ANNALS OF THE NEW YORK ACADEMY OF SCIENCES

Volume 758

EDITORIAL STAFF

Executive Editor
BILL M. BOLAND

Administrative Editor
SHEILA KANE

Managing Editor
JUSTINE CULLINAN

The New York Academy of Sciences
2 East 63rd Street
New York, New York 10021

THE NEW YORK ACADEMY OF SCIENCES
(Founded in 1817)

BOARD OF GOVERNORS, July 1994–June 1995

JOSHUA LEDERBERG, *Chairman of the Board*
HENRY M. GREENBERG, *President*
MARTIN L. LEIBOWITZ, *President-Elect*

Honorary Life Governor
WILLIAM T. GOLDEN

HENRY A. LICHSTEIN, *Treasurer*

Governors-at-Large

ELEANOR BAUM	BARRY R. BLOOM	D. ALLAN BROMLEY
EDWARD COHEN	SUSANNA CUNNINGHAM-RUNDLES	BILL GREEN
SANDRA PANEM	RICHARD A. RIFKIND	DOMINICK SALVATORE
DAVID E. SHAW	WILLIAM C. STEERE, JR.	SHMUEL WINOGRAD

CYRIL M. HARRIS, *Past Chairman*   HELENE L. KAPLAN, *General Counsel* [ex officio]

RODNEY W. NICHOLS, *Chief Executive Officer* [ex officio]

DNA: THE DOUBLE HELIX
PERSPECTIVE AND PROSPECTIVE AT FORTY YEARS

ANNALS OF THE NEW YORK ACADEMY OF SCIENCES
Volume 758

DNA: THE DOUBLE HELIX
PERSPECTIVE AND PROSPECTIVE
AT FORTY YEARS

Edited by Donald A. Chambers

The New York Academy of Sciences
New York, New York
1995

GENENTECH, INC.
460 Pt. San Bruno Blvd.
South San Francisco, CA 94080

Copyright © 1995 by the New York Academy of Sciences. All rights reserved. Under the provisions of the United States Copyright Act of 1976, individual readers of the Annals are permitted to make fair use of the material in them for teaching and research. Permission is granted to quote from the Annals provided that the customary acknowledgment is made of the source. Material in the Annals may be republished only by permission of the Academy. Address inquiries to the Executive Editor at the New York Academy of Sciences.

Copying fees: For each copy of an article made beyond the free copying permitted under Section 107 or 108 of the 1976 Copyright Act, a fee should be paid through the Copyright Clearance Center, Inc., 222 Rosewood Drive, Danvers, MA 01923. For articles of more than 3 pages, the copying fee is $1.75.

∞ The paper used in this publication meets the minimum requirements of American National Standard for Information Sciences–Permanence of Paper for Printed Library Materials, ANSI Z39.48-1984.

The illustration on the designed-cover editions of this volume shows part of a sculpture entitled "Bronze Helix" designed by Charles Reina. This welded solid bronze piece is placed in the Grace Auditorium lobby at the Cold Spring Harbor Laboratory, where it was presented to Dr. James Watson to honor his 25th anniversary as laboratory director.

Library of Congress Cataloging-in-Publication Data

DNA : the double helix : perspective and prospective at forty years / edited by Donald A. Chambers.
    p. cm. – (Annals of the New York Academy of Sciences ; v. 758)
    Includes bibliographical references and index.
    ISBN 0-89766-905-3 (alk. paper). – ISBN 0-89766-906-1 (pbk. : alk. paper)
    1. Nucleic acids–Congresses. 2. Molecular biology–Congresses. I. Chambers, Donald A. II. Series.
Q11.N5 vol. 758
[QP620]
500 s–dc20
[574.87'3282]
                                                             95-8848
                                                                CIP

CCP
Printed in the United States of America
ISBN 0-89766-905-3 (cloth)
ISBN 0-89766-906-1 (paper)
ISSN 0077-8923

ANNALS OF THE NEW YORK ACADEMY OF SCIENCES

Volume 758
June 30, 1995

DNA: THE DOUBLE HELIX
PERSPECTIVE AND PROSPECTIVE AT FORTY YEARS[a]

Editor and Conference Chairman
DONALD A. CHAMBERS

CONTENTS

Dedication	xi
Preface. *By* DONALD A. CHAMBERS	xiii
Conference Participants	xv
Forty Years of DNA. *By* DONALD A. CHAMBERS	1

The Historical Papers [in facsimile]

Molecular Structure of Nucleic Acids: A Structure for Deoxyribose Nucleic Acid (J. D. WATSON and F. H. C. CRICK); Molecular Structure of Deoxypentose Nucleic Acids (M. H. F. WILKINS, A. R. STOKES, and H. R. WILSON); and Molecular Configuration in Sodium Thymonucleate (R. E. FRANKLIN and R. G. GOSLING)..	12
Photos from the Conference	18

Part I. The Double Helix: Perspective

Introduction. *By* DONALD A. CHAMBERS	24
The Aperiodic Crystal of Heredity. *By* GUNTHER S. STENT	25
Photos from the Past	33

Part II. The Pathway to the Double Helix

Introduction: "On the Shoulders of Giants" *By* IRVING M. KLOTZ ..	46
A Fifty-Year Perspective on the Genetic Role of DNA. *By* MACLYN MCCARTY	48

---

[a] This volume represents the proceedings of a conference entitled **DNA: The Double Helix—Forty Years: Perspective and Prospective**, sponsored by the New York Academy of Sciences, the University of Illinois at Chicago, and Green College, University of Oxford, and held in Chicago, Illinois on October 13–16, 1993.

DNA in the Decade before the Double Helix.
By ROLLIN D. HOTCHKISS 55
Linus Pauling: Chemist and Molecular Biologist. By ALEXANDER RICH   74
Historic Reflections on the Clinical Roots of Molecular Biology.
By PAUL HELLER 83

Part III. The Structure and Synthesis of DNA

Introduction. By ROBERT V. STORTI   94
The Nucleic Acids: A Backward Glance. By ALEXANDER RICH   97
Gene Regulatory Proteins and Their Interaction with DNA.
By AARON KLUG 143
Genetics of Retroviruses. By HOWARD M. TEMIN 161
In Memoriam: Howard Temin, the Fierce Scholar.
By DAVID BALTIMORE 166

Part IV. Banquet Program

In Honor of James D. Watson, Francis Crick, and Maurice Wilkins.
By DONALD A. CHAMBERS 171
Greetings. By RODNEY W. NICHOLS 174
Greetings. By JOSHUA LEDERBERG 176
Intellectual Dawns. By CRISPIN TICKELL 180
What the Double Helix Has Meant for Basic Biomedical Science:
A Personal Commentary. By JOSHUA LEDERBERG 182
Values from a Chicago Upbringing. By JAMES D. WATSON 194
DNA: A Cooperative Discovery. By FRANCIS H. C. CRICK 198
DNA at King's College, London. By MAURICE H. F. WILKINS 200
"The Night Before Crickmas": A Poem and Deliverance.
By ROLLIN D. HOTCHKISS 205
Photos from the Banquet 208

Part V. Molecular, Cellular, and Integrative Biology

Introduction. By R. JOHN SOLARO 211
Mammalian Learning and Memory Studied by Gene Targeting.
By SUSUMU TONEGAWA 213
Circuits. By FRANÇOIS JACOB 218

The NK-2 Homeobox Gene and the Early Development of the
Central Nervous System of *Drosophila*. By MARSHALL NIRENBERG,
KOHZO NAKAYAMA, NORIKO NAKAYAMA, YONGSOK KIM,
DERVLA MELLERICK, LAN-HSIANG WANG, KEITH O. WEBBER,
and RAJNIKANT LAD 224

Part VI. DNA and Molecular Medicine

Introduction. By LORD WALTON OF DETCHANT 243

The Molecular Basis for Phenotypic Diversity of Genetic Disease.
By DAVID J. WEATHERALL 245

A Molecular Switch for the Consolidation of Long-Term Memory:
cAMP-Inducible Gene Expression. By CRISTINA M. ALBERINI,
MIRELLA GHIRARDI, YAN-YOU HUANG, PETER V. NGUYEN,
and ERIC R. KANDEL 261

Molecular Analysis of Duchenne Muscular Dystrophy: Past, Present,
and Future. By KAY E. DAVIES, JONATHON M. TINSLEY,
and DEREK J. BLAKE 287

Transgenic Mouse Models of Disease: Altering Adipose Tissue
Function *in Vivo*. By SUSAN R. ROSS, REED A. GRAVES,
LISA CHOY, VERONICA SOLEVEVA, and BRUCE M. SPIEGELMAN ... 297

Recombinant DNA Technology and Oral Medicine.
By HAROLD C. SLAVKIN 314

Part VII. DNA, Oncogenes, and Cancer

Introduction. By SIR RICHARD DOLL 329

The Molecular Basis of Oncogenes and Tumor Suppressor Genes.
By ROBERT A. WEINBERG 331

A Nuclear Tyrosine Kinase Becomes a Cytoplasmic Oncogene.
By DAVID BALTIMORE, RUIBAO REN, GENHONG CHENG,
KONSTANTINA ALEXANDROPOULOS, and PIERA CICCHETTI 339

Recombinant Toxins: New Therapeutic Agents for Cancer.
By IRA H. PASTAN, LEE H. PAI, ULRICH BRINKMANN,
and DAVID J. FITZGERALD 345

Part VIII. Recombinant DNA and Biotechnology

Introduction. By RICHARD L. DAVIDSON 355

DNA: Template for An Economic Revolution. By DAVID A. JACKSON 356

The Molecular Biology of Thyroid Hormone Action.
By RALFF C.J. RIBEIRO, JAMES W. APRILETTI, BRIAN L. WEST,
RICHARD L. WAGNER, ROBERT J. FLETTERICK, FRED SCHAUFELE,
and JOHN D. BAXTER 366

Human and Mouse T-Cell Receptor Loci: Genomics, Evolution,
Diversity, and Serendipity. *By* LEROY HOOD, LEE ROWEN,
and BEN F. KOOP 390

Part IX. The Double Helix: Prospective

Introduction. *By* DONALD A. CHAMBERS 413

Where Will Genome Analysis Lead Us Forty Years On?
By SIR WALTER BODMER 414

The World We Have Lost. *By* HORACE FREELAND JUDSON 427

The Biomedical Revolution at 40 Years: An Overview of the
Conference. *By* DONALD A. CHAMBERS, KENNETH B. M. REID,
and RHONNA L. COHEN 441

Index of Contributors 459
Subject Index 461

Financial assistance was received from:

Supporters
- AMERSHAM CORPORATION
- AMICON, INC.
- BECKMAN INSTRUMENTS, INC.
- GLAXO RESEARCH INSTITUTE
- HELENE CURTIS, INC.
- MARION MERRELL DOW INC.
- PFIZER INC
- PHARMACIA BIOTECH INC.
- THE UPJOHN COMPANY

Contributors
- ABBOTT LABORATORIES
- DuPONT MERCK PHARMACEUTICAL COMPANY
- GENETICS INSTITUTE, INC.
- GIBCO BRL/LIFE TECHNOLOGIES, INC.
- MERCK RESEARCH LABORATORIES
- PARKE DAVIS PHARMACEUTICAL RESEARCH
- SMITHKLINE BEECHAM PHARMACEUTICALS

Others
- COLD SPRING HARBOR LABORATORY PRESS
- EPA CEPHALOSPORIN FUND
- SCIENTIFIC SERVICE OF THE CONSULATE GENERAL OF FRANCE
- OXFORD UNIVERSITY PRESS

The New York Academy of Sciences believes it has a responsibility to provide an open forum for discussion of scientific questions. The positions taken by the participants in the reported conferences are their own and not necessarily those of the Academy. The Academy has no intent to influence legislation by providing such forums.

Dedication

This book is dedicated to all the scientists, past and present, whose curiosity, commitment and passion have made these papers possible and, in particular, to the memory of Oswald Avery, Rosalind Franklin, Fred Griffith, André Lwoff, Barbara McClintock, Jacques Monod, Max Delbrück, Linus Pauling and Howard Temin.

Group photograph taken at the conference: *Kneeling*: M. Meselson. *Seated left to right*: I. Klotz, L. Hood, J. Baxter, A. Rich, J. Watson, G. Stent, D. Chambers, J. Lederberg, D. Baltimore, F. Jacob, R. Doll, H. Temin. *Standing left to right*: S. Tonegawa, D. Jackson, H. Judson, M. Nirenberg, I. Pastan, R. Hotchkiss, P. Heller, J. Solaro, C. Tickell, E. Kandel, K. Davies, R. Storti, H. Slavkin, D. Weatherall, W. Bodmer, S. Ross. *Not present for photo*: H. Bourne, R. Davidson, W. Gilbert, A. Klug, M. McCarty, R. Mulligan, H. Varmus, J. Walton, R. Weinberg.

Preface

As Jim Watson's intellectual foundation was a product of the influence of the University of Chicago, mine was developed initially at the Bronx High School of Science (in the classrooms of Milton Kopelman, Edward Frankel, and Clarence Berger) and then at Columbia College of Columbia University. It was during this formative period that I learned the value of reading the original literature, first in terms of the great books, and then in science. From that time forth, not only was I attracted to the qualities of curiosity, passion, and pleasure that life as a laboratory scientist demands and gives back, but also I developed a retrospective interest in the intellectual history of the biomedical sciences. Fifteen years ago, the kernel of the concept of this meeting arose, but it was only during the fall of 1990, during a visit to Green College, Oxford University as a Visiting Scholar that I revisited the idea of mounting a celebration of the fortieth anniversary of the double helix. Such a meeting would draw upon the thematic elements of the past, present, and future of the biomedical revolution that the elucidation of the double helical structure of DNA begot. Upon my return to Chicago, I spoke with officials of the New York Academy of Sciences and was given an enthusiastic go-ahead provided that we could get the appropriate speakers and demonstrate that sufficient interest existed to warrant such a conference. Immediately thereafter, the University of Illinois at Chicago and Green College became co-sponsors.

To put together a meeting of these dimensions required a tremendous commitment of work, help, and understanding from many groups as well as from persons to whom I owe an immense measure of gratitude. I therefore extend my thanks to the Warden and Fellows of Green College for their hospitality, discussion, and support; to the faculty, staff and students of the Department of Biochemistry and the Center for Molecular Biology of Oral Diseases at UIC, for their indulgence, understanding, and concern, particularly during the months that immediately preceded the conference; to my bosses Dean Gerald Moss of the University of Illinois College of Medicine, Dean Allen Anderson of the UIC College of Dentistry, to the Provost of UIC, David Brosky, the Chancellor of UIC, James Stukel, and the President of the University of Illinois, Stanley Ikenberry, for their commitment of university support; to the members of the Conference and Communications departments of the New York Academy of Sciences (Geraldine Busacco, Lynn Serra and Renée Wilkerson, and Ann Collins) and to the organizing committees at UIC (Walter Devoreen, Lucia Gee, Abdul Khan, Marion Ostrega, Barbara Poltzer, Ann Ross, and Patricia Wager) for efforts above and beyond. I also thank Roberta Dupuis-Devlin of the Office of Publications Services at UIC for the pictures taken at the conference and banquet that appear in this volume. The publication of this volume of the *Annals* stands apart from the meeting. In this endeavor I owe much thanks and appreciation to Bill Boland, Sheila Kane, and Justine Cullinan for their efforts in seeing the book through the press.

There are six people who deserve special citation: to my uncle, Irving Bieber, M.D., who was alive at the conception of the meeting, but not at its reality, for his love, inspiration, and wisdom in helping me achieve my "impossible dream": to David Soifer, Ph.D., a friend and colleague of more than 20 years, who chaired the NYAS committee that oversaw this conference and made the distance between New York City and Chicago seem like a city block long; to Kenneth Reid, Ph.D., Professor of Immunology and Fellow of Green College, Oxford, and friend, who labored hard and long in his role as Green College coordinator; to my children Karen and David for their love and support; and to Rhonna Cohen, D.D.S., Ph.D., Associate Professor and Program Coordinator of the Center for Molecular Biology of Oral Diseases, UIC, whose commitment to this project exceeds my ability to adequately describe and thank. Finally, the ingredients for a successful conference include the speakers, the audience, the support groups and the sponsors, both academic and industrial. To all, I thank you for joining in this labor of love.

DONALD A. CHAMBERS

Conference Participants

DAVID BALTIMORE, born in New York City, received his undergraduate degree from Swarthmore College and his Ph.D. from The Rockefeller University. In 1968, he joined the faculty of the Massachusetts Institute of Technology and in 1982 was named the first Director of the Whitehead Institute, to which he has recently returned. From 1990–1991, he was President of The Rockefeller University. In 1975, at the age of 37, Dr. Baltimore, with Dr. Howard Temin, was awarded the Nobel Prize in Physiology or Medicine for their simultaneous discovery of the enzyme, reverse transcriptase. Dr. Baltimore is a member of the National Academy of Sciences, the American Academy of Arts and Sciences, the Institute of Medicine, and the Royal Society. Research in Dr. Baltimore's laboratory, which covers the three areas of cancer-inducing viruses, the immune system, and infectious disease, seeks to define the biochemical events underlying change in gene expression and gene regulation in mammalian cells.

JOHN D. BAXTER, born in Lexington, Kentucky, received his B.A. degree from the University of Kentucky (where he also set track records for the 880-yard and mile events) and his M.D. from Yale University. In 1970, he joined the faculty of the University of California at San Francisco, where he is Professor of Medicine, Director of the Metabolic Research Unit, and Chief of the Section of Endocrinology of the Moffitt/Long Hospitals. Dr. Baxter, who was the first to report nuclear binding of glucocorticoid receptors, was also first to describe the human mineralocorticoid receptor, clone the growth hormone gene, and produce growth hormone by recombinant DNA technology. He has authored more than 250 scientific publications, started the journal, *DNA*, is one of the four editors of the text, *Endocrinology and Metabolism*, and has founded four successful biotechnology companies.

WALTER BODMER, Director-General of the Imperial Cancer Research Fund (ICRF), received his B.A. and Ph.D. degrees from Cambridge University and was recruited to the faculty of the Department of Genetics of Stanford University by its chairman, Professor Joshua Lederberg (who taught Sir Walter, then a graduate student, the principles of molecular biology), rising to the rank of professor. In 1970, Bodmer returned to the U.K. as Professor of Genetics at Oxford, and, in 1979, joined the ICRF as Director of Research. Sir Walter's research is directed at somatic cell genetics, immunogenetics, and the cancer problem. He has been awarded honorary degrees by eight universities, including Oxford, is a member of the Royal Society, a foreign member of the National Academy of Sciences and the American Philosophical Society, past-president of the British Association for the Advancement of Science, and was knighted in 1986. Sir Walter, with Robin McKie, is the author of *The Book of Man* (1994).

HENRY R. BOURNE received his A.B. in History and Literature from Harvard College and his M.D. from Johns Hopkins University. In 1971, he joined the faculty of the University of California in San Francisco, where he remains today; from 1983–1991 he was Professor and Chairman of the Department of Pharmacology at UCSF. Professor Bourne is an outstanding investigator of cellular signal transduction and is a member of the National Academy of Sciences and the American Academy of Arts and Sciences.

DONALD A. CHAMBERS was born in New York City, trained at Columbia University, and was a research fellow at the Harvard Medical School. He has served on the faculties of the University of California at San Francisco and the University of Michigan. Dr. Chambers at present is Professor and Head of the Department of Biochemistry and Director of the Center for Molecular Biology of Oral Diseases at the University of Illinois at Chicago and Honorary Visiting Fellow of Green College, Oxford University. Dr. Chambers' research interests focus on the molecular biology of epithelial tissues, wound healing, and neuroimmune interactions as well as the intellectual development of the biomedical sciences. In 1990, he was named UIC Inventor of the Year.

FRANCIS CRICK was born in Northampton, England and received his B.Sc. in Physics from University College, London. During World War II, he was engaged in war research. After the war, he moved to the Cavendish Laboratories of Cambridge University from which he received his Ph.D. in 1954. In 1951, while engaged in thesis research, Crick met James D. Watson and thus began their collaboration, culminating with the publication of the *Nature* papers in 1953, and the award of the Nobel Prize in 1962. Sir Francis went on to direct his attention to elucidating the genetic code and currently, as the J.W. Kieckhefer Distinguished Research Professor at the Salk Institute, he has turned his thought to neurobiology and the nature of consciousness, which he has described in a recent book, *The Astonishing Hypothesis*, published in 1993. Sir Francis also published his autobiography, *What Mad Pursuit*, in 1989. [Photograph by Ed Campodonico. Reproduced by courtesy of the Cold Spring Harbor Laboratory.]

RICHARD L. DAVIDSON received his Ph.D. from Case-Western Reserve University and came to the University of Illinois at Chicago in 1980 from the Harvard Medical School to become the Benjamin Goldberg Professor and Head of the Department of Genetics. Dr. Davidson's research focuses on pigment cell biology and he is the founding editor of the journal *Somatic Cell and Molecular Genetics*.

KAY DAVIES received her B.A. and D.Phil. degrees from Oxford University. After a number of years away from Oxford, during which time she located the gene for Duchenne muscular dystrophy, leading to the development of prenatal diagnosis of the disease (Nature **300:** 69–71, 1982), she returned to Oxford in 1984. In 1994, she was appointed to the Chair of Genetics at Oxford, becoming the first woman to hold this position. Professor Davies is the editor of *Human Molecular Genetics* and *Nature, Genetics* and is on the executive council of the Human Genome Organization. At present, she and her research group are involved in the molecular analysis of Duchenne muscular dystrophy, spinal muscular atrophy, the fragile X syndrome and X-linked retinitis pigmentosa. In the 18 years since she received her D.Phil. degree, Dr. Davies has published close to 200 papers.

RICHARD DOLL was born in Hampton, U.K., and received his medical training at St. Thomas's Hospital Medical School. In 1969, he was appointed Regius Professor of Medicine at Oxford and in 1979, became the founding Warden of Green College, Oxford. Sir Richard is the recipient of a number of international awards and prizes for his work on the epidemiology of cancer and the etiology of lung cancer. He was the first to show that cigarette smoking causes cancer of the lung. In addition, he has been awarded honorary degrees by many universities, including Oxford and Harvard. Currently, he remains active in studies of cancer and the theoretical basis of epidemiology and meta-analysis.

WALTER GILBERT did his undergraduate work in chemistry and physics at Harvard College and received his Ph.D. in Mathematics from Cambridge University in 1957. He first joined the Physics faculty at Harvard University and in 1960 turned his attention to problems of molecular biology, initially working on mRNA, ribosomes, and protein synthesis with J.D. Watson. At present, he is the Carl M. Loeb University Professor at Harvard University. Professor Gilbert is a member of the National Academy of Sciences and the Royal Society and he has been the recipient of a plethora of awards, including the Louise and Bert Freedman Award of the New York Academy of Sciences, the Louisa Gross Horwitz Prize of Columbia University, the Lasker Award, and, in 1980, the Nobel Prize for Chemistry. In addition to his scientific investigations, Professor Gilbert is a leading figure in the biotechnology industry, having founded the company Biogen.

PAUL HELLER was born in Komatau, Austria (now the Czech Republic) and received his M.D. in 1938 from Charles University in Prague. In 1938, he was taken prisoner by the Nazis and was interned in German concentration camps until 1945, when, at the time of the liberation of Buchenwald, he met the American journalist Edward R. Murrow, starting a lifelong friendship. Murrow arranged for Dr. Heller to come to this country to complete his medical training. In 1954, Dr. Heller joined the faculty of the University of Illinois College of Medicine and the West Side Veterans Administration Hospital, rising to Professor of Medicine and Chief of the Department of Medicine at the Westside VA Hospital, where he was named one of five Senior Medical Investigators. At present, Dr. Heller continues an active professional life as Professor Emeritus of Medicine by teaching and writing. His research was centered on the hemoglobinopathies, which serve as the first paradigms for molecular medicine. In 1983, Professor Heller pioneered the use of 5-azacytodine and DNA-methylation as potential therapy for sickle cell disease and the thalassemias.

LEROY E. HOOD, born in Missoula, Montana, received both B.S. and Ph.D. degrees from the California Institute of Technology (Caltech) as well as an M.D. from Johns Hopkins. In 1970, Professor Hood joined the faculty at Caltech, became the Bowles Professor of Biology in 1977, chaired the Division of Biology from 1980–1989 and was both the Director of the Caltech Cancer Center and the NSF Science and Technology Center for Molecular Biotechnology. In 1992, he left Caltech for the University of Washington to become the William Gates II Professor and Chairman of the Department of Molecular Biotechnology, the first such department in the country. Of the many awards Professor Hood has received are the California Scientist of the Year (1985), the Lasker Award, the American College of Physicians Award, as well as honorary degrees from five universities. Professor Hood is a member of the National Academy of Sciences and the American Academy of Arts and Sciences. His research interests center on the genetics, evolution and organization of multigene systems, the genetics and evolution of antibody diversity, and the development of microchemical instrumentation for molecular analysis.

ROLLIN DOUGLAS HOTCHKISS was born in South Britain, Connecticut and received his B.S. and Ph.D. degrees from Yale University. In 1935, he joined The Rockefeller Institute, initially working with Oswald Avery. During that period, his work was directed at quantifying the transforming principle and he began studies that showed that this principle could be extended to the transformation of genetic drug resistance by bacteria. Professor Hotchkiss went on to study the mechanism of action of antibiotics and continued his earlier studies on bacterial DNA, bacterial drug resistance, bacterial metabolism, and physiology. He is the winner of numerous awards and honorary degrees and is a member of the National Academy of Sciences and the American Academy of Arts and Sciences and past-president of the Harvey Society. The Cold Spring Harbor Laboratory has recognized his contributions to molecular biology by naming its dining hall after him.

STANLEY O. IKENBERRY, President of the University of Illinois, received his Ph.D. from Michigan State University and came to the University of Illinois in 1979 from Pennsylvania State University, where he was executive vice-president. President Ikenberry is regarded as "Mr. Higher Education" in the state of Illinois. He has been chairman of the American Council of Education and a member of the executive committee of the Association of American Universities. Dr. Ikenberry has announced his retirement as president effective July 1, 1996 and will return to the study of the problems of universities at the end of the century.

DAVID A. JACKSON was born in New York City, received a B.A. from Harvard College and a Ph.D. in Molecular Biology at Stanford University. From 1969–1972, he was a postdoctoral fellow with Professor Paul Berg at Stanford. During that time he was the first author on the first paper from the Berg laboratory to describe successful technology resulting in recombinant DNA (Proc. Natl. Acad. Sci. USA **69**: 2904–2909, 1972). From 1972–1981, he was a faculty member in the Department of Microbiology of the University of Michigan and in 1980 was a founder and scientific director of the Genex Corporation, the biotechnology company that developed aspartame. At present, Dr. Jackson is Scientific Director for Virology and Cancer Research at the DuPont-Merck Pharmaceutical Corporation.

FRANÇOIS JACOB is one of the pioneers and major figures in biochemical genetics. With Jacques Monod, he developed the organizing concept for molecular genetics, the operon (J. Molec. Biol. **3**: 8–56, 1961), and with Monod and Changeaux, the bioregulatory enzymatic concept of allostery (J. Molec. Biol. **6**: 306–309, 1963). Professor Jacob, who has published his autobiography (*The Statue Within*, 1987), was born in Nancy, France, was a member of the resistance during World War II, and received his M.D. and D.Sc. degrees after the war. He became Head of the Cell Genetics Unit of the Pasteur Institute in 1960 and was Chairman of the Board from 1982–1988. Professor Jacob is a member of many learned societies, including the Académie des Sciences (Paris) and a foreign member of the National Academy of Sciences, the American Academy of Arts and Sciences, and the Royal Society. In 1965, Professor Jacob was awarded the Nobel Prize in Physiology or Medicine with Professors Jacques Monod and André Lwoff.

HORACE FREELAND JUDSON, the author of *The Eighth Day of Creation, The Makers of the Revolution in Biology*, (published by Simon and Schuster, New York in 1979; Penguin edition, 1995) was born in New York City and graduated from the University of Chicago. His graduate work in English was done at the University of Chicago and at Columbia University. From 1963–1973, he was a correspondent for *Time* magazine, during which time he began his work on *The Eighth Day of Creation*. In 1981, he was appointed Henry R. Luce Professor in the Writing Sciences at Johns Hopkins University and Professor of the History of Science, and in 1987 was selected a fellow of the John D. and Catherine T. MacArthur Foundation. Currently, Mr. Judson is a Senior Research Scholar and Visiting Professor in the History of Sciences at Stanford University, where he is working on a project of research and writing in the history of recent immunology in its social context.

ERIC R. KANDEL, born in Vienna, Austria, received his B.A. from Harvard College and the M.D. from New York University. In 1965, he joined the Departments of Physiology and Psychiatry at NYU and in 1974 was appointed Professor and Director of the Center for Neurobiology and Behavior at the College of Physicians & Surgeons of Columbia University. In 1983, he was named University Professor at Columbia University. Dr. Kandel, a member of the National Academy of Sciences, the American Academy of Arts and Sciences, and the National Institute of Medicine, has been awarded numerous prizes for his work including the National Medal of Science and the Lasker Award as well as honorary degrees from seven universities. Dr. Kandel is senior author of the definitive text in neurosciences, *Principles of Neural Science*. Dr. Kandel's research relates to the molecular nature of behavior and learning, in which he pioneered the use of the snail, *Aplysia*, as a model system.

IRVING KLOTZ received both undergraduate and graduate degrees from the University of Chicago and is the Morrison Professor Emeritus of Chemistry at Northwestern University. He has received numerous awards including, most recently, the William Rose Prize of the American Society of Biochemistry and Molecular Biology and is a member of the National Academy of Sciences. Dr. Klotz's research is directed at physical biochemistry, molecular interactions and bioenergetics. He is the author of several books of general interest, including *Diamond Dealers and Feather Merchants*.

AARON KLUG was born in and initially trained in South Africa. He received his Ph.D. from Cambridge University and subsequent honorary degrees from seven universities. From 1954–1957, he did research on the structure of viruses with Rosalind Franklin at Birkbeck College, London, and in 1962, he joined the MRC Laboratory of Molecular Biology in Cambridge, of which he is the present Director. Sir Aaron has received numerous prizes including the Louise Gross Horwitz Prize of Columbia University and the Copley Medal of the Royal Society. He was awarded the Nobel Prize in Chemistry in 1982 and knighted in 1988. Dr. Klug works on the structural biology of DNA and DNA–protein interactions.

JOSHUA LEDERBERG was born in Montclair, New Jersey and was educated at Columbia University and Yale University. At Yale, he continued work on bacterial genetics and recombination initiated in Francis Ryan's laboratory at Columbia, with Edward Tatum. In 1958 Dr. Lederberg was awarded the Nobel Prize for Medicine or Physiology jointly with Tatum and George Beadle. He has served as Professor and Chair of the Department of Genetics at the University of Wisconsin and then at Stanford University Medical School. From 1979–1990, Professor Lederberg was President of the Rockefeller University. He is a member of the National Academy of Sciences, and the Royal Society and was President of the New York Academy of Sciences in 1993–94. Professor Lederberg has been awarded numerous honorary degrees and received the U.S. National Medal of Science in 1989.

MACLYN McCARTY, Professor Emeritus at The Rockefeller University, was born in South Bend, Indiana and received his A.B. degree from Stanford University and his M.D. from the Johns Hopkins University. In 1941, he joined the laboratory of Dr. Oswald Avery at the Rockefeller Institute as a postdoctoral fellow and was a coauthor along with Avery and MacLeod of the paper identifying the transforming principle as DNA; this study appeared in the *Journal of Experimental Medicine* in 1943, 10 years before the Watson–Crick paper in *Nature*. Dr. McCarty, holder of numerous awards, remains active today as editor of the *Journal of Experimental Medicine*.

MATTHEW MESELSON did his undergraduate work at the University of Chicago and graduate work at the California Institute of Technology, from which he received his Ph.D. in Physical Chemistry in 1957 and shortly thereafter joined the faculty. In 1960, he became a member of the faculty of Harvard University, where he remains today as the Thomas Dudley Cabot Professor of the Natural Sciences. Professor Meselson has received a number of distinguished awards, including honorary degrees from the University of Chicago and Columbia University, the Presidential Award of the New York Academy of Sciences, and a MacArthur Foundation Fellowship; he is also a member of the National Academy of Sciences and the Royal Society (London). In 1958, with Franklin W. Stahl, he published their classic investigation showing that DNA replication occurs through a semi-conservative mechanism, a series of experiments that have been called the best biological experiments ever devised.

RICHARD C. MULLIGAN received his undergraduate training at the Massachusetts Institute of Technology and took his Ph.D. in Biochemistry in the laboratory of Professor Paul Berg at the Stanford University School of Medicine in 1980. In 1981 he joined the faculty of MIT, where he is Professor of Biology and Member of the Whitehead Institute for Biomedical Research. Dr. Mulligan is the recipient of a MacArthur Foundation Fellowship and is a leader in the development of gene transfer and gene therapy.

RODNEY W. NICHOLS, an applied physicist and science policy analyst, is Chief Executive Officer of the New York Academy of Sciences. Prior to joining the Academy in 1992, he was Scholar-in-Residence at the Carnegie Corporation of New York (1990–1992), and Vice-President and Executive Vice President of The Rockefeller University (1970–1990). Earlier he was a research manager in industry and in the federal government. A Harvard graduate, he has advised federal agencies including the NIH and NSF, and he served on the Executive Committee of the Carnegie Commission on Science, Technology, and Government. A member of the Council on Foreign Relations and the board of advisors to *Foreign Affairs*, he has represented the United States in international negotiations on arms control and on technology for developing countries. Elected a Fellow of the American Association for the Advancement of Science and of the New York Academy of Sciences, Mr. Nichols was awarded the Secretary of Defense Medal for Distinguished and Meritorious Civilian Service.

MARSHALL W. NIRENBERG was born in New York City, receiving his B.S. and M.S. from the University of Florida and his Ph.D. from the University of Michigan. In 1959, he began his association with the NIH as a postdoctoral fellow, where he remains as Chief of the Laboratory of Biochemical Genetics of the National Heart, Blood and Lung Institute. Among his many honors, he is a member of the National Academy of Sciences, the American Academy of Arts and Sciences, and the Institute of Medicine; he has been awarded honorary degrees from 13 universities, including Harvard, Yale, the University of Chicago, and the Universities of Florida, Michigan and Pennsylvania, and he has received the Louise Gross Horwitz Prize of Columbia University, the Lasker Award, and the Presidential Medal. In 1968, Dr. Nirenberg was awarded the Nobel Prize in Physiology or Medicine for his role in "cracking the genetic code." In 1972, Dr. Nirenberg changed his research focus to problems of neurobiology.

IRA W. PASTAN received both B.S. and M.D. degrees from Tufts University. After doing a medical residency at Yale University, he joined the National Institutes of Health, where he is now Chief of the Laboratory of Molecular Biology at the National Cancer Institute. The winner of a number of prizes and named lectureships, including the Pierce Immunotoxin Award, Dr. Pastan is also a member of the National Academy of Sciences. His research has centered on the regulation of genetic expression in animal cells and bacteria. Most recently, he has focused on multidrug resistance and the development of new drugs for cancer treatment.

ALEXANDER RICH was born in Hartford, Connecticut and received both A.B. and M.D. degrees from Harvard University. From 1949–1954 he was a postdoctoral fellow of Linus Pauling at Caltech and then became Chief of the Section of Physical Chemistry of the National Institute of Mental Health. In 1958, he joined the MIT faculty, where at present he is the William Thompson Sedgwick Professor of Biophysics. Professor Rich has received numerous awards, including the Presidential Award of the New York Academy of Sciences, the Lewis S. Rosenstiel Award from Brandeis University, and the James R. Killian Faculty Achievement Award from MIT. He is a member of the National Academy of Sciences, the American Academy of Arts and Sciences, and the Institute of Medicine. Professor Rich's research has focused on structural biology and he has done pioneering work on the structure of the polysome, tRNA, as well as the discovery of Z-DNA.

SUSAN ROSS did her undergraduate work at the University of Pennsylvania and received a Ph.D. from Princeton University. In 1982, she became a faculty member in the Department of Biochemistry at the University of Illinois at Chicago and was named a University Scholar in 1992. In 1994 she became a member of the Cancer Center and the Department of Microbiology at the University of Pennsylvania. Her research has utilized transgenic mice to study the nature of mouse mammary virus interactions and, more recently, the developmental biology of the adipocyte.

HAROLD SLAVKIN was born in Chicago, Illinois and received both an undergraduate degree and a D.D.S. from the University of Southern California (USC). In 1968, he joined the USC faculty, where he was named the George and Mary Lou Boone Professor of Craniofacial Molecular Biology and Director of the Center for Craniofacial Molecular Biology. In 1995, he became Director-designate of the National Institute for Dental Research. His research is directed towards the genetics and developmental biology of the structures of the oro-facial complex. In 1983, Professor Slavkin successfully cloned the amelogin gene.

R. JOHN SOLARO trained as a pharmacist and received his Ph.D. in Physiology from the University of Cincinnati. He has been a faculty member of the Medical College of Virginia and the University of Cincinnati, coming in 1987 to UIC to become Professor and Head of the Department of Physiology and Biophysics. Dr. Solaro's research is concerned with the developmental physiology and molecular biology of heart muscle. In 1992, he was named a University Scholar of the University of Illinois.

GUNTHER STENT was born in Berlin, Germany, and immigrated to the United States, where he received a Ph.D. degree from the University of Illinois in Urbana-Champaign. He did postdoctoral training with Max Delbrück at Caltech and then at the University of Copenhagen and the Pasteur Institute. In 1952, he joined the faculty of the University of California, Berkeley, where he rose to the position of Professor and Chairman of the Department of Molecular and Cell Biology. Professor Stent is a member of the National Academy of Sciences and the American Academy of Arts and Sciences. From 1948–1968 his research focused on molecular biology in the control of the structure and replication of the genetic material and regulation of gene expression. He authored one of the early texts of molecular biology, *The Molecular Biology of Bacterial Viruses* (1963). In 1969, he refocused scientifically to the study of neurobiology. Professor Stent also has a profound interest in the history and philosophy of science and has written and edited several books in that context, including *Phage and the Origins of Molecular Biology* (1969), *Paradoxes of Progress* (1978), and *Morality as a Biological Phenomenon* (1978, 1981).

ROBERT STORTI, a native of Rhode Island, received his Ph.D. from Indiana University and did postdoctoral training with Dr. Alex Rich at the Massachusetts Institute of Technology. Since 1978, he has been a faculty member of the University of Illinois at Chicago, where at present he is Professor of Biochemistry. His research focuses on the molecular biology of muscle development. In 1987, he was named to the first class of University Scholars at the University of Illinois.

JAMES J. STUKEL received his engineering degree from Purdue University and M.S. and Ph.D. from the University of Illinois at Urbana-Champaign (UIUC). He joined the UIUC engineering faculty in 1968, was made professor in 1975, and Associate Dean of the College of Engineering in 1984. In 1985, he came to the University of Illinois at Chicago (UIC) as Vice-Chancellor for Research and Dean of the Graduate School. From 1986–1990 he served as Executive Vice Chancellor and Vice Chancellor for Academic Affairs and in 1991 became Chancellor. His research focused on electrohydrodynamics of multi-phase systems and environmental and energy policy analysis, and he has published 90 papers in that area. As Chancellor of the UIC, Dr. Stukel oversaw a campus of 25,000 students that contains the largest medical school in the U.S., with an annual budget soon to reach $1 billion. In March 1995, Dr. Stukel was named President of the University of Illinois.

HOWARD M. TEMIN received his undergraduate degree from Swarthmore College and his Ph.D. from the California Institute of Technology. He joined the faculty of the University of Wisconsin in 1960, where he remained for his entire professional life, rising to the rank of professor in 1969. A member of the National Academy of Sciences and other distinguished organizations, he has received countless awards and prizes including the National Science Medal in 1992. In 1975, Dr. Temin was awarded the Nobel Prize jointly with Dr. David Baltimore for their independent discovery of the enzyme, reverse transcriptase. Professor Temin died in 1994, and his lecture at this conference was his last public address.

CRISPIN TICKELL, the third Warden of Green College, Oxford, was educated at Christ Church, Oxford and joined the British Diplomatic Service in 1954. He served as Ambassador to Mexico from 1981–83 and Permanent Representative to the United Nations from 1987–90. Sir Crispin was a Visiting Fellow of the Center of International Affairs of Harvard University, during which time he wrote a widely acclaimed book, *Climatic Change and World Affairs*. In addition, he was a Visiting Fellow of All Souls College, Oxford, and has been President of the Royal Geographic Society and Chair of the International Institute for Environment and Development. In 1990, Sir Crispin was named Special Advisor to the Prime Minister on environmental affairs. During his tenure as Warden of Green College, Sir Crispin established the Green College Center for Environmental Policy and Understanding and currently serves as its director.

SUSUMU TONEGAWA was born in Nagoya, Japan, and received a B.S. degree in Chemistry from Kyoto University and a Ph.D. in Biology from the University of San Diego. From 1969–1970, he was a postdoctoral fellow of Dr. Renato Dulbecco at the Salk Institute. In 1971, he joined the Basel Institute for Immunology in Basel, Switzerland, and in 1991, became a member of the faculty of the Massachusetts Institute of Technology as Professor of Biology. Professor Tonegawa is a member of the National Academy of Sciences and the American Academy of Arts and Sciences and has received numerous distinguished prizes. In 1987, he was awarded the Nobel Prize in Physiology or Medicine for his work on the somatic generation of antibody diversity. Most recently, he has turned his attention to neurobiology.

HAROLD E. VARMUS did his undergraduate training in liberal arts at Amherst College and received his M.D. from Columbia University. In 1970, he moved to the University of California at San Francisco, where he began a collaboration on the study of oncogenes with Professor J. Michael Bishop which culminated in their both being awarded the Nobel Prize in Medicine or Physiology in 1989. Dr. Varmus is a member of the National Academy of Sciences, the American Academy of Arts and Sciences, and the Institute of Medicine. In 1993, Dr. Varmus accepted the position of Director of the National Institutes of Health.

JOHN NICHOLAS WALTON, Lord Walton of Detchant, was born in Northumberland, in the U.K., and received his medical training from the University of Durham, followed by a graduate fellowship in neurology at the Massachusetts General Hospital. In 1958, he was appointed Professor of Neurology at the University of Newcastle-upon-Tyne, where from 1971 to 1981 he served as the Dean of Medicine. From 1983–1989, Lord Walton was the second warden of Green College, Oxford. He has served as President of the British Medical Association, the Royal Society of Medicine, and the World Federation of Neurology, also receiving numerous distinguished awards and honorary degrees. In addition to publishing his autobiography, *The Spice of Life* (1992), he is also the editor of *Brain's Diseases of the Nervous System* and *The Oxford Companion to Medicine*. Currently, Lord Walton is very active in the House of Lords, where he serves as a spokesman and interpreter of medicine and science.

JAMES DEWEY WATSON was born in Chicago, Illinois, receiving his B.S. from the University of Chicago in 1947, at the age of 19, and three years later his Ph.D. from Indiana University, where he studied with Salvador Luria. In 1951 Watson began the collaboration with Francis Crick to determine the structure of DNA, which culminated in the award of the Nobel Prize in 1962 jointly with Crick and Maurice Wilkins. Watson went on to join the faculty of Harvard University, where he turned his attention to ribosomes and protein synthesis and introduced Walter Gilbert to molecular biology. In 1965, he published the first edition of his classic text, *The Molecular Biology of the Gene*, and has collaborated with others to write *Recombinant DNA* (2nd ed., 1992), *The Molecular Biology of the Cell* (3rd ed., 1993), and *Phage and the Origin of Molecular Biology* (1966). In 1968, Watson became Director of the Cold Spring Harbor Laboratory, and has built that laboratory to become one of the stellar research institutions of the world. In 1988, Watson was named Associate Director of the Human Genome Project of the NIH, and from 1989–1992, he served as the first director of the National Center for Human Genome Research of the NIH. During the 1993–1994 academic year, Watson was Visiting Professor at Balliol College, Oxford.

DAVID J. WEATHERALL is the Regius Professor of Medicine of Oxford University. He received his medical degree from the University of Liverpool, did postdoctoral research in genetics and hematology at the Johns Hopkins Medical School, and remained a faculty member at the University of Liverpool until 1974. In 1974, Sir David was appointed the Nuffield Professor of Medicine at Oxford University and, in 1992, the Regius Professor. He was knighted in 1987. In 1988, he founded the Institute of Molecular Medicine at Oxford and serves as its director. Sir David has won a plethora of distinctions including the Royal Medal of the Royal Society, the Gold Medal of the Royal Society of Medicine, and the Johns Hopkins Centennial Medal. He is a member of the Royal Society and a foreign member of the National Academy of Sciences and the Institute of Medicine as well as President of the International Society for Hematology and the British Association for the Advancement of Science. Sir David's research concerns itself with molecular medicine and the thalassemias.

ROBERT A. WEINBERG was born in Pittsburgh, Pennsylvania, and received both B.S. and Ph.D. degrees from the Massachusetts Institute of Technology. In 1973, he joined the MIT biology faculty and is at present Professor of Biology and Member of the Whitehead Institute. Professor Weinberg is a member of the National Academy of Sciences and the American Academy of Arts and Sciences. Among his many awards are the Robert Koch Medal, the Warren Triennial Prize of the Massachusetts General Hospital, the Bristol-Myers Award for Distinguished Achievement in Cancer Research and the Sloan Prize of the General Motors Cancer Research Foundation. His research concerns itself with the elucidation of mechanisms of cancer genes.

MAURICE HUGH FREDERICK WILKINS was born in New Zealand and received his Ph.D. from Cambridge University in 1940. During World War II, he did research on the atomic bomb as part of the Manhattan Project in the United States. After the war, he returned to the U.K. and joined the faculty of Kings College, London, where he turned his focus to biophysics and structural biology and his X-ray diffraction studies of the structure of DNA. Professor Wilkins, the recipient of many awards and honorary degrees, was awarded the Nobel Prize in 1962 with Watson and Crick. He is a fellow of the Royal Society, a foreign member of the National Academy of Sciences and the American Academy of Arts and Sciences, and past-president of the British Society for Social Responsibility in Science. At present, he continues to teach at Kings College, London and is writing a scientific autobiography.

Forty Years of DNA

DONALD A. CHAMBERS

*Department of Biochemistry, and
Center for Molecular Biology of Oral Diseases
University of Illinois at Chicago
1819 W. Polk Street
Chicago, Illinois 60612*

From October 14 until October 17, 1994, a scientific fantasy became a reality in Chicago, Illinois, when forty scientists who were the protagonists and leading figures of the biomedical revolution of the twentieth century met to mark the discovery of the structure of DNA in a symposium entitled *The Double Helix: Forty Years, Prospective and Perspective*. The conference, held under the auspices of the New York Academy of Sciences, the University of Illinois at Chicago, and Green College of the University of Oxford brought together more than 1,000 participants, consisting of senior scientists, university and high school faculty, historians, postdoctoral fellows, and graduate, undergraduate and high school students, in an electric atmosphere in which the audience, as interactive "flies on the wall," participated in a review and appraisal of the last forty years of biomedical progress. The goal of this conference, the largest of the five held in this fortieth anniversary year of the publication of the papers in *Nature* describing the double-helical structure of DNA by Watson, Crick, Wilkins, Franklin, and their collaborators, was to bring together both the historical figures of the age and the more recent leaders of the biomedical revolution in an assessment of the last forty years of those developments in biomedical science that were spawned by the elucidation of the structure of DNA. This volume attests to the success of the project.

The history of the discovery of the double helix has been documented by both Watson[1] and Crick[2] as well as by historical observers, in particular by Judson[3] and Olby.[4] FIGURES 1A, B, and C show Watson, Crick, Wilkins, and Franklin in 1953, at the time of the publication of their papers describing the structure of DNA as a double helix in which the two strands of the DNA were held together by hydrogen bonds, joining the complementary purine–pyrimidine base pairs coupled with additional base-stacking forces to form a molecular structure in which the bases were found inside the molecule and the sugar phosphates on the outside.[a] The experimental data upon which the structure was based were two-fold: (*1*) the X-ray diffraction studies from the

---

[a] These papers are reproduced in this volume on unfolioed pages 13–17.

FIGURE 1A. James Watson and Francis Crick in front of a model of DNA. (From Watson.[1] Reproduced by permission of the author and through the courtesy of the Cold Spring Harbor Laboratory Archives.)

group at Kings College led by Wilkins and Franklin, and (2) model building guided by both the known structures of the purine and pyrimidine bases described by Todd[5] and by the rules of base pairing, initially stated by Chargaff[6] (FIG. 2). Over the period since 1953, the original structure has largely stood the test of time, although some modifications have occurred, as described by Rich and Klug in this volume.

Although hard to imagine in 1993, it was not immediately accepted that DNA was the genetic material. Studies relating DNA to genetic information had their origin in the work of Avery (FIG. 3), MacLeod, and McCarty,[7] who in turn had revisited the earlier studies of Griffith[8] on transformation in bacteria, and biochemically isolated and purified the transforming principle and showed it to be DNA (see the papers by McCarty and Hotchkiss in this volume). Notwithstanding the clear demonstration by these workers that removal of protein did not increase the specific activity of the transforming principle or that the transforming principle was resistant to proteases but not nucleases, a sizeable number of scientists led by Alfred Mirsky (FIG. 4)

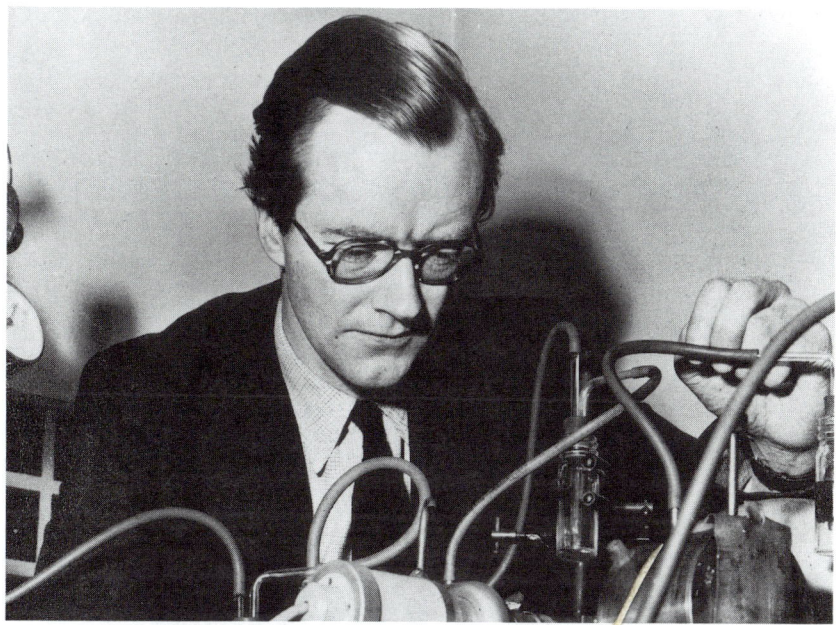

FIGURE 1B. Maurice H. F. Wilkins. (Reproduced by permission.)

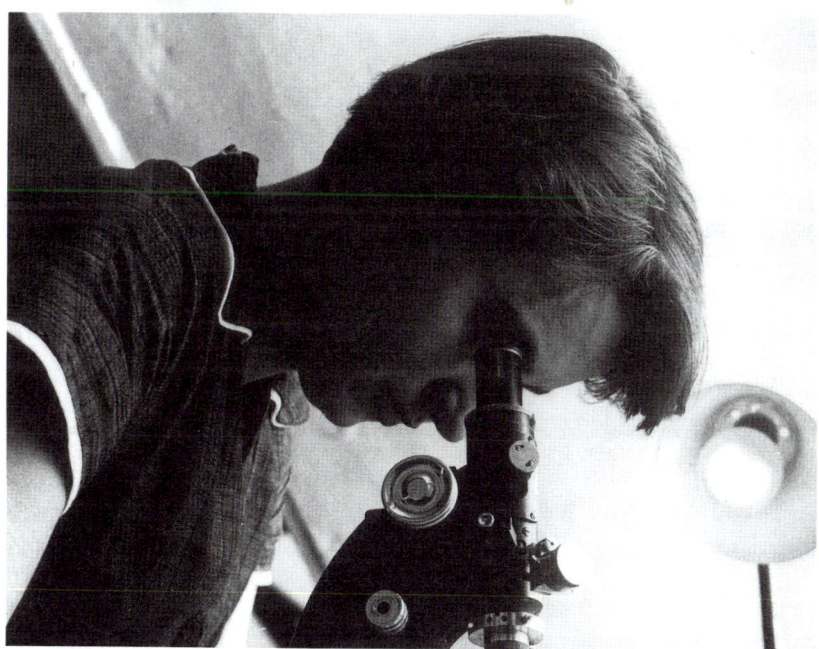

FIGURE 1C. Rosalind Franklin. (Photo by Henry Grant. Reproduced through the courtesy of the Laboratory of Molecular Biology of the Medical Research Council.)

FIGURE 2. Dr. Erwin Chargaff. (Reproduced by permission of the archives at Columbia University.)

questioned whether a chemical structure differing only in four bases was complex enough to be the code of life. In terms of how pure such a molecule would have to be to convince the skeptics that the genetic substance was DNA and not protein, Lederberg argued that Avery and his colleagues were battling Avogadro's number (Lederberg, this volume). Indeed, the entire concept of DNA's being the genome has been questioned by Stent as being premature for acceptance in the 1940s (Stent, this volume).

What, then, changed the tide of opinion? As early as 1939, Max Delbrück (FIG. 5), initially a physicist and student of Niels Bohr, became excited about the potential for what we now call molecular genetics and the deciphering of the code of life. He suggested the study of bacterial viruses as model systems. After World War II, a plethora of physicists, including Crick and Szilard, followed Delbrück's example and switched from weapons research to biology. Thus began the studies of the phage group, which gathered every summer at Cold Spring Harbor to exchange ideas (Stent, this volume). A major breakthrough in the acceptance of DNA as the genetic material came with the Nobel Prize experiments of the phage group members Hershey and Chase (themselves working at Cold Spring Harbor, FIG. 6), in which they showed

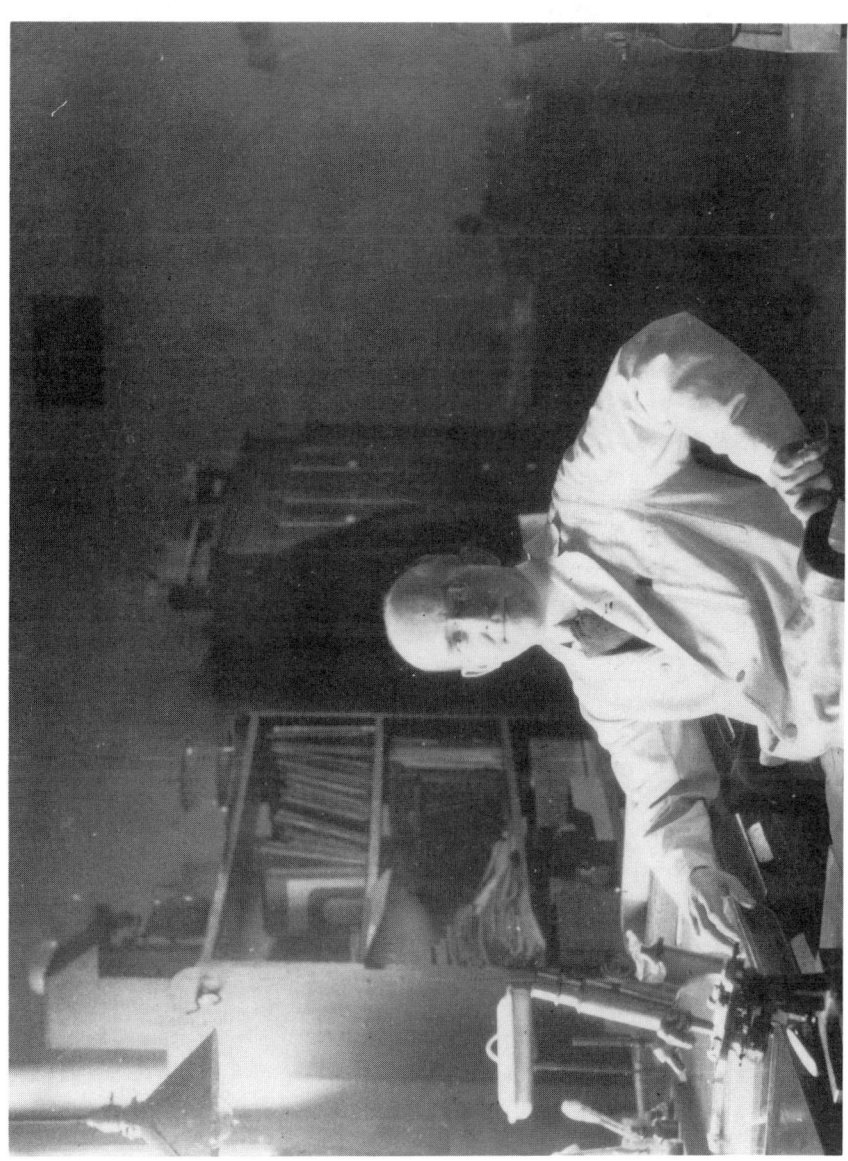

FIGURE 3. Oswald Avery in his laboratory at The Rockefeller Hospital. (Reproduced by permission of The Rockefeller University Archives.)

FIGURE 4. Alfred Mirsky. (Reproduced by permission of The Rockefeller University Archives.)

FIGURE 5. Max Delbrück at the California Institute of Technology. (Reproduced by permission of Caltech.)

FIGURE 6. Martha Chase and Alfred Hershey at Cold Spring Harbor. (Reproduced by permission of Karl Maramorosch.)

that when a phage infected a bacterium and reproduced, only the nucleic acid of the phage entered the bacterium, with the phage protein remaining on the outside.[9] Thus, the atmosphere was now receptive for the new observations of Watson, Crick, Wilkins and Franklin.

The rapidity of acceptance of the structure of DNA has been ascribed to its inherent beauty, as put forth by the coy statement in the Watson-Crick paper that "it has not escaped our notice. . . ."[10] However, we mustn't lose sight of the fact that Watson and Crick and their phage group colleagues were never far from the "big picture." In fact, FIGURE 7 shows a page of Watson's laboratory notebook from 1952, where, in the context of the cell cycle, he sets forth the concept which later became the central dogma of molecular biology: DNA → RNA → protein, demonstrating that even at that time, when he was consumed with the structure of DNA, structure–function relationships of DNA in cell physiology were of great importance. The ability of the phage group to define and focus on issues central to biological mechanisms coupled with their free exchange of ideas was probably a crucial condition of their success (FIG. 8). It was not by chance that the members of this same group went on to define the nature of the genetic code and mutagenesis, discover messenger RNA, postulate transfer RNA, and set the stage for building a molecular construct for neurobiology. Since the late 1960s, despite the arguments of Sydney Brenner and others that the golden age of molecular biology was at an end and that it was now time to turn the intellectual efforts

November 1952

A Hypothetical Scheme of the Interrelationship between the Nucleic Acids and Proteins

Consequences of scheme

1. RNA synthesis and DNA synthesis should not occur at the same time. Protein synthesis and DNA synthesis will occur simultaneously.

2. Nuclear RNA synthesis will occur only in anaphase cells.

3. The total Mg^{++} concentration will increase toward metaphase and decrease during interphase.

4. The content of nucleolar RNA may possibly remain constant during interphase. Synthesis of nucleolar RNA occurs on the chromosomes during prophase - metaphase.

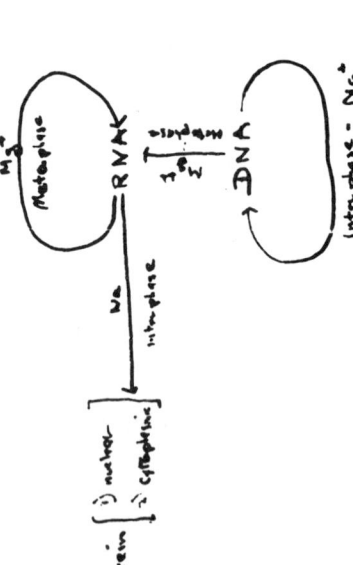

FIGURE 7. Reproduction of a page of James Watson's laboratory notebook. (Reproduced by permission of James Watson.)

FIGURE 8. Members of the phage group, including, from the left, François Jacob, Max Delbrück, Matthew Meselson, Ronald Rolfe, Gunther Stent, and Sydney Brenner. (Reproduced by permission of Gunther Stent.)

to the next frontier, the neurosciences, molecular biology has continued to flourish. Unpredictable consequences of new developments have given us recombinant DNA technology (leading to a biotechnology industry), new DNA diagnostics, "designer genes," gene therapy, and the sequencing of the human genome (Bodmer, this volume). The growth of knowledge encompassed by the term *molecular biology* can be seen by contrasting the size of the first edition of Watson's *Molecular Biology of the Gene*, published in 1965, with that of the current fourth edition, which appeared in 1987.[10,11] From a small, compact single-authored book, it has grown to a two-volumed, multi-authored, encyclopedia-like text. Such rapid growth has not been without its price. I remember that when the first edition appeared, I raced to the bookstore to buy it because I wanted to see how Watson prioritized "his science."

As I read then, and as I reread now, a distinct charm shines through, a feeling of being at the edge of the frontier. Here was a new genre, a book that was not the usual compendium of facts to be mastered, but rather, an outline of concepts and the experiments that led to their development. Even the chapter headings—such as "Cells Obey the Laws of Chemistry" and "The Importance of Weak Chemical Interactions"—seemed to yield important truths. I found the book itself a *tour de force*, and my experience reading it a stroke of illumination, like reading a biological *Hamlet*. By contrast, although the original principles are still maintained in the current edition, its size and magnitude have resulted in the loss of its original charm. Much of the same criticism can be leveled at the whole field of molecular biology, as Horace Judson does in his essay in this volume, "The World We Have Lost." We are witnessing the retreat of the smaller research group of an earlier age, driven by curiosity and passion, where everyone knew one another, yielding to the monoliths of today, funded by government and industry, and succumbing to application and profit.

Today, as the golden age of biology continues to flourish, molecular biology has become integral to understanding the biological process. Whereas in the early part of the twentieth century, anatomy was the mother science of biology and medicine, it has been displaced by the biochemical sciences, the science of biological molecules and their physiological regulation, a new microanatomy of molecular interactions. Traditional biochemistry has evolved by combining molecular biology with structural biology to produce a science capable of understanding physiology and pathology in terms of molecular structure and function. Of such are the contents of this volume.

REFERENCES

1. WATSON, J. D. 1968. The Double Helix. A Personal Account of the Discovery of the Structure of DNA. Atheneum. New York. (*See also* The Double Helix, A Norton Critical Edition. G. S. Stent, Ed. W. W. Norton, 1980.)
2. CRICK, F. H. C. 1979. What Mad Pursuit. Penguin. London.
3. JUDSON, H. F. 1979. The Eighth Day of Creation. Simon and Schuster. New York.
4. OLBY, R. 1974. The Pathway to the Double Helix. MacMillan. London.
5. TODD, A. & D. M. BROWN. 1952. Nucleotides: Part X. Some observations on the structure and chemical behavior of the nucleic acids. J. Chem. Soc.: 52–58.
6. CHARGAFF, E. 1950. Chemical specificity of nucleic acids and mechanism of their enzymatic degradation. Experientia **6**: 201–209.
7. AVERY, O., C. MACLEOD & M. MCCARTY. 1943. Induction of transformation by a desoxyribonucleic acid fraction isolated from Pneumococcus type III. J. Exp. Med. **79**: 137–158.
8. GRIFFITH, F. 1928. The significance of pneumococcal types. J. Hygiene **27**: 141–144.
9. HERSHEY, A. D. & M. CHASE. 1952. Independent functions of viral protein and nucleic acid in the growth of bacteriophage. J. Gen. Physiol. **36**: 39–56.

10. WATSON, J. D. & F. H. C. CRICK. 1953. Molecular structure of nucleic acids: A structure for deoxyribose nucleic acid. Nature **171**: 737–738.
11. WATSON, J. D. 1965. The Molecular Biology of the Gene, first ed. W. A. Benjamin. Menlo Park, CA.
12. WATSON, J. D., N. H. HOPKINS, J. W. ROBERTS, J. A. STEITZ & A. M. WEINER. 1987. Benjamin/Cummings Co. Menlo Park, CA.

The Historical Papers

[The following three papers are reproduced with permission from Nature.*]*

equipment, and to Dr. G. E. R. Deacon and the captain and officers of R.R.S. *Discovery II* for their part in making the observations.

[1] Young, F. B., Gerrard, H., and Jevons, W., *Phil. Mag.*, **40**, 149 (1920).
[2] Longuet-Higgins, M. S., *Mon. Not. Roy. Astro. Soc., Geophys. Supp.*, **5**, 285 (1949).
[3] Von Arx, W. S., Woods Hole Papers in Phys. Oceanog. Meteor., **11** (3) (1950).
[4] Ekman, V. W., *Arkiv. Mat. Astron. Fysik. (Stockholm)*, **2** (11) (1905).

MOLECULAR STRUCTURE OF NUCLEIC ACIDS

A Structure for Deoxyribose Nucleic Acid

WE wish to suggest a structure for the salt of deoxyribose nucleic acid (D.N.A.). This structure has novel features which are of considerable biological interest.

A structure for nucleic acid has already been proposed by Pauling and Corey[1]. They kindly made their manuscript available to us in advance of publication. Their model consists of three intertwined chains, with the phosphates near the fibre axis, and the bases on the outside. In our opinion, this structure is unsatisfactory for two reasons: (1) We believe that the material which gives the X-ray diagrams is the salt, not the free acid. Without the acidic hydrogen atoms it is not clear what forces would hold the structure together, especially as the negatively charged phosphates near the axis will repel each other. (2) Some of the van der Waals distances appear to be too small.

Another three-chain structure has also been suggested by Fraser (in the press). In his model the phosphates are on the outside and the bases on the inside, linked together by hydrogen bonds. This structure as described is rather ill-defined, and for this reason we shall not comment on it.

We wish to put forward a radically different structure for the salt of deoxyribose nucleic acid. This structure has two helical chains each coiled round the same axis (see diagram). We have made the usual chemical assumptions, namely, that each chain consists of phosphate diester groups joining β-D-deoxyribofuranose residues with 3′,5′ linkages. The two chains (but not their bases) are related by a dyad perpendicular to the fibre axis. Both chains follow right-handed helices, but owing to the dyad the sequences of the atoms in the two chains run in opposite directions. Each chain loosely resembles Furberg's[2] model No. 1; that is, the bases are on the inside of the helix and the phosphates on the outside. The configuration of the sugar and the atoms near it is close to Furberg's 'standard configuration', the sugar being roughly perpendicular to the attached base. There is a residue on each chain every 3·4 A. in the z-direction. We have assumed an angle of 36° between adjacent residues in the same chain, so that the structure repeats after 10 residues on each chain, that is, after 34 A. The distance of a phosphorus atom from the fibre axis is 10 A. As the phosphates are on the outside, cations have easy access to them.

The structure is an open one, and its water content is rather high. At lower water contents we would expect the bases to tilt so that the structure could become more compact.

The novel feature of the structure is the manner in which the two chains are held together by the purine and pyrimidine bases. The planes of the bases are perpendicular to the fibre axis. They are joined together in pairs, a single base from one chain being hydrogen-bonded to a single base from the other chain, so that the two lie side by side with identical z-co-ordinates. One of the pair must be a purine and the other a pyrimidine for bonding to occur. The hydrogen bonds are made as follows : purine position 1 to pyrimidine position 1 ; purine position 6 to pyrimidine position 6.

If it is assumed that the bases only occur in the structure in the most plausible tautomeric forms (that is, with the keto rather than the enol configurations) it is found that only specific pairs of bases can bond together. These pairs are : adenine (purine) with thymine (pyrimidine), and guanine (purine) with cytosine (pyrimidine).

In other words, if an adenine forms one member of a pair, on either chain, then on these assumptions the other member must be thymine ; similarly for guanine and cytosine. The sequence of bases on a single chain does not appear to be restricted in any way. However, if only specific pairs of bases can be formed, it follows that if the sequence of bases on one chain is given, then the sequence on the other chain is automatically determined.

It has been found experimentally[3,4] that the ratio of the amounts of adenine to thymine, and the ratio of guanine to cytosine, are always very close to unity for deoxyribose nucleic acid.

It is probably impossible to build this structure with a ribose sugar in place of the deoxyribose, as the extra oxygen atom would make too close a van der Waals contact.

The previously published X-ray data[5,6] on deoxyribose nucleic acid are insufficient for a rigorous test of our structure. So far as we can tell, it is roughly compatible with the experimental data, but it must be regarded as unproved until it has been checked against more exact results. Some of these are given in the following communications. We were not aware of the details of the results presented there when we devised our structure, which rests mainly though not entirely on published experimental data and stereochemical arguments.

It has not escaped our notice that the specific pairing we have postulated immediately suggests a possible copying mechanism for the genetic material.

Full details of the structure, including the conditions assumed in building it, together with a set of co-ordinates for the atoms, will be published elsewhere.

We are much indebted to Dr. Jerry Donohue for constant advice and criticism, especially on interatomic distances. We have also been stimulated by a knowledge of the general nature of the unpublished experimental results and ideas of Dr. M. H. F. Wilkins, Dr. R. E. Franklin and their co-workers at

This figure is purely diagrammatic. The two ribbons symbolize the two phosphate—sugar chains, and the horizontal rods the pairs of bases holding the chains together. The vertical line marks the fibre axis

King's College, London. One of us (J. D. W.) has been aided by a fellowship from the National Foundation for Infantile Paralysis.

J. D. WATSON
F. H. C. CRICK

Medical Research Council Unit for the
Study of the Molecular Structure of
Biological Systems,
Cavendish Laboratory, Cambridge.
April 2.

[1] Pauling, L., and Corey, R. B., *Nature*, **171**, 346 (1953); *Proc. U.S. Nat. Acad. Sci.*, **39**, 84 (1953).
[2] Furberg, S., *Acta Chem. Scand.*, **6**, 634 (1952).
[3] Chargaff, E., for references see Zamenhof, S., Brawerman, G., and Chargaff, E., *Biochim. et Biophys. Acta*, **9**, 402 (1952).
[4] Wyatt, G. R., *J. Gen. Physiol.*, **36**, 201 (1952).
[5] Astbury, W. T., Symp. Soc. Exp. Biol. 1, Nucleic Acid, 66 (Camb. Univ. Press, 1947).
[6] Wilkins, M. H. F., and Randall, J. T., *Biochim. et Biophys. Acta*, **10**, 192 (1953).

Molecular Structure of Deoxypentose Nucleic Acids

WHILE the biological properties of deoxypentose nucleic acid suggest a molecular structure containing great complexity, X-ray diffraction studies described here (cf. Astbury[1]) show the basic molecular configuration has great simplicity. The purpose of this communication is to describe, in a preliminary way, some of the experimental evidence for the polynucleotide chain configuration being helical, and existing in this form when in the natural state. A fuller account of the work will be published shortly.

The structure of deoxypentose nucleic acid is the same in all species (although the nitrogen base ratios alter considerably) in nucleoprotein, extracted or in cells, and in purified nucleate. The same linear group of polynucleotide chains may pack together parallel in different ways to give crystalline[1–3], semi-crystalline or paracrystalline material. In all cases the X-ray diffraction photograph consists of two regions, one determined largely by the regular spacing of nucleotides along the chain, and the other by the longer spacings of the chain configuration. The sequence of different nitrogen bases along the chain is not made visible.

Oriented paracrystalline deoxypentose nucleic acid ('structure B' in the following communication by Franklin and Gosling) gives a fibre diagram as shown in Fig. 1 (cf. ref. 4). Astbury suggested that the strong 3·4-A. reflexion corresponded to the internucleotide repeat along the fibre axis. The ∼ 34 A. layer lines, however, are not due to a repeat of a polynucleotide composition, but to the chain configuration repeat, which causes strong diffraction as the nucleotide chains have higher density than the interstitial water. The absence of reflexions on or near the meridian immediately suggests a helical structure with axis parallel to fibre length.

Diffraction by Helices

It may be shown[5] (also Stokes, unpublished) that the intensity distribution in the diffraction pattern of a series of points equally spaced along a helix is given by the squares of Bessel functions. A uniform continuous helix gives a series of layer lines of spacing corresponding to the helix pitch, the intensity distribution along the nth layer line being proportional to the square of J_n, the nth order Bessel function. A straight line may be drawn approximately through

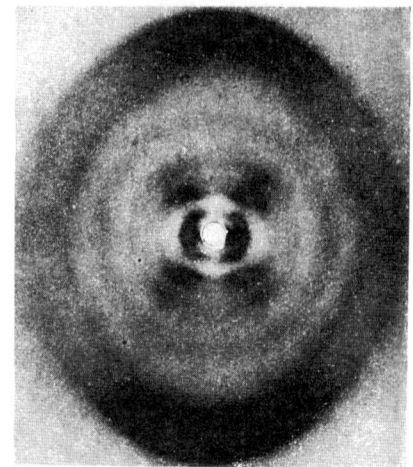

Fig. 1. Fibre diagram of deoxypentose nucleic acid from *B. coli*. Fibre axis vertical

the innermost maxima of each Bessel function and the origin. The angle this line makes with the equator is roughly equal to the angle between an element of the helix and the helix axis. If a unit repeats n times along the helix there will be a meridional reflexion (J_0^2) on the nth layer line. The helical configuration produces side-bands on this fundamental frequency, the effect[5] being to reproduce the intensity distribution about the origin around the new origin, on the nth layer line, corresponding to C in Fig. 2.

We will now briefly analyse in physical terms some of the effects of the shape and size of the repeat unit or nucleotide on the diffraction pattern. First, if the nucleotide consists of a unit having circular symmetry about an axis parallel to the helix axis, the whole diffraction pattern is modified by the form factor of the nucleotide. Second, if the nucleotide consists of a series of points on a radius at right-angles to the helix axis, the phases of radiation scattered by the helices of different diameter passing through each point are the same. Summation of the corresponding Bessel functions gives reinforcement for the inner-

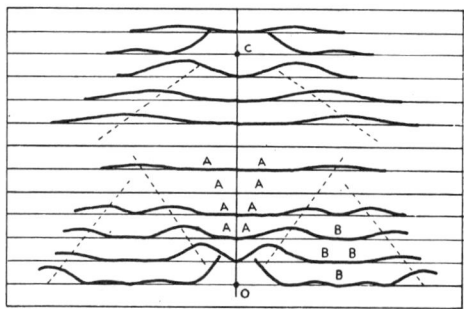

Fig. 2. Diffraction pattern of system of helices corresponding to structure of deoxypentose nucleic acid. The squares of Bessel functions are plotted about 0 on the equator and on the first, second, third and fifth layer lines for half of the nucleotide mass at 20 A. diameter and remainder distributed along a radius, the mass at a given radius being proportional to the radius. About C on the tenth layer line similar functions are plotted for an outer diameter of 12 A.

most maxima and, in general, owing to phase difference, cancellation of all other maxima. Such a system of helices (corresponding to a spiral staircase with the core removed) diffracts mainly over a limited angular range, behaving, in fact, like a periodic arrangement of flat plates inclined at a fixed angle to the axis. Third, if the nucleotide is extended as an arc of a circle in a plane at right-angles to the helix axis, and with centre at the axis, the intensity of the system of Bessel function layer-line streaks emanating from the origin is modified owing to the phase differences of radiation from the helices drawn through each point on the nucleotide. The form factor is that of the series of points in which the helices intersect a plane drawn through the helix axis. This part of the diffraction pattern is then repeated as a whole with origin at C (Fig. 2). Hence this aspect of nucleotide shape affects the central and peripheral regions of each layer line differently.

Interpretation of the X-Ray Photograph

It must first be decided whether the structure consists of essentially one helix giving an intensity distribution along the layer lines corresponding to $J_1, J_2, J_3 \ldots$, or two similar co-axial helices of twice the above size and relatively displaced along the axis a distance equal to half the pitch giving $J_2, J_4, J_6 \ldots$, or three helices, etc. Examination of the width of the layer-line streaks suggests the intensities correspond more closely to J_1^2, J_2^2, J_3^2 than to $J_2^2, J_4^2, J_6^2 \ldots$ Hence the dominant helix has a pitch of ~ 34 A., and, from the angle of the helix, its diameter is found to be ~ 20 A. The strong equatorial reflexion at ~ 17 A. suggests that the helices have a maximum diameter of ~ 20 A. and are hexagonally packed with little interpenetration. Apart from the width of the Bessel function streaks, the possibility of the helices having twice the above dimensions is also made unlikely by the absence of an equatorial reflexion at ~ 34 A. To obtain a reasonable number of nucleotides per unit volume in the fibre, two or three intertwined coaxial helices are required, there being ten nucleotides on one turn of each helix.

The absence of reflexions on or near the meridian (an empty region AAA on Fig. 2) is a direct consequence of the helical structure. On the photograph there is also a relatively empty region on and near the equator, corresponding to region BBB on Fig. 2. As discussed above, this absence of secondary Bessel function maxima can be produced by a radial distribution of the nucleotide shape. To make the layer-line streaks sufficiently narrow, it is necessary to place a large fraction of the nucleotide mass at ~ 20 A. diameter. In Fig. 2 the squares of Bessel functions are plotted for half the mass at 20 A. diameter, and the rest distributed along a radius, the mass at a given radius being proportional to the radius.

On the zero layer line there appears to be a marked J_{10}^2, and on the first, second and third layer lines, $J_9^2 + J_{11}^2, J_8^2 + J_{12}^2$, etc., respectively. This means that, in projection on a plane at right-angles to the fibre axis, the outer part of the nucleotide is relatively concentrated, giving rise to high-density regions spaced c. 6 A. apart around the circumference of a circle of 20 A. diameter. On the fifth layer line two J_5 functions overlap and produce a strong reflexion. On the sixth, seventh and eighth layer lines the maxima correspond to a helix of diameter ~ 12 A. Apparently it is only the central region of the helix structure which is well divided by the 3·4-A. spacing, the outer parts of the nucleotide overlapping to form a continuous helix. This suggests the presence of nitrogen bases arranged like a pile of pennies[1] in the central regions of the helical system.

There is a marked absence of reflexions on layer lines beyond the tenth. Disorientation in the specimen will cause more extension along the layer lines of the Bessel function streaks on the eleventh, twelfth and thirteenth layer lines than on the ninth, eighth and seventh. For this reason the reflexions on the higher-order layer lines will be less readily visible. The form factor of the nucleotide is also probably causing diminution of intensity in this region. Tilting of the nitrogen bases could have such an effect.

Reflexions on the equator are rather inadequate for determination of the radial distribution of density in the helical system. There are, however, indications that a high-density shell, as suggested above, occurs at diameter ~ 20 A.

The material is apparently not completely paracrystalline, as sharp spots appear in the central region of the second layer line, indicating a partial degree of order of the helical units relative to one another in the direction of the helix axis. Photographs similar to Fig. 1 have been obtained from sodium nucleate from calf and pig thymus, wheat germ, herring sperm, human tissue and T_2 bacteriophage. The most marked correspondence with Fig. 2 is shown by the exceptional photograph obtained by our colleagues, R. E. Franklin and R. G. Gosling, from calf thymus deoxypentose nucleate (see following communication).

It must be stressed that some of the above discussion is not without ambiguity, but in general there appears to be reasonable agreement between the experimental data and the kind of model described by Watson and Crick (see also preceding communication).

It is interesting to note that if there are ten phosphate groups arranged on each helix of diameter 20 A. and pitch 34 A., the phosphate ester backbone chain is in an almost fully extended state. Hence, when sodium nucleate fibres are stretched[3], the helix is evidently extended in length like a spiral spring in tension.

Structure in vivo

The biological significance of a two-chain nucleic acid unit has been noted (see preceding communication). The evidence that the helical structure discussed above does, in fact, exist in intact biological systems is briefly as follows :

Sperm heads. It may be shown that the intensity of the X-ray spectra from crystalline sperm heads is determined by the helical form-function in Fig. 2. Centrifuged trout semen give the same pattern as the dried and rehydrated or washed sperm heads used previously[6]. The sperm head fibre diagram is also given by extracted or synthetic[1] nucleoprotamine or extracted calf thymus nucleohistone.

Bacteriophage. Centrifuged wet pellets of T_2 phage photographed with X-rays while sealed in a cell with mica windows give a diffraction pattern containing the main features of paracrystalline sodium nucleate as distinct from that of crystalline nucleoprotein. This confirms current ideas of phage structure.

Transforming principle (in collaboration with H. Ephrussi-Taylor). Active deoxypentose nucleate allowed to dry at ~ 60 per cent humidity has the same crystalline structure as certain samples[3] of sodium thymonucleate.

We wish to thank Prof. J. T. Randall for encouragement; Profs. E. Chargaff, R. Signer, J. A. V. Butler and Drs. J. D. Watson, J. D. Smith, L. Hamilton, J. C. White and G. R. Wyatt for supplying material without which this work would have been impossible; also Drs. J. D. Watson and Mr. F. H. C. Crick for stimulation, and our colleagues R. E. Franklin, R. G. Gosling, G. L. Brown and W. E. Seeds for discussion. One of us (H. R. W.) wishes to acknowledge the award of a University of Wales Fellowship.

M. H. F. WILKINS
Medical Research Council Biophysics
Research Unit,

A. R. STOKES
H. R. WILSON
Wheatstone Physics Laboratory,
King's College, London.
April 2.

[1] Astbury, W. T., Symp. Soc. Exp. Biol., 1, Nucleic Acid (Cambridge Univ. Press, 1947).
[2] Riley, D. P., and Oster, G., Biochim. et Biophys. Acta, 7, 526 (1951).
[3] Wilkins, M. H. F., Gosling, R. G., and Seeds, W. E., Nature, 167, 759 (1951).
[4] Astbury, W. T., and Bell, F. O., Cold Spring Harb. Symp. Quant. Biol., 6, 109 (1938).
[5] Cochran, W., Crick, F. H. C., and Vand, V., Acta Cryst., 5, 581 (1952).
[6] Wilkins, M. H. F., and Randall, J. T., Biochim. et Biophys. Acta, 10, 192 (1953).

Molecular Configuration in Sodium Thymonucleate

SODIUM thymonucleate fibres give two distinct types of X-ray diagram. The first corresponds to a crystalline form, structure A, obtained at about 75 per cent relative humidity; a study of this is described in detail elsewhere[1]. At higher humidities a different structure, structure B, showing a lower degree of order, appears and persists over a wide range of ambient humidity. The change from A to B is reversible. The water content of structure B fibres which undergo this reversible change may vary from 40–50 per cent to several hundred per cent of the dry weight. Moreover, some fibres never show structure A, and in these structure B can be obtained with an even lower water content.

The X-ray diagram of structure B (see photograph) shows in striking manner the features characteristic of helical structures, first worked out in this laboratory by Stokes (unpublished) and by Crick, Cochran and Vand[2]. Stokes and Wilkins were the first to propose such structures for nucleic acid as a result of direct studies of nucleic acid fibres, although a helical structure had been previously suggested by Furberg (thesis, London, 1949) on the basis of X-ray studies of nucleosides and nucleotides.

While the X-ray evidence cannot, at present, be taken as direct proof that the structure is helical, other considerations discussed below make the existence of a helical structure highly probable.

Structure B is derived from the crystalline structure A when the sodium thymonucleate fibres take up quantities of water in excess of about 40 per cent of their weight. The change is accompanied by an increase of about 30 per cent in the length of the fibre, and by a substantial re-arrangement of the molecule. It therefore seems reasonable to suppose that in structure B the structural units of sodium thymonucleate (molecules on groups of molecules) are relatively free from the influence of neighbouring

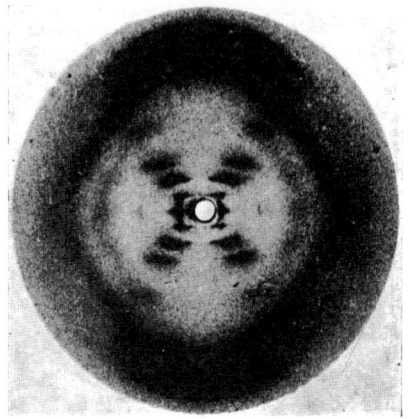

Sodium deoxyribose nucleate from calf thymus. Structure B

molecules, each unit being shielded by a sheath of water. Each unit is then free to take up its least-energy configuration independently of its neighbours and, in view of the nature of the long-chain molecules involved, it is highly likely that the general form will be helical[3]. If we adopt the hypothesis of a helical structure, it is immediately possible, from the X-ray diagram of structure B, to make certain deductions as to the nature and dimensions of the helix.

The innermost maxima on the first, second, third and fifth layer lines lie approximately on straight lines radiating from the origin. For a smooth single-strand helix the structure factor on the nth layer line is given by:

$$F_n = J_n(2\pi rR) \exp i\, n(\psi + \tfrac{1}{2}\pi),$$

where $J_n(u)$ is the nth-order Bessel function of u, r is the radius of the helix, and R and ψ are the radial and azimuthal co-ordinates in reciprocal space[2]; this expression leads to an approximately linear array of intensity maxima of the type observed, corresponding to the first maxima in the functions J_1, J_2, J_3, etc.

If, instead of a smooth helix, we consider a series of residues equally spaced along the helix, the transform in the general case treated by Crick, Cochran and Vand is more complicated. But if there is a whole number, m, of residues per turn, the form of the transform is as for a smooth helix with the addition, only, of the same pattern repeated with its origin at heights mc^*, $2mc^*$ ... etc. (c is the fibre-axis period).

In the present case the fibre-axis period is 34 A. and the very strong reflexion at 3·4 A. lies on the tenth layer line. Moreover, lines of maxima radiating from the 3·4-A. reflexion as from the origin are visible on the fifth and lower layer lines, having a J_5 maximum coincident with that of the origin series on the fifth layer line. (The strong outer streaks which apparently radiate from the 3·4-A. maximum are not, however, so easily explained.) This suggests strongly that there are exactly 10 residues per turn of the helix. If this is so, then from a measurement of R_n the position of the first maximum on the nth layer line (for $n\, 5\leqslant$), the radius of the helix, can be obtained. In the present instance, measurements of R_1, R_2, R_3 and R_5 all lead to values of r of about 10 A.

Since this linear array of maxima is one of the strongest features of the X-ray diagram, we must conclude that a crystallographically important part of the molecule lies on a helix of this diameter. This can only be the phosphate groups or phosphorus atoms.

If ten phosphorus atoms lie on one turn of a helix of radius 10 A., the distance between neighbouring phosphorus atoms in a molecule is 7·1 A. This corresponds to the P . . . P distance in a fully extended molecule, and therefore provides a further indication that the phosphates lie on the outside of the structural unit.

Thus, our conclusions differ from those of Pauling and Corey[4], who proposed for the nucleic acids a helical structure in which the phosphate groups form a dense core.

We must now consider briefly the equatorial reflexions. For a single helix the series of equatorial maxima should correspond to the maxima in $J_0(2\pi rR)$. The maxima on our photograph do not, however, fit this function for the value of r deduced above. There is a very strong reflexion at about 24 A. and then only a faint sharp reflexion at 9·0 A. and two diffuse bands around 5·5 A. and 4·0 A. This lack of agreement is, however, to be expected, for we know that the helix so far considered can only be the most important member of a series of coaxial helices of different radii ; the non-phosphate parts of the molecule will lie on inner co-axial helices, and it can be shown that, whereas these will not appreciably influence the innermost maxima on the layer lines, they may have the effect of destroying or shifting both the equatorial maxima and the outer maxima on other layer lines.

Thus, if the structure is helical, we find that the phosphate groups or phosphorus atoms lie on a helix of diameter about 20 A., and the sugar and base groups must accordingly be turned inwards towards the helical axis.

Considerations of density show, however, that a cylindrical repeat unit of height 34 A. and diameter 20 A. must contain many more than ten nucleotides.

Since structure B often exists in fibres with low water content, it seems that the density of the helical unit cannot differ greatly from that of dry sodium thymonucleate, 1·63 gm./cm.³ [1,5], the water in fibres of high water-content being situated outside the structural unit. On this basis we find that a cylinder of radius 10 A. and height 34 A. would contain thirty-two nucleotides. However, there might possibly be some slight inter-penetration of the cylindrical units in the dry state making their effective radius rather less. It is therefore difficult to decide, on the basis of density measurements alone, whether one repeating unit contains ten nucleotides on each of two or on each of three co-axial molecules. (If the effective radius were 8 A. the cylinder would contain twenty nucleotides.) Two other arguments, however, make it highly probable that there are only two co-axial molecules.

First, a study of the Patterson function of structure A, using superposition methods, has indicated[6] that there are only two chains passing through a primitive unit cell in this structure. Since the $A \rightleftharpoons B$ transformation is readily reversible, it seems very unlikely that the molecules would be grouped in threes in structure B. Secondly, from measurements on the X-ray diagram of structure B it can readily be shown that, whether the number of chains per unit is two or three, the chains are not equally spaced along the fibre axis. For example, three equally spaced chains would mean that the nth layer line depended on J_{3n}, and would lead to a helix of diameter about 60 A. This is many times larger than the primitive unit cell in structure A, and absurdly large in relation to the dimensions of nucleotides. Three unequally spaced chains, on the other hand, would be crystallographically non-equivalent, and this, again, seems unlikely. It therefore seems probable that there are only two co-axial molecules and that these are unequally spaced along the fibre axis.

Thus, while we do not attempt to offer a complete interpretation of the fibre-diagram of structure B, we may state the following conclusions. The structure is probably helical. The phosphate groups lie on the outside of the structural unit, on a helix of diameter about 20 A. The structural unit probably consists of two co-axial molecules which are not equally spaced along the fibre axis, their mutual displacement being such as to account for the variation of observed intensities of the innermost maxima on the layer lines ; if one molecule is displaced from the other by about three-eighths of the fibre-axis period, this would account for the absence of the fourth layer line maxima and the weakness of the sixth. Thus our general ideas are not inconsistent with the model proposed by Watson and Crick in the preceding communication.

The conclusion that the phosphate groups lie on the outside of the structural unit has been reached previously by quite other reasoning[1]. Two principal lines of argument were invoked. The first derives from the work of Gulland and his collaborators[7], who showed that even in aqueous solution the —CO and —NH_2 groups of the bases are inaccessible and cannot be titrated, whereas the phosphate groups are fully accessible. The second is based on our own observations[1] on the way in which the structural units in structures A and B are progressively separated by an excess of water, the process being a continuous one which leads to the formation first of a gel and ultimately to a solution. The hygroscopic part of the molecule may be presumed to lie in the phosphate groups $((C_2H_5O)_2PO_2Na$ and $(C_3H_7O)_2PO_2Na$ are highly hygroscopic[8]), and the simplest explanation of the above process is that these groups lie on the outside of the structural units. Moreover, the ready availability of the phosphate groups for interaction with proteins can most easily be explained in this way.

We are grateful to Prof. J. T. Randall for his interest and to Drs. F. H. C. Crick, A. R. Stokes and M. H. F. Wilkins for discussion. One of us (R. E. F.) acknowledges the award of a Turner and Newall Fellowship.

ROSALIND E. FRANKLIN*
R. G. GOSLING

Wheatstone Physics Laboratory,
King's College, London.
April 2.

* Now at Birkbeck College Research Laboratories, 21 Torrington Square, London, W.C.1.

[1] Franklin, R. E., and Gosling, R. G. (in the press).
[2] Cochran, W., Crick, F. H. C., and Vand, V., *Acta Cryst.*, **5**, 501 (1952).
[3] Pauling, L., Corey, R. B., and Branson, H. R., *Proc. U.S. Nat. Acad. Sci.*, **37**, 205 (1951).
[4] Pauling, L., and Corey, R. B., *Proc. U.S. Nat. Acad. Sci.*, **39**, 84 (1953).
[5] Astbury, W. T., Cold Spring Harbor Symp. on Quant. Biol., **12**, 56 (1947).
[6] Franklin, R. E., and Gosling, R. G. (to be published).
[7] Gulland, J. M., and Jordan, D. O., Cold Spring Harbor Symp. on Quant. Biol., **12**, 5 (1947).
[8] Drushel, W. A., and Felty, A. R., *Chem. Zent.*, **89**, 1016 (1918).

Photos from the Conference

Walter Gilbert speaking to students.

Luncheon and poster viewing.

PHOTOS FROM THE CONFERENCE

Maclyn McCarty giving autographs.

Ira Pastan with students.

John Baxter (*center, with glasses*) listening to a conference attendee.

François Jacob attending to students.

Marshall Nirenberg surrounded by students.

Walter Gilbert talking to students.

PHOTOS FROM THE CONFERENCE

Harold Slavkin and Maclyn McCarty.

James Watson, Walter Gilbert, and Gunther Stent.

David Baltimore, Sir Walter Bodmer, and Richard Mulligan.

From left to right: Marshall Nirenberg, Eric Kandel, Susumu Tonegawa, and Alexander Rich.

Rollin Hotchkiss with students.

Kay Davies (*with back to camera*) with Richard Mulligan and Sir David Weatherall (*back to camera*).

PHOTOS FROM THE CONFERENCE

Joshua Lederberg signing an autograph.

David Jackson at the lectern.

James Watson giving his autograph.

PART I. THE DOUBLE HELIX: PERSPECTIVE

Introduction

DONALD A. CHAMBERS

Department of Biochemistry, and
Center for Molecular Biology of Oral Diseases
University of Illinois at Chicago
1819 W. Polk Street
Chicago, Illinois 60612

Today, in view of the logarithmic growth of science and the number of scientists confronting the challenges and complexities of new frontiers, considerations of the past, of the development of the scientific intellect, are often neglected or left to nonscientists. One of the aims of this volume, and of the conference preceding it, is to address in small part how the past begot the present.

Professor Gunther Stent, a member of the original phage group, leads off by providing an historical perspective on the growth of the field of molecular biology. Of particular import to Professor Stent and historians of science is the question of the extent to which science is shaped by personalities and *vice versa*. Stent asks why, for example, although known to the scientific community, the contributions of Avery, MacLeod, and McCarty (see McCarty, this volume) in the elucidation of DNA as the transforming principle were undervalued. He suggests that this "lack of appreciation" is a function of "prematurity in science," a concept he first put forth in 1972. Joshua Lederberg provides an alternative explanation in his discussion elsewhere in this volume.

The Aperiodic Crystal of Heredity

GUNTHER S. STENT

Department of Molecular and Cell Biology
205 Life Sciences Addition
University of California
Berkeley, California 94720

I learned in my history class at Hyde Park High School, down on the South Side, at 62nd and Stony Island, that the Renaissance began on May 29, 1453, the day Constantinople fell to the Turks. On that date, I thought, everybody suddenly realized that the Middle Ages were over and that the time had come to rediscover the arts and sciences of classical antiquity.

So, I've always been astounded by the uncanny coincidence that molecular biology began exactly five hundred years—almost to the day—after the onset of the Renaissance, on April 25, 1953, the day an article appeared in *Nature* by two young persons, James Watson (formerly a student at Hyde Park High's rival, South Shore High, at 75th and Constance) and Francis Crick, reporting the discovery of the DNA double helix.[1] On that date, I've always thought, everybody suddenly realized that the era of classical genetics was over and that the time had come to think about hereditary transactions in terms of information-carrying macromolecules.

Just as the Renaissance sprang from the confrontation of the Christian West with the Muslim East, so molecular biology sprang from the confrontation of genetics with chemistry. The first known theory of heredity was put forward in the fifth century B.C. by Hippocrates, who held that children possess the qualities of their fathers because the semen contains tiny samples from all parts of the paternal body. Less than a century later, Aristotle showed the inadequacy of the Hippocratic view by pointing out that fathers produce offspring that develop paternal traits (such as gray hair) that are manifest only in the post-reproductive stage of life. He proposed instead that the fathers' semen provides the *plans* according to which the unformed blood of the mother is shaped into the offspring. Thus Aristotle recognized that biological inheritance is attributable to the transmission through the generations of *information* for the embryonic development of the individual, even though, androcentrist that Aristotle was, he couldn't imagine that the mother might also have some information to pass on.

The Renaissance, which has reawakened interest in the physical sciences, produced few novel insights into heredity. It was only in the mid-nineteenth century, when Gregor Mendel provided his radical theory that discrete hereditary elements, later called "genes," are responsible for hereditary traits, that the new era dawned in which the mechanisms governing the self-reproduction of humans and their fellow creatures became ultimately revealed.

Mendel's revolutionary insights remained unnoticed by the community of biologists because his discovery was *premature*, in the sense that his contemporaries could not connect the concept of discrete hereditary elements with their notions of cell biology, and that the statistical methodology by means of which Mendel interpreted his results was foreign to their way of thinking. His 1865 paper was rediscovered only at the turn of the twentieth century, by which time chromosomes, mitosis, and meiosis had been found, and Mendel's theory could now be accounted for in terms of structures visible under the microscope. Moreover, by then the application of statistics to biology had become commonplace.

Upon the rediscovery of Mendel's work, genetics developed into one of the most important frontiers of biological research. Thanks in large part to the work of Thomas Hunt Morgan and his associates, it came to be known that genes are arranged in a linear order on the chromosomes and that they undergo sudden permanent changes, or *mutations*, which result in a change in the particular trait determined by the gene, such as the switch from red eye color to white.

The rise of genetics opened the way for great advances in the understanding of life. On the theoretical plane, genetics provided a firm basis for understanding evolution. It could now be seen that gene mutation, being the prime source of biological novelty, is the motor that drives evolution. And it was realized that what Darwin's mechanism of natural selection actually selects are organisms carrying novel genes, or novel combinations of old genes, that confer greater fitness in the struggle to be the most prolific. On the applied plane, genetics brought tremendous practical benefits. In agriculture, it became possible to design rational breeding procedures by means of which economically superior varieties of crop plants and livestock could be produced. And in medicine, the recognition of the role of genes in many human diseases provided a rationale for taking measures for their prevention or relief.

But throughout the first half of the twentieth century, while genetics became the queen of the biological sciences, the physical nature of its central concept, the gene, had remained shrouded in mystery. As late as 1950, in an essay entitled "The Development of the Gene Theory," Morgan's one-time associate, H. J. Muller, by then one of the elder statesmen of genetics and the leading guru of the gene, wistfully declared that no one knew of what the gene is made, how it manages to impose its character on the organism that carries it, or how it reproduces itself faithfully in cell division.[2]

Despite his wide-ranging discussions and speculations about the chemical nature of the gene, Muller did not mention that six years earlier, Oswald Avery and his colleagues, C. M. MacLeod and Maclyn McCarty, had found that genes are embodied in DNA.[3] Muller failed to consider that DNA has anything to do with genes because Avery's discovery, just like Mendel's, had been "premature": The then-current view of the molecular nature of DNA made it well-nigh inconceivable that DNA could be the carrier of hereditary information.[4] However, unlike Mendel's discovery, which remained virtually unknown until its rediscovery, publication of Avery's discovery caused an immediate sensation and was certainly known to Muller long before he wrote his essay.

Muller's essay appeared as one of 26 articles contributed by the world's leading geneticists to a volume commemorating the golden jubilee of the rediscovery of Mendel's paper. Only two of the articles referred to Avery's discovery. Avery's colleague at Rockefeller, Alfred Mirsky, whose essay was devoted to the chemistry of DNA, cautioned that there was some doubt about the validity of Avery's claim.[5] And Joshua Lederberg, who wrote about bacterial genetics, referred to Avery's work as "the most outstanding and best investigated" example of "infective heredity." But he didn't mention DNA.[6]

Why could the implications of Avery's discovery not be connected with the canonical knowledge of the 1940s? Ever since DNA had been identified in the cell nucleus by Friedrich Miescher in 1868, it had been suspected of exerting some function in hereditary transactions. This suspicion grew stronger in the 1920s, when DNA was found to be a major component of the chromosomes. But the prevailing view of the molecular nature of DNA made it nearly inconceivable that DNA could be the carrier of hereditary information.

First, until well into the 1930s, DNA was generally considered to be merely a tetranucleotide, composed of one unit each of deoxy-adenylic, -guanylic, -thymidylic, and -cytidylic acids. Second, even when it was finally realized by the early 1940s that the molecular weight of DNA is actually much higher than allowed by the tetranucleotide hypothesis, it was still widely believed that the tetranucleotide was the basic repeating unit of the large DNA polymer, in which the four different kinds of nucleotides recur in regular sequence.

DNA was viewed as a monotonous macromolecule, which, like other biopolymers such as starch or cellulose, is always the same, no matter what its biological source, and the ubiquitous presence of DNA in the chromosomes was generally explained in purely physiological or structural terms. It was usually to the chromosomal protein that the informational role of the genes had been assigned, since the great differences in the specificity of structure that exist between various proteins in the same organism, or between similar proteins in different organisms, had been appreciated since the beginning of the century.

The conceptual difficulty of assigning a genetic role to DNA had not escaped Avery and his coworkers. At the conclusion of their paper they stated: "If the results of the present study of the transforming principle are confirmed, then nucleic acids must be regarded as possessing biological specificity, the chemical basis of which is as yet undetermined."[3]

The mystery of the nature of the gene, and the possibility that the mechanisms of its self-replication and governance of cell function might be explainable only in terms of hitherto unknown principles of physics and chemistry, attracted some physicists to genetics. The eventually most influential of these physicists was Max Delbrück. In 1935, aged twenty-nine, Delbrück published a speculative paper, in which he pointed out that one of the most striking aspects of the gene is its long-term stability.[7] To account for that stability, he proposed that the gene is a molecule whose constituent atoms are fixed in their mean positions and electronic states. Whenever an atom happens to acquire an energy greater than the activation energy required to change its particular state, a discontinuous, saltatory change would occur, corresponding to a mutation.

Ten years later, the ideas developed in Delbrück's little-known paper were popularized in the widely read little book, *What is Life?*, written by one of the two inventors of quantum mechanics, Erwin Schrödinger.[8] He resurrected the Aristotelian notion of the gene as an *information carrier* and postulated that its information is stable because the chromosome in which it is embedded is a one-dimensional, *aperiodic crystal*. Schrödinger suggested that the aperiodic crystal consists of a long sequence of a few repeating, isomeric elements, whose exact pattern of succession represents the hereditary information. He illustrated the vast combinatorial possibilities of such a crystal by means of an example in which he used the three symbols of the Morse code— dot, dash, and space—as its isomeric elements. Thus Schrödinger became the first person to put forward the notion of a linear genetic code.

Meanwhile, Delbrück had begun to attack the gene problem experimentally. He took up the study of bacterial viruses, or *phages*, in 1938, as a postdoctoral fellow at Caltech. Although phage particles are too small to be seen under ordinary microscopes and structurally and chemically are very simple— half protein, half DNA—they are nevertheless endowed with the capacity for self-reproduction. As Delbrück showed, each phage particle infecting a bacterial host cell gives rise to a hundred or so identical progeny phages within half an hour. Thus the central problem of gene replication could be put in simple terms: How does the parental phage particle manage to produce a crop of a hundred progeny particles during that half hour? Two years later, Delbrück's meeting Salvador Luria and Alfred Hershey brought into being the Phage Group, whose members were united by a single common goal—to solve the mystery of the gene. In 1947, Luria, by then a professor at Indiana University, took on the nineteen-year-old James Watson as his graduate student and initiated him as a member of the Phage Group.

One of the most influential of the experiments done by members of the Phage Group was Hershey's demonstration that only the phage DNA enters the cell upon infection of its bacterial host.[9] The phage protein remains outside, devoid of any further function in the reproductive drama about to ensue. Thus it was concluded that the genes of the parent phage carrying the information necessary for directing the synthesis of progeny reside in the phage DNA.

In contrast to Avery's experimentally much more clear-cut result, Hershey's reconfirmation that DNA is the genetic material had an immediate and profound impact. Henceforth, genetic thought was mainly focused on DNA, because ideas about the molecular structure of DNA had changed significantly in the meantime. More refined chemical analyses by Erwin Chargaff of DNA samples from diverse biological sources had shown that, contrary to the demand of the tetranucleotide hypothesis, the relative contents of the four nucleotide bases—adenine, guanine, thymine, and cytosine—vary over a wide range.[10] Hence DNA is not a monotonous macromolecule and could embody genetic information in the form of an arbitrary sequence of the four different kinds of nucleotides. In other words, the DNA molecule would correspond to Schrödinger's one-dimensional aperiodic crystal of heredity, with the four different nucleotides being the repeating, isomeric elements of the genetic code.

Chargaff also noted that, notwithstanding the wide variations in relative content of adenine, guanine, thymine, and cytosine in DNA samples from different biological sources, it was always the case that adenine and thymine are present in equal proportions, as are guanine and cytosine. Chargaff wrote that this equivalence rule is "noteworthy . . . but whether [it] is more than accidental cannot yet be said."[10] The equivalence rule did turn out to be definitely more than accidental and would provide strong chemical support for Watson and Crick's proposed DNA double helix.

Concurrent with the rise of the Phage Group there had also taken place a movement into biology by another set of physical scientists, whose interest was focused on the analysis of the three-dimensional structure of biological macromolecules by X-ray crystallography. One of these crystallographers, W. T. Astbury, obtained the first X-ray diffraction pictures of DNA in 1945, from which he inferred that the purine and pyrimidine bases of successive nucleotides form a dense stack perpendicular to the long axis of the DNA molecule.[11]

Six years later, Linus Pauling achieved the first great triumph of the structural analysis of biological macromolecules when he discovered the α-helix as the secondary structure of proteins.[12] Pauling's success was due in part to a novel approach to structure determination, in which guesswork and model-building played a much greater role than they did in the more straightforward analytical procedure of more conventional crystallographers.

Learning of Pauling's success and a chance meeting with Maurice Wilkins, who was already carrying out X-ray crystallographic analyses of DNA in London, inspired James Watson, who had obtained his Ph.D. under Luria in the previous year, to have a try at working out the structure of the DNA molecule. Watson joined John Kendrew at the Cavendish Laboratory in Cambridge to gain the necessary skills in X-ray crystallography for his project. There, Watson met Francis Crick, a Ph.D. student to whom it had also occurred that the three-dimensional structure of DNA would be likely to provide important insights into the nature of the gene. Watson and Crick then began a collaboration, whose story is known to one and all through Watson's memoir, *The Double Helix*.[13]

The discovery of the DNA double helix opened up enormous vistas to the imagination. It was to provide the highroad to understanding how biological information is encoded in the gene as a string of molecular symbols and how that information is replicated for transmission from one generation to the next. By now, at the close of the 20th century, it has become obvious that Watson and Crick's paper was the watershed of our century's biology. Genetics was only the first biological specialty to undergo its molecular-biological *aggiornamento*. Eventually, no life science discipline, pure or applied, from evolution and microbiology through physiology, embryology, neurobiology, and psychology to medicine, nutrition, pharmacology and agriculture, failed to be profoundly transformed by the advent of molecular biology. In fact, molecular biology has become so all-pervasive that it seems to be disappearing as a discrete disciplinary entity, as biochemistry disappeared some years ago.

Was there anything unique about this epochal contribution of Watson and Crick? Would molecular biology have developed differently if there hadn't been any Watson and Crick to start it off in the grandiose way they did? In his review of *The Double Helix*, Chargaff posed this question and, by way of an answer, asserted that "it is not the men who make science, it is science that makes the men. What A does today, B and C and D could surely do tomorrow."[14] Thus he insinuated that if Watson and Crick had not existed, we would have had the double helix anyhow. Yet, Chargaff thought that such ready replaceability is not the rule for the men who make art, of one of whom it can be said that "*Timon of Athens* could not have been written . . . had Shakespeare . . . not existed."

Chargaff's comparison of Watson and Crick with Shakespeare carries little philosophical or historical weight. Obviously, we would not have had *Timon*, had Shakespeare not lived, any more than we would have had Watson and Crick's April 1953 *Nature* paper had they not lived. As for the *content* of Watson and Crick's paper, however, we can be fairly sure that even without them, a double helical structure of DNA would have been described before too long by other people—for instance, as has often been suggested, by Maurice Wilkins and Rosalind Franklin. But then neither is the *content* of *Timon* unique, in that even without Shakespeare, plays not only would have been but *have* been written that show how a man may make his response to the injuries of life, how he may turn from lighthearted benevolence to passionate hatred towards his fellow-men.

One could reasonably argue that *Timon*'s content *is* uniquely Shakespeare's, because playwrights E, F, and G, although they may have communicated more or less the same insights as the Great Bard, would not have done it in quite the same exquisite form. But then we cannot take for granted that doctors B, C, and D, who eventually would have found the structure of DNA would have communicated it in quite the same exquisite form as Watson and Crick and published a paper that produced the same revolutionary effect on 20th century biology. On the basis of my personal acquaintance with the proto-molecular biologists of that time, I believe that had Watson and Crick not existed, the insights they provided in one single package would have come out much more gradually, dribbling out over a period of many months or years.

Peter Medawar summed it all up in his review of *The Double Helix*: "The great thing about Watson and Crick's discovery was its completeness, its finality. . . . If Watson and Crick had been seen groping toward an answer. . . . If the solution had come out piecemeal instead of in a blaze of understanding, then it still would have been a great episode in biological history."[15] But it would not have been the dazzling achievement that it, in fact, was.

REFERENCES

1. WATSON, J. D. & F. H. C. CRICK. 1953. A structure for deoxyribonucleic acid. Nature **171**: 737–738.
2. MULLER, H. J. 1950. The development of the gene theory. *In* Genetics in the 20th Century. L. C. Dunn, Ed. :77–99. Macmillan. New York.
3. AVERY, O. T., C. M. MACLEOD & M. MCCARTY. 1944. Studies on the chemical nature of the substance inducing transformation of pneumococcal types. J. Exp. Med. **79**: 137–159.
4. STENT, G. S. 1972. Prematurity and uniqueness in scientific discovery. Scientific American **227**(December): 84–93.
5. MIRSKY, A. E. 1950. Some chemical aspects of the nucleus. *In* Genetics in the 20th Century. L. C. Dunn, Ed. :127–153. Macmillan. New York.
6. LEDERBERG, J. 1950. Genetic studies with bacteria. *In* Genetics in the 20th Century. L. C. Dunn, Ed. :263–289. Macmillan. New York.
7. TIMOFEEF-RESSOVSKY, N. W., K. G. ZIMMER & M. DELBRÜCK. 1935. Über die Natur der Genmutation und Genstruktur. Nach. Ges. Wiss. Goettingen. Math-Phys. Kl. Fachgruppe VI. **13**: 190–245.
8. SCHRÖDINGER, E. 1945. What is Life? Cambridge University Press. New York.
9. HERSHEY, A. D. & M. CHASE. 1952. Independent functions of viral protein and nucleic acid in growth of bacteriophage. J. Gen. Physiol. **36**: 39–56.
10. CHARGAFF, E. 1950. Chemical specificity of the nucleic acids and mechanisms of their enzymatic degradation. Experientia **6**: 201–209.
11. ASTBURY, W. T. 1947. Nucleic Acids. Symp. Soc. Exp. Biol. **1**: 66.
12. PAULING, L., R. B. COREY & H. R. BRANSON. 1951. The structure of proteins: Two hydrogen-bonded helical configurations of the polypeptide chain. Proc. Natl. Acad. Sci. USA **37**: 205–211.
13. WATSON, J. D. 1968. The Double Helix. Atheneum. New York. [A Critical Edition, G. S. Stent, Ed. Norton. New York, NY. 1980.]
14. CHARGAFF, E. 1968. A quick climb up Mount Olympus. Science **159**: 1448–1449.
15. MEDWAR, P. 1968. Lucky Jim. New York Review of Books. March 28: 3–5.

Photos from the Past

Meeting of the RNA tie club in Portugal Place, Cambridge, England, 1955. *Rear*: Francis Crick, Leslie Orgel. *Front*: Alexander Rich, James Watson. (Photograph courtesy of A. Rich.)

Francis Crick and James Watson in the Cavendish. (From *The Double Helix* by J. D. Watson. Photo by A. C. Barrington Brown courtesy of the Cold Spring Harbor Laboratory Archives.)

Francis Crick and James Watson in front of King's College Chapel. (From *The Double Helix* by J. D. Watson. Photo courtesy of the Cold Spring Harbor Laboratory Archives.)

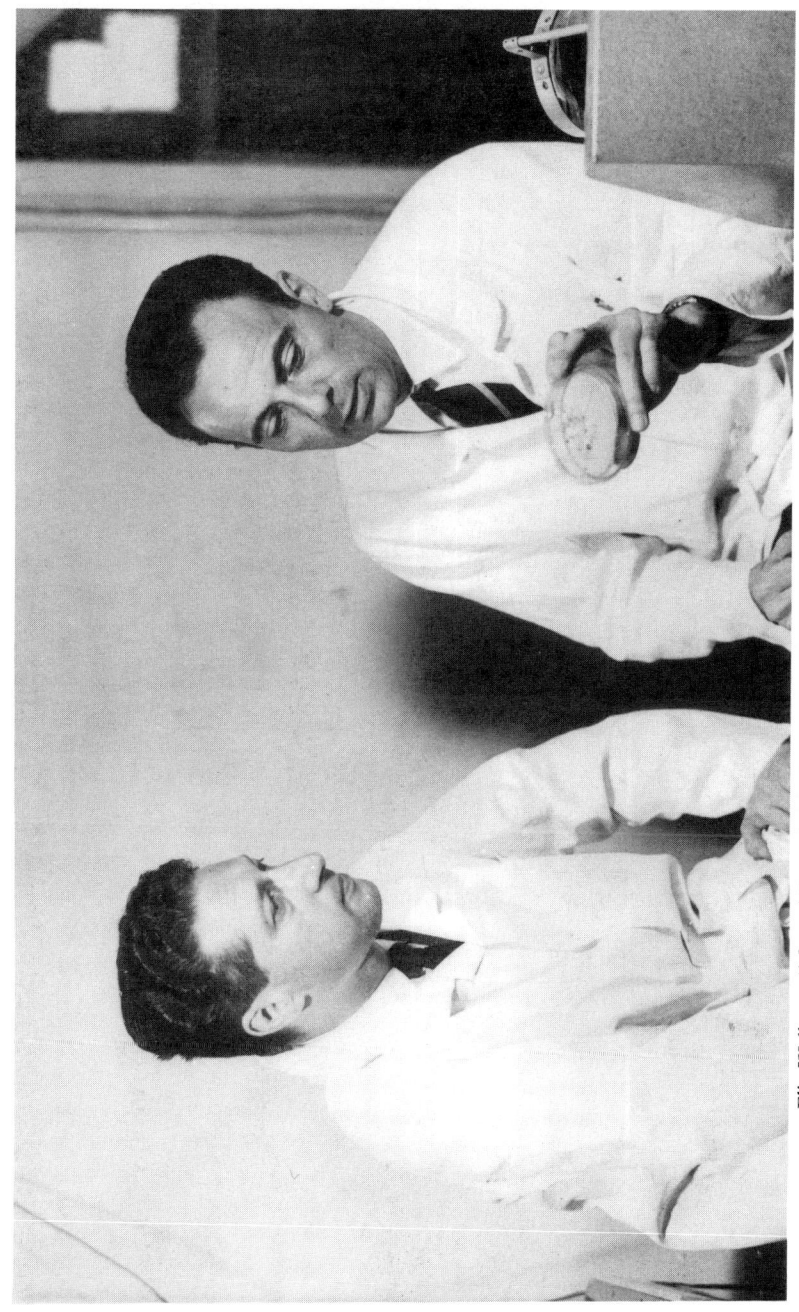

Elie Wollman (*left*) and François Jacob (*right*) looking at a petri dish. (Photo courtesy of F. Jacob.)

Meeting at the Cavendish Laboratory, Cambridge, England, 1957. *From left to right*: Donald Caspar, Pauline Cowan, Alexander Rich, Francis Crick, Sydney Brenner. (Photo courtesy of A. Rich.)

PHOTOS FROM THE PAST

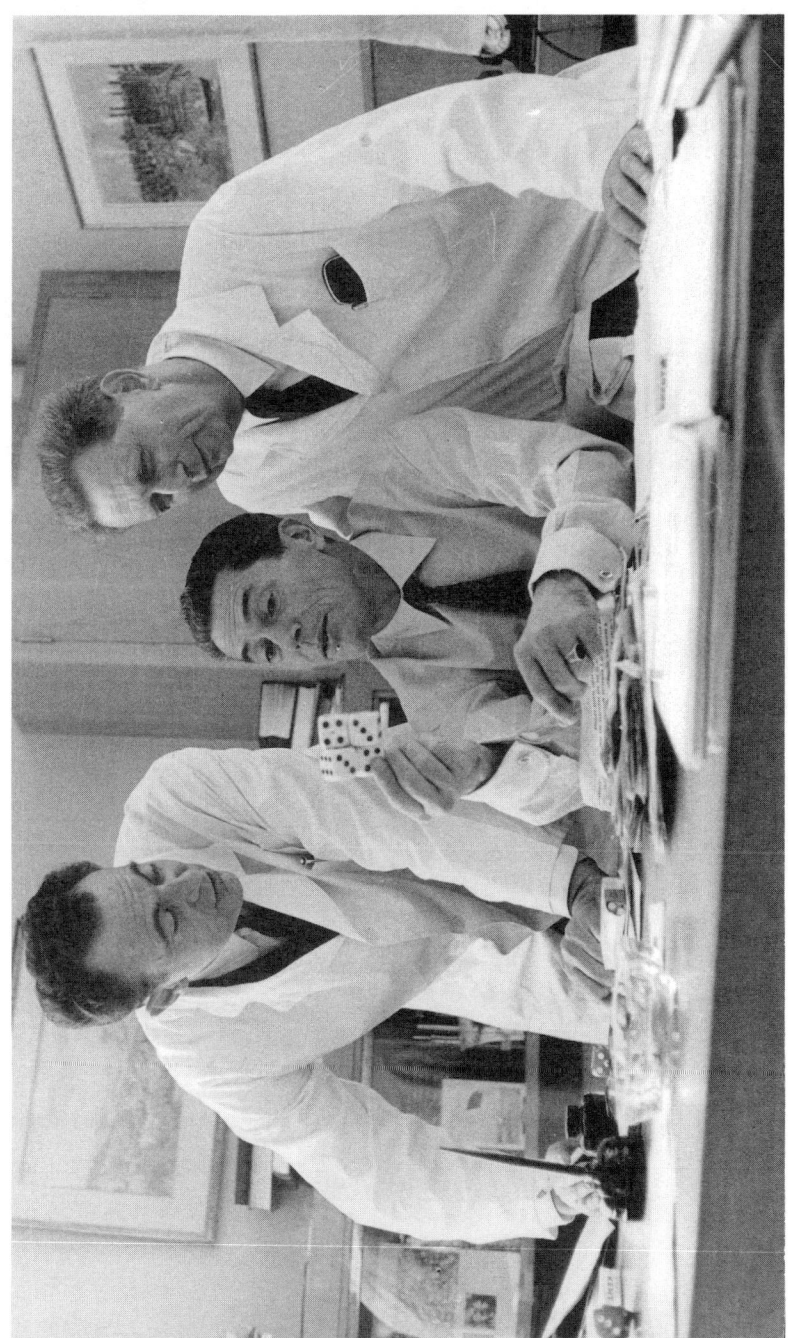

From left to right: François Jacob, Jacques Monod, André Lwoff. (Photo courtesy of F. Jacob.)

Salvador Luria, June 1951. (Photo courtesy of Karl Maramorosch.)

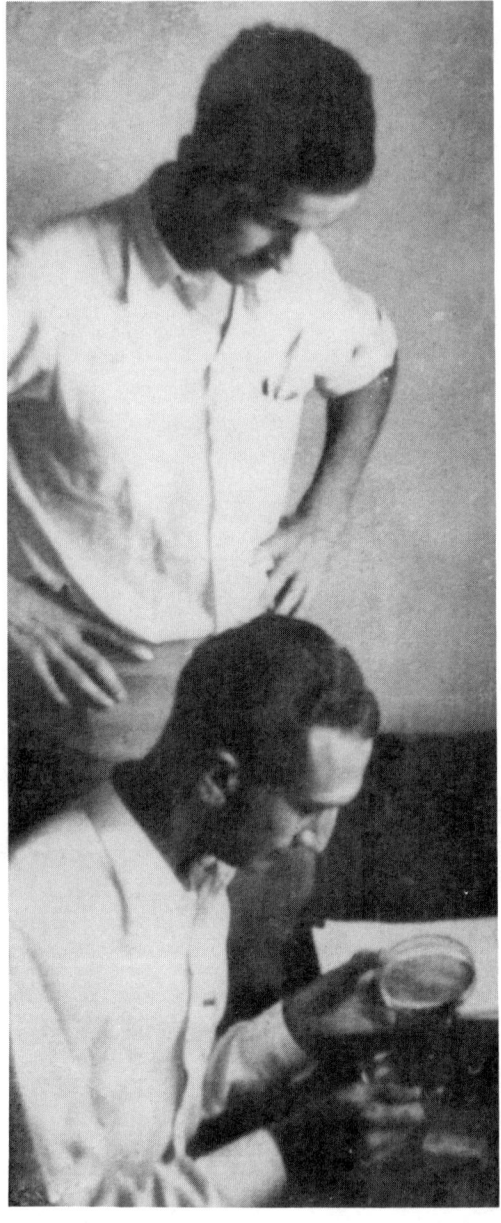

Salvador Luria and Max Delbrück (*seated*). (Photo courtesy of the Cold Spring Harbor Laboratory Archives.)

PHOTOS FROM THE PAST

Joshua Lederberg. (Photo courtesy of Karl Maramorosch.)

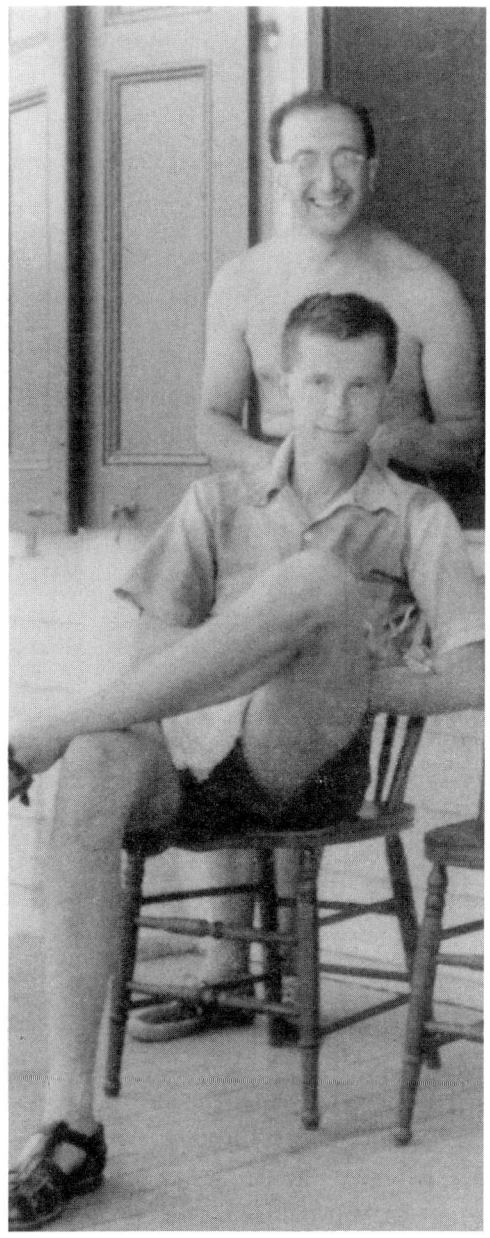

Salvador Luria and Max Delbrück (*seated*). (Photo by Karl Maramorosch courtesy of the Cold Spring Harbor Laboratory Archives.)

From left to right: Jacques Monod, André Lwoff, and Lou Siminovitch.
(Photo by Karl Maramorosch.)

Rollin Hotchkiss and Max Delbrück. (Photo by Karl Maramorosch.)

PHOTOS FROM THE PAST

Leo Szilard and Alfred Hershey. (Photo by Karl Maramorosch.)

Martha Chase and Alfred Hershey at Cold Spring Harbor. (Photo by Karl Maramorosch.)

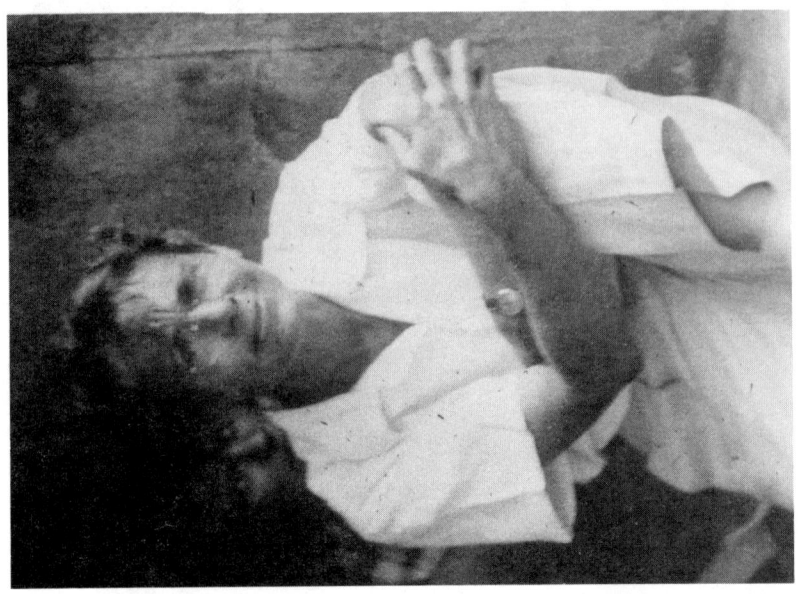

Barbara McClintock at Cold Spring Harbor in the early 1940s. (Photos courtesy of the Cold Spring Harbor Laboratory Archives.)

PHOTOS FROM THE PAST

Meeting that took place in Scotland around 1965 to discuss the future of molecular biology. The meeting was deliberately informal and unrecorded. Several key molecular geneticists attended, including (*from left to right at the top of the staircase*) Sol Spiegelman, Salvador Luria, and André Lwoff and Francis Crick (*at the bottom of the staircase*). Also seen here are Leo Sachs, Renato Dulbecco, Michael Staker, Harry Ruben, Bill Paul, James Darnell, Guido Pontecorvo, Lionel Crawford, and Martin Pellock. (Photo courtesy of Sir Walter Bodmer.)

Participants at the first EAMDA (European Alliance of Muscular Dystrophy Association) meeting held in Holland in 1984 and organized by Ysbrand Poortman (*on extreme right*). This was one of many meetings that promoted exchange of ideas and resources between cytogeneticists, clinicians, and molecular biologists. Such cooperation eventually led to the cloning of the DMO gene.) (Photo courtesy of Kay Davies.)

PHOTOS FROM THE PAST

James Watson and Francis Crick, honored at the fortieth anniversary of the discovery of the double helix. (Photo by Ed Campodonico courtesy of the Cold Spring Harbor Laboratory Archives.)

James Watson (*right*) congratulating Francis Crick. (Photograph by Ed Campodonico courtesy of the Cold Spring Harbor Laboratory Archives.)

PART II. THE PATHWAY TO THE DOUBLE HELIX

Introduction: "On the Shoulders of Giants"

IRVING M. KLOTZ

Department of Chemistry
Northwestern University
Evanston, Illinois 60208-3113

As we all know, it was Isaac Newton who once remarked "If I have seen farther than others, it is because I have stood on the shoulders of giants." In this section, we will hear from some of our giants.

Any so-called "introduction" of these individuals would be superfluous. What might be appropriate are a few words on the prevailing conceptual attitudes in their fields at the time of their momentous discoveries.

With respect to Maclyn McCarty's paper on the chemical nature of the transforming principle, I want to quote some comments of a contemporary young biochemist who heard one of the earliest talks on this subject, in 1945. Fritz Schlenk writes, "I remember the excellent presentation given by Maclyn McCarty; unfortunately it did not arouse much excitement. . . . Proteins preoccupied the affection of most investigators at that time to such an extent that their surmised role as carriers of genetic information was not readily abandoned." In fact somewhat earlier in this century, skepticism about nucleic acids was even sharper. For example, the geneticist William Bateson, the inventor of the term *genetics*, wrote as follows: "It is inconceivable that particles of chromatin . . . can possess those powers that must be assigned to [genes]. The supposition that particles of chromatin, indistinguishable from each other and indeed almost homogeneous under any known test, can . . . confer all the properties of life surpasses the range of even the most convinced materialis[t]." (At that time, Bateson must have been referring to the infant Francis Crick.)

Maclyn McCarty presents a retrospective of this field from the vantage point of half a century later.

Repeatedly in science, the possible presence of a minuscule amount of impurity in a substance has interjected uncertainty into judgments of the significance of the striking properties of the material. Years before the discovery of the transforming principle by Avery, MacLeod and McCarty, Sumner had encountered similar skepticism in regard to his claims that the enzyme urease was intrinsically a catalytic entity. It took years of careful, painstaking work before the concept and its generality became irrefragable. One must also recognize that in some instances, impurities indeed have been the basis of exceptional properties—for example, silicates in polywater, or boron in graphite blocks for the nuclear pile.

It also required years of intensive studies of the biology of bacteria to remove all skepticism about the nucleic acid nature of genetic elements. A major contributor to the development of bacterial genetics and metabolism in that period, and thereafter, was Rollin Hotchkiss. In his characteristic whimsical style he will present his retrospective of that era using the metaphor of investment counseling.

Shortly before the turn into the twentieth century, the chemist A. W. H. Kolbe opined, "atoms arranged in space . . . the sober chemical world has no taste for such hallucinations." The complete reversal of this viewpoint in our century is due primarily to Linus Pauling, the preeminent chemist of this century. One of Pauling's many great triumphs is the recognition of the α-helix of proteins. In contrast to the skeptical reception of the nucleic acid nature of the transforming principle, the α-helix as a crucial element of protein structure was accepted essentially immediately. As Ian MacArthur, from the laboratory of one of Pauling's major competitors, said in 1953, "straight away this 'α-helix' has an air of conviction." As we are all aware, the α-helix provided a far-reaching construct in the unravelling of relationships between molecular structure and biological behavior.

Unfortunately, at the last moment, Professor Pauling had to withdraw from this symposium. Providentially, Alex Rich is present at this meeting and is the most appropriate stand-in for Pauling, not only because of his own major contributions to nucleic acid and protein structures, but also because of his long association with the Pauling laboratory. The young Alexander Rich, a freshly-hatched M.D., arrived in Pasadena in 1949, just at the onset of the "giant leaps forward" (adopting Chairman Mao's expression) in protein and nucleic acid structure, and Rich has been at the very center of the field ever since.

McCarty, Hotchkiss and Rich have reviewed the origins of molecular biology from the bottom up, from atoms to biological behavior. It behooves us also to be aware of a perspective from the opposite direction, to see the insights into molecular biology that arose from clinical observations. Paul Heller, an eminent hematologist, has spent a lifetime translating hemoglobinopathies into molecular terms. He will present a retrospective from the vantage point of clinical medicine.

All of us here have stood on the shoulders of these giants. It is to them and their colleagues and contemporaries that we owe the enormous extension in the range of our vision in molecular and cellular biology.

A Fifty-Year Perspective on the Genetic Role of DNA

MACLYN McCARTY

The Rockefeller University
1230 York Avenue
New York, New York 10021

The term "fifty years" in the title was chosen to indicate the central position of the year 1943 in the completion of the research and the development of the first report on the identification of the chemical nature of the pneumococcal transforming substance. The laboratory work came to nearly a complete halt in the summer of that year so that our efforts could be devoted to the preparation of the manuscript, which was ultimately submitted for publication on November 1.

By the fall of 1942 we were aware that our laboratory results from a variety of different experimental lines seemed to be telling us that the biological activity of our transforming extracts was associated with the DNA. Accordingly, we devoted the next several months, extending into the Spring of 1943, to the task of preparing several large lots of transforming DNA by the purification procedures that we had developed, and subjecting each of the preparations to a panel of analytical procedures. This was not a trivial effort, since we used type III pneumococcal cells from at least 200 liters of culture for each preparation and carried out a lengthy series of purification steps before undertaking the analyses. It is probably worthwhile to pause and ask why we felt it was necessary to repeat this process so many times before reporting our findings. The answer, I believe, lies in the state of knowledge of the nucleic acids at that time.

Substances of this class had been known for more than seventy years since the work of Miescher, who first derived the material that he termed "nuclein" from the nuclei of cells from purulent material. He described many of the properties of his nuclein, including its high content of phosphorous. It took a few years for this substance to become better defined, in part as a result of the presence of protein in the nuclein preparations, and it was not until 1889 that a clear separation of the protein and phosphorus-containing components was achieved by Richard Altman, who first applied the name nucleic acid to the latter.

Much of the biochemical work on the composition of the nucleic acids emanated from the laboratory of Albrecht Kossel, who identified the purine and pyrimidine bases of the nucleic acids and detected the presence of carbohydrate as a constituent of the molecule. It remained for Phoebus A. Levene, in his extensive studies of nucleic acids beginning in the early 1900s, to identify the sugars as ribose and deoxyribose. The latter was not published until 1929, so that at the time Griffith first reported his discovery of the phenomenon of pneumococcal transformation, the name deoxyribonucleic acid did not exist, not even in the original "desoxy" version. Levene, among his many other contributions, was also responsible for recognition of the nucleotide as the primary structural unit of nucleic acids.

At the same time that the nucleins were becoming better defined, cell biologists were getting a clearer picture of the cell nucleus and the existence of chromatin and the chromosomes. Walther Flemming had suggested in 1882 that chromatin might be identical with nuclein. It is not surprising, therefore, that a number of investigators entertained the idea that nuclein might be the substance that carried hereditary information, and some of these even focused on the nucleic acid component. One of these was an American, Albert Mathews, who first commented on the possible genetic role of nucleic acid while working on the composition of nuclein with Kossel in Marburg. He later retracted his views on this point, however, when continued chemical studies on nuclei from various sources led him to conclude that the nucleic acids from all sources were pretty much the same, while the associated proteins were highly variable.

Another example came from E. B. Wilson, who wrote in his book *The Cell in Development and Heredity* in 1900 that "There is . . . considerable ground for the hypothesis that in a chemical sense this substance [nucleic acid] is the most essential nuclear element handed on from cell to cell." He also retracted this view in a later publication, possibly influenced by Mathews' data on the chemical uniformity of nucleic acid, but also by histochemical evidence that had been interpreted as showing that nucleic acid disappeared from the nucleus during certain phases of the cell cycle.

Somewhat later the eminent chemist Emil Fischer contributed some remarkable speculations of his own on the genetic role of the nucleic acids. In a reprinting of Fischer's book *Aus meinem Leben* that appeared in 1987, Bernard Witkop prepared a prologue that quotes a passage from a paper published by Fischer in the *Berichte* in 1914. This quote, which is in discussion of his work on the synthesis of methylated purines, runs as follows in Witkop's English translation:

> With the synthetic approaches to this group we are now capable of obtaining numerous compounds that resemble, more or less, natural nucleic acids. How will they affect various living organisms? Will they be rejected or metabolized or will they participate in the construction of the cell nucleus? Only the experiment will give us the answer. I am bold enough to hope that, given the right conditions, the latter may happen and that artificial nucleic acids may

be assimilated without degradation of the molecule. Such incorporation should lead to profound changes of the organism, resembling perhaps permanent changes or mutations as they have been observed before in nature.

These visionary comments seem to be proposing genetic engineering with synthetic nucleic acids, and were clearly too advanced to draw much attention at the time. I know of nothing to suggest that Fischer ever returned to these speculations in the few remaining years of his life.

These thoughts about a genetic role for the nucleic acids were the exceptions, and they had disappeared by the 1930s, when the notion that they had a rather monotonous and unvarying structure certainly became the dominant reason for the widespread tendency to discount the possibility that they carried the genetic message. This point of view was further strengthened by Levene's popular tetranucleotide theory. Even though he did not state it explicitly in these terms and there was very little in the way of experimental evidence to support it, the theory gave rise to a picture of the nucleic acid molecule as an array of tetranucleotides in which the four nucleotides appeared in the same repeating order.

Even without this restricted view of nucleic acid structure, there was ample reason to be more impressed with the potential of proteins for diversity, with their greater number of constituent units and large variation in physical properties. Thus, the emphasis on the probable protein nature of the gene was almost universal, the commonly expressed conclusion being that if genes were composed of a known substance, there was only protein to be considered. One of the rare exceptions that we found to this point of view was in a paper from Jack Schultz, who, in writing in 1941 on the nucleoprotein nature of the gene, pointed out that there was too little information to conclude that the nucleic acids were monotonously uniform and that much more work had to be done before one could exclude the possibility of specificity of this chromosomal component.

We were acquainted with the generally negative attitude toward a possible genetic role for nucleic acids from about the time that we first began to suspect that the transforming principle might be DNA. We had also heard verbally from more than one source that it couldn't be DNA because "nucleic acids are all alike." This, I believe, is the explanation for the extended efforts that we made to assure ourselves that the evidence we had accumulated to implicate DNA was indeed correct. Avery was more reluctant to challenge the prevalent dogma than MacLeod and I were, but by April 1943, when it was time to prepare our annual report to the Board of Scientific Directors of the Rockefeller Institute, he was at least willing to take the first step.

Up to that time, the annual reports had not even mentioned that DNA had been identified as a component of the transforming extracts, an observation that had first been made in January 1941. In preparing the 1943 report, the full background was provided for the first time, together with a description of the current experiments and the evidence suggesting that the active material was DNA. The genetic interpretation of the transformation phenomenon was stressed, and, with the proviso "if the present studies are confirmed," the report went on to discuss the implications of DNA involvement in transformation. Thus we went semi-public with this information for the first time.

In May, Avery wrote the famous letter to his brother, Roy, explaining why he would not be able to move to Nashville on reaching emeritus status that summer as originally planned. He went further in this letter than he had in the annual report, in chattier, less formal prose, but the same reservations and uncertainties were included. After an extended description of the research, which had apparently not been communicated to Roy before, he went on as follows:

> We are now planning to prepare new batch & get further evidence of purity & homogeneity by use of ultracentrifuge & electrophoresis. This will keep me here for a while longer. If things go well I hope to go up to Deer Isle, rest a while—Come back refreshed & try to pick up loose ends in the problem & write up the work. If we are right, & of course that's not yet proven, then it means that nucleic acids are <u>not</u> merely structurally important but functionally active substances in determining the biochemical activities and specific characteristics of cells—& that by means of a known chemical substance it is possible to induce <u>predictable</u> and <u>hereditary</u> changes in cells. This is something that has long been the dream of geneticists. The mutations they induced by x-ray and ultraviolet are always unpredictable, random, and chance changes. If we're proven to be right—and of course that's a big <u>if</u>—then it means that both the chemical nature of the <u>inducing stimulus</u> is known & the chemical structure of the <u>substance produced</u> is also known—the former being thymus nucleic acid—the latter Type III polysaccharide and both are thereafter reduplicated in the daughter cells—and after innumerable transfers and without further addition of the inducing agent, the same active & specific transforming agent can be recovered far in excess of the amount originally used to induce the reaction. Sounds like a virus—may be a gene. But with mechanisms I am not now concerned—one step at a time—& the first is, what is the chemical nature of the transforming substance? Someone else can work out the rest. . . .

As a footnote to this letter, I should point out that Avery did not move to Nashville until five years later, in the summer of 1948. The "new batch" that he referred to in the letter was already well under way. The procedure was modified only in that all steps, except the extraction of the heat-killed type III pneumococci and the enzymatic treatment of the extracts, were carried out in the cold. Not surprisingly, it turned out to be the most active preparation we had ever obtained, and all of the analyses, including the ultracentrifugal and electrophoretic studies lived up to expectations. Thus, by the end of June, when Avery was preparing to head for Deer Isle, we were ready to begin work on a manuscript. He suggested that over the summer I prepare a draft of the body of the paper, including materials, methods, and description of the experiments, while he drafted the introduction and discussion sections. The stage was set for a full-time effort to complete the manuscript in the fall.

In addition to working on the paper that summer, I had some experiments to carry out to confirm certain observations that we had made during the previous months. For example, in preparing our purified DNA for elementary analysis, we had dialyzed the samples against distilled water and subsequently found that the material had lost all activity in the transforming system. This was a surprise since we had known for some time that solutions of the purified DNA in physiological saline were stable at 4°C for months. In comparing solutions of the same preparations in distilled water and in saline with daily titrations, I confirmed that in distilled water all activity was lost in a few days, while there was no detectable change in the activity of the saline solution over a period of weeks. The loss of activity in the absence of salt was not accompanied with any loss in the viscosity of the solution. Although we did not know the explanation for this phenomenon, it was clearly something to be avoided.

A similar situation was encountered with regard to drying the purified DNA. At an earlier stage of the research, we had been in the habit of drying crude extracts by lyophilization in order to store them for future fractionation experiments. The full activity of these crude extracts appeared to be retained indefinitely after drying. However, we had encountered evidence that this was not true for our purified product. On quantitative studies of preparations of purified product dried either by lyophilization or from alcohol and ether, I found that while the activity was preserved in a sample of the dried material, when the material was redissolved in saline immediately after drying was completed, substantial loss of activity was seen the next day upon titration of another sample, and after three days all activity was gone. Here again there was no apparent change in physical properties nor any clue as to the basis for the loss in activity. The idea was to include information on the effects of distilled water solution and drying in the manuscript as a warning to those trying to repeat the studies.

Some additional work also seemed to be needed to support the statement that the transforming substance was recoverable from transformed cells in amounts far in excess of that originally used to induce the transformation. It seemed certain that this might be so, but we could find no record of an actual experiment demonstrating this point. It was a simple exercise to isolate a single colony of transformed type III pneumococci and grow a few liters of culture from which to prepare the usual extract. Of course, these cells proved to be as good a source of transforming DNA as the native type III strain, but it was necessary to document the point.

I also had an opportunity during that summer to continue these experiments that I had initiated to obtain a purified DNase. The literature had indicated that mammalian pancreas was the richest source of this enzyme, and at this point I was using commercial pancreatin, which was a convenient source but not very amenable to fractionation procedures. However, by that August we had learned something about the enzyme, such as the requirement of magnesium ion for activation, and had crude preparations that would rapidly destroy the transforming activity of our purified DNA when used at concentrations as low as 0.01 mg/ml. This was far from good enough, however, since the rather crude material had proteolytic as well as other enzymatic activities. Thus we decided that we were not in a position to include any of this information in the paper. It took me almost two years to obtain a suitable purified DNase and nail down its effect on transforming DNA.

When Avery returned from Deer Isle in the fall, we began the process of finalizing the manuscript. I had had some previous experience with his meticulous care in achieving accuracy and felicitous phrasing in his writing and editing, and so I knew what to expect. On this occasion, we repaired to a small room in the Institute library for our writing, in order to avoid interruptions even by the telephone. There, over a period of a few weeks, we hammered out a draft that approached final form, reaching this point in the middle of October—or almost exactly fifty years ago. After the final touches were completed, Avery delivered it by hand to Peyton Rous, the editor of the *Journal of Experimental Medicine* on November 1.

Before the close of 1943 the full staff of the Rockefeller Institute had a preview of our conclusions, when Avery presented the work at the regular Friday afternoon meeting on December 15. There are those who remember this as the first occasion that the identification of the transforming principle as DNA was first challenged by Alfred Mirsky. As one intimately concerned with the matter, I can assure you that this was not the case. Avery's talk was received warmly, without substantive discussion of the details of the research.

At the end of 1943 we were thus pretty well convinced that our evidence showed that the current view that nucleic acids lacked the potential for biological specificity must be incorrect, and we also thought it likely that its property of transferring hereditary information could not possibly be limited to the bacterial world. Of course, we had no idea what the structural basis for the biological specificity might be and could see that determining this was an important challenge for the future. It should be noted that at that time the structural basis for the specificity of enzymes and antibodies was also not well established, so that in this respect things were not much further ahead for proteins.

We knew that the rough, unencapsulated strains of pneumococcus had as much DNA as the type III strains, but that it was totally lacking in transforming activity in our system. It seemed necessary to assume that only a small fraction of the DNA in the transforming extracts was related to the synthesis of a capsule, and that the rest was concerned, as in the rough strains, with all of the other functions of the bacterial cell. We did not get around to putting it this way in print until the report on the effect of purified DNase on the transforming substance, but these points had been part of laboratory discussions. We were thus aware of the challenge of determining the basis for the specificity of the nucleic acids, but nothing in our experience suggested an experimental approach to this problem.

It should be remembered that all of the research on pneumococcal transformation from the time of its discovery by Griffith until almost ten years afterwards was associated with the clinical problem of pneumococcal pneumonia and pursued for its possible bearing on the control of this disease. The situation was different during the final years of the research, since we no longer had any illusions that the identification of the transforming principle would have direct applicability to the problem of pneumonia and it seemed obvious that it must have broader biological implications. Once we became convinced that we were dealing with DNA, however, we could imagine that this information might prove useful in medical research in the future. On the other hand, it is clear that at that time we did not even dream of the kind of developments that would occur in the next fifty years and make possible the application of DNA biology to a vast array of medical problems.

Since I did not pursue these matters intensively after 1946 but followed my original training by moving to different problems concerned with the pathogenesis of disease, I have of necessity observed these startling developments pretty much as an outsider. It has nevertheless been a real pleasure to follow each step along the way to the current flourishing state of molecular biology, and I appreciate the opportunity to participate in this conference honoring the architects of that major step that took place forty years ago.

DNA in the Decade before the Double Helix

ROLLIN D. HOTCHKISS[a]

The Rockefeller University[b]
New York, New York 10021

The acronym DNA belongs—as appropriate for acronyms—to an agency of government, an ancient and powerful one, deoxyribonucleic acid. We celebrate here the discovery that DNA is a cooperatively organized and genuinely bipartisan governing agency, and at this point we shall be reflecting a little on how it was elected to its present eminence.

The biological action of DNA revealed by Avery, MacLeod, and McCarty in 1944,[1] and further consolidated by Maclyn McCarty in the next two years,[2] was a challenging "breakthrough." It was an altogether unexpected discovery. For these workers—three M.D.s—were investigating an unprecedented change of growth habit in bacteria—the Griffith "transformation"—ringed about with uncertainties. Exploring it required the development of some new accessory tools and techniques. Clearly, their results furnished a primary stimulus for the question taken up some years later by Watson and Crick, whose model for DNA[3] we so rightly honor at this festival.

I had the privilege of participating right in the Avery laboratory in the further development that immediately followed the classical papers, from 1946 onward. I am grateful for the invitation to recapitulate here some of the early steps and hurdles that a few of us had to negotiate in surveying for those broad highway bridges that now connect the territories of biochemistry and genetics. I shall be telling a rather personal tale of the decade from 1945 to about 1954—moving from one well-known piece of work toward the one that we celebrate here. I am supposing that, because history has a foreshortening effect, some of that intervening period will be unfamiliar by now, though of course I cannot offer you new scientific conclusions from that era that you do not already know by implication. But I believe that in that period we were able to heighten the importance, and sharpen the definition, of what became the problem of DNA structure.

---

[a] Current address: Research Professor in Biology, SUNY at Albany, 1400 Washington Avenue, Albany, New York 12222.
[b] Professor Emeritus.

DEVELOPING THE AVERY-McCARTY FINDING

The Avery, MacLeod, and McCarty paper of 1944[1] was "complete" in the sense that it admirably fulfilled its own carefully restrained claims. Nevertheless its implications were challenging, and that meant that many scientists reflecting on the report worried about one or the other of two broad groups of questions that it portended: Can it really be DNA, *and only DNA*, that is accomplishing the change?—and: does the heritability of bacterial capsules have anything to do with genetics? Most scientists, according to territorial imperatives, were more troubled by one, and less by the other, of these two lines of questions. Of course there were numerous secondary questions, such as: can such big molecules as DNA get inside bacterial membranes? I have reported on the questions and interpretations raised in the literature of that time (also under the auspices of the New York Academy of Sciences[4]), and suggested that most interested biologists were challenged, but could not readily perceive what to do next to explore the Avery phenomenon. This paralysis was seen by Gunther Stent as due to a general unpreparedness for what he has characterized as a "premature" finding.[5]

From 1946 to 1948, while Avery proceeded with his retirement from the Rockefeller Institute and McCarty moved on to the genetics and etiology of the streptococcal diseases, I enjoyed generous access to all of their laboratory records and materials. During the first year, there was also the advantage of the ongoing participation by Harriett Taylor. She and I kept the fairly complicated transformation system alive and well and were able to simplify its requirements and make it more efficient and reproducible.[6] With Avery's continued interest and participation, we entertained and explored many working hypotheses and questions about the system. Harriett was a brilliant and enthusiastic biologist; we inevitably had rich opportunities for hours of discussion, reinforcing each other reciprocally in biology and chemistry, and those influences were to continue for the next several years.

After a year Harriett Taylor moved to Paris, where as Ephrussi-Taylor she continued a fine analysis of pneumococcal encapsulation. That aspect of transformation was also explored by MacLeod, with coworkers Robert Austrian and Marjorie Krauss, at New York University. Andre Boivin and coworkers in Toulouse soon reported similar antigen transformations for certain *E. coli* strains, but their strains and system were not confirmed or successfully followed up. By 1950, Alexander, from Columbia Medical School with Leidy and soon, Zamenhof,[7] took up capsular antigen transfer with *Hemophilus*. These are the few investigators who dealt with DNA-produced bacterial transformation during the decade leading up to the Watson-Crick paper of 1953.

TABLE 1. What We Needed to Know about DNA Transformation in 1945

Is "DNA" a genuine coherent category?
Is transforming DNA much like calf thymus DNA?
Yet is it different from thymus DNA?
Are DNAs usually found as nucleoproteins?
Can DNA also contain RNA?
Is nucleoprotein or RNA present in transforming DNA?
Does DNase act *only* upon DNAs?

How could DNAs differ chemically?
Is DNA really a repeating tetranucleotide?
Are bases all in 1:1 ratios?
Are DNA polynucleotide chains interrupted?
Can DNA have a secondary structure?

What is the genetic significance of bacterial transformation?
Is transformation an induction?
Do bacteria (with no sex cycle) actually have genes?
Is transformation a directed mutation?
Is transformation a gene transfer?
Are only *surface* antigens altered by DNA?
Can other species than pneumococci be transformed?
Are all traits modifiable by appropriate DNA?
Can traits be displaced, or "recombined" by DNA?
Can DNA transformation be quantitatively measured?
Can transformation be carried out in defined media?

THE QUESTIONS AT ISSUE

Avery had particularly asked me, as early as 1943 or 1944, how one could eliminate the concern that small amounts of active protein might still be present in the transforming DNA. So this was one important part of my 1946 investigation of the transforming agent, but I also began to approach some of the genetic and chemical concerns outlined above. TABLE 1 gives several specific questions just about as I framed them, I believe, and not modified by hindsight!

DNA as a "natural product" had been an object of curiosity since 1870, though most samples studied in chemical detail had come from one of two favorite sources—fish sperm or calf thymus. Therefore, in 1944 we could still hardly be sure that there was one single set of criteria defining the many occurrences of DNA, such as those located by Feulgen staining. A few persons even held the lingering misconception that DNA was the nucleic acid of *animals*, and RNA was *plant* nucleic acid!

As TABLE 1 indicates, we had the related concern whether the DNases were dependable defining reagents. As already brilliantly developed by Maclyn McCarty, it did look as if that were true—and time has shown it to be—yet time has also shown that it was then reasonable to fear that DNase might turn out to be one of the phospho-esterases that can act upon ribo- as well as deoxyribo-phosphate esters. I need to remind you that the chemist Alfred Marshak had reported phosphorus isotope-labeled intermediates containing both DNA and RNA.[8] That work would not be disregarded as easily today as it was in those years now that we know of many DNA-RNA hybrids formed in transcription. A possible involvement of RNA along with DNA in transformation was even one idea we had considered during those months of collaboration in the Avery laboratory.

Besides chemical analysis (to which I shall return) a different set of questions troubled biologist colleagues. Geneticists especially—who lovingly taught genetics in terms of the beautiful classic minuet of the chromosomes of sexual forms—honestly wondered whether the asexual bacterial cells, not channeled for multiplication by way of a simplified sexual zygote, had, or even needed, genes in order to duplicate their cellular constituents, "autocatalytically," for example. Furthermore, there were neoclassic examples of non-Mendelian inheritance in paramecia and other ciliates. That stimulating teacher, Tracy Sonneborn, gave his students a thorough exposure to bacterial transformation, but, as a productive investigator of epigenetic phenomena, he felt obligated to make the serious speculation in the literature that transforming DNA might be a contact inducer of carbohydrate synthesis.[9]

Sonneborn well exemplifies an important and broader point I want to emphasize: it is essentially a *duty* of the specialist-expert to critically consider alternative hypotheses that might explain such new discoveries. The role of this loyal opposition in science is well-established and honorable. The relative layman or nonexpert, on the other hand, has to rely principally on "common sense" and thereby is more ready simply to accept new ideas that seem plausible. I shall return to this point later on.

Thus, during the decade I am recalling here, there were many welcome and often pleasant opportunities to argue with various specialists about interpretations, or various contaminants, or experimental loopholes we were trying to eliminate. I will have to skim past most of them here.

CHEMISTRY OF DNA AND TRANSFORMATION

By 1948, I had developed a way to perform quantitative base analysis of DNA by paper chromatography.[10] I show you here what is surely the first (semi)quantitative chromatogram ever made of total bases from DNA hydrolysates (FIG 1).[11] Erwin Chargaff and coworkers were at this point just begin-

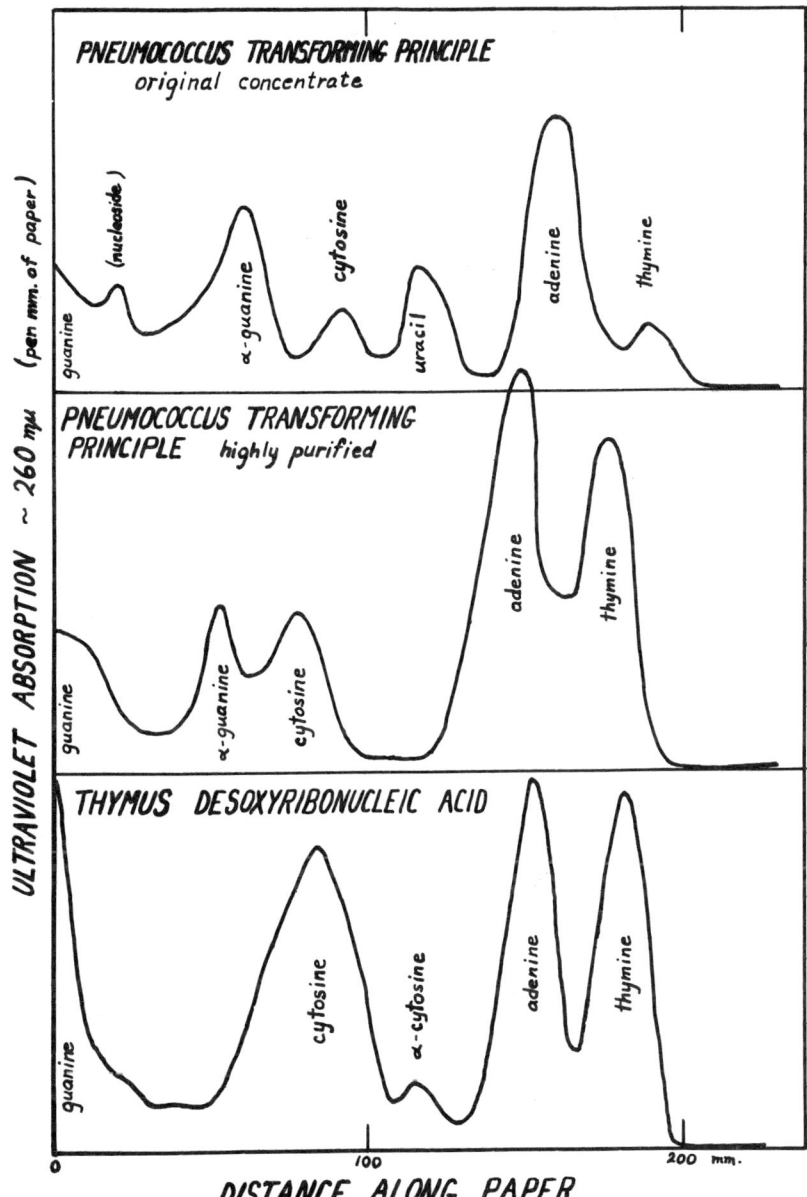

FIGURE 1. Semiquantitative chromatogram with *n*-butyl alcohol for base recovery[10] from hydrolysates of DNA: crude original (Stage 1) and highly purified (RNAse-treated and deproteinized Stage 4) pneumococcal DNA preparations, and calf thymus DNA. Acid hydrolysis: in 6N HCl at 118°. These hydrolysates were also used for chemical analyses reported in TABLE 2. (Experiment of February 1948; modified from Hotchkiss.[11])

TABLE 2. Chemical Properties of Transforming DNA as Progressively Purified

DNA Preparation	Stage 1	Stage 2	Stage 3	Stage 4	Calf Thymus
Treatment:	alc.pptn	RNase	Deprotn.	alc.pptn	Standard
Pptn. (times)	2	3	4	9	4
activity (%)	23	88	93	100	n.a.
Ratios:					
adenine:P	0.16	0.11	0.17	0.22	
uracil:P	0.09	0.0	0.0	0.0	
thymine:P	0.04	0.16	0.26	0.33	
cytosine:P	0.18	0.12	0.21	0.24	
total N:P	1.97	1.66	1.88	1.72	1.72
P:UV	3.6	6.8	4.5	4.4	4.2
DOP:UV	0.013	0.074	0.081	0.086	0.093
DNA Hydrolysate in Strong HCl 118°C:					Pure Adenine
α-NH$_2$:N	0.020	0.015	0.011	0.005	0.015
gly N:α-NH$_2$				1.01	1.02
				1.03	1.03
	(proteins gave quick release of 0.33 to 0.8 of their N as α-NH$_2$ N, and 0.0009 to 0.023 of that as glycine N)				

NOTE: α-NH$_2$, alpha amino acid N; activity, inverse of concentration giving half maximal transformation yield; gly, glycine (nitrogen); DOP, deoxypentose (by diphenylamine); deprotn, deproteinization (Sevag); N, total nitrogen; P, total phosphorus; pptn, reprecipitation with alcohol; UV, 260-mμ absorption of a standard concentration.

ning independently also to use UV spectrophotometry to perform similar, and eventually more reliable analyses of DNA.[12] My results showed that purified transforming DNA did indeed contain all four expected bases, *including thymine—but no discernable uracil*. The bottom chromatogram indicates that thymus DNA might be different, and also reveals the first demonstration of an "extracurricular" base, which I suggested (correctly, it turned out) was the novel base, 5-methyl-cytosine.

But we are largely indebted to Chargaff and his several collaborators in those next years for evidence of compositional differences in DNAs, and for amply tabulating the constant and variable ratios of bases in various DNAs. Later (about 1959) this modulation into biospecific forms was supported by physical evidence signalled by Marmur, Doty, Sueoka and others. Sequential differentiation could not be shown until much later.

But what about the often-suggested presence of small amounts of protein, which was supposed by some to be the true transforming agent? Precise micro determination of total and alpha-amino acid showed (TABLE 2) that the amount of amino acid was very small and surely far less than 1 percent of the total nitrogen. As we had always known, because of the similar nitrogen content of proteins and nucleic acids (16 percent), the classic nitrogen-to-phosphorus ratio is only insensitively increased in parallel to supposed minority protein (e.g., from 1.72 to 1.81 for a 5 percent contamination). But we found a sensitive criterion: classical calf thymus DNA and, in fact, the simple pure DNA component adenine, also slowly "hydrolyze," decomposing to the amino acid, glycine. By precise glycine analyses, I soon showed that this constituted *the entirety* of the amino acid coming from purified DNA, within the limits of the method used (TABLE 2), and therefore that the content of protein in this nucleic acid was vanishingly small.[13]

Something had been clear to any experienced chemist from the beginning: given the crudeness of the early base analyses before 1940 (depending on separating out and weighing some fairly insoluble derivative), those early estimates had to be low and unreliable. Calculations of any integral base ratios had relied upon the organic chemists' successful assumptions that served for usual low-molecular compounds, but could not be reliable for big molecules. By the 1940s Hammarsten, Avery, Chargaff, and Levene himself no less, knew that DNA was a high-molecular material. Thus the famous "tetra-nucleotide hypothesis" stemming from the turn of the century was merely an *arithmetic* convenience, and any unitary ratios for the four bases reported at that time could only be crude approximations.

To make matters worse, the literature contained a gross error. In Table XVIII in the much-cited textbook *The Nucleic Acids*[14] the authors Levene and Bass purported to show the fit of two sets of DNA base analyses (from 25 years earlier) to the proportions calculated for a simple tetranucleotide. One set of "Found" data, from Levene's own laboratory, is no support, accounting for little more than 80 percent of the bases, and deviating by 13.5 to 19 percent from the "Calculated" values. But the other set, cited from Steudel as "Found," *were actually* that worker's *values calculated* in theory for a tetranucleotide salt! This shocking error I demonstrated by reprinting photocopies from the original papers in my earlier review;[4] Olby had also discovered and mentioned it,[15] but it seems to have passed with otherwise little notice.

I knew Phoebus A. Levene personally, and Lawrence Bass superficially; I attribute this misquotation of Steudel only to careless zeal, but it collapses the 1931 presentation of the old "tetranucleotide hypothesis" to a single set of very poorly fitting analyses. Yet it must have influenced the views of many people in the era of the 1940s who likely consulted the 1931 monograph more often than they would the early original sources themselves!

DNA A DENATURABLE MACROMOLECULE

Another feature of DNA chemistry that I want to recall is the growing recognition in that decade that DNA had a denaturable secondary or tertiary su-

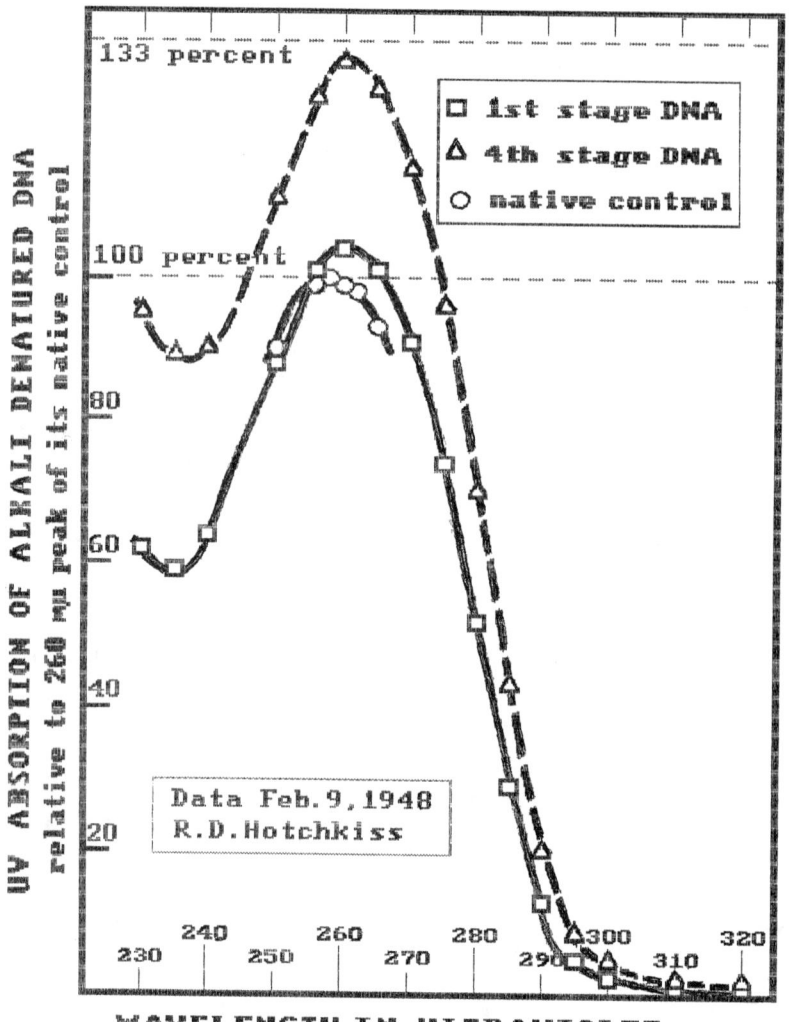

FIGURE 2. UV spectrum of intact pneumococcal DNA, crude and purified, showing the hyperchromic shift produced by alkali (1/60 N NaOH). *Upper curves*: Stage 4 and Stage 1, alkaline. *Bottom curve*: a short section of neutral Stage 4 as native control for both DNAs (experiment of 02/09/48). "Stage" of purification as in FIGURE 1, TABLE 2, and text.

perstructure. The first workers to report on it were probably Gulland and Jordan: briefly put, their 1947 acid-base titration curves of DNA[16] showed that by the time DNA had been titrated to an alkaline pH, it had somehow changed, exposing more titratable buffering groups, so that back titration with acid followed a different ("hysteresis") curve. They suggested some kind of phase change.

My own experience with DNA denaturation began with spectrophotometric work in which I found that alkali treatment of intact transforming DNA produced a substantial increase in UV absorption—what came to be called the hyperchromic effect. Quantitative data (FIG. 2) showed a slight hyperchromicity (9 percent at 260 mμ) for a crude transforming preparation still containing RNA. When processed beyond the usual stage (stage 3 of my series would be standard[1] pneumococcal DNA), a highly active preparation (stage 4) four times as active as the first, contained no uracil or protein and its hyperchromic response in alkali was about 31 percent. My data (FIG. 2) of 1948, and before, were the earliest known to me showing this effect; possibly the very first observations made, though the odds may be against it, ultraviolet spectrophotometers having become common by that time. Anyhow, it is perhaps irrelevant, as well as conjectural, for I did not publish such data until 1957[17] in a handbook article. Rene Thomas[18] has recently described his similar discovery in 1950, which lay undocumented for another year or two.

During 1948–49 I came to rely on spectrophotometry for monitoring the nativeness of DNA preparations.[17] Calf thymus DNA (FIG. 3) gave the typical hyperchromic shift of about 33 percent. When I learned of the parallel use of the hyperchromicity of DNA *hydrolysis* by Kunitz[19] to assay and crystallize the enzyme DNase, I proved that the effects were connected, by demonstrating that alkali denatured DNA no longer showed hydrolytic hyperchromicity. By 1950 such denaturation effects[18] were being described in work from Frick, Tsuboi, Rene Thomas and others. Semiquantitative effects of these agents upon biological activity were also described by Zamenhof.[7] Clearly, the evidence by 1952 indicated that DNA had a denaturable superstructure (TABLE 3). Somewhat later, the role of *heat* denaturation of DNA was taken up by Marmur and Lane, and Marmur with Doty and Schildkraut,[20] and soon by others.

TRANSFORMATION AS A GENETIC PROCESS

Over the same period we were clarifying the biological and genetic significance of bacterial transforming agents. Earlier quantitation of transformation had relied on finding limiting dilutions of DNA at which it generally failed to give detectable qualitative transformation in a sample population.[1,2,6,7,25] With our drug-resistance markers, we could determine the actual *number of*

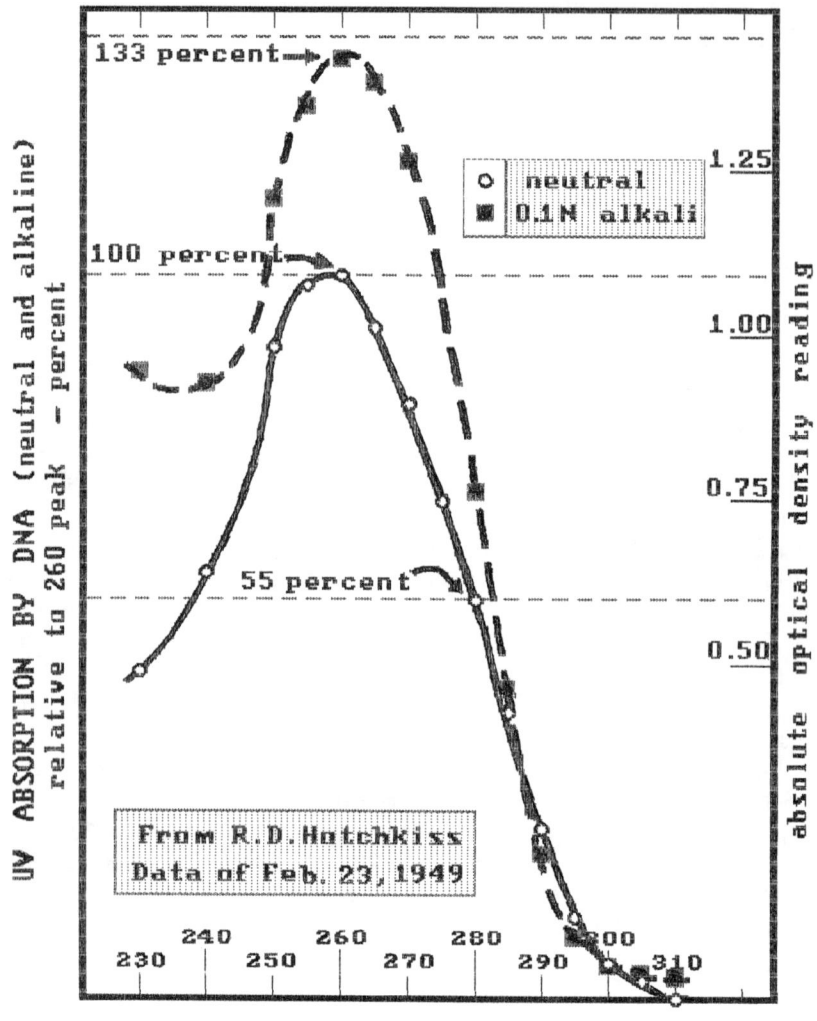

FIGURE 3. UV spectrum of a purified calf thymus DNA showing the full curves of neutral (native) and alkaline (denatured) DNA. Corrected for additive blank and dilution; a similar treatment with DNase gives a very similar hyperchromic increase due to hydrolysis at wavelengths near 260 mμ.[17]

TABLE 3. What Had Been Learned About DNA by 1954[a]

Chemical Composition of Transforming DNA

Transforming agent does not give qualitative color reactions for protein, and activity is not affected by proteinases.

(Avery, MacLeod & McCarty, 1944)

The DNA contains the usual four bases: AGCT; uracil (hence RNA) appears to be absent. The base frequencies seem different from those of thymus DNA.

(Hotchkiss, 1949)

Hydrolysates contain only minute amounts of amino acid; all of that amino acid is glycine, a product of slow decomposition of adenine (only negligible protein could be present).

(Hotchkiss, 1949, 1952)

DNAs in general have widely different base compositions, the principal regularities being that A = T and G = C.

(Chargaff laboratory, 1948, 1950)

Crystallized DNase (Kunitz) has the same specificity and ability to inactivate transformation as McCarty's DNase.

(McCarty, 1946; Kunitz, 1950; Hotchkiss, 1949)

Native DNA Is a Denaturable Macromolecule

DNA irreversibly changes buffering state at alkaline pH.

(Gulland & Jordan, 1947)

Alkali releases hyperchromicity of native DNA about 33 percent.

(Hotchkiss, [1948 and 1957]; similar results reported by K. Tsuboi [1950], L. F. Cavalieri [1952], G. Frick [1952], J. Shack & J. M. Thomsett [1952, 1953], and R. Thomas [1953, 1954])

DNase degradation of DNA releases hyperchromicity of DNA.

(Kunitz, 1950)

[Later work on heat denaturation and reduction of transforming activity: Marmur & Lane (1960); Doty, Marmur & Schildkraut (1960); Roger & Hotchkiss (1961); Ginoza & Zimm (1961).]

[a] This listing (and those in other tables) are not intended as bibliographic sources, but to associate the names of investigators with the dates and history discussed here. Source references will in many cases be found by consulting the main references in this article.

cells surviving in the drug after a standard exposure to DNA, and so record what has always been for me the most explicit "molecular biology": an actual *biological manifestation* of the uptake of *a single DNA molecule* by an individual cell! This quantitative technique made it possible to show that levels of penicillin resistance attained by spontaneous mutation in a donor strain were recapitulated exactly in steps of transformation by DNA from those strains.[21] Thus in short, we could relate unitary steps in *transformation* of a cell marker—

TABLE 4. What Had Been Learned about DNA Genetics by 1954

Bacterial surface antigens can be introduced by DNA.	Avery, MacLeod & McCarty, 1944 Boivin et al., 1947 Austrian & MacLeod, 1949 Alexander & Leidy, 1951
Transformed traits are carried by DNA from the transformants.	Avery, MacLeod & McCarty, 1944
Defective (mutant) capsular antigens can be transferred, or, if recombined, make a normal capsule.	MacLeod & Krauss, 1947 Ephrussi-Taylor, 1949 Ephrussi-Taylor, 1951
Antibiotic resistance mutations are transferred by DNA.	Hotchkiss, 1951 Leidy, Alexander & Redman, 1954
Resistance mutants allow selection & assay of transformed cells,	Hotchkiss, 1951
so efficiency and time course of gene entry and expression can be quantitated.	Hotchkiss, 1951, 1952, 1954
DNA from spontaneous mutants transmit those same mutations in successive independent unit steps.	Hotchkiss, 1951
Some pairs of markers are linked and frequently transferred together into single transformed cells.	Hotchkiss & Marmur, 1954
Such pairs exhibit linkage and recombination for all allelic configurations of markers.	Hotchkiss & Marmur, 1954 Hotchkiss, 1955
DNA uptake is a cyclic function of cell division.	Hotchkiss, 1954
A sugar dehydrogenase enzyme trait is transferred by DNA—even by DNA from unadapted donor cells.	Marmur & Hotchkiss, 1955

no longer of a surface antigen—to the steps of *mutation* in the history of the strain carrying that DNA. It was found that cell unit traits carried by the DNA were generally introduced singly into the recipient cells. Some of the findings reported in 1951 and the next few years are indicated in TABLE 4.

Work by Demerec had indicated that the drug-resistance markers were the result of individual spontaneous mutational events similar to those in classical genes of higher organisms.[22] The evidence was accumulating that bacteria were controlled by something very much like genes, and that some of these at least were carried on DNA molecules.

We thought for a short time, incorrectly, that individual DNA molecules bore individual marker traits. This might have given molecular support for one genetic concept, that of genes as "beads on a string" (the chromosome). Luckily, with Julius Marmur we were soon to find that *not all* markers were independent. In transferring a mannitol-using trait of pneumococcus we found that its DNA determinant frequently traveled along with our streptomycin-resistance marker into some transformants.[23] This association was exhibited by all allelic combinations of the two markers, so that it could displace a selective resistance marker from, as well as add it to, recipient cells. Those recipients in which the association broke down were genetically exchanging cell DNA with segments of added DNA, and at *unique* marker-specific sites! So, DNA was fulfilling a classical expectation for genetic linkage and recombination. Most significantly, two biochemically unrelated markers, both carried on DNA, seemed to be connected to each other by nothing more nor less than a string of more DNA!

The mannitol marker furthermore gave us a biochemical marker we identified as controlling a cell enzyme, mannitol-phosphate dehydrogenase.[24] This determinant was convincingly like a gene in one more respect: it was equally effective coming from the DNA, regardless of whether or not the donor cells were in a mannitol-adapted or an unadapted state.

Another interesting example of genetic recombination in transforming DNA was developed by Ephrussi-Taylor, who recovered two or three types of spontaneous variants of Type III encapsulated pneumococci. She showed in 1951 that certain pairs of these strains with altered, defective capsules would interact: cell with DNA, in either direction, to produce stable, normally encapsulated transformants.[25] This again had some of the earmarks of classic recombination — one that reconstituted a normal DNA from two modified ones.

GENERAL IMPLICATIONS: TRANSFORMATION AND PHAGE

This genetic-chemical work highlighted the significance of DNA structure in itself, and also through a parallel with bacteriophage chemistry that would converge with our story and confirm it. Alfred Hershey had pointed out (with Rotman in 1949) that certain traits of bacteriophages exhibited linkage when they infected *E. coli* cells. Our researches on bacteria did influence the phage work, as I have detailed before;[4] Hershey expressed interest in my observations as they stood in 1949, and I sent him a summary of my data to use in a bacteriological society roundtable. He reported with thanks that ". . . you have cleared up most of the doubts [about protein being present]. Some people may cling to the virus theory a little longer, perhaps. . . ." Hershey in 1949 was referring to a postulate by some skeptics that some sort of virus contamination in our DNA could explain its activity. It is fascinating that, three years later, his work would show the converse: that it was the DNA component of coliphage that accounted for its activity! In 1952 Hershey with Martha Chase[26] demonstrated that it was the phosphorus and not the sulfur constituents of phage that entered infected cells. This rather famous work was widely interpreted to mean that the DNA, and not the protein, was the "genetic" material of phage. While our bacterial DNA was easily a hundred-fold better-defined than ^{32}P-labeled phage — think, for example, of phosphoproteins! – it conveyed bacterial markers only *piecemeal*. Hershey's finding had the advantage of demonstrating that the *entire genome* of phage was present in its phosphorus-bearing component(s).

James Watson has testified that Hershey's result was instrumental in motivating him to bring the question of DNA structure to Crick, the Cavendish Laboratory, and eventually to all of us.

Therefore, those of us working in bacterial transformation can feel a satisfaction and pleasant sense of fulfillment in the knowledge brought to our subject by the triumphs of later bacteriophage and structural work as they opened up such new possibilities for DNA management and transfer as those represented in the modern papers of this symposium.

THE CONTINUING THREAD OF DNA TRANSFORMATION

Cutting off this reminiscence at about 1953–1954 risks unraveling some of the threads of history it contains. Just as early bacterial transformation and bacteriophage work brought a chemical basis and an incentive for studying DNA molecular structure, so did the successful double helix model in return stimulate the later growth of the DNA molecular genetics from which it arose. Much of the latter has now become a familiar mainstream topic, but a few more comments should be added to preserve a sense of that continuity.

Although transformation in the decade 1944–54 was limited essentially to the few laboratories and workers mentioned, DNA-mediated gene transfer continued at an accelerated pace. Many who greatly contributed to molecular and genetic study of DNA began their work around 1954, surely drawn by the broadening prospect. Arnold Ravin in association with Ephrussi-Taylor, and also Rene Thomas; Julius Marmur and Maurice Fox in our Rockefeller Institute laboratory had started by about 1953; Sanford Lacks began in 1955; Harriett Bernheimer commenced her productive studies of the natural capsule variants of pneumococcus with Austrian and MacLeod about 1960. The notable work of Goodgal and Herriott on *Hemophilus* transformation began about 1956. *Bacillus subtilis* transformation came first in 1958, Spizizen's work[27] eventually attracting geneticists Ganesan, Stocker, Anagnostopolous and others into this line of work.

The growth of the field after 1954 is obvious from a masterly review of transformation by Pierre Schaeffer,[28] who led a transformation group after a first project with us at Rockefeller in 1954. Assembled in 1962, his bibliography included the following (approximate) count of papers:

DNA transformation references in Schaeffer review

Period:	to 1944	1944–48	1949–53	1954–58	1959–62
Years	(15–20)	5	5	5	4
Papers	7	8	37	103	214

The vast majority of these papers were directly concerned with transformation, but the listings above also included (as I cannot here) a proportion of related work in bacteriophage and DNA chemistry.

SOME PERSONAL REFLECTIONS ON THE DNA REVOLUTION

Throughout those years, and since, it remained my conscious concern that we were engaged in matching together the brilliant *formal* conceptions of classical genetics (the genes, linkage groups, mutation, and recombination) with a *material* reality of biochemistry–the molecule, its structure, and chemical reactions. Therefore, I often expressed (e.g., in Ref. 29) the warning that the inclination to say that "the gene *is* made of DNA" is to mix two vocabularies (like saying "mother-love *is* the hormone estradiol," or whatever). If we show respect for the great genetic formalisms, I believe we should still avoid the verb "to be" and express such thoughts as, for example, in "DNA *exhibits* . . . or carries . . . the properties of the gene" and "gene mutation *is accounted for* by changes in the DNA."

The decade on which I have focused can be looked upon as creating the "molecular" parallel to the preceding revolution, which had related the (formal) linkage group to the (material) chromosome. Of course, this latter was still continuing with the further discoveries by Barbara McClintock; furthermore, the "viral genome as chromosome" concept was just beginning to be fruitful, and about to converge with ours at the new macromolecular level.

Perhaps my participation in preparing for the Watson-Crick model may lend some interest to my reactions after their 1953 announcement. Staying close to my theme of the chemical nature of genetic material, I will only add a point or two to the many retrospections others have made. Although the double helical model was an answer to my dreams, yet I had those "problems of an expert": alternatives to consider!

In spite of much concern to show the absence of protein in the transforming DNA, I had always assumed it to act in concert with specific cell proteins. In fact, as I have detailed elsewhere,[29] in 1952 I'd had my own excitement–a shipboard romance with my very own *complementary* model for gene replication, whose components I imagined as specific DNA and specific protein! (Only later did I learn that Pauling and Delbrück[30] had already considered such a model.) Biochemically, the replication apparatus did, of course, involve proteins, and was complex all right. Hence, my "expertise" already knew that the charms of complementarity in the helix model were available to *any duplex*, and I could not feel as some biologists have claimed to, that the replication model proved the structural model!

So, I was thrown back to the question of how realistic the helix itself was.

The bonding and binding in the helix model I could not judge nor question. The consequent complementary structure opening and closing was exciting, and I could happily greet the direct evidence when Mazia and Plaut, then Herbert Taylor, then Meselson and Stahl, began to provide it.

Finally, the "puzzling" evidence of the RNA viruses: having long and often deliberated over RNA itself, I never was sympathetic to the irreversibility of "DNA controls RNA" in the messenger step of the "Central Dogma," and the absolute primacy of DNA. This came to a head in 1957 in a huge broad-based symposium held by the New York Academy of Sciences, in a discussion of tobacco mosaic virus, where I commented on gene nature as follows[31]:

> ... Perhaps the confusing relations between RNA and DNA may be illuminated by the speculation that, as a genetic determinant, RNA was replaced during biochemical evolution by the more molecularly and metabolically stable DNA. Cell lines have preserved the RNA entities which, evolutionwise, were primary to DNA and may have allowed them to store their information in DNA and thereby become subservient to it metabolically. This secondary position would in a sense have been forced on RNA because of its lower stability and perhaps, because of its failure to become organized in such an elegant apparatus as the chromosomes. Viruses, as products of retrograde evolution by loss of function, may have had a choice of either RNA or DNA when specializing to get themselves made in the ample environment of the host cell.

This somewhat prescient opinion, presented before a large and sophisticated audience, was one I felt very serious about, and also published in another symposium.[32] It has frustrated me considerably, to see how long it was to be before reverse transcription and other realities made my speculation into an axiom for those interested in the origin, replication, and evolution of living cells.

THE PUBLIC AS AUDIENCE FOR SCIENCE

This personal memoir about a solidifying phase in the DNA revolution brings to mind some points about the philosophical background that always underlies such developments. I have pointed out that professional specialists are obliged to be more critical of new science than the nonspecialists. My experiences suggested a more debatable point: I felt that the 1940s and '50s were a time when the audience for science at several frontiers was "opening up" and widening. As an investigator whose work was becoming of more interest to a larger and broader public, I felt a growing obligation to strive to satisfy the needs of *that* audience. As concern for a broader "relevance" has grown, I saw that it can become difficult to serve, or strike a balance between, both our public and our professional goals. And more so, if either goal becomes an objective of selfish motives. I wonder how many of us, how earnestly, have on occasion tried to persuade a broad public and demonstrate an easy plausibility, rather than, as earlier, to defend in detail the validity of our conclusions before a sophisticated set of experts.

How often then I responded to seductive invitations to speak and hobnob at various assemblies, exposing my early results to get responses I was eager to learn! Soon, an eager editor would be pressing me for my symposium report. The weak link in this chain was the low residual energy I had for afterward writing another more detailed report for the legendary "refereed journal." Having sung my song, I often became a "troubadour with writer's cramp" and would do the next experiments and be just about ready to sing again when the next symposium invitation came along.

Yet, in 21 symposium articles (and many more public lectures) representing that decade, I tried to give some of the best of my thoughts and insights. And quite a few of those (and presumably the poor ideas too!) never reached the 8 (out of 29!) papers that were sent to the much-vaunted "refereed journals"! But I believe, and I have had testimony, that ideas transmitted in the more informal committee or symposium atmosphere, just as in the lecture hall, are often a useful part of the general dissemination of knowledge, and uniquely stimulatory for further advances. Certainly for DNA, the mass of ambient verbiage has been at times huge, and I often found it stimulating and informative. Since convention dictates that we not regularly refer to, nor index, these "current contents," our historians are offered, as it were, mainly our right-brain concepts, and must struggle harder to get hold of the contributions of our more intuitive left-brains.

Now reviewing that decade represented by my 21 symposium reports and 8 articles, I think some of my reports may have been better for being "refereed" *in vivo* by experts and laymen at meetings, and completed *afterward*! It might even be asked, whether these days a lonely, harassed, and busy referee can do as much for a paper as does a partly expert audience. Of course the referee can ascertain that you probably own at least a dozen test tubes! . . . and that you have read – or at least cited – some well-recognized references. How self-fulfilling, however, that he will probably frown upon referring to more than an occasional symposium paper!

Thus, when new scientific ground is being broken, it usually appears in both the (sanctioned) refereed journals, and the (suspect) symposia. Only after history has judged it to be important, can the emotion and sentiment associated with discovery be openly admitted – on just such occasions as this!

DNA has exposed me to another "relevant diversion" besides the symposium: the advisory committee. Both were for me immensely educational, and I believe, important functions. During sixty to seventy committee-man–years, almost half of them for cancer research institutes, I know many of us tried to give our best insights. I do not in the least regret those chances to serve, and I am glad to have lived in a period in which serving human health has come to have almost as much respect as national military service. And—offered for little or no pay—to have presumably done more to enrich life!

Luckily, we did have all those meetings and symposia to move us onward in our bridge-building, in that decade between the '40s and '50s. I hope I have convinced you that it was an exciting and rewarding time to be active. For a number of us, it contained challenging opportunities, cooperative fellowship, clever friends, a growing interest from a wider public, and respect without too much hero-worship. And isn't that what makes for good science?

REFERENCES

1. AVERY, O. T., C. M. MACLEOD & M. MCCARTY. 1944. J. Exp. Med. **79:** 137–158.
2. MCCARTY, M. & O. T. AVERY. 1946. J. Exp. Med. **83:** 89–96; 97–104.
3. WATSON, J. D. & F. H. C. CRICK. 1953. Nature **171:** 964–969.
4. HOTCHKISS, R. D. 1979. Ann. N.Y. Acad. Sci. **325:** 321–342.
5. STENT, G. S. 1972. Sci. Amer. **227:** 84–93; Adv. Biosci. **8:** 433–449.
6. HOTCHKISS, R. D. & H. EPHRUSSI-TAYLOR. 1951. Fed. Proc. **10**(1): 200.
7. ZAMENHOF, S. 1952. *In* Phosphorus Metabolism. D. McElroy & B. Glass, Eds. Vol. **2:** 301–328. Johns Hopkins Press. Baltimore, MD.
8. MARSHAK, A. & A. C. WALKER. 1945. Science **101:** 94–95.
9. SONNEBORN, T. M. 1943. Proc. Natl. Acad. Sci. USA **29:** 329–343.
10. HOTCHKISS, R. D. 1948. J. Biol. Chem. **175:** 315–332.
11. HOTCHKISS, R. D. 1949. Colloq. Internat. C.N.R.S. **8:** 57–65.
12. CHARGAFF, E. 1950. Experientia **6:** 201–209.
13. HOTCHKISS, R. D. 1952. *In* Phosphorus Metabolism. D. McElroy & B. Glass, Eds. Vol. **2:** 426–436. Johns Hopkins Press. Baltimore, MD.
14. LEVENE, P. A. & L. W. BASS. 1931. The Nucleic Acids. Chemical Catalog Co. New York, NY.
15. OLBY, R. 1974. The Path to the Double Helix :87. University of Washington Press. Seattle, WA.
16. GULLAND, J. M., D. O. JORDAN & H. F. W. TAYLOR. 1947. J. Chem. Soc. 1131–1141.
17. HOTCHKISS, R. D. 1957. *In* Methods in Enzymology. S. P. Colowick & N. O. Kaplan, Eds. Vol **3:** 708–715. Academic Press, New York, NY.
18. THOMAS, R. 1993. Gene **135:** 77–79.
19. KUNITZ, M. 1950. J. Gen. Physiol. **33:** 349–377.
20. MARMUR, J. & D. LANE. 1960. Proc. Natl. Acad. Sci. USA **46:** 453–461; DOTY, P., J. MARMUR, J. EIGNER & C. SCHILDKRAUT. *ibid.*: 461–476.
21. HOTCHKISS, R. D. 1951. Cold Spring Harbor Symp. Quant. Biol. **16:** 457–461.
22. DEMEREC, M. 1945. Proc. Natl. Acad. Sci. USA **31:** 16–24.
23. HOTCHKISS, R. D. & J. MARMUR. 1954. Proc. Natl. Acad. Sci. USA **40:** 55–60.
24. MARMUR, J. & R. D. HOTCHKISS. 1955. J. Biol. Chem. **214:** 383–396.
25. EPHRUSSI-TAYLOR, H. 1951. Cold Spring Harbor Symp. Quant. Biol. **16:** 445–456.
26. HERSHEY, A. D. & M. CHASE. 1952. J. Gen. Physiol. **36:** 39–56.
27. SPIZIZEN, J. 1958. Proc. Natl. Acad. Sci. USA **44:** 1072–1078.

28. SCHAEFFER, P. 1963. *In* The Bacteria. I. C. Gunsalus & R. Y. Stanier, Eds. Vol. **5**: 87–153. Academic Press. New York, NY.
29. HOTCHKISS, R. D. 1966. *In* Phage and the Origins of Molecular Biology. J. Cairns, G. S. Stent & J. D. Watson, Eds.: 180–200. Cold Spring Harbor Laboratory. Cold Spring Harbor, NY.
30. PAULING, L. & M. DELBRÜCK. 1940. Science **92**: 77–79.
31. HOTCHKISS, R. D. 1957. Spec. Publ. N.Y. Acad. Sci. **5**: 226–227.
32. HOTCHKISS, R. D. 1959. *In* Proceedings of the Third Canadian Cancer Conference :3–12. Academic Press. New York, NY.

Linus Pauling: Chemist and Molecular Biologist

ALEXANDER RICH

Department of Biology
Massachusetts Institute of Technology
Cambridge, Massachusetts 02139–4307

Linus Pauling, who was widely regarded as the greatest chemist of the twentieth century, was born in 1901 and died in 1994. In his autobiographical sketches, Pauling said that he became interested in chemistry at age 13, and that by 18 had mastered most of conventional chemistry. At that age he already understood the qualitative nature of the material changes that constitute chemistry and was ready to look more closely into why things occur. He took his undergraduate training at Oregon State University in Corvallis preparing to be a chemical engineer. However, it became increasingly clear that his interests lay in understanding more fundamental problems about the nature of chemical change. He regarded it as his good fortune to be accepted as a graduate student at the California Institute of Technology in 1922. There he was strongly influenced by Arthur A. Noyes and Richard Tolman, while taking his Ph.D. training with Roscoe Dickinson in determining the structure of molecules using X-ray diffraction techniques.

After finishing his Ph.D. in 1926, Pauling went abroad as a Guggenheim Fellow to work in Arnold Sommerfeld's laboratory in Munich. In his fellowship application he proposed to apply the recently discovered quantum mechanics to a study of the structure of molecules and the nature of the chemical bond. He described himself as very lucky to be entering science at a time when new developments were making possible a fundamental approach to these problems.

Pauling had the ability to see things in a general way, where many others could only see the particulars. This talent was strikingly illustrated in 1928 when he began looking into the molecular structure of ionic minerals. These are made of units that have both positive and negative charges. Rather than looking at their three-dimensional structures one at a time, he asked himself instead why they have the structures they have. He formulated a set of simple rules of coordination based on ionic charges and radii. In addition, he introduced a postulate of local neutralization of charges. This allowed the structure of silicates as well as many other minerals to be understood. With these rules it was possible to predict the structure of minerals using rational chemical principles, thereby clarifying an entire field of structural chemistry.

One of Pauling's long-term interests, which originated while he was still a teenager, was to understand the nature of the chemical bond. In 1931 he published one of his early papers on the subject, applying the methods of quantum mechanics. In this he was able to rationally explain the tetrahedral coordination of carbon as well as the square and octahedral coordination of transition metals. He describes the background of the paper:

> During the last four years the problem of the nature of the chemical bond has been attacked by theoretical physicists, especially Heitler and London, by the application of quantum mechanics. This work has led to an approximate theoretical calculation of the energy of formation and other properties of very simple molecules, such as H_2, and has also provided a formal justification for the rules set up in part by G. N. Lewis for the electron pair bond. In the following paper it will be shown that many more results of chemical significance can be obtained from the quantum mechanics equation, permitting the formulation of an extensive set of rules for the electron-pair bond supplementing those of Lewis. Those rules provide information regarding the relative strengths of bonds formed by different atoms, the angles between bonds, free rotation, or lack of free rotation about bond axes, the relation between the quantum numbers of bonding electrons and the number and spatial arrangement of bonds, and so on. A complete theory of the magnetic moments of molecules and complex ions is also developed, and it is shown that for many compounds involving elements of the transition group this theory together with the rules of electron pair bonds leads to a unique assignment of electron structures as well as a definite determination of the type of bonds involved.

Thus, Pauling began his program of rationalizing the chemical bond. By 1935 Pauling felt he had an essentially complete understanding of the nature of the chemical bond which he derived by applying quantum mechanics. A great deal of his understanding of the subject is summarized in *The Nature of the Chemical Bond*, published in 1938. A major theme of this book is that chemistry can only be understood in three dimensions. The book showed that the properties of substances depend in part on the type of bonds between its atoms as well as on the atomic arrangement and distribution of bonds. Pauling systematically illustrated these principles in a variety of cases, thus marrying both theoretical as well as qualitative chemistry. He provided a large number of observations supporting his generalizations about the nature of chemical bonding. His Nobel Prize in Chemistry was awarded on the basis of much of this work.

One of Pauling's great contributions was to bring his insight about the nature of chemical processes into the field of biology. He started to learn something about biology in the late 1920s when Thomas Hunt Morgan came to Caltech, bringing with him a number of younger members of the new biology division. In 1931 Pauling had become interested enough to present a seminar describing the crossing-over of chromosomes. From this beginning he began increasingly to include work on biological molecules. Together with Charles Coryell he measured the magnetic susceptibility of hemoglobin as a function of the state of oxygenation. In the mid-1930s he met with Karl Landsteiner, who stimulated his interest in the nature of antigen–antibody reactions.

It is difficult to realize the primitive nature of the knowledge about proteins in the early 1930s. Most researchers regarded proteins as colloids—somewhat amorphous materials—rather than as molecules, as we understand them today. Pauling played a key role in helping to transform this view. Alfred Mirsky came to Caltech from Rockefeller for one year in 1936, and Pauling published a paper with him which contained a general theory of protein structure. They described proteins as consisting of a polypeptide chain folded in various ways and held together by hydrogen bonds and a number of other tertiary interactions. Their paper described protein denaturation, which was not well understood, as an unfolding of this structure in such a way that there remained a chain that had largely lost its tertiary interactions. This is essentially a modern statement of the view of protein structure.

Pauling and his associates carried out many studies of hapten–antibody reactions. He tried to develop a proposal about how antibodies are formed, but this work was too early to take into account the role of genetic selection. Nonetheless, the basic ideas with which he described molecular complementarity are valid today. This complementarity was mediated by the weak forces such as hydrogen bonding, salt links, dipolar forces, and van der Waals interactions. It is this description of the nature of macromolecular interactions that laid the foundation for the modern understanding of this field. It was an approach that Linus Pauling used in many different areas.

Pasqual Jordan, a German physicist, wrote a paper suggesting that identical molecules interacted with each other and were held together by fluctuating resonance interactions. Pauling discussed this at length with Max Delbrück, and in 1940 they published a paper on the nature of intermolecular forces operating in biological processes. They pointed out that Jordan's proposal was inadequate since the forces generated were not able to hold the molecules together. They stated that biological specificity must instead result from the existence of molecules with complementary structures. They used the example of gene duplication since this could be explained in terms of two mutual complementary molecules. However, at that time it was generally believed that basic proteins in the nucleus constituted the genes.

The manner in which Pauling elucidated the basic structural motifs of proteins is an interesting example of his ability to bring together information from a variety of fields to make predictions. The methodology was based on an extensive knowledge of stereochemical details—bond lengths, angles, van der Waals radii, hydrogen bond lengths, and so on. In addition, Pauling obtained clues by looking at X-ray diffraction patterns of the complex structures that he was trying to solve. In the case of proteins this meant looking at fiber diffraction patterns. His method consisted of developing a model to fit the pattern, using very good stereochemistry to minimize energy. A key issue is finding the right clue in the diffraction data of a complex system with which to construct a model. One can clearly see the manner in which he used this methodology in his studies of protein structure.

In 1937, he thought he knew enough about the structure of amino acids and peptides to try to figure out how they are put together. He looked at the diffraction pattern of fibrous proteins that had a reflection at 5.1 Å on the meridian. At that time he could not build a peptide model that would predict this spacing, and he concluded that this must be related to the fact that he did not have enough experimental information about peptide stereochemistry.

Robert Corey joined him in the laboratory and over the next decade or so they and their colleagues solved over a dozen crystallographic structures of amino acids and some peptides. By 1948 enough evidence had accumulated, so he then returned to the problem. In reviewing the data, he concluded that he had not learned anything in the intervening period. The bond angles and lengths were obtained in a somewhat more accurate form, but the change was not significant. Furthermore, the planarity of the peptide bond was verified, a fact that he had put forth more than a decade earlier. In 1948 he was a visiting professor at Oxford and at one point was ill with a cold and stayed indoors playing with models. He found that he could coil a peptide chain around in a helical fashion with a nonintegral number of residues per turn. It made hydrogen bonds and formed a compact structure. The only problem was that it predicted a 5.4 Å spacing as a helical repeat, but it did not predict the 5.1 Å spacing. At that point he felt strongly that the model was correct, and he decided to ignore the 5.1 Å spacing (which was later shown to arise from a coiled coil). This is an example of Pauling's intuitive understanding of a molecular structure and the manner in which you can use certain data and ignore others. He delayed almost two years before publishing the α-helix. However, at that time he and Corey also published the β-pleated sheet structure deduced by the same methods.

The correctness of the α helix was confirmed quite rapidly. Looking at the model, Max Perutz discerned that there should be a 1.5 Å repeat along the helix axis. A survey of several proteins showed the existence of this, clearly implying that the α helix was a correct structure.

In this same period Pauling was thinking at length about complementarity, and he discovered a very simple interpretation for how enzymes might operate. In 1948 he said:

> I think that enzymes are molecules that are complementary in structure to the activated complexes of the reactions that they catalyze, that is, to the molecular configuration that is intermediate between the reacting substances and the

Photo taken at a meeting on the structure of proteins held in Pasadena, California in September 1953. *Foreground, left to right*: David Harker, Linus Pauling (facing the camera), Gordon Sutherland. *Background, facing camera, left to right*: Maurice Wilkins, Alexander Rich, Murray King. (Photograph courtesy of Alexander Rich.)

products of reaction for these catalyzed processes. The attraction of the enzyme molecule for the activated complex would thus lead to a decrease in its energy, and hence to a decrease in the energy of activation of the reaction and to an increase in the rate of the reaction.

This was a very simple statement, and it seemed intrinsically plausible once he articulated it. It was the first clear description of how enzymes work, and it was based on his knowledge of molecular complementarity. Subsequent structural work with enzymes bound to their substrates amply demonstrated the accuracy of his prediction. Furthermore, this approach is widely used today in designing monoclonal antibodies with enzymatic activity.

In his thinking about the nature of the gene and how it replicates, Pauling developed the idea that the gene probably consists of two complementary parts, each of which can produce a replica of the other. In a lecture delivered in May 1948 at the Sir Jesse Boot Foundation in England, he declared:

> The detailed mechanism by means of which a gene or a virus molecule produces replicas of itself is not yet known. In general the use of a gene or virus as a template would lead to the formation of a molecule not with identical structure but with complementary structure. It might happen, of course, that a molecule could be at the same time identical with and complementary to the template on which it is moulded. However, this case seems to me to be too unlikely to be valid in general, except in the following way. If the structure that serves as a template (the gene or virus molecule) consists of, say, two parts, which are themselves complementary in structure, then each of these parts can serve as the mould for the production of a replica of the other part, and the complex of two complementary parts thus can serve as the mould for the production of duplicates of itself. In some cases the two complementary parts might be very close together in space, and in other cases more distant from one another—they might constitute individual molecules, able to move about within the cell.

Despite this clear picture of how the gene may be duplicated, Pauling failed to discover the structure of DNA. Instead he proposed an incorrect triple-stranded model of DNA structure with phosphates in the middle and bases on the outside. He had used as a clue for this proposal the manner in which phosphate groups form a helical array in certain minerals. However, it was the wrong clue in this case, and the structure was incorrect, as he inferred shortly thereafter. Watson and Crick, who developed the duplex model for DNA, used the methods that Pauling had developed, paying careful attention to bond angles, distances, and hydrogen bonding interactions. The clues that they used were associated with the diffraction pattern produced by DNA molecules, ultimately leading them to the correct structure. They have frequently acknowledged their debt to Pauling for providing the method that enabled them to uncover the structure.

In the mid-1940s Pauling went to a meeting organized by Vannevar Bush.

Linus Pauling at the Institute of Molecular Biology in Moscow in 1975. *From left to right*: Alexander Rich, Linus Pauling, Alexander Bayev, Vladimer Engelhardt, and Walter Gilbert. (Photograph courtesy of Alexander Rich.)

There he met William Castle, a professor of medicine at Harvard Medical School and a hematologist. He spoke with Pauling about the sickling of red blood cells in sickle cell anemia. This sickling occurred when the cells were deoxygenated and birefringent rods could be seen assembling within the cell. Pauling concluded that this could only occur if there was a change in the surface of the hemoglobin molecule which produced a complementarity with another part of the surface in order to build this stable association. He asked Harvey Itano and other associates to look at the hemoglobin of sickle cell anemia. Fortunately, the mutation in sickle cell anemia is an ionic one, changing glutamic acid to valine. By using electrophoresis they discovered that the protein was different. In their paper in 1949 they described this as a "molecular disease." Molecular diseases are commonplace since many diseases are associated with alterations in proteins. It took Pauling's intuitive understanding of the nature of protein interactions to define the first of these molecular diseases.

Another example of intuitive understanding was his work with Zuckerkandl in 1962. By looking at changes in the amino acid sequence of hemoglobin in different animals, they could correlate these with the evolutionary period at which these animals diverged from each other. This suggested that mutations accumulate in a fairly regular manner with time in a particular protein and therefore can be used to measure evolutionary time. With this insight, Pauling was able to unify paleontology, geology, and molecular genetics, thus initiating the field of molecular evolution. The idea was startling when presented, but is now so widely accepted that it seems almost self-evident.

Pauling was a compassionate and warm man who cared deeply for people who were oppressed for their political beliefs. In the post-World War II period he was openly criticized by some politicians. Later, after the development of huge nuclear arsenals, he became an active campaigner for peace, stimulated in this direction by his wife Ava Helen.

Another instance in which Pauling's intuition played a key role was related to his conviction that radiation, even in small amounts, was harmful to the genome. On the strength of this conviction, he crusaded actively for a cessation of nuclear testing, starting a campaign to obtain signatures of scientists opposed to such testing. The petition was very successful in focusing attention on the issue. Today, it is generally understood that radiation from radioactive fallout or X-rays is harmful. However, when Pauling began his campaign, the suggestion was regarded as outrageous and even subversive. He was subpoenaed to testify before the Senate Internal Security Subcommittee for his support of a propaganda campaign against nuclear testing. In 1962 Pauling won the Nobel Peace Prize for his efforts in helping to bring about a test ban treaty.

By the 1970s Pauling's interest began to focus on the antioxidants vitamin

E and especially vitamin C. His analysis led him to the conclusion that we would be healthier if we took supplements of these vitamins. He recommended that increased doses of vitamin C would decrease the incidence of the common cold and later still espoused the use of vitamin C for decreasing the frequency of cancers. This proposal also aroused considerable controversy, and a large number of studies were done. Some of these studies, but not all, have shown protective effects against cancer. A number of other studies have shown that increased intake of antioxidants has important preventive activity in cardiovascular diseases. Perhaps as much as anyone, he stimulated a now universal awareness of the relationship between diet and health. The debate on the role of antioxidants in general and on vitamin C in particular continues to this day.

In his later years, Pauling reflected on the overall impact of his work and the work of others. He said that biology could now be understood without any *élan vital*. Furthermore, he said,

> . . . my other experiences during the last fifty years, involving the ever-increasing understanding of the world on the basis of rational principles, have led me to reject all dogma and revelation, all authoritarianism. It is possible that the greatest contribution of the new world view that has resulted from the progress of science will be the replacement of dogma, revelation, and authoritarianism by rationality—even greater than the contribution to medicine or that to technology.

Pauling was widely recognized for his outstanding contributions to humankind in many different arenas. The esteem with which he was regarded was illustrated to me in a vivid way in 1951, when, as a postdoctoral fellow of Pauling's, I visited Albert Einstein in Princeton. Einstein's comment to me was "Ah, that man is a real genius!"

Historic Reflections on the Clinical Roots of Molecular Biology

PAUL HELLER[a]

Department of Medicine[b]
University of Illinois
College of Medicine at Chicago
VA Medical Center[c]
Chicago, Illinois 60612

One of the outstanding highlights in the history of biology and medicine during the first half of the twentieth century preceding the discovery of the double helix was the gradual, perhaps too gradual recognition of genetic mechanisms underlying biologic and pathologic phenomena, thus building the foundation of molecular biology and pathology. The decisive event in this development was the landmark discovery of sickle cell hemoglobin by Pauling and his associates[1] as the cause of a serious, genetically determined disease. It took almost 45 years before the rather puzzling clinical and laboratory observations in a few patients which fascinated and frustrated several clinical investigators found their explanation. The story of the discovery of sickle cell hemoglobin has been told several times, most thoroughly by Conley,[2] and another repetition carries the risk of a sophomoric exercise unless a new emphasis is added to the analysis of the facts.

On the long road from the first observation[3] in 1904 of a patient with an apparently previously undescribed disease associated with strange-looking red blood cells, to the clarification of the pathogenic mechanism forty-five years later, several roadblocks had to be overcome which were created not only by the complexity of the object of investigation, but also by the investigators themselves, who mainly registered their observations without interpretation. One reason for this failure is obvious, namely the limited and limiting state of knowledge at the time and the underdeveloped methodologic potential for further inquiries by experimentation. But another, perhaps more decisive factor that prevailed during those years and hindered the development of medical science, was the absence of a dialogue between the physician making clinical observations which are sometimes puzzling, and the basic scientist, who might be able to solve such puzzles.

---

[a] Address for correspondence: 1522 Dobson Street, Evanston, Illinois 60202.
[b] Professor Emeritus.
[c] VA Senior Medical Investigator, Emeritus.

The following is a story with a gratifying ending, which was achieved after physicians and basic scientists started to talk to each other. It begins with a 20-year-old black student from Grenada, who had arrived in Chicago in the fall of 1904 to enroll in Chicago's College of Dentistry,[4] and to seek medical advice. He had been suffering from ulcers of both legs, and a few weeks after his arrival reported to Dr. Herrick's clinic at the Presbyterian Hospital with symptoms and signs that we now consider typical of sickle cell anemia. The blood picture was also characteristic of this disease (FIG. 1). Dr. Herrick and his associate, Dr. Irons, were puzzled and remained so during the entire three years the patient lived in Chicago and had several admissions because of symptoms, which we would now designate as pain crises, with or without fever and respiratory infections. At the end of these three years, he acquired a degree in dentistry and left Chicago without informing his physicians.[4]

Dr. Herrick was a prominent clinician in the mold of the famous European academic physicians of that time, some of whom he occasionally visited. Their reputation was mainly based on their astute analysis of clinical findings, which in many instances led to the description and classification of new diseases, some of them with the eponym of their discoverer. One of the prominent features of the medical literature were case reports, preferably legitimized by sur-

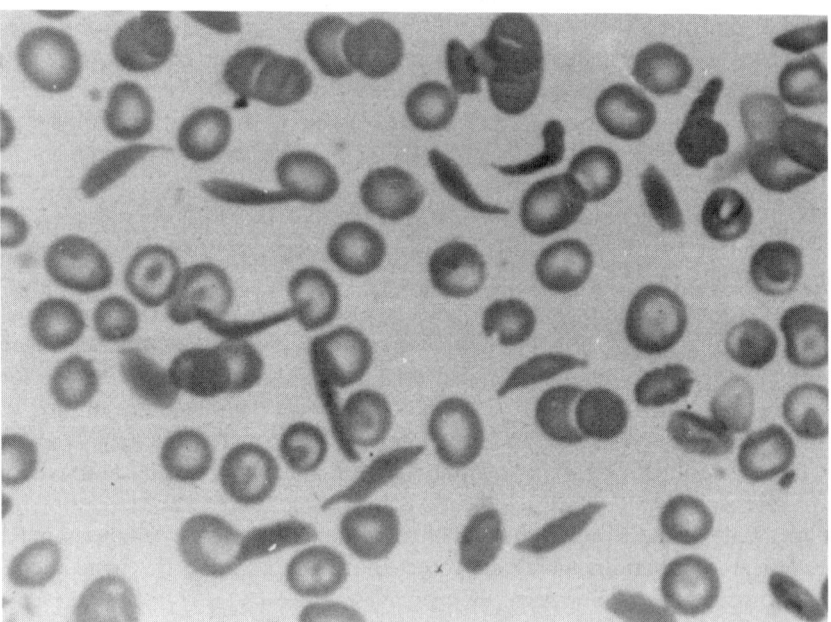

FIGURE 1. Blood smear of a patient with sickle cell anemia. The sickle-shaped cells have become irreversibly sickled after having gone through several sickling–unsickling cycles in venous and arterial blood. The other cells have not yet reached that stage.

gical or autopsy findings. Such a report about Dr. Herrick's patient and his strange red blood cells was in order, but because the results of the hematological examination and the clinical findings were bewildering to him and he could not make a diagnosis, he decided[3,4] to defer publication until similar patients were detected and some hints to the nature of the disease could be elicited. Neither happened, although there must have been forty to fifty such patients in the Chicago area at that time, a figure suggested by the census of 1900, which counted around 32,000 blacks in Chicago, and by the now known incidence of sickle cell anemia in the black population of 0.14 to 0.16%. High pediatric mortality and insufficient alertness of physicians to a hitherto unknown disease probably prevented its detection. In the absence of other experiences or reports, Dr. Herrick's early intense interest probably slackened and he left the care of the patient to Dr. Irons. There also were no efforts to stay in correspondence with the patient, who, on the basis of the recent FBI-like detective investigation by Drs. Savitt and Goldberg,[4] had established a successful dental practice in Grenada, but died rather suddenly in 1916 at the age of 32.

Finally, in 1910 Dr. Herrick presented the findings of "peculiar elongated and sickle-shaped red blood corpuscles in a case of severe anemia" to the Association of American Physicians, followed by a publication with the same title and content,[3] consisting essentially of the findings of 1904, with only a few sketchy follow-up notes. It is not clear which circumstances finally moved Dr. Herrick to this action—perhaps only the desire to complete unfinished business. His diminished interest in the new disease, after the initial enthusiasm, seems to be evident from the fact that he did not mention sickle cell anemia at all in his memoirs published in 1949,[3a] the year when the discovery of sickle cell hemoglobin in Pauling's laboratory had excited the medical world, and a host of other clinical and experimental studies of sickle cell anemia were available. The final publication in 1910 mainly reflects Dr. Herrick's frustration in his attempts to explain this disease. All possible diagnoses were considered from syphilis, which he thought to be most likely, to aniline poisoning. He took it for granted that the patient had a *secondary* anemia. Only in the last short sentence, rather disconnected from the core of the paper, does he mention the possibility that a "change in the composition of the corpuscle itself may be the determining factor."

It is perhaps surprising that Dr. Herrick, who was a scholarly, imaginative clinician and whose special interest was the coronary circulation and the patterns of cardiac pain,[3b] did not speculate on the possibility that the "elongated and sickle-shaped red blood corpuscles" might have difficulties crossing the capillary circulation and might be responsible for some of the clinical manifestations of his patient, especially the leg ulcers and the abdominal and muscular pain. He left such a postulate to later clinical investigators. At any rate, the delay in publication is regrettable; if the story of the patient had been pub-

TABLE 1. Chronology of Significant Developments

Date	Highlight
1904 (published in 1910)	"Peculiar elongated and sickle shaped red blood corpuscles in a case of severe anemia." (Herrick)
1917	Sickling in sealed chamber; unrecognized hint concerning heredity. (Emmel)
1923	Sickle cell anemia is hereditary with variable gene expression ("sickle cell trait"). (Huck)
1923–1949	Many clinical studies. (Sydenstricker, Diggs, and others)
1927	Hypoxia causes sickling; red cell ghosts do not sickle. (Hahn and Gillespie)
1940	Sickled cells in fixed blood samples; birefringence. (Sherman)
1945	Conversation of Castle and Pauling
1949	Homo- and heterozygosity for a mutant gene. (Neel)
1949	"Sickle cell anemia, a molecular disease." (Pauling, Itano, Singer, and Wells)

lished six years earlier, other physicians might also have been alerted earlier and the time table of significant observations might have been shifted to earlier dates.

After Dr. Herrick's report, significant developments were rather slow, as the intervals between relevant publications suggest (TABLE 1). Several investigators made interesting observations, but most of them remained unexplained and uninterpreted. As a matter of historic inquiry and analysis, admittedly as such afflicted with the stigma of hindsight, the question may be raised whether this discrepancy between observation and its interpretation might have been, at least to some degree, avoidable by the initiation of collaborative efforts.

In 1917, Victor Emmel, an anatomist, wanted to test the hypothesis that sickle cells are not released from the bone marrow as such, but develop in the circulation from their biconcave disk shape afterwards.[5] He observed the behavior of the erythrocytes of a patient with sickle cell anemia by incubating them in what was considered at that time "a tissue culture preparation," namely a microscopic slide with the blood sample under a cover slip sealed with petrolatum. Needless to say, he confirmed his hypothesis, and with it invented the sickle cell test, which became the diagnostic test of choice in the clinical and research laboratory for many years. Among his control blood samples was one from the patient's healthy father whose red cells sickled on incubation. Nevertheless, in his report, Dr. Emmel neither considered the possibility of heredity nor speculated about the possible reason for the change in the shape of the red cells in the sealed chamber. Six years later Huck[6,7] recognized the possible significance of Emmel's discovery of sickling in the "culture" preparation of the blood of the patient's healthy father. He detected sickling in several family members of the patient under his observation and postulated that the disease was inherited as a dominant trait, with the difference between asymptomatic and sick individuals with positive tests resulting from variable gene expression. This concept remained the prevalent general understanding prior to Neel's study,[8] published in 1949, in which the sickle cell test was used to examine the blood of members of many families, leading to the conclusion that individuals with the sickle cell trait were heterozygous for an abnormal and normal gene. Beet,[9] working in Northern Rhodesia, came to the same conclusion on the basis of the findings in a single family. A few weeks later this hypothesis was convincingly confirmed by Pauling and his associates.[1]

With regard to the importance of hypoxia, it was only in 1927 that the experiment crucial to the elucidation of the effect of hypoxia was performed by Hahn and Gillespie[10] with a relatively simple technique that permitted the exposure of the blood sample to varying degrees of oxygen tension. These experiments clearly established the dependence of sickling on diminished oxygenation and thus on the presence of deoxyhemoglobin. Sickling of those cells that were not yet irreversibly sickled (see the legend to FIGURE 1) was promptly reversed by exposure of the sickled cells not only to oxygen, but also to carbon monoxide. Most importantly, red cell ghosts (erythrocytes with hemoglobin eluted) did not sickle. It is quite remarkable that subsequently these important findings did not receive the appropriate attention. Hypoxia was certainly recognized as a requirement for sickling, but sickling and deoxyhemoglobin seemed to have been accepted without question as independent sequelae of hypoxia, as seems evident from the rare mentioning of deoxyhemoglobin in publications between 1927 (Hahn and Gillespie) and 1949 (Pauling). During these years, many important clinical studies were reported, as related by Conley,[2] but efforts in experimental research were mainly centered on the improvement of the technique of Hahn and Gillespie, especially on the avoidance of leakage of the gas mixture before it reached the red cells. Beck and Hertz[11] devised a method, somewhat modified by Sherman[12] five years later, in which blood after exposure to varying oxygen concentrations was introduced directly into a fixative without the possibility of entrance of room air, and the degree of sickling could be microscopically quantified. Sherman also showed with this method the difference in the number of sickle cells between arterial and venous blood and the absence of sickling in the blood of individuals with the sickle cell trait. His study subsequently received major attention, not so much because of the reported quantitative data, which generally were considered to be definite proof of the importance of hypoxia, but more so because of an almost *post scriptum* remark in the comments section of the paper, as follows: "In this connection we wish to report without interpretation the observation that under the polarizing microscope characteristic sickle cells exhibit a definite birefringence which disappeared after aeration of the cells and the consequent return to the normal discoid form." It was admittedly beyond Dr. Sherman's competence to himself seek an explanation of this phenomenon. Although he does not mention deoxyhemoglobin at all in the description and discussion of his experiments, he must have thought of the possibility that hemoglobin itself has a causative role. In his essay on sickle cell anemia, Conley[2] mentions that Sherman (who worked, as did Conley, at the Johns Hopkins Hospital) attempted without success to detect an immunologic difference between the hemoglobin of normal individuals and those with sickle cell anemia and the sickle cell trait. Although astute clinicians observed the gradual condensation of hemoglobin in erythrocytes during sickling[13] interpreted as the formation of paracrystalline structures,[14] the awareness of the importance of hypoxia, which in 1945 certainly was fully recognized by all students of sickle cell anemia, did not lead any investigator to examine the medium of oxygen exchange itself. It is certainly true that anyone having such an idea might not have known how to perform such studies, but this is probably not the only reason for the absence of any such interest. Another can be found in the historic-cultural climate of hematology at the time. Sickle cell anemia was a hematologic disease and hematology was almost exclusively considered to be a morphologic discipline. Hematologic investigators before Pauling apparently found it difficult, because of the power of training, tradition, and professional bias, to intellectually transcend morphologic considerations. This fixation seemed to have delayed any productive thought and plan of examining hemoglobin itself. One could well imagine, without the stigma of a hindsight exercise, that certain discoveries that were made after those of Pauling and associates could actually have been made before with relatively simple techniques, available at the time (e.g., gelling of concentrated deoxyhemoglobin solutions[15] or the diminished solubility of sickle cell hemoglobin in high molar phosphate buffer[16,17] or even Harris' famous tactoids,[18] which are sickle-shaped formations of concentrated deoxygenated hemoglobin itself without cells. With regard to the possibility of an abnormal globin's being a pathogenic factor, it is of interest to remember the relatively simple experiment of Hoerlein and Weber in 1948 in an effort to recognize the cause of the dominant form of methemoglobinemia.[19,20] They isolated globin from the brownish hemolysate and combined it with normal heme and detected that it was the globin portion which was responsible for the methemoglobinemia, thus establishing this genetic disorder as the first discovered hemoglobinopathy one year prior to Pauling's publication. This hemoglobin was later given the designation hemoglobin M.

To return to the suspected magic significance of the phenomenon of birefringence for the explanation of sickling: William B. Castle, the eminent clinical and experimental hematologist, had been interested in sickle cell anemia for several years, especially its hemolytic and vaso-occlusive aspects. He and Ham[21] recognized that both phenomena were ultimately attributable to intracellular changes, which lead to the elongated shape of the red cells, thus rendering them more mechanically fragile and increasing the viscosity of blood and hindering its free flow through the capillary circulation. Castle was intrigued by the birefringence observed by Sherman and suspected that this was an indicator of the basic nature of the disease and was due, as he expressed it, to "some type of molecular alignment," but he was careful not to commit himself to the postulate that hemoglobin itself was directly involved.[22] He knew that the exploration of this phenomenon was beyond his qualifications and none of his Bostonian friends among basic scientists seemed interested. Besides, there was a war on, which had a limiting effect on scientific activities not essential to the military effort. Finally, in 1945, Castle had an opportunity to discuss the matter with Pauling, who at that time was already well known for his fundamental studies on the relationship between the chemistry and the function of hemoglobin. Pauling and Castle were members of an important committee which had been assembled to advise the President on postwar science policy. Incidentally, it was this committee which advised the creation of the National Science Foundation and the National Institutes of Health. The circumstances around Dr. Pauling's learning about sickle cell anemia were different in the recollection of the two scientists. Dr. Castle, as I knew him, was a great champion of the collaboration of clinical and basic scientists in the solution of clinical-biological problems and was a master of clarity, logic, and persuasion; he later related[22] that he had suggested to Dr. Pauling, who at that time had no knowledge of sickle cell anemia, that he might be interested in the fascinating question of sickle cells showing birefringence only when they were deoxygenated,[22] and Dr. Pauling was most receptive. Dr. Castle remembers that the discussion took place on the train from Denver to Chicago. In Dr. Pauling's version,[23] he overheard Dr. Castle during an intermission at the meeting talking informally to the members of the committee about sickle cell anemia and became very interested. This difference in recollection apparently has never been resolved, but in whichever way the now legendary conversation took place, Dr. Pauling responded enthusiastically and recognized immediately, and I quote from his later report,[23] that

> ... the relation of sickling to the presence of oxygen clearly indicated that the hemoglobin molecules in the red cells are involved in the phenomenon of sickling and that the difference between sickle cell anemia red corpuscles and normal red corpuscles could be explained by postulating that the former contain an abnormal kind of hemoglobin which when deoxygenated has the power of combining with itself into long rigid rods which then twist the red cell out of shape.

His prediction turned out to be correct. The "abnormal kind" of hemoglobin was soon discovered by the use of free-boundary electrophoresis after Harvey Itano, a young physician and Ph.D. candidate in chemistry[1] joined Dr. Pauling's laboratory in September 1946.[d] The electrophoretic demonstration

---

[d] As Brecher has pointed out,[29] Pauling initially suggested that studies of the magnetic momentum of sickle cell hemoglobin might reveal characteristic differences from normal hemoglobin. This not having been the case, Itano and Singer chose electrophoresis.

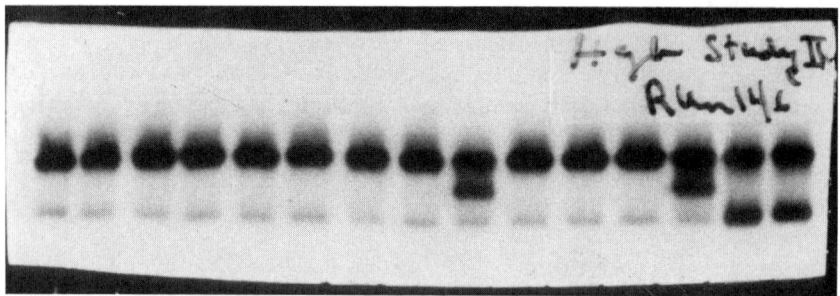

FIGURE 2. Starch gel electrophoregrams of hemolysates performed in the clinical laboratory. The third and seventh from the right show heterozygosity for Hb S (sickle cell trait) and the two on the right end show heterozygosity for Hb C (C-trait; not further discussed in text). The others are normal.

of normal and sickle cell hemoglobin (FIG. 2), later designated as Hb A and Hb S in individuals with the sickle cell trait, also established the reality of the inheritance pattern of the abnormal gene as it was postulated by Neel[8] a few weeks before Pauling's publication.

This legendary story of the discovery of hemoglobin S has been told many times, and I hope to be forgiven for repeating it here, not only because it is by itself such an important revelation, but also in order to emphasize how productive the dialogue between physician and basic scientist can be. Perhaps also this story can celebrate, together with the 40th anniversary of the discovery of the double helix, the Pauling–Castle conversation as the beginning of a new climate in medical research. I know that Dr. Castle considered Pauling's help in bringing this development about to be a great contribution, perhaps more important than his discovery of the cause of pernicious anemia. In a letter to Dr. Pauling in 1955, quoted by Conley,[2] he stated: "Never has a chance remark of mine turned out so well as my mention to you some years ago during our railroad journey from Denver to Chicago of the phenomenon of birefringence when sickle cells are deoxygenated that had been observed by Sherman." Since then the clinical and basic investigator have become increasingly one, either as an individual by training in both disciplines or as a team. The magnificent elucidation of the molecular basis of thalassemia by the application of DNA technology, which has taken place over several years by virtue of the initiative of many investigators and research teams, is perhaps the most brilliant example of this cooperation.[24]

I would like to conclude with a few remarks bearing on the significance of the discoveries of molecular biology for patients. As physicians we have to accept the frequently voiced concern that the knowledge of the molecular mechanism of sickle cell anemia has not helped a single patient during the past 40 years. There are good indications that this may change. Recent developments in gene technology may become the basis for gene therapy, a term

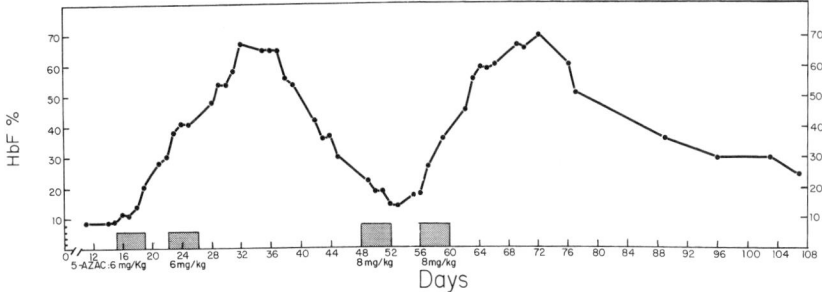

FIGURE 3. Pattern of fetal hemoglobin elevations in the baboon after subcutaneous injection of 5-azacytidine, 6 mg/kg or 8 mg/kg, as indicated by the bars.

used with increasing frequency in several fields of medicine in the hopefully realistic anticipation of future events. Apart from these efforts, the recent attempts to use drugs to increase the level of fetal hemoglobin in erythrocytes of patients with sickle cell anemia and homozygous beta-thalassemia are really a form of gene therapy.[25-27] The very sound rationale behind these efforts is based on the clinically and experimentally well-established favorable effect of fetal hemoglobin on sickling and on the unbalanced globin chain synthesis in thalassemia, but its discussion is beyond the scope of this presentation. Suffice it to say, the claim that the use of these agents is gene therapy is based on their effect either by the direct action on the gene itself, as is the case with 5-azacytidine[25,26] and butyrate,[27] or by an indirect effect through changing the dynamics of hemopoietic stem cells, as is the case with hydroxyurea.[28] After the discovery in our laboratory that 5-azacytidine increases rather dramatically the level of fetal hemoglobin in baboons[25] (FIG. 3), probably because it inhibits methylation of DNA in crucial areas of the regulating sequences of the beta-globin–gene complex, it has been clinically used in patients with life-threatening disease,[26] especially in thalassemia, with remarkable success; its toxicity and possible carcinogenesis forbid wide application, however. And, although multi-center cooperative studies on the clinical application of hydroxyurea and butyrate are currently in progress and their results are eagerly awaited, let us hope instead that the more direct form of replacing the abnormal gene with a normal one will become increasingly feasible and perhaps at the end of this century will no longer be a utopian concept.

REFERENCES

1. PAULING, L., H. A. ITANO, S. J. SINGER & I. C. WELLS. 1949. Sickle cell anemia, a molecular disease. Science 110: 543–548.
2. CONLEY, L. C. 1980. Sickle cell anemia–The first molecular disease. In Blood, Pure and Eloquent. M. M. Win, Ed.: 318–371. McGraw-Hill. New York.

3. HERRICK, J. B. 1910. Peculiar elongated and sickle-shaped red blood corpuscles in a case of severe anemia. Arch. Intern. Med. **6:** 517–521.
3a. HERRICK, J. B. 1949. Memories of Eighty Years. University of Chicago Press. Chicago.
3b. HERRICK, J. B. 1912. Certain clinical features of sudden obstruction of the coronary arteries. Tr. Assoc. Am. Physicians **27:** 100–116.
4. SAVITT, T. L. & M. F. GOLDBERG. 1989. Herrick's 1910 case report of sickle cell anemia, the rest of the story. J. Am. Med. Assoc. **261:** 266–272.
5. EMMEL, V. E. 1917. A study of the erythrocytes in a case of severe anemia with elongated and sickle-shaped red blood corpuscles. Arch. Intern. Med. **20:** 586–598.
6. HUCK, J. G. 1923. Sickle cell anemia. Bull. Johns Hopkins Hosp. **34:** 335–344.
7. TALIAFERRO, W. H. & J. G. HUCK. 1923. The inheritance of sickle-cell anaemia in man. Genetics **8:** 594–598.
8. NEEL, J. V. 1949. The inheritance of sickle cell anemia. Science **110:** 64–66.
9. BEET, E. A. 1949. The genetics of the sickle-cell train in a Bantu tribe. Ann. Eugenics **14:** 279–284.
10. HAHN, E. V. & E. B. GILLESPIE. 1927. Sickle cell anemia. Report of a case greatly improved by splenectomy. Experimental study of sickle-cell formation. Arch. Intern. Med. **39:** 233–254.
11. BECK, J. S. P. & C. S. HERTZ. 1935. Standardizing sickle cell method and evidence of sickle cell trait. Am. J. Clin. Pathol. **5:** 325–332.
12. SHERMAN, I. J. 1940. The sickling phenomenon, with special reference to the differentiation of sickle cell anemia from the sickle cell trait. Bull. Johns Hopkins Hosp. **67:** 309–324.
13. DIGGS, L. W. & J. BIBB. 1939. The erythrocyte in sickle cell anemia. Morphology, size, hemoglobin content, fragility and sedimentation rate. J. Am. Med. Assoc. **112:** 695–701.
14. PONDER, E. 1948. Hemolysis and Related Phenomena. Grune and Stratton. New York.
15. SINGER, K. & L. SINGER. 1953. Studies on abnormal hemoglobins. VIII. The gelling phenomenon of sickle cell hemoglobin: Its biologic and diagnostic significance. Blood **8:** 1008–1023.
16. PERUTZ, M. F., A. M. LIQUORI & F. EIRICH. 1951. X-ray and solubility studies of the haemoglobin of sickle cell anemia patients. Nature **167:** 929.
17. YAKULIS, V. J. & P. HELLER. 1964. An elution test for the visualization of hemoglobin S in blood smears. Blood **24:** 198–201.
18. HARRIS, J. W. 1950. Studies on the destruction of red blood cells. VIII. Molecular orientation in sickle cell hemoglobin solutions. Proc. Soc. Exp. Biol. Med. **75:** 197–201.
19. HOERLEIN, H. & G. WEBER. 1948. Ueber chronische familiaere Methaemoglobinaemie und eine neue Modifikation des Methaemoglobins. Dtsch. Med. Wochenschr. **73:** 476–478.
20. HELLER, P. 1969. Hemoglobin M–an early chapter in the saga of molecular pathology. Ann. Int. Med. **70:** 1038–1041.
21. HAM, T. H. & W. B. CASTLE. 1940. Relation of increased hypotonic fragility and of erythrostasis to the mechanism of hemolysis in certain anemias. Trans. Assoc. Am. Physicians. **55:** 127–132.
22. Quoted by STRAUSS, M. B. 1964. Of medicine, men and molecules: Wedlock for divorce? Medicine **43:** 619–624.
23. PAULING, L. 1955. Abnormality of hemoglobin molecules in hereditary hemolytic anemias. In Harvey Lectures: 1953–1954. Series **49:** 216–241. Academic Press. New York.

24. WEATHERALL, D. 1990. The thalassemias. *In* Hematology. W. J. Williams, E. Beutler, A. J. Erslev & M. Lichtman, Eds.: 510–539. McGraw-Hill. New York.
25. DESIMONE, J., P. HELLER, L. HALL & D. ZWIERS. 1982. 5-Azacytidine stimulates fetal hemoglobin synthesis in anemic baboons. Proc. Natl. Acad. Sci. USA **79:** 4428–4431.
26. LOWRY, C. H. & A. W. NIENHUIS. 1993. Treatment with 5-azacytidine of patients with endstage beta-thalassemia. N. Engl. J. Med. **329:** 845–848.
27. PERRINE, S. P., G. D. GINDER, D. V. FALLER, *et al.* 1993. A short-term trial of butyrate to stimulate fetal-globin-gene expression in the beta-globin disorders. N. Engl. J. Med. **329:** 81–86.
28. CHARACHE, S., G. H. DOVER, M. A. MOYER & J. W. MOORE. 1987. Hydroxyurea-induced augmentation of fetal hemoglobin production in patients with sickle cell anemia. Blood **69:** 109–116.
29. BRECHER, G. 1981. A red cell discovery of the 20th century. Blood Cells **7:** 481–483.

PART III. THE STRUCTURE AND SYNTHESIS OF DNA

Introduction

ROBERT V. STORTI

Department of Biochemistry
The University of Illinois at Chicago
Chicago, Illinois 60612

The 1953 publication in *Nature* of the paper entitled "Molecular Structure of Nucleic Acid" by Watson and Crick[1] proposed a model for the structure of DNA that consisted of a double helix of complementary, paired, so-called Watson and Crick strands. This model raised three important questions concerning the function of DNA. The first question concerned the replication of DNA; that is, how the genetic information encoded in the DNA molecule is copied or replicated during cell division so that an exact copy or genome equivalent of DNA is transmitted to each daughter cell. The second question raised by the model concerned the genetic information encoded in the DNA and how this information is decoded and processed for protein synthesis. Finally, the third question raised at that time concerned the nature of the code. In the decade that followed the events of 1953, each of these questions became the subject of intense investigation. What evolved from this work is the paradigm for information flow that came to be known as the central dogma.

As shown in FIGURE 1, the central dogma describes the process by which DNA is the template for transmitting the genetic code in a two-step process to make protein. The first step, *transcription*, involves the transfer of the genetic information from DNA to an intermediate carrier, the so-called messenger RNA. The second step in this process, *translation*, is the decoding of the genetic code by transfer RNA and, together with the ribosomal machinery, the synthesizing of protein. Each of the speakers in this session has made significant contributions in elucidating various aspects of this pathway of information flow. Our first speaker, Dr. Alexander Rich, is known for his early work on protein and nucleic structure, in particular his studies on Watson and Crick base pairing at atomic resolution. Dr. Rich is also known for his early studies on polysome function in protein synthesis, studies on transfer RNA structure and function and, more recently, his work on Z-DNA.

Our second speaker, Dr. Aaron Klug, pioneered crystallographic techniques for studying macromolecular nucleic acid complexes and has applied these techniques to studying complex structures such as viruses, chromatin and nucleosome structure, and transfer RNA structure. More recently he has been studying protein-DNA recognition in gene regulation.

In 1961 the Jacob and Monod model of gene regulation proposed the syn-

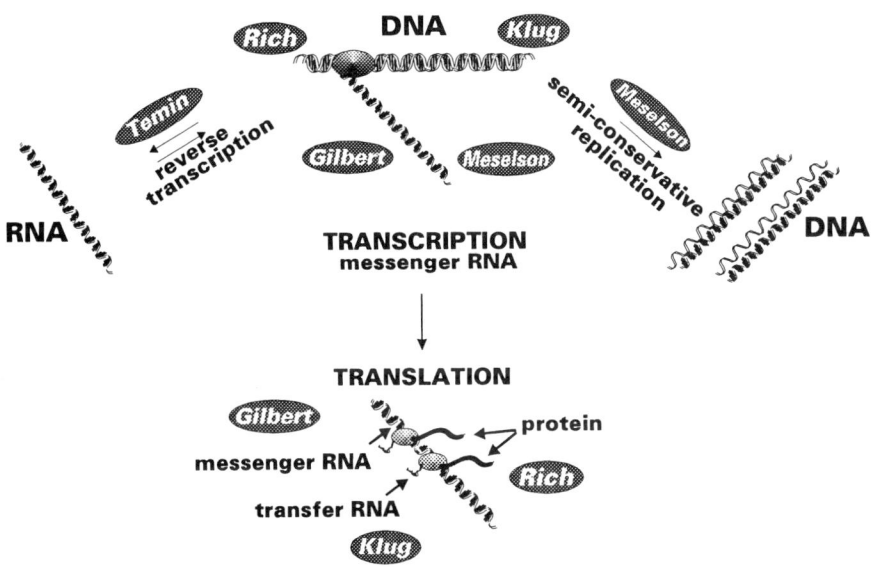

FIGURE 1. The "central dogma" of the function of DNA.

thesis of an intermediate carrier of genetic information, mRNA, and the regulation of its synthesis by operator and repressor control regions. The synthesis of mRNA and its function in protein synthesis were demonstrated in a series of experiments in the early 1960s by our third speaker, Dr. Walter Gilbert. In addition, his later work on gene regulation and the *lac* repressor provided the framework for subsequent studies on prokaryotic and eukaryotic gene control.

The central dogma also defines the process by which the complementary strands of DNA act as templates for their own replication by a process known as semi-conservative replication. This self-templating process was predicted by Watson and Crick in a follow-up paper in *Nature*, "Genetic Implications of the Structure of DNA."[2] This hypothesis was experimentally validated by our fourth speaker, Dr. Matthew Meselson, in the classic Meselson and Stahl density-gradient centrifugation experiments published in 1958.[3] In addition to his studies on replication, Dr. Meselson was also involved in the early experiments on mRNA synthesis and more recently on gene regulation.

The central dogma as outlined in the mid 1950s left open the possibility of an alternate route for DNA replication; namely, that the synthesis of RNA from DNA could be reversed and that RNA could serve as a template for DNA synthesis. This process was demonstrated in experiments conducted by our last speaker, Dr. Howard Temin. In collaboration with Dr. David Baltimore, Dr. Temin showed that replication of the retrovirus genome occurred through an RNA intermediate using the viral encoded reverse transcriptase. These studies have more recently been extended to show that some cellular DNAs can also replicate using an RNA intermediate.

The central dogma has been the framework for molecular biology research for the last 40 years. During this time we have developed an understanding of how DNA is organized into genes and intragenic DNA and how it is replicated. Many of the details concerning gene regulation and the mechanics of transcription and translation have been described and the speakers in this section will discuss how this new information has revolutionized our thinking of biological problems.

REFERENCES

1. WATSON, J. D. & F. H. C. CRICK. 1953. Molecular structure of nucleic acid: a structure for deoxyribose nucleic acid. Nature **171:** 737–738.
2. WATSON, J. D. & F. H. C. CRICK. 1953. Genetic implications of the structure of DNA. Nature **171:** 964–967.
3. MESELSON, M. & STAHL, F. W. 1958. The replication of DNA in *Escherichia coli*. Proc. Natl. Acad. Sci. USA **44:** 671–682.

The Nucleic Acids
A Backward Glance

ALEXANDER RICH

Biology Department
Massachusetts Institute of Technology
Cambridge, Massachusetts 02139

A meeting held to celebrate an anniversary provides a useful interlude between the more familiar meetings addressed solely to the most recent findings. It provides an opportunity to look back at the evolution of a discipline and perhaps to reflect on the manner in which knowledge was gained and how research work evolved. Here, I present an overview of the evolution of my research work in which I try to emphasize how discoveries were made in the context of the currently existing knowledge.

It is difficult to realize how little we knew 40 years ago of what we call molecular biology. Today textbooks of molecular biology bulge with facts, and we have available a large variety of powerful experimental techniques. Very little of this was available 40 years ago. Biochemistry texts of the day had no discussion of information transfer at the molecular level. The nucleic acids were relegated to a minor chapter, and there was considerable uncertainty about the general outlines of biological systems. Most scientists thought genes were proteins. However, some scientists, knowing the 1944 work of Avery and his colleagues,[1] on the pneumococcal transforming factor, believed genes were made of DNA. Some also thought that DNA might make RNA, which, in turn, might make proteins, but the experimental basis for this assumption was tenuous. With the proposal of the double-helical structure for DNA by Watson and Crick,[2] a concrete system was put forth that led directly to a reasonable inference about how genes could be replicated. Beyond that, all was obscure. Virtually nothing was known of the physical nature of RNA, and only misinformation was available about the nature of protein synthesis. It was believed by many that proteins were assembled by proteolytic enzymes working in reverse![3] This just illustrates the state of ignorance that was widespread at the time.

Remarkably, within about 15 to 20 years of the discovery of the double helix, we had obtained a reasonably coherent picture of the mechanism of information transfer in biological systems: how DNA can replicate, how RNA is assembled, and how protein synthesis occurs. Many persons participated in this revolution in our understanding of biological systems at the molecular

FIGURE 1. Postdoctoral days at Caltech in Pasadena, California, 1954. *Left to right*: Jack Dunitz, Giovanni Giacometti, James Watson, Alexander Rich, Jane Rich.

level. Here, I will try to comment on the evolution of our knowledge, illustrating these advances by the contributions that came out of my own laboratories with many different collaborators.

WHAT FOLLOWS AFTER THE DNA DOUBLE HELIX?

In 1953 I was working as a postdoctoral fellow with Linus Pauling (FIG. 1). After some initial work characterizing abnormal hemoglobins and developing double-refraction of flow instruments, I began to focus on X-ray diffraction analysis, using both single-crystal and fiber-diffraction techniques. I had been doing some work on DNA fibers early in 1953, but the work was severely hampered because of the primitive fiber-diffraction equipment available at that time.

Linus Pauling had proposed an incorrect triple-stranded structure for DNA. However, it quickly became apparent in the spring of 1953 that the structure of the double helix formulated by Watson and Crick[2] was a more likely candidate. I had spent some time working on their structure after it was published, trying to understand how the strands might come apart during DNA replication. Toward that end, I tried to see whether it was possible to make the double helix go left-handed as well. A careful analysis of the structure (i.e., left-handed B-DNA) led me to conclude that it was not likely to be stable because of unfavorable van der Waals contacts.

> *It is probably impossible to build this structure with a ribose sugar in place of the deoxyribose, as the extra oxygen atom would make too close a van der Waals contact.*
> — *Watson and Crick (1953)*

FIGURE 2. Watson and Crick[2] had not anticipated that an RNA double helix could be formed by adopting a slightly different conformation than that proposed for the DNA double helix.

Jim Watson arrived at Caltech later in 1953 where he was a postdoctoral fellow with Max Delbrück. We spoke a great deal about DNA as well as RNA and decided to work together to study fibers of RNA to try and learn something about its molecular structure. Very little was known about RNA at that time. The 3′–5′ linkage had been established only recently for both DNA and RNA.[4] However, the question of whether nature used the 2′ hydroxyl of RNA as a way of building a chain was still an open issue in the literature. Many felt that RNA was a multiply-branched polymer. The best available evidence generated from the work of Watson and Crick suggested that the double helix could not be built by RNA strands since the 2′ hydroxyl might yield unfavorable van der Waals contacts (FIG. 2).

We obtained RNA from a number of sources and made fibers using methods similar to those that had been successful with DNA. These fibers were negatively birefringent and showed a small degree of orientation.[5] The data available at that time suggested that the base ratios of the RNA molecules were not complementary so that it seemed unlikely that they could be forming a double helix. Nonetheless, some features of the diffraction pattern suggested the possibility of double helix formation.[6] In retrospect, what we were looking at were samples that usually contained a large amount of ribosomal RNA, some of which had double-helical segments in them. Although we discussed the question of how DNA could make RNA, which could then make proteins, the problem remained unsolved.[6] The concept of messenger RNA was several years in the future. We also discussed the genetic role that RNA must be playing, especially in the plant viruses that contain only RNA as their genetic element.

By mid-1954, I left Caltech to join the National Institutes of Health, where I had been appointed as head of a Section on Physical Chemistry. Jack Dunitz joined me in that move, and later on I was successful in inducing

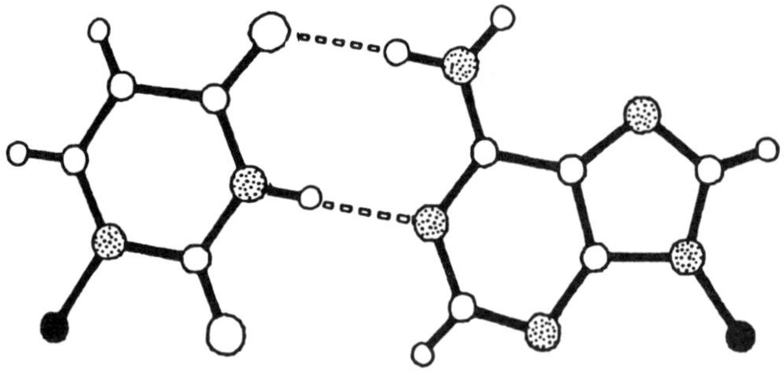

Uracil   **Adenine**

Rich and Davies (1956)

FIGURE 3. The diffraction pattern produced by the 1:1 mixture of poly-A plus poly-U was very similar to that produced by the DNA duplex, although there were subtle differences as well.[8] It clearly suggested that the bases were pairing together in a manner similar to that of the adenine·thymine base pair of DNA. Nitrogen atoms are *stippled*; oxygen atoms are represented by *large circles*.

David Davies to return from England, where he had gone after completing a postdoctoral fellowship at Caltech. Later still, I persuaded Gary Felsenfeld to return from England after his postdoctoral work there so that he joined us at NIH.

THE FIRST HYBRIDIZATION TO FORM A DOUBLE HELIX: POLY-A + POLY-U

The pace of research on RNA changed rather dramatically with the discovery of the enzyme polynucleotide phosphorylase by Marianne Grunberg-Manago and Severo Ochoa.[7] This made possible the production of long polymers containing ribonucleotides. In 1955 Ochoa lectured at the NIH, and he was kind enough to provide me with samples of polyadenylic acid (poly-A) and polyuridylic acid (poly-U).

When these two ribonucleic acid polymers were dissolved together, there was a rapid increase in viscosity, and from this solution glassy fibers could be drawn which were highly birefringent and produced a helical diffraction pattern similar, but not identical to that of double-helical DNA.[8] In particular, the first layer line was strong in this pattern compared to the weak intensity found in DNA fibers. David Davies and I interpreted this pattern in terms of a double helix in which adenine paired with uracil in a manner similar to that which had been postulated in DNA for thymine pairing with adenine (FIG. 3). We suggested that this form of an RNA double helix might be the one in

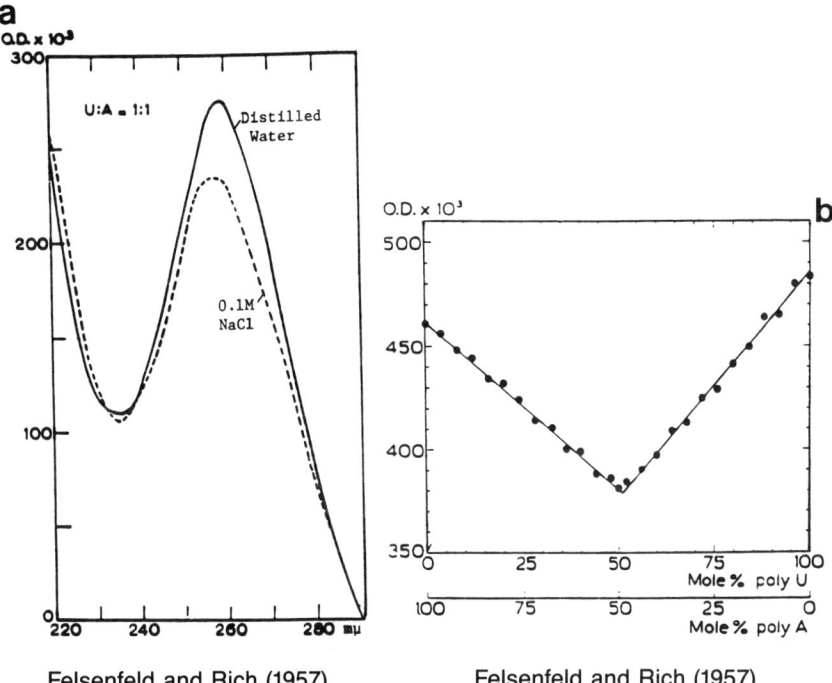

Felsenfeld and Rich (1957)   Felsenfeld and Rich (1957)

FIGURE 4. When poly-A and poly-U were mixed together, a strong hypochromism was observed in the ultraviolet band.[9] (a) The ultraviolet band at 259 mμ showed a strong drop when the denatured mixture in distilled water is compared with that in 0.1 M NaCl. (b) By measuring the change in optical density at 259 mμ as a continuous function of mole ratio, it could be seen that the maximal drop occurred sharply at a 1:1 mixture. The sharpness of the drop made it possible to estimate the equilibrium constant for the reaction. (From Felsenfeld and Rich.[9] Reproduced by permission.)

which RNA carries out its implied molecular duplication in the plant and smaller animal viruses.

Associated with an increase in viscosity on mixing the two polymers, there was also a drop in the optical density of the polymer seen at 260 mμ[9] (FIG. 4a), an observation also made by Warner.[10] By carefully analyzing the drop in optical density at one wavelength for various mole ratios of these two polymers, Felsenfeld and I were able to show that the largest drop in optical density occurred with a 1:1 mixture of the polymers[9] (FIG. 4b).

Shortly after observing the double-helical diffraction pattern produced by a fiber of poly-A plus poly-U, I was walking through a long corridor at NIH and bumped into Herman Kalckar, who worked in the same building.

"Herman," I said to him. "I have just found that mixing poly-A and poly-U will produce a double helix."

Kalckar replied, somewhat incredulously, "You mean without an enzyme?"

Kalckar's astonishment reflected the prevailing view at the time that an enzyme was required to build the DNA double helix. Arthur Kornberg had already shown that DNA polymerase was able to incorporate deoxynucleotides into a DNA double helix,[11] and it was generally believed that this could come about only through enzymatic activity. What we had shown in our experiments was that an enzyme was not required to create the double-helical structure. Instead, the driving force was the stability of the duplex molecule, compared to the unstructured form of the individual components. This stability was associated with stacking of the bases (which produced the observed hypochromism; FIG. 4).

This experiment had significance in two respects. It represented the first hybridization of two nucleic acid molecules to form a double helix. This reaction, in which single-stranded nucleic acids can be induced to form a double helix, was destined to become one of the major methods whereby nucleic acids were analyzed, an eventuality that was somewhat anticipated in the first publication: "this method for forming a two-stranded helical molecule by simply mixing two substances can be used for a variety of studies directed towards an understanding of the formation of helical molecules using specific interactions."[8] A full realization of hybridization came about only some four years later, when it was shown that it could be carried out to form DNA–DNA duplexes as well as DNA–RNA hybrid duplexes.

The second important conclusion from this work was that the RNA backbone would accommodate a double helix, although its form would differ somewhat from that of the DNA double helix. The relationship of the RNA double helix to the DNA double helix in its two different forms was not apparent at the time. Only several years later was it clear that the A form of the DNA double helix was very similar to the form adopted by the RNA double helix.

TRIPLE-STRANDED NUCLEIC ACIDS

While exploring the changes in optical density hypochromism with continuous changes in mole ratio of polyadenylic acid and polyuridylic acid, it was noted that occasionally the curve at higher mole ratios of poly-U was bowed, that is, had a lower value when there was more poly-U present than poly-A. It was also discovered that the addition of divalent cations would transform the curve dramatically so that it then formed a new minimum which occurred at 67 mole percent of poly-U and 33 mole percent of poly-A (FIG. 5a); thus, a 2:1 complex formed.[12] This had also been seen in the ultracentrifuge. The sedimentation constant of the 2:1 complex increased 50% over that seen with the 1:1 complex, suggesting that the mass of the molecule had

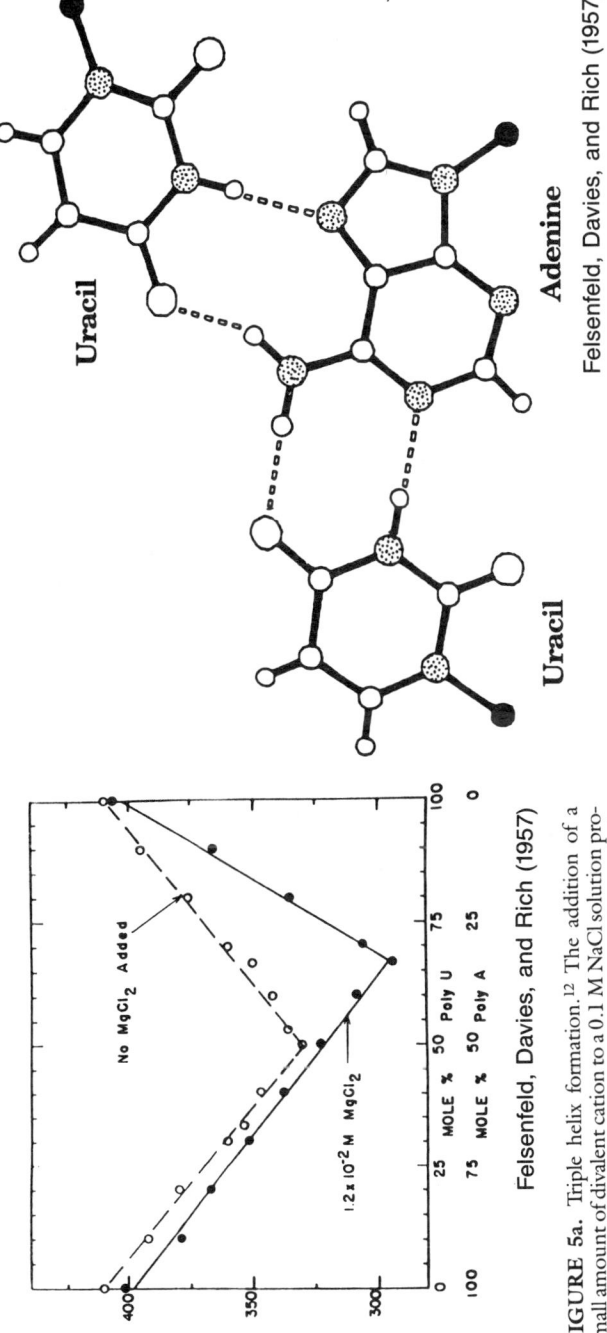

FIGURE 5a. Triple helix formation.[12] The addition of a small amount of divalent cation to a 0.1 M NaCl solution produced a new minimum in optical density absorption at a ratio of 2U:1A. This suggested that the divalent cations stabilized the entrance of the third strand into the major groove by minimizing electrostatic repulsion between phosphate groups. (From Felsenfeld et al.[12] Reproduced by permission.)

FIGURE 5b. The triple helix did not show a significant increase in the diameter of the molecule. The fact that the contribution to hypochromism for the second strand was equal to that of the first strand suggested that the bases were parallel to each other. This was interpreted as indicating a triplex interaction of the type shown here.

increased 50% even though there was no change in the hydrodynamic diameter of the molecule, so that it was clear that the third strand was fitting into the major groove of the duplex. X-ray diffraction photographs of triple-stranded complexes showed no change in the diameter of the molecule. These were interpreted as indicating the formation of a triplex in which the third uracil strand formed two hydrogen bonds with the adenine using the amino group of adenine and the imidazole N7, as indicated in FIGURE 5b.[12] A similar type of hydrogen bonding was seen two years later by Karst Hoogsteen in the single-crystal analysis of 9-methyl adenine and 1-methyl thymine,[13] and it is now referred to as Hoogsteen base pairing.

Several cations were shown to be capable of stabilizing the three-stranded complex. The fact that the third strand produced a hypochromic effect similar to that seen on the addition of the first two strands strongly suggested that the second strand of poly-U assumed a configuration similar to that of the first poly-U strand, that is, helically wound with the plane of the uracil residues normal to the helix's axis.

RNA duplexes and triplexes were not solely limited to purine–pyrimidine complexes. An early study was carried out on the association of polyadenylic acid with polyinosinic acid (poly-I).[14] The hypoxanthine base in inosinic acid is similar to that of guanine, except that it lacks the 2-amino group. Looking at the spectrum of poly-A and poly-I, it could be seen that significant hypochromism occurred when the two strands were mixed together (FIG. 6a). Under these conditions, a small amount of hypochromism was seen at one hour and considerably more at five hours. When this was analyzed by means of a continuous variation curve, at one minute the optical density drop was maximal at a 1:1 complex; at five hours the drop showed a 2:1 complex with two moles of inosinic acid to one mole of adenylic acid. This was interpreted as first forming a 1:1 double helix containing hypoxanthine and adenine, with further addition to make a 2:1 complex, with the second hypoxanthine hydrogen bonded to the adenine amino group and N7 in a manner similar to that which had been proposed for poly-A, plus two poly-Us (FIG. 6b). X-ray diffraction analyses of the fibers showed a well-organized pattern, with a helical scattering radius of about 9.5 Å, consistently with the phosphate groups' being organized on the outer edge of the double or triple helix.[14]

The polynucleotides available at that time consisted of poly-A, poly-U, poly-I, and poly-C; it was very difficult to make a ribo-G polymer. Studying the reactions of these polynucleotides showed that a variety of two- and three-stranded complexes could be formed. It was generally assumed that the third strand was stabilized by at least two hydrogen bonds in formulating these structures. However, the detailed nature of the interactions was not known since the fiber diagrams were only capable of providing general features of the molecular organization, not detailed features. Single-crystal diffraction analysis would be required to uncover the fine details of these structures.

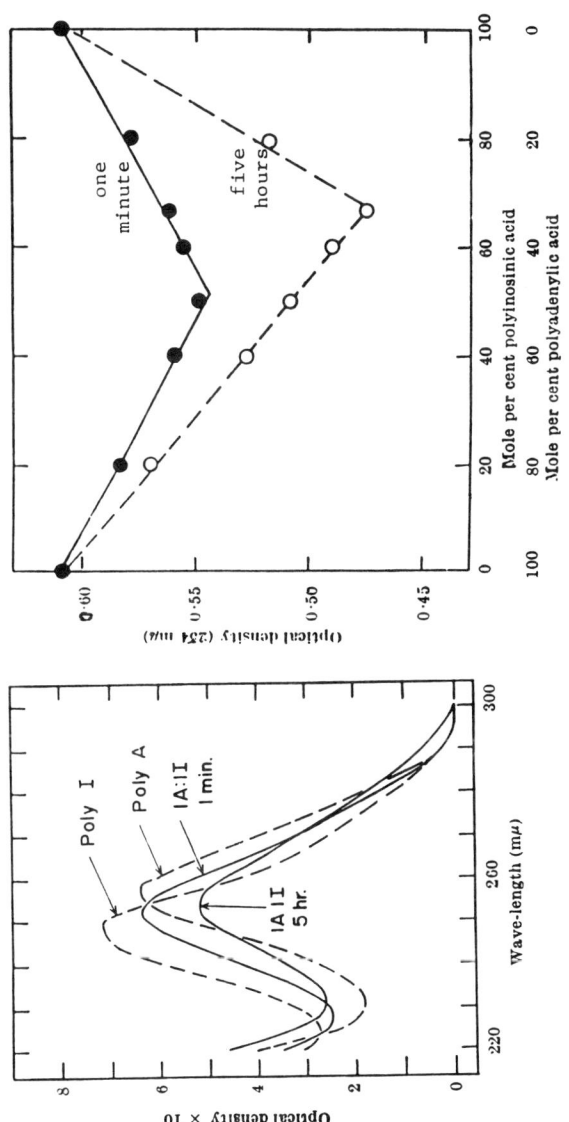

FIGURE 6a. Spectroscopic analysis of the interaction of polyadenylic acid with polyinosinic acid.[14] Reaction conditions were chosen so that the hypochromism observed in the ultraviolet developed gradually over time (*left*). By scanning the hypochromism at 259 mμ, it could be shown that a 1:1 complex formed at 1 hour, while at 5 hours a 1A:2I complex was observed (*right*).

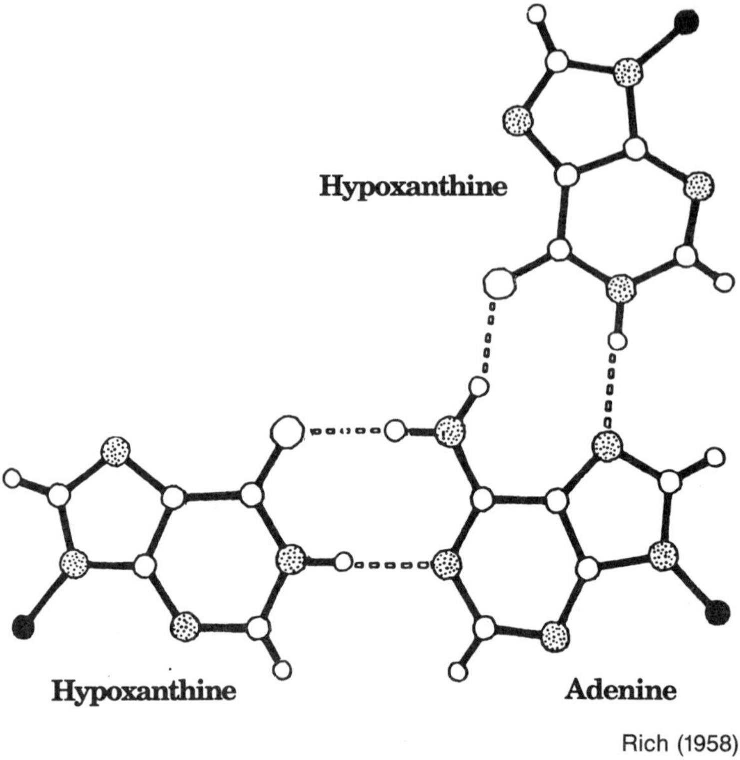

FIGURE 6b. The well-ordered diffraction pattern showed the formation of a helical molecule with a diameter slightly larger than that of the poly-A plus poly-U system. The three-stranded complex was interpreted as shown with the second strand poly-I bonding to adenine N7 and N6 positions in a manner analogous to that seen in poly-A plus 2 poly-U. (This figure and FIGURE 6a are from Rich.[14] Reproduced by permission.)

It is interesting to reflect on the fact that following the discovery of three-stranded nucleic acids in 1957, very little research was carried out to follow up this observation except for studies dealing with polynucleotide interactions of the type described above. Recently, however, there has been a renaissance in this field and many types of triplexes are now known. This renewed interest was stimulated in part by the use of newer technologies with defined segments of DNA to form triplexes. It was also stimulated by the work of M. Frank-Kamenetskii and his colleagues that showed that homopurine–homopyrimidine segments of DNA would, under the influence of negative torsional strain, undergo partial strand separation so that the liberated single strand of homopyrimidines could then fold back on an intact segment of homopurine–homopyrimidine duplex to form a stable triplex.[15] (This has been called H-DNA.) In addition, the use of recombinant methods has made it possible to introduce plasmids containing segments of DNA or RNA that will form triplexes with selective regions of the genome. This, in turn, is used as a method of regulating gene expression. A great deal of this work is now being done in the biotechnology industry.

HOW IS RNA MADE FROM DNA?

In the middle and late 1950s molecular biologists were faced with a series of puzzling dilemmas. Indirect evidence seemed to suggest that RNA had a central role in protein synthesis, and this gave rise to the belief that RNA makes protein. The question of how RNA was made remained a puzzle. People generally felt that there must be an enzyme that could make RNA from DNA, but this enzyme initially proved difficult to purify, even though a number of workers were active in the area.

In 1959 I discussed a number of ideas in a paper on the relations between DNA and RNA.[16] It seemed that there were two possibilities, as illustrated in FIGURE 7. At the time, we believed DNA went through a cycle in which double strands underwent chain separation. Thus, DNA went from a double-

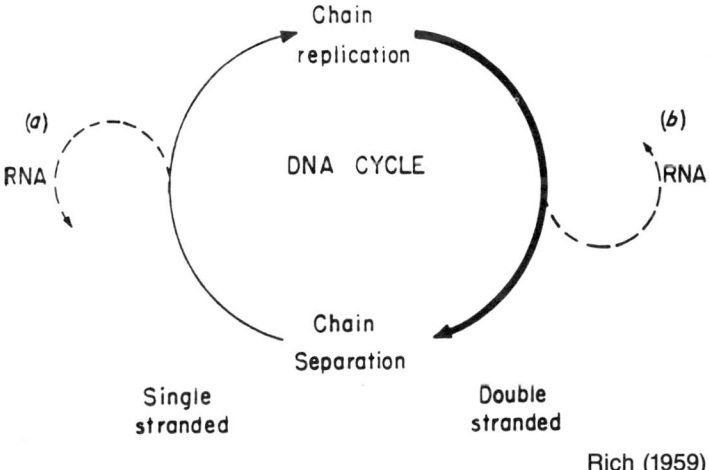

FIGURE 7. The central cycle in this diagram illustrates the manner in which DNA is replicated; that is, individual strands go through a cycle in which they are initially double-stranded, but undergo separation during replication to produce a new DNA duplex. Two alternatives are illustrated for RNA production, either from (*a*) a single-stranded DNA template or (*b*) a double-stranded DNA template. The stimulus for thinking about a double-stranded DNA template arose because of the variety of triplex molecules that had been discovered. The conclusion of the analysis was that (*a*) was the most probable mechanism since (*b*) did not appear to have the requisite specificity.[16] (From Rich.[16] Reproduced by permission.)

TABLE 1. Possible Types of Nucleotide Polymerase Enzymes That Could Exist if Single-Stranded Templates were Used[a]

Enzyme	Primer	Substrate	Product
A	Single-chain DNA	Desoxy ATP Desoxy GTP Desoxy CTP Desoxy TTP	DNA polynucleotide chain
B	Single-chain DNA	ATP (ADP) GTP (GDP) CTP (CDP) UTP (UDP)	RNA polynucleotide chain
C	Single-chain RNA	ATP (ADP) GTP (GDP) CTP (CDP) UTP (UDP)	RNA polynucleotide chain
D	Single-chain RNA	Desoxy ATP Desoxy GTP Desoxy CTP Desoxy TTP	DNA polynucleotide chain

[a] Adapted from Rich.[16]

NOTE: Enzyme A (DNA polymerase) had been identified at the time (1959) of the original publication of this table, while enzyme B (RNA polymerase) was rapidly being purified. Enzyme C was not discovered for many years, and enzyme D (reverse transcriptase) took more than 10 years to be discovered.

stranded to a single-stranded form during replication and then became double stranded again. RNA could be made either by having a third strand added to the DNA duplex in analogy to the many types of polynucleotide triplex structures that we had been investigating. Alternatively, an RNA strand could be made from a single-strand DNA template. In discussing the latter possibility, it seemed reasonable to describe four different types of nucleotide polymerase enzymes, i.e., those that have single strands as templates and have polynucleotide chains as a product (TABLE 1).[16] In this table, which originally appeared in 1959, four types of enzymes (A–D) were described, two which have single DNA strands as templates and two with RNA single strands. The products of these would be either DNA or RNA strands. One form of enzyme A was already known in the DNA polymerase that Arthur Kornberg was studying, which was active with a single DNA chain and had as its product a single DNA polynucleotide. Enzyme B was a possible candidate for the synthesis of RNA strands, analogous to the example cited in FIGURE 7. However, two other types of polymerase enzymes were described, one in which a single chain of RNA is a template for an RNA polynucleotide chain. I speculated that this type of reaction is likely to be found in the replication of certain RNA viruses. The fourth enzyme D would produce a DNA chain from a single chain of

RNA. Enzyme D is, of course, a description of reverse transcriptase, which was discovered over ten years later. In 1959 it was stated that although we have no experimental information about such an enzyme, it cannot be discounted entirely.[16]

In considering the possibility that a double-stranded DNA might act as a template for the synthesis of RNA, the reactions known at that time in forming triple-helical complexes suggested that this type of mechanism would be improbable as a basis for RNA synthesis. That is, looking at the necessary structural requirements, it was not obvious that the four RNA bases could be specified by the DNA base pairs. The conclusion of that article in 1959 was that a single-stranded template was most likely.[16]

This left open the question of whether such a system could physically work. That is, could a hybrid molecule be formed and be stable enough to work in the transfer of genetic information from DNA and RNA? Putting the question more concretely, if a hybrid helix were formed, could the 2′ hydroxyl present on one chain accommodate a double helix in which the other strand lacks the 2′ hydroxyl group? There was no experimental evidence in 1959 with which to address that question.

AN RNA STRAND HYBRIDIZES TO A DNA STRAND

In 1960 it became possible to address this question experimentally, however. By that time two observations gave some indirect support for the idea that a single strand was an adequate template for replication. Kornberg and his associates found that the best primer for the replication of DNA was a single strand of DNA, rather than the double-stranded complementary polymer.[17] The second important discovery was that of Robert Sinsheimer, who showed that the DNA from the small virus ΦX 174 is itself single-stranded.[18] This single strand of DNA thus contained all the biological information necessary to replicate the virus. Both of these discoveries reinforced the concept that the biological production of RNA might proceed through an analogous mechanism whereby a single strand of DNA serves as the template for the production of a complementary RNA strand.

The question of structural compatibility was addressed early in 1960. H. Gobind Khorana had developed a chemical method for synthesizing polydeoxythymidylic acid (poly-dT).[19] Using a sample that he graciously provided to me, I fractionated it and then studied its interaction with polyriboadenylic acid (poly-A). Dissolving the solutions in dilute neutral sodium chloride, the optical density of the mixture had a clear hypochromicity, as shown in FIGURE 8a.[20] Analysis of the reaction with continual changes of mole ratios led to an optical density minimum at 1:1 in dilute salt concentrations (FIG. 8b). At higher salt concentrations a 2:1 minimum formed. These

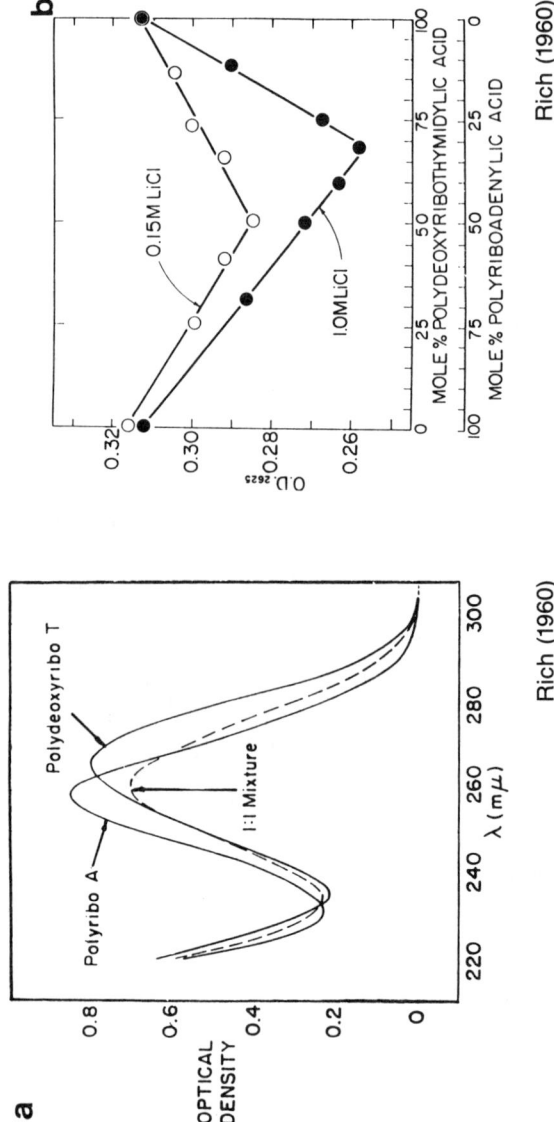

FIGURE 8. Formation of a DNA–RNA hybrid is seen in the ultraviolet spectra of poly-A and poly-dT.[20] (a) The ultraviolet spectrum of the two polymers alone is shown in a dilute salt solution, together with the hypochromism of a 1:1 mixture. (b) The ultraviolet absorption of the mixture of polymers is seen to have a minimum of 1:1 in a dilute salt concentration and a minimum of 2 dT:1 rA in a high-salt solution. The stability of the 1:1 complex in the dilute salt solution suggested that it would be a stable intermediate that could be used by an enzyme to make an RNA strand using a single DNA strand as template. (From Rich.[20] Reproduced by permission.)

results were somewhat analogous to those that had been observed with the poly-A plus poly-U system. This study was complemented by an ultracentrifugation analysis that demonstrated the formation of the hybrid helix. An analysis of the reactivity of poly-dT clearly showed that the polymer reacted in a manner that was expected in terms of the thymine hydrogen bonding properties. The conclusion of these studies was that a DNA–RNA hybrid can form a stable structure, and it could be used to transmit information from DNA to RNA.[20]

This first hybridization of a DNA and an RNA strand is one that is still widely used today in the isolation of messenger RNA. Thus, poly-dT usually attached to the column is used to isolate messenger RNA by forming a hybrid helix with the poly-A tail of messenger RNA.

It is interesting that at the same time the RNA–DNA hybridization was published, Julius Marmur, Paul Doty, and their associates published papers describing their experiments in hybridizing DNA strands to each other.[21,22] They demonstrated that single strands of DNA obtained by thermal denaturation could be slowly cooled and annealed together to reform the double helix. This meant that the strands could form a double helix not only when all the bases were the same on the individual strands, as was the case with the hybridized RNA–RNA duplex in 1956 or the DNA–RNA hybrid, but could also form a double helix when the bases varied along the polynucleotide chain. In the latter case, slow cooling was essential to allow the bases to make appropriate contact. A year later Ben Hall and Sol Spiegelman showed that RNA–DNA hybrids could also be formed from bacteriophage-encoded RNA by the same slow annealing process.[23] The naturally occurring RNAs would also form a stable hybrid helix with single-stranded DNA molecules.

These hybridization experiments provided one of the major experimental tools for exploring the relationship between nucleic acids. Hybridization continues to be widely used today, involving both DNA–DNA and DNA–RNA complexes.

In the same year (1960) investigators were beginning to isolate early preparations of what we now call RNA polymerase. In order to get incorporation of a radioactive ribonucleotide, DNA was required as well as all four of the nucleoside triphosphates. These experiments were carried out by J. Hurwitz, A. Stevens, S. Weiss, and their collaborators. The detailed role of DNA was not clearly understood until 1961. At that time it was apparent that one could use denatured DNA (that is, single-stranded DNA) to direct the synthesis of RNA. Experiments by Hurwitz demonstrated that if one used polydeoxy-T as a primer, the synthesis of polyribo-A followed.[24] Interestingly, Hurwitz also obtained the polydeoxy-T polymer from H. Gobind Khorana.[19] This was the same polymer that I had used in the previous year to demonstrate the stability of the hybrid helix. The enzymatic reaction clearly indicated that a single strand of DNA containing thymine would synthesize its complement containing only adenine. At that same time, experiments by both the Hurwitz and the Stevens laboratories involving the use of a d(AT) polymer primer yielded RNA containing only uracil and adenine.[25,26] By 1961 the enzymatic evidence had clearly indicated that the transfer of information from DNA to RNA went through the formation of a single strand of DNA acting to assemble its complement in an RNA strand.

THOUGHTS ON EVOLUTION AND THE ORIGIN OF LIFE

In 1961 I was asked to contribute an article to a volume dedicated to Albert Szent-Györgi. I decided to write about the evolution of biochemical information transfer and the origin of life.[27] This was prompted by the current prevailing ideas regarding the origin of life. The prominent view at the time, largely supported by the writings of the Russian biochemist Oparin, was that life began through the creation of primitive protein molecules that provided the necessary environment for developing cells and self-replication. In this view of evolution, amino acids assembled at random eventually formed molecules that had enzymatic activity which would function to develop the machinery of life. The greatest weakness in a theory of this type is that it does not really explain the most difficult step, namely the evolution of nucleic acid–controlled protein synthesis. It seemed far more reasonable to suggest that life began with nucleic acids.

Experimental work had already begun on the reactions whereby components of polynucleotide chains could be made, starting with elementary organic molecules, and I could imagine a process whereby these eventually led to the development of nucleic acid chains. It seemed reasonable to believe that such primitive polynucleotide chains could act as a template for synthesizing its complement to build a two-stranded molecule. In view of the different chemical and biological activities of the two nucleic acid polymers, it also seemed reasonable that this parent polynucleotide molecule was initially an RNA polymer, rather than DNA (FIG. 9). In this proposal these molecules would act as a rather inefficient catalyst for assembling their complements and facilitating their polymerization into double-stranded RNA molecules. This statement might be regarded as a very early statement of what is now known as the "RNA world," that is, an early system in which RNA contains information that is replicated and directs the synthesis of primitive RNA enzymes. The reason for choosing RNA rather than DNA stemmed from the conviction stated earlier,[16] namely, that since we know the RNA molecule is able to carry genetic information and it is also involved in protein synthesis, it seemed reasonable to believe that this would be involved in the origin of life. DNA was thus regarded as a specialized derivative molecule that evolved in a form that carried out only the molecular replicating cycle that is an inherent part

> *We postulate that the primitive polynucleotide chains are able to act as a template, or as a somewhat inefficient catalyst for promoting the polymerization of the complementary nucleotide residues to build up an initial two-stranded molecule.*
>
> *It may be reasonable to speculate that the hypothetical stem or parent polynucleotide molecule was initially an RNA-like polymer . . .*
>
> *—Rich (1962)*

FIGURE 9. In an article on evolution and the origin of life, it seemed reasonable in 1962 to believe that life began by an RNA catalytic activity to build an RNA duplex.[27] This may be the first statement describing what is known today as the "RNA world."

of the transmission of genetic information. DNA is less reactive metabolically, perhaps because of the absence of the hydroxyl group, and this may have had a selective advantage in an evolving biochemical system.

Other points developed in this evolution essay dealt with various mechanisms for the selection of purines and pyrimidines.[27] It was pointed out that a nucleic acid replicating system could have contained three base pairs instead of two since the base pair isoguanine and isocytosine could fit into the same DNA duplex with a unique system of hydrogen bonding, assuming that the bases were maintained in the keto form. It is interesting that Steve Benner and associates in 1990 found that nucleoside triphosphates with isoguanine and isocytosine could be specifically incorporated into the DNA duplex by DNA polymerase.[28]

At that time in 1962, the concept of messenger RNA had just been developed, but the mechanism whereby RNA polymerase made messenger RNA was not fully understood. One of the things that seemed apparent to me was that if RNA copies were made of both complementary chains of DNA, one of them would be involved in coding for proteins, while the other might be a component of a control or regulatory system (FIG. 10). This is perhaps

> *Messenger RNA may be made* in vivo *as complementary copies of one or both strands of DNA. If both strands are active, then the DNA would produce two RNA strands which are complementary to each other. Only one of these might be active in protein synthesis, and the other strand might be a component of the control or regulatory system.*
> — *Rich (1962)*

FIGURE 10. In 1962 RNA polymerase had been only recently purified and messenger RNA discovered a short time previously. However, it was not certain how the enzyme worked. This speculation represents an early statement of what is now called anti-sense RNA.[27]

the first statement of what is now known as "anti-sense" RNA, suggesting that it might have a role in biological systems.

It was characteristic of writing in these early days in the development of molecular biology that very large features of biological systems were unknown. It was therefore possible to speculate and anticipate features that could be investigated only several years in the future.

PROTEINS ARE MADE ON POLYSOMES

By 1962 evidence had accumulated suggesting that a rapidly synthesized RNA subfraction called messenger RNA was made by DNA and combined with ribosomes to carry out protein synthesis.[29,30] Through the work of Paul Zamecnik and Mahlon Hoagland and associates, it was known that amino acids were activated by transfer RNA (called soluble RNA at the time) in protein synthesis.[31] Evidence had been found for a triplet code, and thus three nucleotides specified one amino acid.[32] At that time I began to wonder about the relationship between the length of the message and the size of the ribosome. In working with rabbit reticulocytes that were synthesizing hemoglobin, it seemed apparent that the hemoglobin molecule with about 150 amino acids had to be specified by 450 nucleotides. If the nucleotides were stacked in a system in which each base took up 3.4 Å, then the length of the message should be greater than 1,500 Å long. However, the ribosome itself was only some 230 Å in diameter. Working with my first graduate student, Jonathan Warner, and a postdoctoral fellow, Paul Knopf, we tried to isolate larger structures containing more than one ribosome. This experiment was successful. Using sucrose density gradients, we showed that only the clusters of ribosomes that we called polysomes rather than the single ribosomes, were active in protein synthesis.[33,34] In the case of the hemoglobin synthesis, about five ribosomes were found in clusters. To characterize these, we used electron microscopy, which clearly showed not only the number of ribosomes in each section of a sucrose density gradient, but also (with negative staining) actually showed the messenger RNA running from ribosome to ribosome (FIG. 11).

We then had a dynamic picture of how proteins were synthesized. Ribosomes attached themselves to one end of the message to initiate protein synthesis. The concept of initiator and terminator codon signals was not developed until much later. The ribosome moved along the messenger RNA, and the polypeptide chain grew as the ribosome progressed. At the end, the ribosome released the protein and became detached from the message.[35] This construct could be readily adapted to polycistronic messages,[36] which were

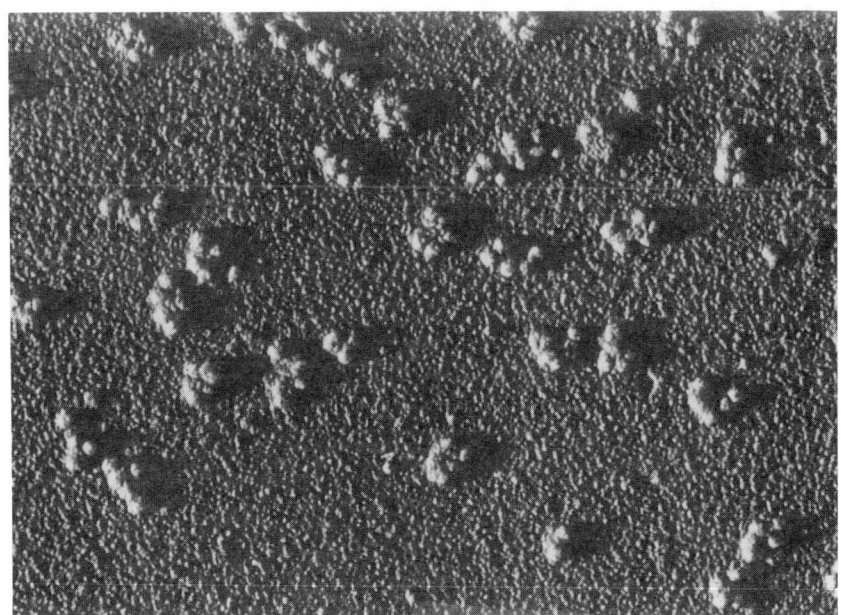

FIGURE 11a. In 1962 it was discovered that protein synthesis takes place on clusters of ribosomes called polysomes.[33] Electron micrograph of polyribosomes synthesizing hemoglobin. They are visualized here with platinum shadowing.

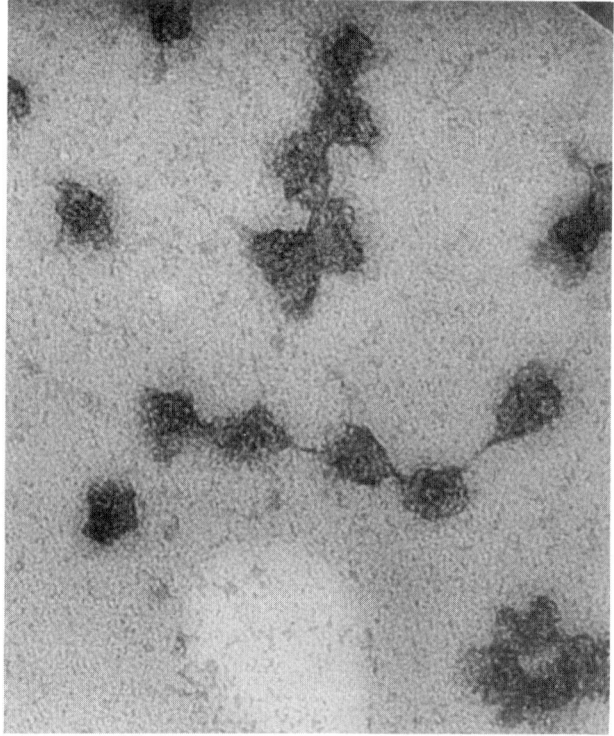

FIGURE 11b. Negative staining reveals a thin strand of messenger RNA passing between the ribosomes. (This figure and FIGURE 11a are from Warner et al.[33] Reproduced by permission.)

then beginning to appear in bacterial systems, and it also made possible a number of additional investigations in the system.[37] At that time many people felt that the concept of ribosomal movement was unusual, and it took a while before it evolved to its present status of being self-evident.

Later investigations demonstrated that there were two transfer RNA molecules firmly bound to the ribosome in protein synthesis.[38] We characterized these as being attached to the amino acyl (A) site or the peptidyl (P) site. The elongation of the peptide chain was then carried out by the peptidyl transferase.

In other experiments, aspects of ribosomal architecture were characterized. By digesting ribosomes active in protein synthesis with peptidases, it was possible to show that about 30 amino acids were shielded from peptidase digestion while they were buried in the ribosome.[39] We postulated a channel some 50 Å long leading from the site of peptide bond formation to the exit position on the ribosome. Likewise, nuclease digestion experiments on the messenger RNA showed that four codons from the peptidyl site were shielded by the ribosome.[40] These early indirect experiments allowed us to learn something about the ribosomal architecture.

TRANSFER RNA IS MADE ON DNA

An illustration of how little was known about RNA was the question of whether transfer RNA was made by DNA or whether it could be made by replication of double-helical RNA. To address that question in 1962, a graduate student, Howard Goodman, carried out experiments in which *E. coli* transfer RNA (called soluble RNA at that time) was hybridized to *E. coli* DNA.[41] RNA–DNA hybridization was first demonstrated with polynucleotides in 1960,[20] and in 1961 Hall and Spiegelman[23] showed that one could use bacteriophage RNA and hybridize it to DNA using a system similar to that which had been devised by Marmur and Doty in their DNA annealing experiments of 1960.[21,22] The first question to be asked is how many tRNA molecules could be added to the DNA before it reached saturation. Radioactive transfer RNA was used. Hybridization was carried out while adding increasing amounts of transfer RNA. The amount of transfer RNA was then measured by digesting the now-hybridized tRNA with ribonuclease and measuring the amount of radioactivity that was resistant. It had already been demonstrated that a DNA–RNA hybrid was resistant to nuclease digestion. In this way a plateau was found at 40 molecules of transfer RNA per *E. coli* genome.[41] This suggested that the triplet code had some degeneracy in that there were fewer than the 64 species that one might find if each triplet were represented with a different tRNA molecule.

These experiments were extended to explore the relationship of *E. coli* tRNA to DNA from related or distant bacterial strains.[41] The results were quite clear. With strains that were genetically closer to *E. coli*, the amount of hybridization was not as great as with *E. coli*, but nonetheless considerable. With strains that were distinct from *E. coli* and not close neighbors, the amount of hybridization was minimal. This method became a tool, not only for measuring the number of genes that code for tRNA, but also the extent to which these genes were conserved in different species.

CHLOROPLASTS AND MITOCHONDRIA CONTAIN DNA

By the early 1960s, it was clear that the inheritance of characteristics in chloroplasts was non-nuclear in origin. The physical basis of this was not known. Some indirect experiments suggested the possibility that chloroplasts might contain DNA, but it had not yet been demonstrated. In 1963, together with some students, we decided to use the then newly discovered method of density gradient centrifugation to see whether it was possible to identify DNA in chloroplasts.[42] Four different organisms were used: spinach and beets, as well as the photosynthetic green algae *Chlamydomonas* and *Chlorella*. DNA was extracted and run in a cesium chloride density gradient, together with a marker. The results showed that in addition to a large peak, α, which was found to come from the nuclear preparation, another two additional peaks, β and γ, were found to appear in these preparations (FIG. 12). When chloroplasts were isolated from a crude mixture, the β peak increased by 10–30-fold. Experiments demonstrated that the DNA was double-stranded, and it clearly indicated that the chloroplasts not only contained DNA, but they also contained DNA with significantly different base ratios since the density at which they banded in cesium chloride was distinct from that found in the nucleus.[42]

In carrying out a preparation from the whole cell, it was observed that the sample contained two different species, β and γ, and that their ratio varied significantly. The origin of the γ peak suggested that it arose in the preparation. Since the preparation involved a procedure of centrifuging particles, the suggestion was made that the γ peak probably represented DNA from mitochondria.

These were the first demonstrations of DNA in organelles. This method was widely adopted, and a large number of studies were carried out subsequently on organelle DNA. Discovery that the DNA had different base ratios from that found in the nucleus provided some support for the concept that these organelles originated with the infection of a eukaryotic cell by microorganisms such as bacteria which subsequently learned to survive in a symbiotic way within the cell. This hypothesis is still discussed today, although it is far from established.

ELECTRONIC COMPLEMENTARITY OF NUCLEIC ACID BASES

In 1965 we tried to see whether the nucleic acid bases had an affinity for each other independent of the geometry of the double helix. We studied the

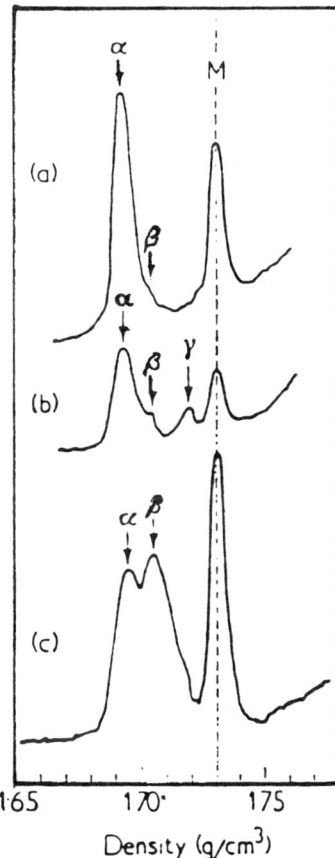

FIGURE 12. In 1963, we used density gradient centrifugation to see whether chloroplasts actually contained DNA.[42] Densitometer scans are shown of uv absorption of DNA in cesium chloride gradients. (a) DNA from the whole cell; (b) DNA from the chloroplast fraction; and (c) the DNA from β region of the chloroplast preparation was centrifuged again. M is added marker DNA; α is the nuclear DNA. In addition to the β peak seen in the chloroplast fraction (b), another peak γ was observed. It was surmised that this represented DNA from mitochondria.[42] Subsequent studies by other investigators verified this supposition. (From Chun *et al.*[42] Reproduced by permission.)

hydrogen bonding specificity of nucleic acid bases in the absence of base stacking and independent of the polynucleotide chain. Infrared studies were carried out in chloroform solutions of soluble adenine and uracil derivatives.[43] The infrared spectrum in chloroform specifically makes it possible to register changes in N-H stretching vibrations when hydrogen bonding occurs. The results of this study revealed an interesting phenomenon. Although both adenine and uracil (or thymine) derivatives self-associate, that is to say, form hydrogen-bonded pairs in solution, they did so with much less affinity than was observed when they hydrogen-bonded with each other. The association constants for forming an adenine–uracil base pair was 15–30 times greater than the association constant for adenine–adenine or uracil–uracil base pairs.[44,45] These experiments were carried out in a chloroform solution in the absence of water. This environment is one that allows the interactions of the bases to be studied independently of other interactions. Furthermore, the hydrophobic

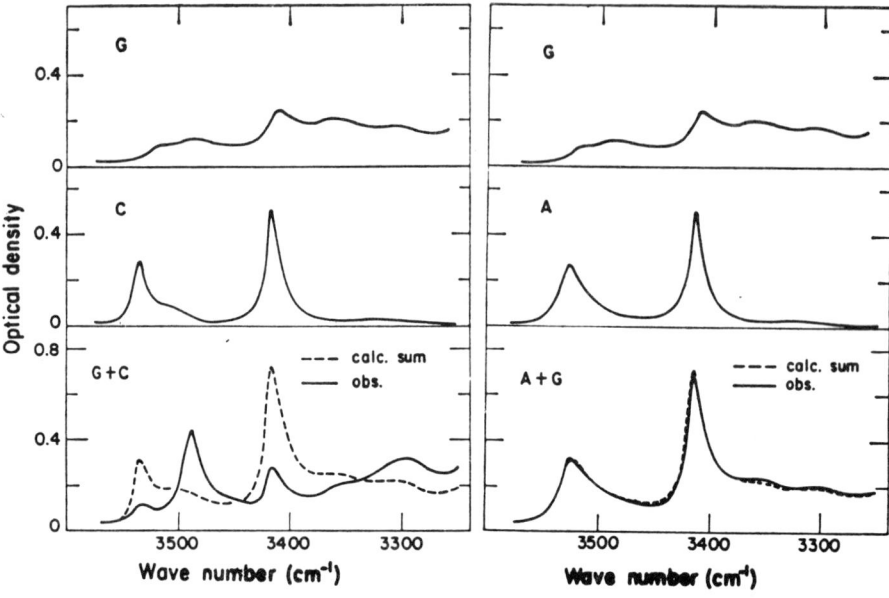

Kyogoku, Lord, and Rich (1966)

FIGURE 13. In 1965 we discovered that chloroform-soluble derivatives of adenine and uracil formed hydrogen-bonded dimers in solution.[43] In 1966 this was extended to guanine and cytosine derivatives.[46] Mixing these derivatives together in solution resulted in the formation of a 1:1 complex with the appearance new N-H stretching vibrations at 3300 cm^{-1} and at 3488 cm^{-1}. However, when guanine and adenine chloroform-soluble derivatives are mixed under the same conditions, they do not interact specifically. Of all the four possible mixtures, only G and C as well as A and U (or T) were found to form strong hydrogen-bonded pairs in chloroform solution. (From Kyogoku et al.[46] Reproduced by permission.)

environment of this solution may be similar in part to the hydrophobicity inside the nucleic acid double helix, where the bases stack and water is excluded.

These experiments were extended to studies of chloroform-soluble guanine and cytosine derivatives, where a similar phenomenon was observed.[46] The self-association with cytosine and of the guanine derivatives was significantly smaller than the association of cytosine with guanine derivatives (FIG. 13). Studies were then carried out to measure the association of all the possible paired combinations, and the results were quite striking. Only the guanosine–cytosine and adenine–uracil (thymine) pairs associated with each other with a higher association constant than any of the other cross pairs (FIG. 13).

This phenomenon was called *electronic complementarity*, and it is probably an expression of the dipole–dipole and dipole-induced dipole interactions which are characteristics of these bases. The interesting phenomenon observed here is that the intrinsic stability of these bases paired with each other was complementary and reinforced the geometrical complementarity that is an inherent feature of the nucleic acid double helix.

SINGLE CRYSTAL X-RAY ANALYSES OF BASE PAIRS

When the structure of DNA was formulated by Watson and Crick in 1953, they suggested that the guanine–cytosine base pair would be held together by two hydrogen bonds, similarly to the adenine–thymine base pair.[2] However, in 1956 Pauling and Corey reviewed the geometry of purines and pyrimidines and their potential for hydrogen bond formation and concluded that the guanine–cytosine base pair in DNA would probably be held together by three hydrogen bonds.[47] Experimental evidence on this point was not available in the X-ray diffraction data from DNA fibers since the resolution of the patterns was very far from being adequate to settle questions on subtleties of this type.

In 1963 my colleagues and I made a co-crystal of 9-ethyl guanine and 5-bromo-1-methyl cytosine.[48] The structure was solved by the heavy atom method, and the results are shown in FIGURE 14. It was clear that the guanine and cytosine residues were held together by three hydrogen bonds ranging in length from 2.86 to 2.95 Å.

This structure provided useful information about the detailed structural geometry of these two bases when they were hydrogen-bonded together. Furthermore, it gave a reliable number for the distance between what are analogous to glycosyl carbon atoms attached to the nitrogens N1 of the cytosine derivative and N9 of the guanine. This distance of 10.8 Å is important since it specifies the distance between the two polynucleotide strands in the DNA molecule.

A number of other purine–pyrimidine complexes were analyzed, including several complexes of adenine and uracil derivatives that all formed Hoogsteen base pairs.[49–52]

THE DOUBLE HELIX AT ATOMIC RESOLUTION

Fiber-diffraction patterns of DNA were interpreted in terms of the double helix formulated by Watson and Crick.[2] These diffraction patterns rarely had reflections beyond 3 Å resolution. Most reflections were 4–5 Å or less in resolution. Thus, they had a relatively small number of reflections compared to the number of unknowns which were being determined by the structure, that is, the positions of the atoms. The diffraction patterns were "interpreted" rather than "solved." What that means is that molecular models were built, and the diffraction pattern that they predicted was then compared with the observed patterns. Then corrections were made in the structure to get a closer agreement. This by no means proved that the structure had that form since it would have to be established that no other conformation would fit the data. Given the limited resolution of the data, this was not a settled issue.

Things remained that way for almost twenty years after the formulation

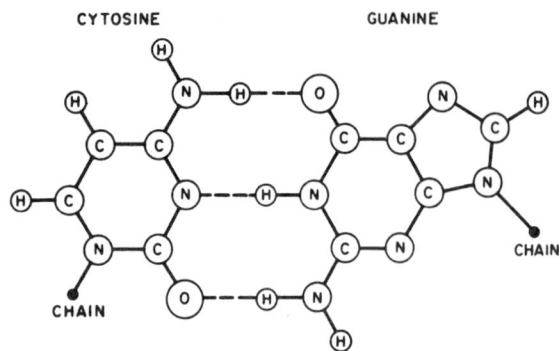

The hydrogen bonding which is believed to exist between guanine and cytosine in DNA.

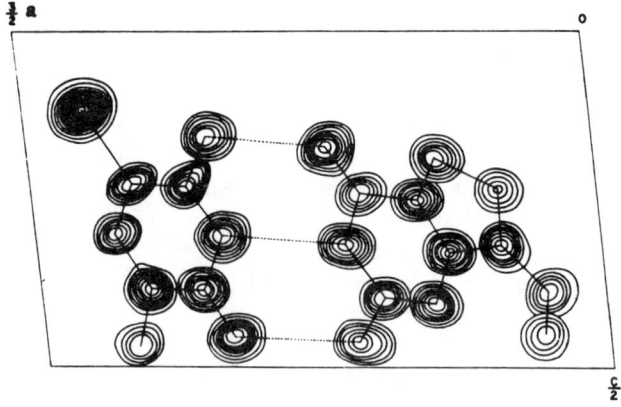

Sobell, Tomita, and Rich (1963)

FIGURE 14. The crystal structure of 9-ethylguanine and 1-methyl-5-bromouracil.[48] This structure solved at atomic resolution clearly indicated the presence of three hydrogen bonds stabilizing the interaction. (From Sobell et al.[48] Reproduced by permission.)

of the double helix in 1953. However, in 1973, together with N. Seeman, J. Rosenberg and others, we solved two structures at atomic resolution in which the double helix was finally visualized. These were the dinucleotide phosphates GpC[53] and ApU.[54] They were solved at 0.7 Å resolution, which made it possible to define all atoms in the unit cell, including solvent molecules and cations.

The structure GpC (FIG. 15a) showed that the molecules had crystallized in a double helical array with three hydrogen bonds between guanine and cytosine residues, just as had been seen earlier in the single-crystal analysis of guanine and cytosine derivatives.[48] The bases were stacked with considerable overlap. In addition, this crystalline fragment made it possible to obtain pre-

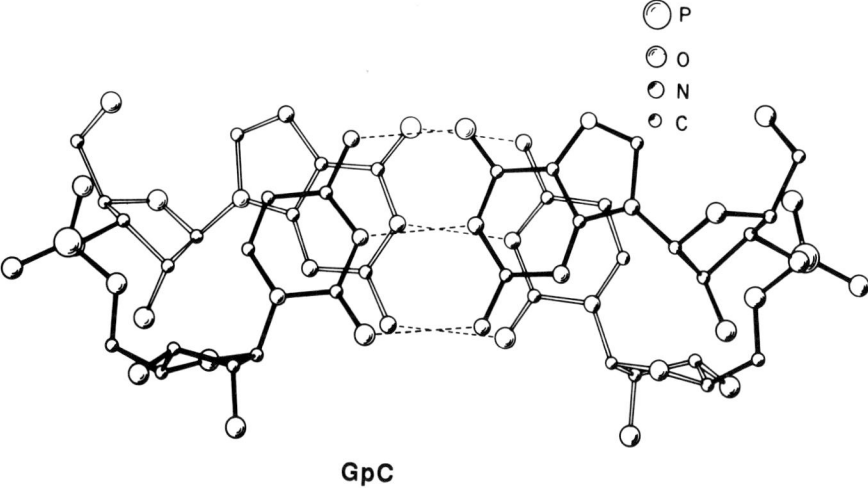

GpC

Day, Seeman, Rosenberg, and Rich (1973)

FIGURE 15a. The double helix was first seen at atomic resolution in two dinucleotide phosphate structures. This figure shows the structure of GpC with the solid bonds indicating the bases closer to the reader. (From Day et al.[53] Reproduced by permission.)

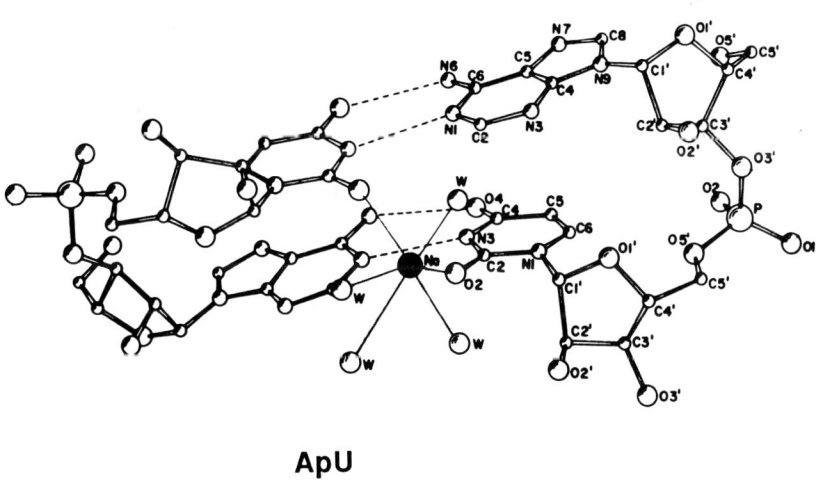

ApU

Rosenberg et al. (1973)

FIGURE 15b. The structure of ApU is shown in perspective, together with a sodium ion found in the minor groove that coordinates the two bases and four water molecules. (From Rosenberg et al.[54] Reproduced by permission.) Both this structure and that in FIGURE 14a had similar conformation and, if extended into a long double helix, formed an RNA duplex with 11 base pairs per turn.

cise information about the detailed geometry about the ribose phosphate backbone.[55,56]

In some ways the structure of GpC might have been anticipated in that GC hydrogen bonding of the Watson–Crick type had been observed in single crystals of purine–pyrimidine complexes. This was not the case with purine–pyrimidine complexes involving adenine and uracil (or thymine) derivatives. The first of these, 9-ethyl adenine and 1-methyl thymine solved by Hoogsteen, yielded base pairing through adenine N6 and N7,[13] just the type of hydrogen bonding postulated when the third strand of poly-U was added to the double-stranded poly-A plus poly-U.[12] However, it was rather puzzling since several other crystalline complexes were solved in our laboratory[49-51] as well as in others between adenine and uracil (or thymine) derivatives and all had Hoogsteen, or reversed Hoogsteen base pairing. None of them had Watson–Crick pairing. To make the situation even more dramatic, a model of DNA had been proposed in which the two strands were held together entirely by Hoogsteen-type pairing between A and T, as well as G and C.[57]

It was of great interest to find that only when the adenine and uracil residues were connected with the sugar phosphate backbone did they assume the form of a double helix in which Watson–Crick base pairing was found between the adenine and uracil derivatives (FIG. 15b).[54] The structure also revealed an intricate pattern of interaction with hydrated sodium ions as well as water molecules.

Preprints of these articles were mailed to several people, including Jim Watson. He replied to me that it was only after seeing the structure of ApU with its Watson–Crick base pairs that he had his "first good sleep in 20 years."

It is also of interest to note that the structures of ApU and GpC were quite similar to each other, aside from the differences in the actual bases. They were very similar to a model of RNA 11, that is, an RNA double helix in which there were eleven base pairs per turn. These molecules had been interpreted as crystallizing in the A form, in which there is a deep major groove and a shallow minor groove. Visualization of the double helix at atomic resolution resolved the issue of the conformation of the RNA double helix. These studies also provided precise geometry for the sugar phosphate backbone of RNA molecules. These data were of great value in interpreting the structure of transfer RNA, which was solved during the same time period.

tRNA CAN BE CRYSTALLIZED

In the mid-1960s it was not at all clear that transfer RNA molecules folded up into a discrete molecular form. Many investigators felt that it might be somewhat unstructured, containing only the secondary-structure double-helical stems and loops that had been identified in the clover-leaf sequence first recognized by Robert Holley and colleagues.[58] It was of considerable interest in 1968 when Sung-Hou Kim and I were able to obtain large single crystals of *E. coli* F-met tRNA.[59] These crystals were hexagonal and had three to six molecules in the asymmetric unit. They were heavily hydrated and diffracted X-rays to only 20 Å resolution. Nonetheless, this was enough to suggest that the molecule had a discrete form. The availability of purified tRNA species around that time made it possible for a number of laboratories to crystallize tRNA during the next couple of years.

By 1969 a different crystal form was obtained for the F-met tRNA, an orthorhombic crystal form with one molecule in the asymmetric unit. It diffracted to 7 Å resolution. It was possible to carry out a Patterson calculation and from this to make a rough estimate of the size of the molecule, 80 ± 10 Å by 25 × 35 Å.[60] It is interesting that these initial approximations are not far removed from the size revealed when its structure was ultimately determined.

A vigorous attempt was made to improve the resolution of the diffraction pattern. For X-ray diffraction studies a resolution of better than 3 Å is required in order to resolve the bases from the sugars and the phosphate residues. Crystals that diffracted at 7 Å resolution were useful principally in stimulating and spurring a more vigorous search for better crystals. However, they were of no value in terms of determining the structure of the molecule. This search was finally successful in 1971, when high-resolution crystals were obtained from yeast phenylalanyl tRNA that crystallized in an orthorhombic lattice with a resolution of 2.7 Å.[61] This diffraction pattern was quite interesting since it showed a distribution of intensities that was characteristic of helical components in the molecule. The existence of crystals with this resolution meant that it was only a matter of time before the three-dimensional structure of the molecule could be determined.

A number of different crystal forms of this molecule were obtained including monoclinic, hexagonal, and cubic.[62] The key to obtaining high-resolution crystals was the addition of the polyamine spermine. It had been known that spermine is found widely in association with the nucleic acids. It is an elongated molecule with four positive charges, and it seemed to play a key role in stabilizing the lattices formed with yeast tRNA^{phe}. At the time we thought this might be a general key to obtaining high-resolution tRNA crystals. Unfortunately, spermine works in many cases, but does not work with all tRNA species.

TRACING THE BACKBONE OF tRNA

Once high-resolution crystals of yeast tRNA^{phe} were obtained, work was pursued in obtaining heavy atom derivatives. Although the technique for obtaining heavy atom derivatives in protein molecules was well-developed, it had never been attempted before with macromolecular nucleic acids. Through a series of extensive trials, a number of isomorphous derivatives were obtained[63] using samarium ions which went to several sites, osmium salts that complexed the 2′ and 3′ hydroxyl groups at the 5′ end of tRNA,[64] and platinum salts. These made possible construction of an electron-density map at successive levels of resolution. At 5.5 Å resolution the outlines of the molecule could be seen, although the surface could not be traced unambiguously.[63] However, at 4 Å resolution, not only could the shape of the molecule be seen clearly, but also the folding of the polynucleotide chain could be traced.[65] This is in general not possible with proteins since the constituent atoms have approximately equal scattering power for X-rays. However, phosphate atoms are somewhat electron-dense, and so the electron-density map showed peaks which were interpreted to be phosphorous positions. Approximately 80 peaks were found in the asymmetric unit, and the chain tracing was carried out by recognizing that the phosphate–phosphate distance between covalently linked nucleotide ranged from 5.5 to 7 Å. A search was carried out, and in this way a tracing of phosphates could be found. The tracing was largely, but not entirely unambiguous. The tracing showed the manner in which the familiar clover-leaf folding, which had been found as a constant component of the secondary structure of transfer RNA sequences, showed up in a striking transformation in the three-dimensional electron-density map. The cloverleaf, as shown in FIG. 16, was transformed in that the acceptor stem was co-linear with the dihydro U stem, and the anti-codon stem was virtually co-linear with the T pseudo U stem. The molecule was composed of two stacking domains more or less organized at right angles to each other in the shape of a letter "L." The T pseudo U loop and D loop met at the corner of the molecule, while the anti-codon loop was found about 80 Å away from the CCA acceptor end of the molecule, this being the distance between the two ends of the L-shaped molecule. The overall dimensions of the molecule were in rough agreement with those that were obtained from the very crude three-dimensional Patterson map at 12 Å resolution.[60]

Detailed examination of the folding revealed that it contained elements that were compatible with the known chemistry of tRNA molecules. For example, when a thiouridine is found at nucleotide 8, ultraviolet light induced a crosslink with cytosine 13. These residues were close enough to form such a linkage.

HIGH-RESOLUTION STRUCTURE OF tRNA REVEALED MANY UNEXPECTED CONFORMATIONS

The structure of yeast tRNA^{phe} was solved at 3 Å in 1974, both for the orthorhombic form that we had been working on,[66] as well as for the mono-

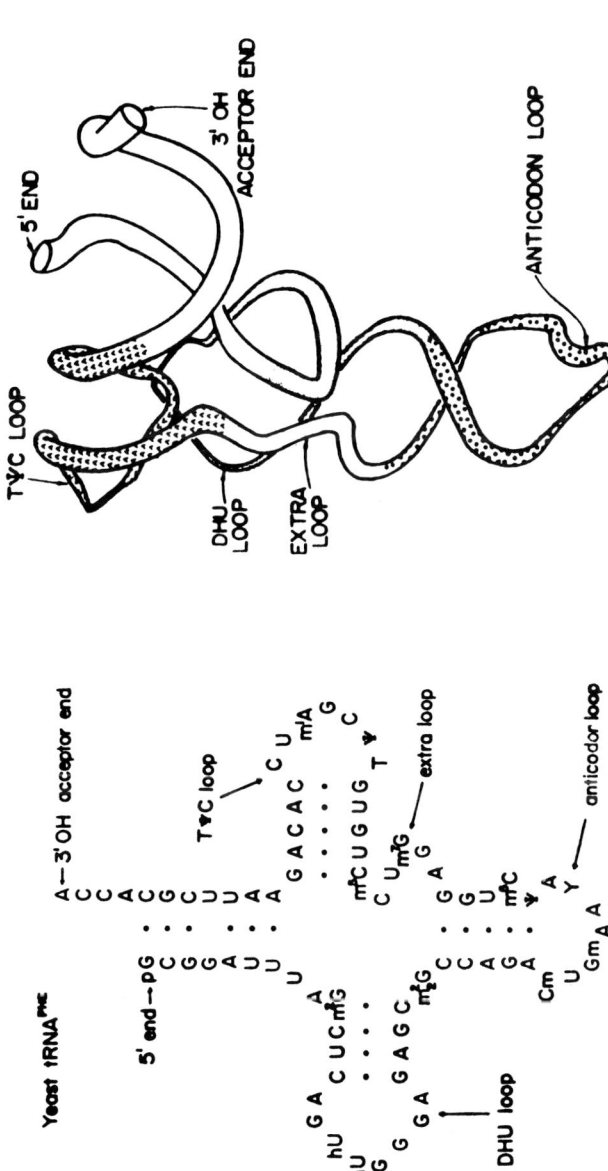

FIGURE 16. At 4 Å resolution, the orthorhombic crystal of yeast tRNA[phe] revealed the electron-dense peaks of the phosphate groups. This made it possible to trace the backbone chain.[65] It revealed that Holley's clover-leaf secondary-structure diagram (*left*) was preserved in the tRNA molecule but transformed into an L-shaped structure (*right*) in which the anti-codon at one end is 76 Å away from the 3' acceptor at the other end. The anti-codon stem stacked on the dihydro U stem and the acceptor stem stacked on the T pseudo U stem. (From Kim *et al*.[65] Reproduced by permission.)

clinic form solved by A. Klug and associates.[67] The two were very similar and confirmed the general folding of the polynucleotide chain that had been seen at 4 Å, but provided a host of tertiary interactions that defined the nature of the L-shaped molecule. The double-helical stems had a form very similar to that which was seen in the single-crystal analyses of atomic-resolution ApU and GpC crystals. In addition, there were nine base–base tertiary interactions in which bases were hydrogen-bonded together, either in pairs or in triplexes (FIG. 17). Of these, only one was of the Watson–Crick type. All the others were novel. Some of them, however, such as the pairing between A9 and A23, were similar to those that had been identified in the adenine–adenine pairing of polyribo adenylic acid solved several years earlier in a fiber–diffraction study.[68] Several of the base pairs in the dihydro U stem had other bases in the major groove which formed base triplexes that held segments of the variable loop close to the stem. What became apparent is that a detailed analysis of the structure clarified and explained a wide variety of chemical reactions that had been seen in tRNA molecules. By looking at the role of conserved bases in maintaining the three-dimensional structure, it became apparent that this was indeed the general structure of transfer RNA molecules.[69]

This tRNA structure has now been seen in a number of single-crystal studies including studies of tRNA bound to aminoacyl tRNA synthetases, which showed that the molecule maintains its form even when bound to a protein carrying out aminoacylation.

One of the interesting features of the tertiary structure pointed out by Gary Quigley was the fact that a similar folding of the polynucleotide chain was found in the anti-codon loop as well as in the T pseudo U loop.[70] This was a folding in which a uracil residue was involved in stabilizing a 180° turn of the polynucleotide chain. It was called a U-turn and involved several stabilizing interactions. After describing the U-turn in tRNA, we commented, "It will be of interest to see if uridine turns will also occur in other RNA structures." The answer to that question required nearly 20 years. In 1994, the structure of the hammerhead ribozyme was published by McKay and his colleagues, and it was found that the U turn was incorporated into this molecule almost precisely as it was found in the tRNA.[71] It is interesting that the uridine turn provides the active site of the ribozyme which facilitates bond cleavage.

Solution of the three-dimensional structure of transfer RNA opened the door to a large number of additional studies. It showed that the anti-codon bases were found in a stacked array, some 76 Å away from the site at which the amino acid is added. This provided some constraints on the mode of action of the tRNA molecule inside the ribosome during protein synthesis.[72] It also set the stage for further studies to elucidate the specificity of aminoacylation and the various modes in which tRNA aminoacyl synthetases interact with tRNA.

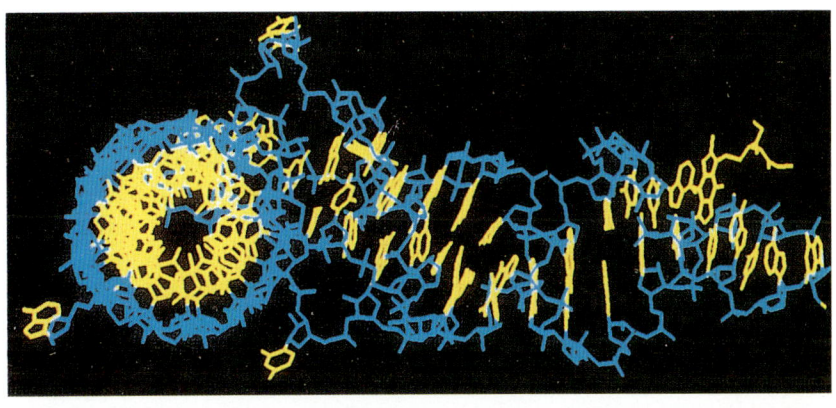

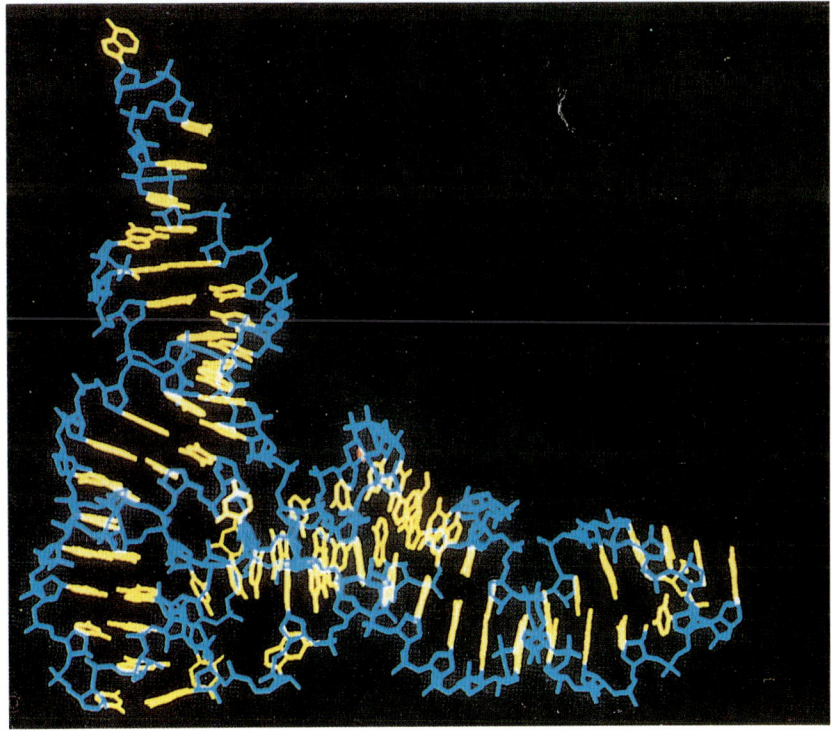

FIGURE 17. At 3 Å resolution the yeast tRNA[phe] structure revealed many complex tertiary-structure interactions.[66] The molecule retained its L-shaped structure as seen on the *left*, where bases are yellow and the backbone is blue. On the right we are viewing the molecule down the acceptor stem with the anti-codon loop at the bottom.

DNA TURNS LEFT-HANDED

When the first dinucleotide phosphate nucleic acid fragments were solved as single crystals and found to contain double-helical conformations, the prediction was made in 1973 that this would be followed by a large number of studies defining nucleic acid conformation at the atomic level from single-crystal diffraction studies. However, it was not until 1979 that the next crystal structure was solved for a DNA hexamer containing the sequence d(CGCGCG).[73] The rate-limiting step in the slow evolution of these studies was the gradual development of a chemical technology which allowed us to synthesize DNA segments. This type of synthesis was still quite difficult in the 1970s, and it was only through our collaboration with Jacques van Boom and his colleagues in Leiden that we were able to obtain purified samples that allowed us to carry out crystallization experiments. Andrew Wang, Gary Quigley and others in the lab discovered that this material crystallized with a resolution of 0.9 Å, and we eagerly anticipated that this crystal might provide us with the first atomic-level view of the DNA double helix. We tried to solve the structure with translation-rotation functions in which we assumed, for example, that the molecule had either an A- or a B-type double helix. This procedure failed, and so we carried out the more lengthy procedure of soaking in heavy-atom derivatives. Barium, cobalt, and copper ions were diffused into the lattice and from this emerged the structure of the molecule.[73] To our astonishment, the molecule was left-handed. This was somewhat surprising although, on looking at the literature more carefully, it became less so. Several years earlier Fritz Pohl and Tom Jovin had discovered that in a high-salt solution poly (dG-dC) has a nearly inverted circular dichroism compared to the low-salt form.[74] It seemed reasonable to believe that this was the form that had been trapped in our crystal lattice. Subsequent experiments showed this to be the case.[75]

In this alternative, left-handed conformation, the double helix had undergone some remarkable transformations (FIG. 18). The two strands were antiparallel and were held together by Watson–Crick base pairs; however, the residues alternated with *syn* and *anti* conformations of the bases. In this structure all the guanine residues were in the *syn* conformation, which had not been seen previously in DNA molecules, but had been anticipated from an earlier analysis.[76]

Although now we are very familiar with the idea of alternative conformations for DNA, especially as it carries out its various metabolic activities, in 1979 this was a somewhat novel concept. It stimulated a great deal of investigations of non-B-DNA alternative conformations, a field of research which has now broadened considerably.

Within a short time period much was learned about the chemistry of left-handed Z-DNA, as it was called. Certain ions stabilized the left-handed con-

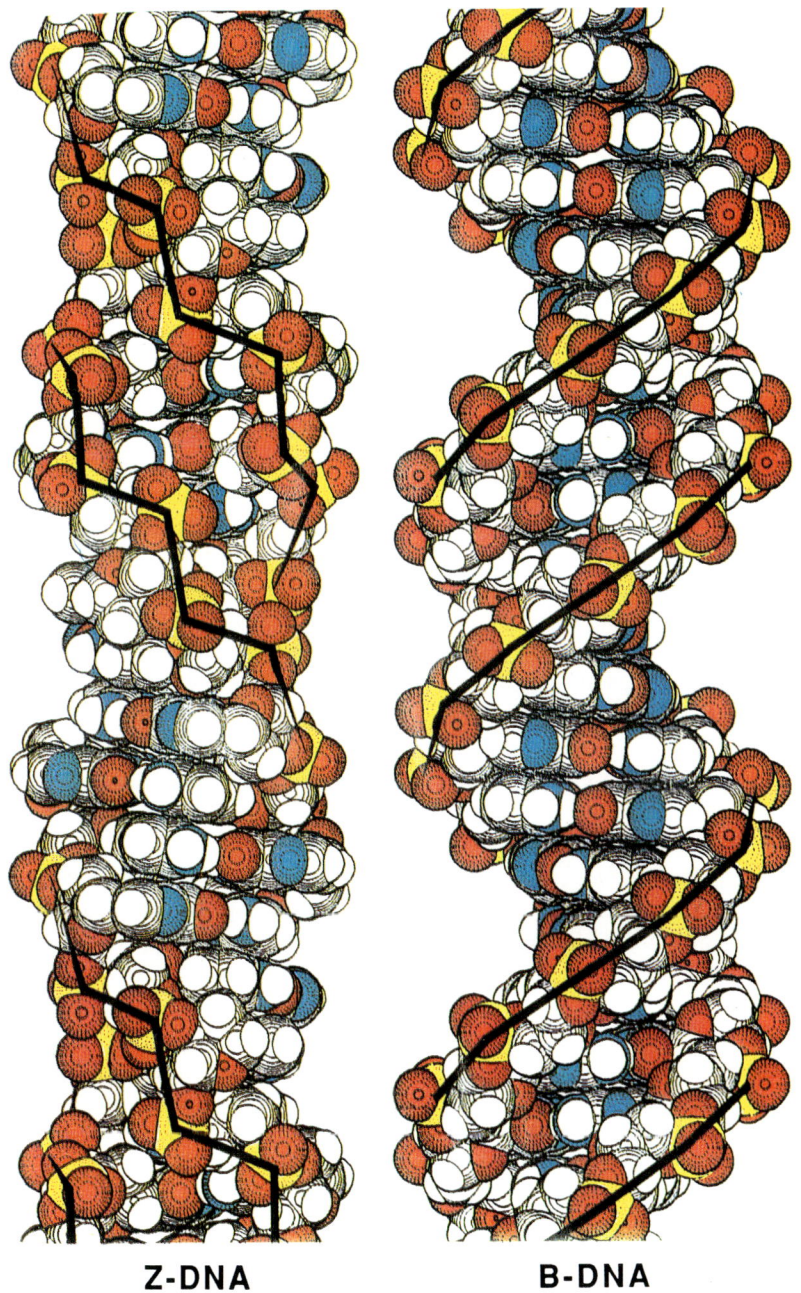

FIGURE 18. Van der Waals diagrams of Z-DNA (*left*) and B-DNA (*right*).[73] The *solid black line* goes from phosphate group to phosphate group along the chain. The zig-zag form of the Z-DNA backbone is evident. Z-DNA has a slightly smaller diameter than B-DNA and it no longer has the wide major groove that is seen in B-DNA. Phosphorus is yellow; oxygen, red; nitrogen, blue; hydrogen, white; and carbon is shown by concentric circles.

formation,[77] and its chemical reactivity was modified considerably compared to that of B-DNA owing to the alternative conformation of the bases.[78]

Z-DNA was found to be immunogenic.[79] Both polyclonal and monoclonal antibodies were prepared that could react with Z-DNA. These proved to be useful objects for experimental study. The most important component for stabilizing *in vivo* Z-DNA is negative supercoiling,[77] which untwists the DNA duplex. If an antibody to Z-DNA is added to the solution, torsional strain is then no longer required since the protein alone is able to maintain it in that conformation.[80]

Although it was relatively easy to study the chemical and physical conditions that are involved in converting B-DNA to Z-DNA, it was much more difficult to define its biological role *in vivo*. There were very early indications that Z-DNA exists *in vivo*. In autoimmune diseases and especially in systemic lupus erythematosus a significant titer of antibody specific to Z-DNA develops in the course of the disease.[81] This suggested the presence of Z-DNA in these systems.

It has been proposed that Z-DNA is involved in a variety of activities, including homologous recombination. However, the most significant demonstration of Z-DNA occurring inside cells is usually associated with the negative supercoiling that develops during transcription. When RNA polymerase travels along DNA, it leaves negative supercoiling behind it in the upstream region of the gene and develops positive supercoiling in the downstream region.[82] In experiments carried out in collaboration with Burghardt Wittig and his associates, we were able to demonstrate the formation of Z-DNA in specific regions of individual genes such as *c-myc*.[83] These experiments were carried out using cells embedded in agarose microbeads. When permeabilized nuclei are prepared from cells in this protected environment, they maintain their ability to carry out DNA replication and transcription, and biotin-labeled antibodies specific to Z-DNA can be diffused into them. With a 10-nanosecond laser pulse at 266 mμ, the antibody is cross-linked to the Z-DNA. After restriction endonuclease digestion, the restriction fragments containing Z-DNA attached to the antibody can be isolated. This has made it possible to map the production of Z-DNA specifically. In the upstream region of the *c-myc* gene, three regions convert to Z-DNA when the gene is accurately transcribed. However, when the gene is downregulated and no longer transcribed, Z-DNA is no longer present.[83] Thus, we have a picture of a dynamic state of DNA in which negative torsional strain associated with transcription flips some regions into Z-DNA, and when transcription ceases, Z-DNA returns to the lower-energy B conformation.

The full story of the biology of Z-DNA is still unfolding. The recent isolation of Z-DNA binding proteins may provide a useful tool for further understanding its activity.

A MECHANISM FOR DNA BENDING

In solution the DNA molecule is somewhat rigid. Important components of this are the phosphate–phosphate repulsive interactions. In 1978 I began to think about the influence of asymmetric neutralization of phosphate charges on the conformation of DNA. It seemed reasonable to believe that if the phosphate groups along one side of the DNA double helix were neutralized by the presence of positive charges, the repulsion of the phosphate groups on the opposite side of the helix would lead to a bending.[84] In my discussion of this with Andrei Mirzabekov, he pointed out that in nucleosomes this is precisely what happens since the positively charged side chains of histones interact with phosphate groups along one side of the DNA, and of course the DNA is coiled around the nucleosome.[85]

In 1994 this prediction was strikingly confirmed by experiments of J. Strauss and L.J. Maher.[86,87] They synthesized DNA selectively using some methyl phosphonates, which do not have a negative charge. By placing the methyl phosphonates so that they would all be found on one side of the DNA duplex, they were able to measure DNA bends that had been predicted many years earlier. In this case, the asymmetric presence of negative charges on the opposite side of the DNA duplex resulted in bending the DNA through phosphate–phosphate repulsion.

It is quite likely that this mechanism for DNA bending is used quite often when proteins with positive charges interact with DNA along one of its edges.

DNA CAN ALSO BE FOUR-STRANDED

The lesson taught by the left-handed Z-DNA structure was that specialized sequences of nucleotides can adopt alternative conformations that differ from regular B-DNA. Z-DNA is favored by alternations of purines and pyrimidines, especially alternating guanine and cytosine residues. As we mentioned above, under supercoiling, large stretches of homopurines–homopyrimidines can be induced to undergo partial strand separation with the formation of triple-stranded DNA.[15] In a similar way the specialized sequences found at the ends of chromosomes in telomeres are also capable of adopting alternative conformations.

At the chromosome end, specialized sequences are found in which one strand contains many repeats of a predominantly guanine-rich sequence and the other strand a complementary cytosine-rich sequence.[88] Telomere sequences vary somewhat although there seems to be a common sequence in metazoa.

The guanine-rich sequences can adopt a four-stranded conformation. The

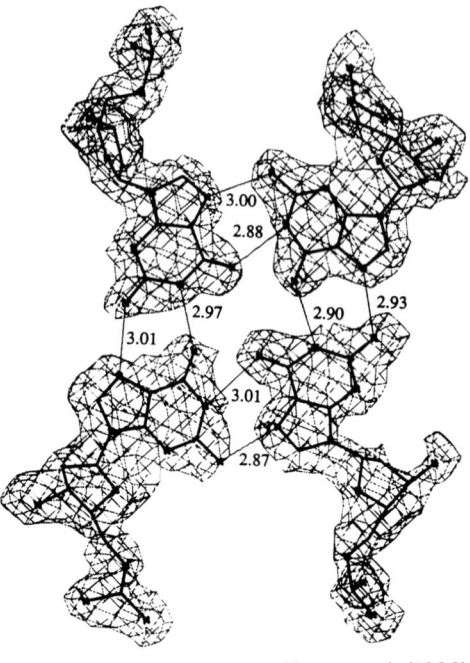

Kang et al. (1992)

FIGURE 19. The structure of the *Oxytrichia* telomere d($G_4T_4G_4$).[93] This forms a four-stranded complex with the four levels of guanine quartets in the center, and loops at either end contain thymine residues. Electron-density map of one guanine quartet shows the eight hydrogen bonds stabilizing the interaction. Hydrogen bonding distances are listed in angstroms. (From Kang et al.[93] Reproduced by permission.)

origin of this discovery goes back to early experiments with polyinosinic acid which suggested that this molecule formed a parallel three- or four-stranded structure, as deduced from the X-ray diffraction fiber studies.[89] The bases in the center were held together by cyclic hydrogen bonding. A striking example of this association was discovered by Davies and colleagues in gels of guanosine monophosphate. Fiber-diffraction patterns were interpreted in terms of four bases of guanine hydrogen bonded in a cyclic manner.[90] The addition of the 2-amino groups of guanine relative to hypoxanthine in polyinosinic acid gave rise to a proposed eight hydrogen bonds per base plane involving the four guanine residues.

The guanine-rich strand of telomeres has been found to have two single-stranded repeats at the 3' end of the molecule.[88] Several investigators suggested that these sequences could fold back on themselves to form four-stranded guanine structures with the intervening sequences between the guanine segments involved in forming loops.[91,92] The *Oxytrichia* telomere has a repeat of d(G_4T_4). We found that one and one-half repeats with the sequence d($G_4T_4G_4$) formed

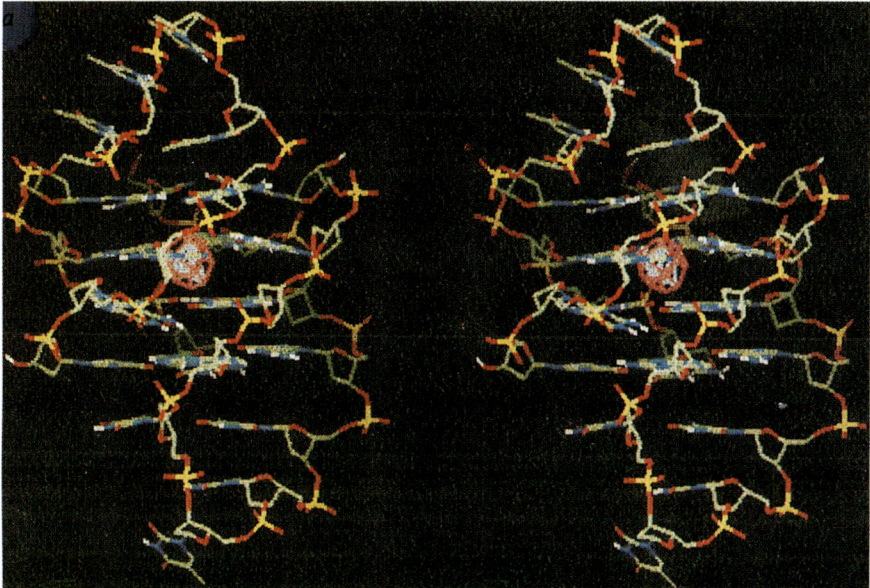

FIGURE 20. Stereoscopic view of the structure of the *Oxytrichia* telomere. The central region contains an omit map with electron-density contours indicating the presence of an axial potassium ion. (From Kang *et al.*[93] Reproduced by permission.)

a single crystal which was solved by X-ray diffraction methods.[93] This provided the first view of an electron-density map of the four guanines in a plane held together by cyclic hydrogen bonding (FIG. 19). Several years earlier it had been suggested that an axial ion, usually potassium, would be found between the planes of the guanine quartets. Such an ion was seen in the crystal structure of the *Oxytrichia* telomere (FIG. 20).

This type of guanine quartet structure is highly polymorphic. Many studies have now been carried out of four-stranded guanine structures. In addition, several proteins have been found that bind specifically to these structures.

It is a curious fact that the cytosine-rich sequences of telomeres also have the ability to form a four-stranded structure. The origin of this goes back to early single-crystal work in 1962 which showed that cytosine residues were capable of forming hemi-protonated base pairs ($C \cdot C^+$) in which one proton added to a cytosine makes it capable of forming a base pair with three hydrogen bonds.[94] A fiber-diffraction pattern of polycytidylic acid was interpreted by Robert Langridge and myself as having two parallel strands held together by $C \cdot C^+$ hydrogen bonds.[95] In 1993 a further advance came with an NMR study by Maurice Guéron and associates of $d(TC_5)$ which showed that there were actually two parallel stranded duplexes of cytosine residues held together by $C \cdot C^+$ hydrogen bonds to form a four-stranded structure.[96] These two duplexes

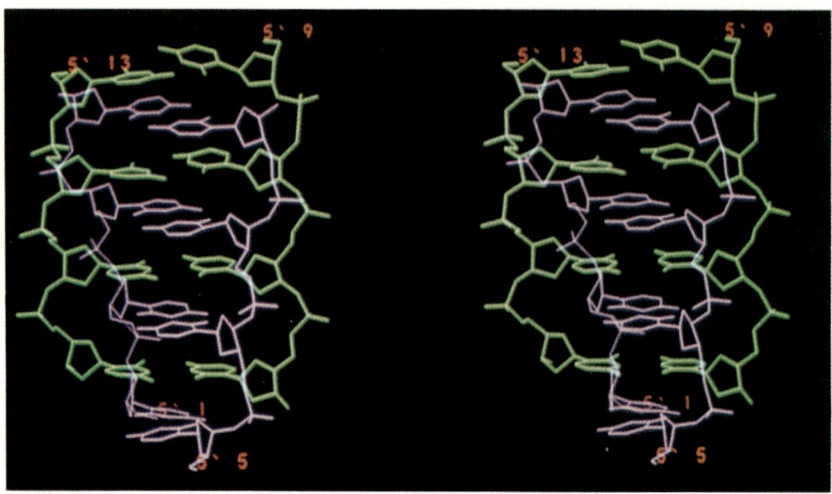

FIGURE 21. The structure of d(C$_4$) was solved at 2.3 Å resolution and it is shown in a stereoscopic figure.[97] There are four strands in the structure consisting of two pairs (green and pink). Each of these pairs is parallel to each other, but the two pairs are arranged in opposite orientation with the bases intercalated. In the center the base pairs are held together by C·C$^+$ hydrogen bonds. (From Chen et al.[97] Reproduced by permission.)

were oriented in the opposite direction, and the bases between them were intercalated. Crystal structure analysis of d(C$_4$),[97] as well as of d(TC$_3$),[98] confirmed the overall arrangement of this intercalated motif, although there were several modifications revealed in the X-ray structures (FIG. 21).

The four-stranded, intercalated cytosine structures are generally stabilized in the acidic region by its requirement for a protonated cytosine to hold the strands together. However, they are frequently stable up to pH 7, so that one has to consider the possibility of their being active in biological systems.

The discovery of four-stranded DNA structures involving both guanine-containing residues as well as cytosine-containing residues is relatively recent, and it is somewhat early to have a clear picture of their biological role. They are, however, examples of the intense polymorphism that DNA expresses. In seeing such stable alternative conformations, one is tempted to believe that nature is likely to utilize them. In the case of the G-4 and C-4 four-stranded structures, we suggested that they may represent a possible system for self-recognition.[97] For example, if, under the stress of negative supercoiling, a segment of DNA with an inverted repeat has at its center a segment containing two clusters of guanine (or cytosine) residues, then a cruciform extrusion could result in these being found in the DNA loops at the end. If two DNA molecules were present with similar extrusions, they could be held together through the formation of four-stranded complexes containing either guanine and cytosine residues. Such complexes could also be stabilized by the presence of specific binding proteins.

Is it possible that these could be used in recognizing identical chromosomes during meiosis or in recognizing identical sequences of DNA to give the alignment necessary as a prelude for homologous recombination? Identity of sequence would be required because extrusions of this type would occur at the same intervals along the DNA strand only if the two sequences were identical. If that were the case, then cooperativity would ensue, and identical sequence recognition could be stabilized by many of these four-stranded interactions. Alternatively, if the two DNA sequences were different, the recognition cruciforms would no longer be in phase and the molecules would fall apart.

This is another example of the manner in which alterations of DNA conformation make us face the question of whether these are also associated with specific biological functions. At the very least, they provide an excellent stimulus for directing research programs.

EPILOGUE

In this brief overview, I have touched upon a number of research topics covered in my laboratory during the past four decades or so of working on nucleic acids. What becomes apparent in this brief discussion is that in the early years very large and basic questions were asked, such as: Can RNA form a double helix? How is DNA made from RNA? How is protein synthesis carried out? What does transfer RNA look like? In more recent years, however, increasingly specialized questions have been addressed, such as the biological implications of having DNA existing in alternative non-B conformations. Does nature utilize them? And if so, how?

The change in the type of question asked, of course, reflects the enormous maturation which the discipline has undergone. Today, it is no longer a question of whether we can explain living organisms at the molecular level. Rather, the question is how detailed our understanding can become and in how comprehensive a manner we can look at all the phenomena at the molecular level and then integrate them into an understanding of the organism as a whole.

In this effort basic studies dealing with the nucleic acids will continue to play an important role, even though the main focus of activity increasingly has to do with the nature of the proteins encoded by the nucleic acids and the manner in which these proteins interact, especially with the nucleic acids, to influence the behavior of biological systems.

ACKNOWLEDGMENTS

The work reviewed here was carried out with a large number of collaborators—graduate students, postdoctoral fellows, and other colleagues. It was supported by several federal granting agencies.

REFERENCES

1. AVERY, O. T., C. M. MACLEOD & M. MCCARTY. 1944. Studies on the chemical nature of the substance inducing transformation of pneumococcal types. J. Exp. Med. **79**: 137–157.
2. WATSON, J. D. & F. H. C. CRICK. 1953. A structure for deoxyribose nucleic acid. Nature **171**: 738–740.
3. This was discussed extensively at an American Chemical Society Symposium on Proteins, New York City, 1951.
4. BROWN, D. M., G. D. FASSMAN, D. I. MAGRATH, A. R. TODD, W. COCHRAN & M. M. SOOLFSON. 1953. Structure of adenylic acids a and b. Nature **172**: 1184–1185.
5. RICH, A. & J. D. WATSON. 1954. Physical studies on ribonucleic acid. Nature **173**: 995–996.
6. RICH, A. & J. D. WATSON. 1954. Some relations between DNA and RNA. Proc. Natl. Acad. Sci. USA **40**: 759–764.
7. GRUNBERG-MANAGO, M., P. J. ORTIZ & S. OCHOA. 1955. Enzymatic synthesis of nucleic acidlike polynucleotides. Science **122**: 907.
8. RICH, A. & D. R. DAVIES. 1956. A new two-stranded helical structure: polyadenylic acid and polyuridylic acid. J. Am. Chem. Soc. **78**: 3548.
9. FELSENFELD, G. & A. RICH. 1957. Studies on the formation of two- and three-stranded poly ribonucleotides. Biochim. Biophys. Acta **26**: 457–468.
10. WARNER, R. C. 1956. Ultraviolet spectra of enzymatically synthesized polynucleotides. Fed. Proc. **15**: 379.
11. KORNBERG, A., I. R. LEHMAN, M. J. BESSMAN & E. S. SIMMS. 1956. Enzymic synthesis of deoxyribonucleic acid. Biochim. Biophys. Acta **21**: 197–198.
12. FELSENFELD, G., D. R. DAVIES & A. RICH. 1957. Formation of a three-stranded polynucleotide molecule. J. Am. Chem. Soc. **79**: 2023–2024.
13. HOOGSTEEN, K. 1959. The crystal and molecular structure of a hydrogen-bonded complex between 1-methylthymine and 9-methyladenine. Acta Crystallogr. **12**: 822.
14. RICH, A. 1958. Formation of two- and three-stranded helical molecules by polyinosinic acid and polyadenylic acid. Nature **181**: 521–525.
15. FRANK-KAMENETSKII, M. D. 1990. DNA supercoiling and unusual structures. In DNA Topology and Its Biological Effects. N. R. Cozzarelli & J. C. Wang, Eds.: 185–215. Cold Spring Harbor Laboratory Press. Cold Spring Harbor, NY.
16. RICH, A. 1959. An analysis of the relation between DNA and RNA. Ann. N.Y. Acad. Sci. **81**: 709–722.
17. LEHMAN, I. R., S. B. ZIMMERMAN, J. ADLER, M. J. BESSMAN, E. S. SIMMS & A. KORNBERG. 1958. Enzymatic synthesis of deoxyribonucleic acid. V. Chemical composition of enzymatically synthesized deeoxyribonucleic acid. Proc. Natl. Acad. Sci. USA **44**: 1191.
18. SINSHEIMER, R. L. 1959. A single-stranded deoxyribonucleic acid from bacteriophage ϕX174. J. Mol. Biol. **1**: 43.
19. TENER, G. M., H. G. KHORANA, R. MARKHAM & E. H. POL. 1958. Studies on polynucleotides. II. The synthesis and characterization of linear and cyclic thymidine oligonucleotides. J. Am. Chem. Soc. **80**: 6223.
20. RICH, A. 1960. A hybrid helix containing both deoxyribose and ribose polynucleotides and its relation to the transfer of information between the nucleic acids. Proc. Natl. Acad. Sci. USA **46**: 1044–1053.
21. MARMUR, J. & D. LANE. 1960. Strand separation and specific recombination in

deoxyribonucleic acids: Biological studies. Proc. Natl. Acad. Sci. USA **46**(4): 453–461.
22. DOTY, P., J. MARMUR, J. EIGNER & C. SCHILDKRAUT. 1960. Strand separation and specific recombination in deoxyribonucleic acids: Physical chemical studies. Proc. Natl. Acad. Sci. USA **46**(4): 461–476.
23. HALL, B. D. & S. SPEIGELMAN. 1961. Sequence complementarity of T2-DNA and T2-specific RNA. Proc. Natl. Acad. Sci. USA **47**: 137.
24. FURTH, J. J., J. HURVITZ & M. GOLDMANN. 1961. The directing role of DNA in RNA synthesis. Biochem. Biophys. Res. Commun. **4**(5): 362–367.
25. FURTH, J. J., J. HURWITZ & M. GOLDMAN. 1961. The directing role of DNA in RNA synthesis: Specificity of the deoxyadenylate deoxythymidylate copolymer as a primer. Biochem. Biophys. Res. Commun. **4**(6): 431–435.
26. STEVENS, A. 1961. Net formation of polyribonucleotides with base compositions analogous to deoxyribonucleic acid. J. Biol. Chem. **236**: PC43–PC45.
27. RICH, A. 1962. On the problems of evolution and biochemical information transfer. *In* Horizons in Biochemistry. M. Kasha & B. Pullman, Eds.: 103–126. Academic Press. New York.
28. BENNER, S. A., J. A. PICCIRILLI, T. KRAUCH & S. E. MORONEY. 1990. Enzymatic incorporation of a new base pair into DNA and RNA extends the genetic alphabet. Nature **343**(6253): 33–37.
29. BRENNER, S., F. JACOB & M. MESELSON. 1961. An unstable intermediate carrying information from genes to ribosomes for protein synthesis. Nature **190**: 576.
30. GROS, F., H. HIATT, W. GILBERT, C. G. KURLAND, R. W. RISEBROUGH & J. D. WATSON. 1961. Unstable ribonucleic acid revealed by pulse labelling of *Escherichia coli*. Nature **190**: 581.
31. HOAGLAND, M. B., P. C. ZAMECNIK & M. L. STEPHENSON. 1957. Intermediate reactions in protein biosynthesis. Biochim. Biophys. Acta **24**: 215–216.
32. CRICK, F. H. C., L. BARNETT, S. BRENNER & R. J. WATTS-TOBIN. 1961. General nature of the genetic code for proteins. Nature **192**: 1227.
33. WARNER, J. R., A. RICH & C. E. HALL. 1962. Electron microscope studies of ribosomal clusters synthesizing hemoglobin. Science **138**: 1399–1403.
34. WARNER, J. R., P. M. KNOPF & A. RICH. 1963. A multiple ribosomal structure in protein synthesis. Proc. Natl. Acad. Sci. USA **149**: 122–129.
35. GOODMAN, H. M. & A. RICH. 1963. Mechanism of polyribosomal action during protein synthesis. Nature **199**: 318–322.
36. KIHO, Y. & A. RICH. 1965. A polycistronic messenger RNA associated with a b-galactosidase induction. Proc. Natl. Acad. Sci. USA **54**: 1751–1758.
37. WARNER, J. R. & A. RICH. 1964. The number of growing polypeptide chains on reticulocyte polyribosomes. J. Mol. Biol. **10**: 202–211.
38. WARNER, J. R. & A. RICH. 1964. The number of soluble RNA molecules on reticulocyte polyribosomes. Proc. Natl. Acad. Sci. USA **51**: 1134–1141.
39. MALKIN, L. I. & A. RICH. 1967. Partial resistance of nascent polypeptide chains to proteolytic digestion due to ribosomal shielding. J. Mol. Biol. **26**: 329–346.
40. KUECHLER, E. & A. RICH. 1970. Position of the initiator and peptidyl sites in the *E. coli* ribosome. Nature **225**: 920–924.
41. GOODMAN, H. M. & A. RICH. 1962. Formation of a DNA-soluble RNA hybrid and its relation to the origin, evolution and degeneracy of soluble RNA. Proc. Natl. Acad. Sci. USA **48**: 2101–2109.
42. CHUN, E. H. L., M. H. VAUGHN & A. RICH. 1963. The isolation and characterization of DNA associated with chloroplast preparations. J. Mol. Biol. **7**: 130–141.

43. HAMLIN, JR., R. M., R. C. LORD & A. RICH. 1965. Hydrogen-bonded dimers of adenine and uracil derivatives. Science **148**: 1734–1737.
44. KYOGOKU, Y., R. C. LORD & A. RICH. 1967. An infrared study of hydrogen bonding between adenine and uracil derivatives in chloroform solution. J. Am. Chem. Soc. **89**: 496–504.
45. KYOGOKU, Y., R. C. LORD & A. RICH. 1967. The effect of substituents on the hydrogen bonding of adenine and uracil derivatives. Proc. Natl. Acad. Sci. USA **57**: 250–257.
46. KYOGOKU, Y., R. C. LORD & A. RICH. 1966. Hydrogen bonding specificity of nucleic acid purines and pyrimidines in solution. Science **154**: 518–520.
47. PAULING, L. & R. B. COREY. 1956. Specific hydrogen-bond formation between pyrimidines and purines in deoxyribonucleic acids. Arch. Biochem. Biophys. **65**(1): 164–181.
48. SOBELL, H. M., K. TOMITA & A. RICH. 1963. The crystal structure of an intermolecular complex containing a guanine and cytosine derivative. Proc. Natl. Acad. Sci. USA **49**: 885–892.
49. KATZ, L., K. TOMITA & A. RICH. 1965. The molecular structure of the crystalline complex ethyladenine: methyl-bromouracil. J. Mol. Biol. **13**: 340–350.
50. KATZ, L., K. TOMITA & A. RICH. 1966. The crystal structure of the intermolecular complex 9-ethyladenine: 1-methyl-5-bromouracil. Acta Crystallogr. **21**: 754–764.
51. TOMITA, K., L. KATZ & A. RICH. 1967. Crystal structure of the intermolecular complex 9-ethyladenine: 1-methyl-5-fluorouracil. J. Mol. Biol. **30**: 545–549.
52. KIM, S. H. & A. RICH. 1967. Crystal structure of the 1:1 complex of 5-fluorouracil and 9-ethylhypoxanthine. Science **158**: 1046–1048.
53. DAY, R. O., N. C. SEEMAN, J. M. ROSENBERG & A. RICH. 1973. A crystalline fragment of the double helix: The structure of the dinucleotide phosphate guanylyl-3',5'-cytidine. Proc. Natl. Acad. Sci. USA **70**: 849–853.
54. ROSENBERG, J. M., N. C. SEEMAN, J. J. P. KIM, F. L. SUDDATH, H. B. NICHOLAS & A. RICH. 1973. Double helix at atomic resolution. Nature **243**: 150–154.
55. SEEMAN, N. C., J. M. ROSENBERG, F. L. SUDDATH, J. J. P. KIM & A. RICH. 1976. RNA double helical fragments at atomic resolution: I. The crystal and molecular structure of sodium adenylyl-3',5'-uridine hexahydrate. J. Mol. Biol. **104**: 109–144.
56. ROSENBERG, J. M., N. C. SEEMAN, R. O. DAY & A. RICH. 1976. RNA double helical fragments at atomic resolution: II. The crystal structure of sodium guanylyl-3',5'-cytidine nonahydrate. J. Mol. Biol. **104**: 145–167.
57. ARNOTT, S., M. H. F. WILKINS, L. D. HAMILTON & R. LANGRIDGE. 1965. Fourier synthesis studies of lithium DNA. Part III: Hoogsteen models. J. Mol. Biol. **11**: 391–402.
58. HOLLEY, R. W., J. APGAR, G. A. EVERETT, J. T. MADISON, M. MARGUISSE, S. H. MERRILL, J. R. PENWICK & A. ZAMIR. 1965. Structure of a ribonucleic acid. Science **147**: 1462.
59. KIM, S.-H. & A. RICH. 1968. Single crystals of transfer RNA: An X-ray diffraction study. Science **162**: 1381–1384.
60. KIM, S.-H. & A. RICH. 1969. Crystalline transfer RNA: The three-dimensional Patterson function at 12-Angstrom resolution. Science **166**: 1621–1624.
61. KIM, S.-H., G. QUIGLEY, F. L. SUDDATH & A. RICH. 1971. High resolution X-ray diffraction patterns of tRNA crystals showing helical regions of the molecule. Proc. Natl. Acad. Sci. USA **68**: 841–845.
62. KIM, S.-H., G. QUIGLEY, F. L. SUDDATH, A. McPHERSON, D. SNEDEN, J. J. KIM, J. WEINZIERL & A. RICH. 1973. X-ray crystallographic studies of polymorphic forms of yeast phenylalanine transfer RNA. J. Mol. Biol. **75**: 421–428.

63. KIM, S.-H., G. QUIGLEY, F. L. SUDDATH, A. MCPHERSON, D. SNEDEN, J. J. KIM, J. WEINZIERL, P. BLATTMANN & A. RICH. 1972. Three-dimensional structure of yeast phenylalanine transfer RNA: Shape of the molecule at 5.5 Å resolution. Proc. Natl. Acad. Sci. USA **69**: 3746–3750.
64. CONN, J. F., J. J. KIM, F. L. SUDDATH, P. BLATTMAN & A. RICH. 1974. Crystal and molecular structure of an osmium bispyridine ester of adenosine. J. Am. Chem. Soc. **96**: 7152–7153.
65. KIM, S.-H., G. J. QUIGLEY, F. L. SUDDATH, A. MCPHERSON, D. SNEDEN, J. J. KIM, J. WEINZIERL & A. RICH. 1973. Three-dimensional structure of yeast phenylalanine transfer RNA: Folding of the polynucleotide chain. Science **179**: 285–288.
66. KIM, S.-H., F. L. SUDDATH, G. J. QUIGLEY, A. MCPHERSON, J. J. KIM, J. L. SUSSMAN, A. H.-J. WANG, N. C. SEEMAN & A. RICH. 1974. Three-dimensional tertiary structure of yeast phenylalanine transfer RNA. Science **185**: 435–439.
67. ROBERTUS, J. D., J. E. LADNER, J. T. FINCH, D. RHODES, R. S. BROWN, B. F. C. CLARK & A. KLUG. 1974. Structure of yeast phenylalanine tRNA at 3Å resolution. Nature **250**: 546.
68. RICH, A., D. R. DAVIES, F. H. C. CRICK & J. D. WATSON. 1961. The molecular structure of polyadenylic acid. J. Mol. Biol. **3**: 71–86.
69. KIM, S.-H., J. L. SUSSMAN, F. L. SUDDATH, G. J. QUIGLEY, A. MCPHERSON, A. H.-J. WANG, N. C. SEEMAN & A. RICH. 1974. The general structure of transfer RNA molecules. Proc. Natl. Acad. Sci. USA **71**: 4970–4974; RICH, A., U. L. RAJBHANDARY. 1976. Transfer RNA: Molecular structure, sequence and properties. Annu. Rev. Biochem. **45**: 805–860.
70. QUIGLEY, G. J. & A. RICH. 1976. Structural domains of a transfer RNA molecule. Science **194**: 796–806.
71. PLEY, H. W., K. M. FLAHERTY & D. B. MCKAY. 1994. Three-dimensional structure of a hammerhead ribozyme. Nature **372**: 68–74.
72. RICH, A. 1974. How transfer RNA may move inside the ribosome. In Ribosomes. M. Nomura, A. Tissieres & P. Lengyel, Eds.: 871–884. Cold Spring Harbor Laboratory. Cold Spring Harbor, NY.
73. WANG, A. H.-J., G. J. QUIGLEY, F. J. KOLPAK, J. L. CRAWFORD, J. H. VAN BOOM, G. VAN DER MAREL & A. RICH. 1979. Molecular structure of a left-handed double helical DNA fragment at atomic resolution. Nature **282**: 680–686.
74. POHL, F. M. & T. M. JOVIN. 1972. Salt-induced co-operative conformational change of a synthetic DNA: Equilibrium and kinetic studies with poly(dG-dC), J. Mol. Biol. **67**: 375–396.
75. THAMANN, T. J., R. C. LORD, A. H.-J. WANG & A. RICH. 1981. The high salt form of poly (dG-dC)·poly (dG-dC) is left-handed Z-DNA: Raman spectra of crystals and solutions. Nucleic Acids Res. **9**: 5443–5457.
76. HASCHEMEYER, A. E. V. & A. RICH. 1967. Nucleoside conformations: An analysis of steric barriers to rotation about the glycosidic bond. J. Mol. Biol. **27**: 369–384.
77. PECK, L. J., A. NORDHEIM, A. RICH & J. C. WANG. 1982. Flipping of cloned d(pGpG)$_n$·d(pCpG)$_n$ DNA sequences from right to left-handed helical structure by salt, Co(III), or negative supercoiling. Proc. Natl. Acad. Sci. USA **79**: 4560–4564.
78. RICH, A., A. NORDHEIM & A. H.-J. WANG. 1984. The chemistry and biology of left-handed Z-DNA. Annu. Rev. Biochem. **53**: 791–846.
79. LAFER, E. M., A. MOLLER, A. NORDHEIM, B. D. STOLLAR & A. RICH. 1981. Antibodies specific for left-handed DNA. Proc. Natl. Acad. Sci. USA **78**: 3546–3550.

80. LAFER, E. M., R. SOUSA & A. RICH. 1985. Anti-Z-DNA antibody binding can stabilize Z-DNA in relaxed and linear plasmids under physiological conditions. EMBO J. **4:** 3655–3660.
81. LAFER, E. M., R. P. C. VALLE, A. MOLLER, A. NORDHEIM, P. SCHUR, A. RICH & B. D. STOLLAR. 1983. Z-DNA specific antibodies in human systemic lupus erythematosus. J. Clin. Invest. **71:** 314–321.
82. LIU, L. F. & J. C. WANG. 1987. Supercoiling of the DNA template during transcription. Proc. Natl. Acad. Sci. USA **84:** 7024–7027.
83. WITTIG, B., S. WOLFL, T. DORBIC, W. WAHRSON & A. RICH. 1992. Transcription of human c-myc in permeabilized nuclei is associated with formation of Z-DNA in three discrete regions of the gene. EMBO J. **11:** 4653–4663.
84. RICH, A. 1978. Localized positive charges can bend double helical nucleic acid. *In* Gene function, 12th FEBS Meeting, Dresden. S. Rosenthal, H. Bielka, C. Coutelle & C. Zimmer, Eds.: 78–81. Pergamon Press. London.
85. MIRZABEKOV, A. D. & A. RICH. 1979. Asymmetric lateral distribution of unshielded phosphate groups in nucleosomal DNA and its role in DNA bending. Proc. Natl. Acad. Sci. USA **76:** 1118–1121.
86. CROTHERS, D. M. 1994. Upsetting the balance of forces in DNA. Science **266:** 1819–1820.
87. STRAUSS, J. K. & L. J. MAHER, III. 1994. DNA bending by asymmetric phosphate neutralization. Science **266:** 1829–1834.
88. BLACKBURN, E. H. 1991. Structure and function of telomeres. Nature **350:** 569–573.
89. RICH, A. 1958. The molecular structure of polyinosinic acid. Biochem. Biophys. Acta **29:** 502–509.
90. GELLERT, M., M. N. LIPSETT & D. R. DAVIES. 1962. Helix formation by guanylic acid. Proc. Natl. Acad. Sci. USA **48:** 2013–2018.
91. WILLIAMSON, J. R., M. K. RAGHURAMAN & T. R. CECH. 1989. Monovalent cation-induced structure of telomeric DNA: The G-quartet model. Cell **59:** 871–880.
92. SUNDQUIST, W. I. & A. KLUG. 1989. Telomeric DNA dimerizes by formation of guanine tetrads between hairpin loops. Nature **342:** 825–829.
93. KANG, C., X. ZHANG, R. RATLIFF, R. MOYZIS & A. RICH. 1992. Crystal structura of four-stranded Oxytricha telomeric DNA. Nature **356:** 126–131.
94. MARSH, R. E., R. BIERSTEDT & E. L. EICHHORN. 1962. The crystal structure of cytosine-5-acetic acid. Acta Crystallogr. **15:** 310–316.
95. LANGRIDGE, R. & A. RICH. 1963. The molecular structure of helical polycytidylic acid. Nature **198:** 725–728.
96. GEHRING, K., J.-L. LEROY & M. GUÉRON. 1993. A tetrameric DNA structure with protonated cytosine·cytosine base pairs. Nature **363:** 561–565.
97. CHEN, L., L. CAI, X. ZHANG & A. RICH. 1994. Crystal structure of a four-stranded intercalated DNA: $d(C_4)$. Biochemistry **33:** 13540–13546.
98. KANG, C. H., I. BERGER, C. LOCKSHIN, R. RATLIFF, R. MOYZIS & A. RICH. 1994. Crystal structure of intercalated four-stranded $d(C_3T)$ at 1.4 Å resolution. Proc. Natl. Acad. Sci. USA **91:** 11636–11640.

Gene Regulatory Proteins and Their Interaction with DNA

AARON KLUG

MRC Laboratory of Molecular Biology
Cambridge, CB2 2QH, England

REGULATION OF GENE EXPRESSION IN HIGHER ORGANISMS—A COMBINATORIAL PRINCIPLE

At any one time most of the genes of a higher organism are silent: only a minority of the genes in a given cell is expressed, that is, being transcribed into messenger RNA, which in turn acts as a template for protein synthesis. The selective expression of any one gene is accomplished primarily through the interaction of protein transcription factors with characteristic DNA sequences included in the control region of the gene, which is most commonly located near to the actual coding region. The binding of a set of such factors, or regulatory proteins, acts as a molecular switch for the activation of RNA polymerase and other components of the transcriptional machinery, which are common to all genes. The supply of a particular combination of such transcription factors ensures that a gene is switched on at the right place and at the right time.

The essential idea that gene expression is controlled by binding of a protein to a regulatory region, or promoter, goes back to François Jacob and Jacques Monod more than thirty years ago, when working on the bacterium *Escherichia coli*. However, whereas transcription control mechanisms in prokaryotes are relatively simple, involving only one or two such factors, it has emerged in the last ten years or so that more complex organisms (eukaryotes), from yeast to man, need, and have, more complex mechanisms for regulating the expression of their genes. Usually, a combination of several (as many as six) transcription factors is necessary to form a transcription complex which can harness and activate the RNA polymerase to initiate transcription at the right starting point.

A little reflection will convince one of the reason why a set of such protein factors is required, rather than a single protein for each gene. If the latter were the case, that protein would have to be coded for by the expression of another gene, which would in turn require another protein transcription factor and so on, leading to an infinite recurrence. However, if a set of proteins is involved, then different combinations can be used for different genes. Thus a smaller number of regulatory proteins can control a large number of genes. In addition, this provides the means for multiple control at the level of the gene, which has the advantage that transcription may be regulated in a quantitative rather than in an all-or-none manner, and also for producing a network of interacting genes, since the protein product of one gene can affect the expression of another. So one has a combinatorial principle at work here operating at the level of a combination of proteins. As we shall see later, for example, in the case of zinc finger proteins, the principle also operates within individual proteins, where different subdomains can be combined to give greater variety or precision of recognition—a microcosm, as it were, of the macroscopic picture.

In this article we are concerned not with the actual process of transcription, or the wider problems of networking, but rather with the way in which a regulatory protein achieves recognition of the right segment of DNA. How does a transcription factor recognize a specific target site against the vast background of DNA in the nucleus?

RECOGNITION OF SPECIFIC DNA SEQUENCES BY PROTEINS

Eukaryotic DNA-binding transcription factors, like prokaryotic, achieve recognition by having embedded in them a discrete structure or domain that serves for binding to DNA. Most of the structures identified so far fall into a number of different types, each type having a characteristic amino-acid sequence and three-dimensional structure.[1,2] The first such structure to be identified was the so-called helix-turn-helix motif, discovered about ten years ago by X-ray crystallographic studies of certain bacterial regulatory proteins. Sequences suggestive of similar structured motifs were also subsequently found in proteins which are involved in establishing cellular patterns in developing embryos of *Drosophila* and mice—the so-called homeodomains. However, it was only in 1987 that the first high-resolution structure of a complex with DNA of one of these prokaryotic proteins was determined. X-ray crystallographic analysis of a number of other helix-turn-helix motif/DNA complexes have provided a precise insight into how this structural motif recognizes specific DNA sequences.

Since then, remarkable progress has been made towards understanding the nature of specific protein–DNA interactions. A number of types of structural motifs for DNA recognition have been identified on the basis of amino acid sequence comparisons associated with biochemical studies, and the three-dimensional structures have been determined for members of most of these types. Each motif represents a different solution to the problem of designing a piece of protein surface to fit a particular segment of a DNA double helix.

At first glance the DNA double helix looks a very uniform structure with the two strands, the phosphate-sugar backbones, spiralling around each other, and held together by the base pairs to form the DNA double helix. The bases on each strand form complementary pairs with the bases on the other strand, using hydrogen bonds to recognize each other. The double helical structure results in two kinds of grooves—the major and the minor—between the chains, each with a specific constellation of chemical groupings on the surface. The local structure of DNA varies in detail along its length and is determined by the particular sequence of base pairs. The width and depth of the major and minor grooves, the pattern of chemical groupings in the grooves, and the flexibility of the double helix are all determined by the base sequences, so that a specific sequence will result in a double helix with distinct characteristics. Every DNA sequence is thus recognizable from the shape of the double helix and the chemical identity of the bases. DNA-binding domains of proteins have surfaces complementary to DNA and hence a shape—often containing an α-helix—that fits well into the major (or minor) groove of DNA. This is often achieved by an α-helix protruding from the bulk of the whole protein, but presented in different ways in different types of proteins. When the pattern of a particular amino acid on the surface of the protein matches the pattern of groups on the surface of the DNA double helix, binding takes place through the formation of hydrogen bonds and van der Waals contacts. One might think of this recognition as a small piece of protein "reading," that is, interacting with, a short sequence of base pairs.

A protein reading-head could thus recognize a short DNA sequence of perhaps three or four base pairs, but such a short sequence occurs too frequently to be unique and uniquely recognizable. High specificity in the selective control of gene expression requires the recognition of a reasonable length of DNA, and Nature has found more than one way of putting together reading heads to achieve this. One design exemplified by several classes of DNA-binding proteins, such as prokaryotic helix-turn-helix proteins, the hormone receptors and basic leucine zipper proteins to be described below, is to combine two monomers so as to bind as a dimer. The use of dimer does not merely extend the length of DNA sequence recognized, but also brings into play the possibility of a variation in spacing between the two half-sites, so increasing the rarity of occurrence of the particular run of bases to be recognized, and hence its degree of uniqueness. Another distinctly different design for sequence-specific recognition is one in which reading heads are directly repeated in tandem, as found in the zinc finger proteins.

THE HELIX-TURN-HELIX AND HOMOEODOMAIN

The prokaryote helix-turn-helix group of DNA-binding proteins is to date the most thoroughly studied. The helix-turn-helix motif consists of two α-helices packed together at an angle of about 120°, with a tight turn at the elbow between them. It is not a stably folded structure on its own, and there is usually a third helix from the rest of the protein which stabilizes it. The second helix in the helix-turn-helix lies in the major groove of the DNA and carries the main amino acid residues responsible for specific binding (FIG. 1a) – hence it is called the recognition helix. This involves primarily direct or water-mediated interactions of the side chains on the recognition helix with the exposed chemical groups on the edges of the base pairs, but the whole motif is also fixed in a particular orientation by hydrogen bonds to phosphates on the DNA backbone.

The prokaryotic proteins all bind to DNA as symmetrical dimers with their two-fold axes coincident with a two-fold axis of the DNA centered on the palindromic sequence of the DNA-binding site. The central base pairs of the latter are not necessarily in contact with the protein, but can influence the conformation of the whole site by virtue of its detailed geometry and/or by its deformability, allowing a better fit between the protein and DNA. This last point is indeed a general one, namely that DNA is not a passive participant in the recognition process. The sequence-dependent features in the double helix constitute another, physical level of information in DNA sequences, secondary to the chemical code.[3]

The helix-turn-helix motif in eukaryotes was first identified by the sequence similarity with prokaryotic helix-turn-helix of a highly conserved 60–amino acid sequence found in many homeotic genes of *Drosophila*, now called the homeodomain. It is, however, different in two respects. First, the recognition helix is substantially longer, containing at its C-terminal end more basic residues which interact with the phosphates, leading to a somewhat different orientation in the major groove of the DNA (FIG. 1b). Secondly, the parent protein of the homeodomain binds as a monomer, although some proteins also contain other DNA binding regions, such as the POU domain. This originates in a conserved sequence found in certain homeotic genes, but which always appears to be accompanied by a homeodomain.[4]

THE BASIC LEUCINE ZIPPER

The basic leucine zipper was discovered in 1988, when it was observed[5] that a number of transcription factors contained a common pattern of repeating Leu residues spaced seven residues apart and named the leucine zipper. A basic region adjacent to the "zipper" directs sequence-specific binding (hence the acronym bLZip). The zipper sequence is in fact the motif responsible for dimerization of the parent protein by forming an α-helical coiled coil of the kind described by Francis Crick thirty-five years ago. The prototype of the basic leucine zipper family was the yeast transcriptional ac-

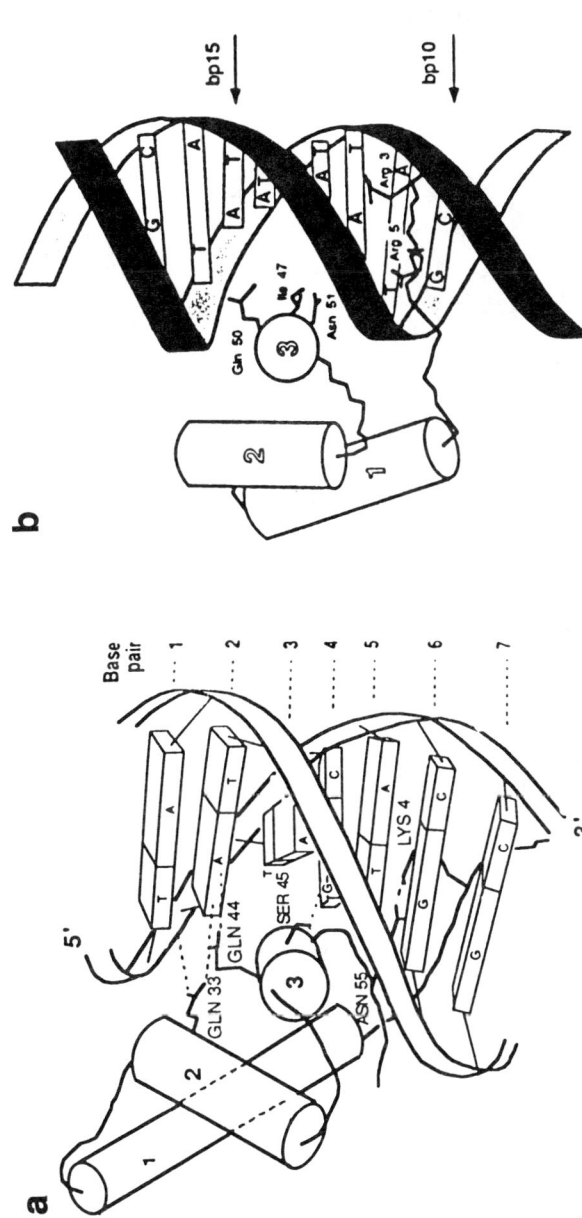

FIGURE 1. Helix-turn-helix protein-DNA interactions. (a) Part of the structure of a complex of the DNA-binding domain of the phage λ repressor and a λ operator site, showing the recognition α-helix, marked 3, inserted into the major groove of the DNA half-site (from Jordan and Pabo[41]). Some of the hydrogen-bond interactions are shown. Helices 2 and 3 constitute the helix-turn-helix motif. (b) Structure of an engrailed homeodomain–DNA complex (from Kissinger et al.[42]). The critical contacts made by the recognition helix in the major groove are shown. Note also the additional contacts made by the N-terminal basic region in the minor groove.

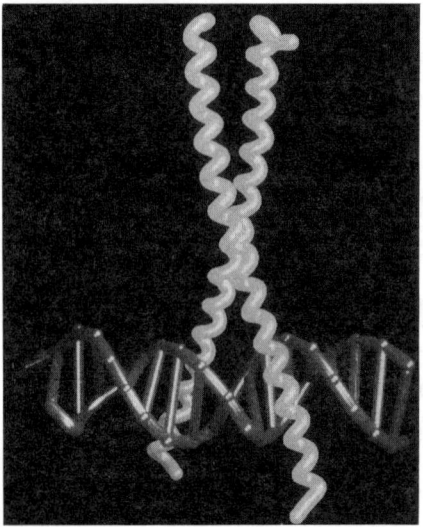

FIGURE 2. Structure of the GCN4 basic leucine zipper–DNA complex (from Ellenberger et al.[6]), showing a smoothed Cα backbone for the protein. The view is at right angles to the dyad of the complex formed by the protein dimer and the palindromic DNA-binding site. The C-termini of the monomers (*at the top*) pack together as a coiled coil, which gradually diverges to allow the basic region residues at the N-terminal ends to insert into the major groove of each DNA half-site.

tivator GCN4, which binds specifically as a dimer to a DNA segment 9 base pairs long, containing two palindromic 4-base pair half-sites separated by a single base pair. In the binding site for the CREB proteins, the same half-sites are separated by two base pairs.

The crystal structure of the GCN4 basic leucine zipper–DNA complex has been determined recently[6] (FIG. 2). The two basic leucine zipper monomers, 56 amino acids long, each form a smoothly curved continuous α-helix, the two packing together at their C-terminal ends as a coiled coil, which gradually diverges to allow the residues in the basic region to follow the major groove of each DNA half-site. The DNA, which is straight, is thus grasped, as it were, by a pair of splayed chopsticks. It is noteworthy that the basic regions, which are disordered in the free proteins, become ordered into helices on contacting the DNA.

Although GCN4 functions as a homodimer, it is an important feature of the basic leucine zipper class of proteins that they form both homo- and heterodimers. Thus the homodimer of the proto-oncogene c-Jun is unstable and binds DNA inefficiently, whereas the Jun-Fos heterodimer is stable and binds Ap-1 DNA sites with much higher affinity. Fos is thus a positive regulator of Jun. The use of heterodimers thus allows a combinatorial mode of action and increases the repertoire of regulatory functions for this family of proteins.

Another group of proteins that have similarities to the basic leucine zipper group are the helix-loop-helix (HLH), which are so far characterized only by sequence regularities.[7] A basic region of about 15 amino acids lies immediately N-terminal to a 15–amino acid segment with helical characteristics, which is separated from a second such putative helix by a region of variable length. This family also functions as dimers and in almost all cases a heterodimer is the active DNA-binding species—thus the protein Myc needs the protein Max to be effective.

β-RIBBON MOTIFS

There are other families of DNA-binding proteins that use α-helices to bind in the major groove, such as the two main classes of zinc finger proteins, which are discussed below, and the dimeric DNA-binding domain of the papilloma virus transcriptional activator, whose structure complexed with DNA has recently been determined.[8] It therefore came as a surprise when it was found[9] that the prokaryotic MetJ repressor interacts with DNA in a quite different fashion. Here a pair of anti-parallel β-strands (a "β-ribbon"), formed between a symmetrical dimer of the protein, interacts with DNA in the major groove (FIG. 3). Two other helices from each monomer are important for the stability of the whole dimer structure, which forms a core from which the β-ribbon protrudes so as to be able to lie in the major groove. The other parts

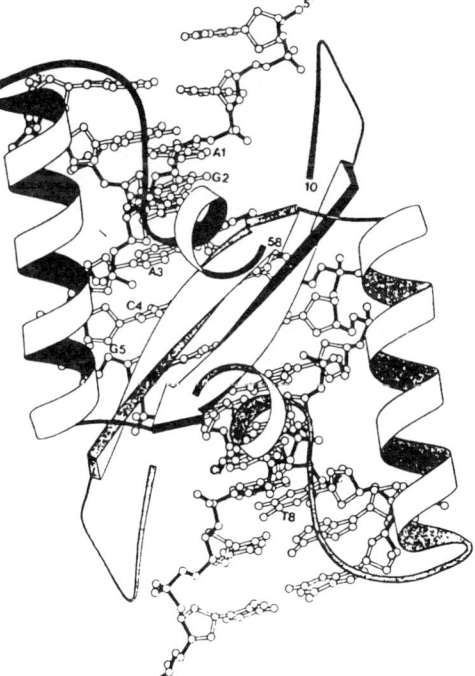

FIGURE 3. Structure of the MetJ repressor β-ribbon-helix motif bound to DNA (from Somers and Phillips[9]). Each monomer of the dimer contributes a strand to the β-ribbon. The two helices that support the ribbon are truncated to make it visible. The view is parallel to the dyad axis.

of the protein make extensive phosphate contacts on the backbones, and so help fix the whole complex.

The β-ribbon recognition element here requires dimerization to produce its folded structure, but there is an earlier example where a β-ribbon formed by a continuous loop in a monomer appears to be the main recognition element. This is the bacterial HU protein, whose structure has been determined crystallographically,[10] but no complex with DNA has yet been solved. Chemical protection studies show interactions in the minor groove.

These examples show that a β-sheet can be used in recognition, but nevertheless it seems clear that an α-helix, with its greater intrinsic stability, is a more suitable element for probing the characteristics of a particular segment of DNA, and accounts for its more widespread use.

THE TATA-BOX BINDING PROTEIN

Another recent case, but one where a large β-sheet is used rather than a ribbon, is that of the TFIID TATA-box binding protein required for the initiation of transcription for all three classes of eukaryotic polymerase. The first step was the solution of the structure of the isolated protein,[11] and with data from chemical and mutagenesis studies, some of the residues involved in DNA binding could be located. The whole molecule is saddle-shaped, being formed of two almost identical halves related by a pseudo dyad, from each of which is appended a β-loop, so that the two form a pair of stirrups, as it were, that could wrap around grooves in the DNA, so it seemed that the saddle could sit astride the helix.

Subsequently, two crystal structures of TATA-box binding proteins bound to DNA were determined.[12,13] Rather than sitting astride the DNA double helix, the TATA-box binding protein sits almost parallel to the axis of the DNA, which arches under the concave surface of the protein. This interaction distorts the minor groove drastically so that the double helix is unwound and also bent through about 80°, suggesting how the melting of the double helix necessary for the initiation of transcription may begin. This result also illustrates the point made above[3] that the physical properties of special DNA sequences can carry another level of information in their interaction with protein. In this case TATA-box binding protein exploits the deformability of the TATA sequence to initiate the unwinding of DNA as a prelude to transcription.[14]

ZINC FINGER PROTEINS–THE FIRST CLASS

The "zinc finger" motif was first identified by Miller *et al.*[15] in the *Xenopus* transcription factor IIIA (TFIIIA) where it is repeated consecutively nine times. It consists of a 30–amino acid sequence containing two His, two Cys, and three hydrophobic amino acids, all at conserved positions. On this basis, together with earlier measurement of zinc content and partial proteolysis data, it was proposed that each zinc finger motif forms a small, independently folded zinc-containing mini-domain, used repeatedly in a modular fashion to achieve sequence-specific recognition of DNA. Since that time zinc finger motifs of the TFIIIA type have turned up in hundreds of proteins, and they appear to be the most widely used of all types of DNA-binding domains. Indeed, they are estimated to constitute as much as 1% of the human genome.[16] This design may truly be called modular, since the multiply-repeated domains all have the same structural framework, but can achieve chemical distinctiveness through variations in certain key amino acid residues.

NMR studies first showed that the structured region of the zinc finger motif comprises about 25 amino acids forming a compact unit with a 12–amino acid helix packed against a β-hairpin,[17] confirming an earlier proposal for the structure of the finger.[18] NMR studies of other such peptides showed isomorphous structures (FIG. 4a), while those of a two-zinc finger peptide showed that the two-structured finger mini-domains folded independently, with a short, flexible linker sequence between them.[19,20] The mode of binding to DNA was first determined from the crystallographic analysis of a three-zinc finger peptide from the early-response protein Zif268 bound to its cograte DNA.[21] The three fingers bind in an equivalent manner to a segment of DNA 9 base pairs long so that each finger contacts three base pairs, with no gaps between the three sites (FIG. 4b). The main interaction for each finger comes from its helical region, which uses amino acid residues three (or six) positions apart in the sequence, and thus facing the same way, to interact with successive (or next nearest) bases on the DNA. These contacts are achieved by the angle at which the recognition helix sits in the major groove. The linkers between fingers are largely extended and make no significant DNA contacts. Hence the linkers seem to play no part in determining the manner in which the fingers interact with DNA, but they do seem to play a wider, secondary role in the binding in solution.[22] There are, of course, other interactions between the body of each finger and backbone phosphates, but the specific recognition is effected by the one-to-one amino acid to base pair interactions from the recognition helix to a triplet of base pairs. This simple pattern suggested a potential for devising a recognition code for the design of zinc finger modules that would recognize specific DNA sequences, and indeed some fingers have already had their specificity changed using a small number of site-directed mutations.[23]

This idea gains credence from a statistical study of more than a thousand zinc finger motifs by Jacobs,[24] who found that amino acids in three positions are highly variable. These positions are precisely those used to make contacts in the Zif268 complex, namely, those falling on the first, second, and third turns of the recognition α-helix. However, the crystal structure of a second zinc finger–DNA complex has been solved in this laboratory,[25] indicating that zinc finger–DNA recognition may be more complex than at first perceived. The DNA-binding domain is that of the *Drosophila* regulatory protein Tramtrack, which has two fingers and binds with relatively high affinity (5 × 10^7 M^{-1}) to a DNA binding site 6 base pairs long.[26] The second finger shows a similar pattern of contacts to those observed in the Zif268 complex, but

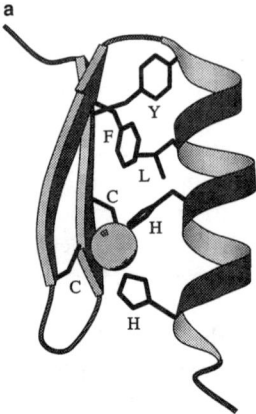

FIGURE 4. Zinc fingers of the first class (TFIIIA-type) and their interaction with DNA. (a) Structure of one of the zinc fingers of the yeast protein SWI5 from a 2-D NMR study in solution.[43] The two substructures forming the finger, the β-sheet and the helix are pinned together by the Zn ion (*grey ball*) ligated to a pair of Cys and a pair of His. The structure is further stabilized by the cluster formed by the three invariant hydrophobic residues, Phe(F), Tyr(Y) and Leu(L), packed together at the top of the domain. (b) The structure of the complex formed by the three zinc fingers of Zif268 and its binding site (adapted from Pavletich and Pabo[21]). Each finger binds in a similar manner to three base pairs in the major groove. The linkers between the three fingers are extended, and the fingers bind consecutively with no gaps in the DNA. Zn ions are represented by *balls*, as in panel **a**.

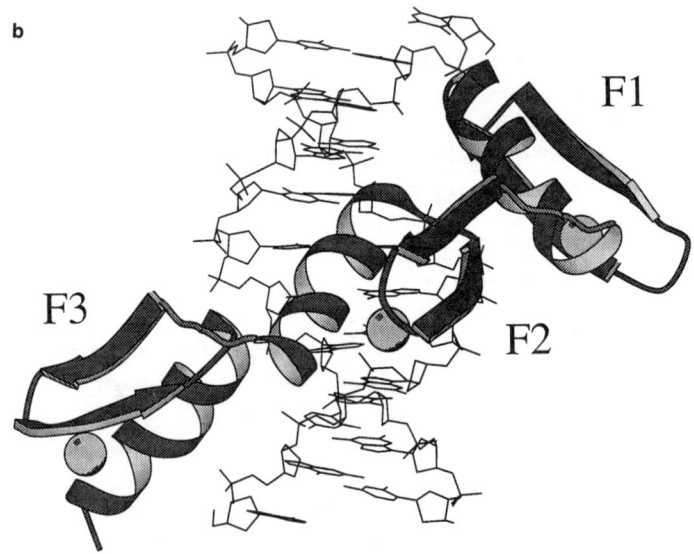

the first finger uses a well-conserved Ser residue, which is not in one of the three canonical positions, to make a hydrogen bond with a thymine, at a point where the DNA helix is distorted. Clearly one must await further structures of complexes to clarify whether there might be additional variations from the simple pattern shown by Zif268.

A SECOND CLASS OF ZINC FINGERS: HORMONE RECEPTOR DNA-BINDING DOMAINS

Shortly after zinc fingers were discovered in TFIIIA, sequence motifs that appeared to be related were found in several other protein or cDNA sequences of molecules that bound DNA. It was therefore at first thought that these might have a rather similar structure to the TFIIIA-type finger domains. The history of the discovery of the first two classes is told in Rhodes and Klug.[27]

The most important and widespread examples were those from members of the superfamily of hormone-activated nuclear receptors, which play a central role in the control of eukaryotic gene expression, and are indeed transcription factors.[28,29] The DNA-binding domains of such receptors all include two motifs in tandem, each about 30 amino acids long, but each motif contains two pairs of Cys rather than a pair of Cys and a pair of His, as in the first class. They do indeed bind Zn^{2+}, but the three-dimensional structure of two such DNA-binding domains, determined in solution using 2-D NMR spectroscopy, showed that the receptor DNA-binding domain is structurally distinct from the TFIIIA-type of zinc finger.[30,31] The two motifs in each domain each fold up into an irregular loop followed by an α helix, but the two together form a single structural unit with their helices crossing at right angles, so that the DNA recognition helix (from the first motif) is supported by the helix from the second motif (FIG. 5a).

Hormone receptors bind to palindromic sites (response elements, RE) on the DNA as dimers, and the DNA-binding domains alone also form dimers, the dimer interface arising from a region of the loop of the second motif of each receptor. These different roles for the two motifs within one structural unit, namely helix recognition and dimerization, were deduced by mapping on to the three-dimensional structure data on site-directed mutagenesis data from a number of laboratories, particularly those of P. Chambon, R. Evans, and G. Ringold (see references in Ref. 31). This combination of structural analysis and biochemical and genetic experiments pointed toward a mechanism of interaction with DNA and a general model was proposed by Härd et al.[30] and Schwabe et al.[31]

The detailed chemistry of the interactions at the protein–DNA interface, however, has had to await a crystal structure of a complex. The first to be determined was that at 2.9 Å resolution by Luisi et al.[32] of the DNA-binding

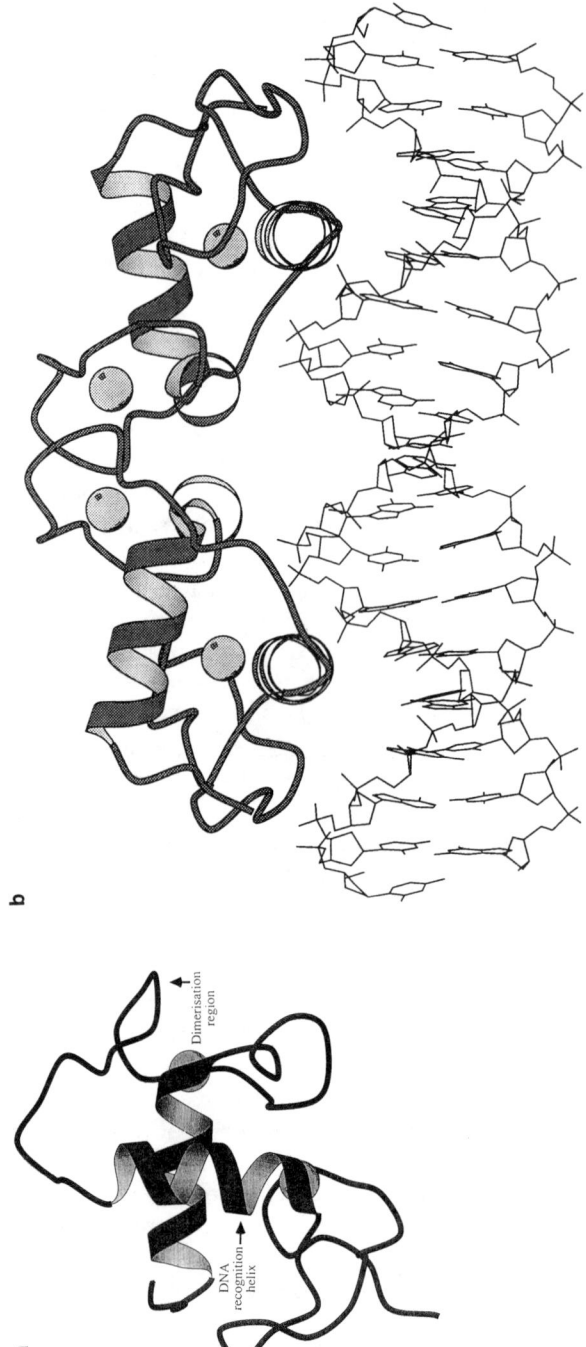

FIGURE 5. Estrogen receptor DNA-binding domain and its interaction with DNA. (a) Structure of the DNA-binding domain of the estrogen receptor by a 2-D NMR study in solution (from Schwabe et al.[31]). A single structural unit is made by two approximately similarly folded motifs, each consisting of an irregular loop followed by a helix. The Zn ion in each motif is shared between the loop and helix. The conformation of the loop in the dimerization region changes when the dimer is formed on binding to DNA (see panel **b**). (**b**) Structure of the complex formed between the DNA-binding domain of the estrogen receptor and its binding site.[34] The protein binds as a dimer making identical contacts to the two half-sites on the DNA. The recognition helices bind deep in the major groove, and there are also contacts from basic residues at the dimer interface to the minor groove lying between the half-sites. The view is perpendicular to the dyad of the whole complex.

domain of the glucocorticoid receptor (GR) complexed with a DNA segment 18 base pairs long containing the glucocorticoid response element (GRE). However, the particular DNA segment (GRE$_{s4}$) used was composed of two half-sites (each of 6 base pairs) separated by a non-native spacing, with four (rather than three) intervening base pairs. As a consequence of this, the two DNA-binding domains do not bind equivalently to DNA. One monomer of the GR.DNA-binding domain dimer appears to bind specifically to the DNA, with its recognition helix deep in the major groove, while the other binds in the same general way, but forms less close nonspecific interactions.

An intriguing feature of the structure of the DNA-binding domain-GRE$_{s4}$ complex is that the protein–protein contacts at the dimer interface can override the potential specific protein–DNA contacts at the second site in the GRE, yet they are not strong enough to hold the protein as a dimer in solution. The explanation for this must lie in the cooperative nature of the interaction of the DNA-binding domain with the DNA, that is, the protein–protein dimer interface might be induced and stabilized by interactions with the DNA backbone.

The specific-nonspecific character of this complex arises because of the presence of an extra base pair between the two half-sites in the GRE. To obtain specific binding at both half-sites, the native spacing of three base pairs is required. The information contained in the spacing of the two half-sites is an integral part of the binding-site "code," which is "recognized" through the formation of a protein–dimer interface. This is especially important for members of the nuclear-receptor family, which can recognize the same half-site sequence, but with different orientations and/or spacings. For these receptors, orientation and spacing are the only means of discrimination.[28,33]

Very recently the crystal structure of a second hormone receptor–DNA complex has been solved (at 2.4Å resolution) by Schwabe et al.,[34] in this laboratory. This is of the DNA-binding domain of the estrogen receptor (ER), which recognizes a different half-site from GR but with the same native separation of three base pairs between half-sites (ERE$_{s3}$). The protein binds as a symmetrical dimer in an equivalent manner to both half-sites (FIG. 5b) and now rather more interactions can be seen than were visible in the crystal structure of the DNA-binding domain–GRE$_{s4}$ complex,[32] which thus probably represents a compromise in the binding of a dimer to a site with an unfavorable spacing between its two halves. The interactions seen in the ER.DNA-binding domain–ERE$_{s3}$ complex are characteristic in number and type of those found in specific interactions in other families of protein–DNA complexes.

OTHER ZINC-BINDING DOMAINS

Clearly the second class of zinc finger DNA-binding domains is a not simple variant of the first TFIIIA class. They differ both in their structure and in the way in which they interact with DNA. Above all, the GR and ER receptors operate as dimers which bind to palindromic DNA sites, whereas the binding of zinc finger of the first class make no use of the symmetry of the DNA structure nor of base sequence. The latter function as independent modules which can be strung together in a directly repeating (tandem) fashion with no restriction on their number.

It should, however, be added that some members of the nuclear receptor family (e.g., thyroid, vitamin D, retinoic acid) also bind as dimers, but as nonsymmetrical dimers to a DNA binding site made of two directly repeated identical "half-sites." Here clearly another interface is brought into play, but again the discrimination depends entirely on the separation between the repeats in the DNA sequence.[33]

The structure of a member of a third class of zinc-binding domain of a distinct structural type has emerged recently.[35] This is of the GAL4 transcriptional activator, representative of a small family so far found only in yeast. NMR results[36,37] show that the two zinc ions and six Cys in the DNA-binding domain form a binuclear cluster with each Zn^{2+} coordinated by four Cys, so that two of the Cys are shared. This cluster holds together two short helices, related by a quasi dyad, one of which inserts into the major groove of the DNA as in the hormone receptors; we thus see the recognition helix supported by a second helix. Again, the molecule binds as a dimer to a palindromic DNA-binding site, with two short half-sites separated by approximately 1.5 turns of the DNA helix so that the DNA-binding domains bind on opposite faces of the DNA.

This GAL4 class of zinc-binding proteins is by no means the final example of proteins in which Zn^{2+} is used structurally to help fold a polypeptide chain into a compact domain that is used for binding nucleic acids. A fourth class is constituted by the $Cys-X_2-Cys-X_4-His-X_4-Cys$ sequences found in the nucleocapsid proteins of retroviruses,[38] which form "stubby" fingers. This general use of zinc was foreshadowed some time ago,[39] but the wide extent to which this happens is becoming increasingly clear, even though the structural information is still limited. A diverse set of families of proteins which bind Zn^{2+} and interact with nucleic acids are becoming uncovered[23] and are loosely grouped under the name of zinc finger proteins. The name is not inappropriate since the zinc-binding domains in all cases known do grip or grasp the double helix.

[NOTE ADDED IN PROOF: The use of one or more zinc ions to stabilize and help fold an autonomous structural entity is not confined to nucleic acid binding domains, although this is where they were first discovered. The use of zinc for folding a short peptide thirty to sixty amino acids long is more general, a possibility noted some time ago.[39] There are now ten classes of zinc binding motifs whose three-dimensional structures have been determined.[44]]

CONCLUDING REMARKS

There is little doubt that there is still more variety to be discovered in the design of proteins for recognizing DNA, but some principles have emerged. In most cases so far the basic element of specific recognition is that between a short stretch of base pairs (three or four) and a small piece of protein, usually part of an α-helix, but sometimes a β-sheet. These specific base contacts are reinforced by a relatively large number of contacts to the phosphates on the DNA backbone(s) which help fix the orientation of the whole structural element. Even where there are differences in detail on the way in which the recognition helix inserts into the major groove of the DNA, there are nevertheless common features among some of the different eukaryotic families.[40]

One such single interaction element is rarely enough to give sufficient binding energy and a high degree of discrimination between DNA target sites. Thus, very often, two such interaction elements are used when the protein binds as a symmetrical dimer. This also brings into play the spacing between the two half-sites, which must be set accurately to give proper binding. Several examples of the importance of spacing have been given above. The dimer design lends itself to still greater versatility by the use of heterodimers.

Another way of putting together interaction elements is by direct repetition, linking them together in tandem, and making no use of the symmetry of the DNA as do the dimers. This is exemplified by the zinc finger proteins of the first class (TFIIIA type). They take advantage of the modular design since there is no restriction on the number of domains that can be deployed (unlike the limit of two in the dimer case), nor on the distances between them. This modular design clearly offers a large number of combinatorial possibilities for sequence-specific recognition of DNA, and it is no wonder that zinc finger proteins of this class are so widely found in nature.

SUMMARY

The selective expression of a gene is achieved through the interaction of protein transcription factors with characteristic DNA sequences located in the regulatory region of the gene, which is usually distinct from the coding region. These proteins contain domains that bind specifically to the DNA sites (or response elements). Some general principles in the design of these DNA-binding domains are described, followed by examples of the different structural classes discovered so far and how they recognize their binding sites.

REFERENCES

1. PABO, C. O. & R. T. SAUER. 1992. Transcription factors: Structural families and principles of DNA recognition. Annu. Rev. Biochem. **61:** 1053–1095.

2. HARRISON, S. C. 1991. A structural taxonomy of DNA-binding domains. Nature **353**: 715–719.
3. TRAVERS, A. A. & A. KLUG. 1990. Bending of DNA in nucleoprotein complexes. In DNA Topology and its Biological Effects. N. R. Cozzarelli & J. C. Wang, Eds.: 57–106. Cold Spring Harbor, NY.
4. HERR, W., R. A. STURM, R. G. CLERC, L. M. CORCORAN, D. BALTIMORE, P. A. SHARP, H. A. INGRAHAM, M. G. ROSENFELD, M. FINNEY, G. RUVKUN & H. R. HORVITZ. 1988. The POU domain: A large conserved region in the mammalian *pit-1*, *oct-1*, *oct-2*, and *Caenorhabditis elegans unc-86* gene products. Genes & Development **2**: 1513–1516.
5. LANDSCHUTZ, W. H., P. F. JOHNSON & S. L. MCKNIGHT. 1988. The leucine zipper: A hypothetical structure common to a new class of DNA-binding proteins. Science **240**: 1759–1764.
6. ELLENBERGER, T. E., C. J. BRANDL, K. STRUHL & S. C. HARRISON. 1992. The GCN4 basic-region leucine zipper binds DNA as a dimer of uninterrupted α-helices: Crystal structure of the protein-DNA complex. Cell **71**: 1223–1237.
7. MURRE, C., P. MCCAW & D. BALTIMORE. 1989. A new DNA binding and dimerization motif in immunoglobulin enhancer binding, *daughterless*, *MyoD*, and *myc* proteins. Cell **56**: 777–788.
8. HEGDE, R. S., S. R. GROSSMAN, L. A. LAIMINS & P. B. SIGLER. 1992. The 1.7Å structure of the bovine papillomavirus-1 E2 DNA-binding domain bound to its DNA target. Nature **359**: 505–512.
9. SOMERS, W. S. & S. E. V. PHILLIPS. 1992. Crystal structure of the *met* repressor-operator complex at 2.8Å resolution reveals DNA recognition by β-strands. Nature **359**: 387–393.
10. WHITE, S. W., K. APPELT, K. S. WILSON & I. TANAKA. 1989. A protein structural motif that bends DNA. Proteins **5**: 281–288.
11. NIKLOV, D. B., S. H. HU, J. P. LIN, A. GASCH, A. HOFFMANN, M. HORIKOSHI, N.-H. CHUA, R. G. ROEDER & S. K. BURLEY. 1992. Crystal structure of the TFIID TATA-box binding protein: A central transcription initiation factor. Nature **360**: 40–46.
12. KIM, J. L., D. B. NIKOLOV & S. K. BURLEY. 1993. Co-crystal structure of TATA-box protein recognizing the minor groove of a TATA element. Nature **365**: 520–527.
13. KIM, Y., J. H. GEIGER, S. HAHN & P. B. SIGLER. 1993. Crystal structure of a yeast TATA box protein/TATA-box complex. Nature **365**: 512–520.
14. KLUG, A. 1993. Opening the gateway. Nature **365**: 486–487.
15. MILLER, J., A. D. MCLACHLAN & A. KLUG. 1985. Repetitive zinc-binding domains in the protein transcriptional factor IIIA from *Xenopus* oocytes. EMBO J. **4**: 1609–1615.
16. HOOVERS, J. M. N., M. MANNENS, R. JOHN, J. BLIEK, V. VAN HEYNINGEN, D. J. PORTEOUS, N. J. LESCHOT, A. WESTERVELD & P. F. R. LITTLE. 1992. High-resolution localization of 69 potential human zinc finger protein genes: A number are clustered. Genomics **12**: 254–263.
17. LEE, M. S., P. G. GIPPERT, K. V. SOMAN, D. A. CASE & P. E. WRIGHT. 1989. Three-dimensional solution structure of a single zinc-finger DNA binding domain. Science **245**: 635–637.
18. BERG, J. M. 1988. Proposed structure for the zinc-binding domains from transcription factor IIIA and related proteins. Proc. Natl. Acad. Sci. USA **85**: 99–102.
19. NEUHAUS, D., Y. NAKESEKO, K. NAGAI & A. KLUG. 1990. Sequence-specific

[¹H]NMR resonance assignments and secondary structure identification for 1-and 2-zinc finger constructs from SW15. FEBS Lett. **262:** 179–184.
20. NAKASEKO, Y., D. NEUHAUS, A. KLUG & D. RHODES. 1992. Adjacent zinc finger motifs in multiple zinc finger peptides from SW15 form structurally independent, flexibly linked domains. J. Mol. Biol. **228:** 619–636.
21. PAVLETICH, N. P. & C. O. PABO. 1991. Zinc finger-DNA recognition: Crystal structure of a Zif268-DNA complex at 2.1Å. Science **252:** 809–817.
22. CHOO, Y. & A. KLUG. 1993. A role in DNA binding for the linker sequences of the first three zinc fingers of TFIIIA. Nucleic Acids Res. **21:** 3341–3346.
23. BERG, J. 1993. Zinc finger proteins. Current Opin. Struct. Biol. **3:** 11–16.
24. JACOBS, G. H. 1992. Determination of the base recognition positions of zinc fingers from sequence analysis. EMBO J. **11:** 4507–4517.
25. FAIRALL, L., J. W. R. SCHWABE, L. CHAPMAN, J. T. FINCH & D. RHODES. 1993. The crystal structure of a two zinc-finger peptide from the *Drosophila* protein Tramtrack complexed with DNA shows an extended mechanism for zinc-fingers to bind to DNA. Nature. Submitted for publication.
26. FAIRALL, L., S. D. HARRISON, A. A. TRAVERS & D. RHODES. 1992. Sequence-specific DNA binding by a two zinc-finger peptide from the *Drosophila melanogaster* Tramtrack protein. J. Mol. Biol. **226:** 349–366.
27. RHODES, D. & A. KLUG. 1993. Zinc fingers. Sci. Amer. **268:** 56–65.
28. EVANS, R. M. 1988. The steroid and thyroid hormone receptor super family. Science **240:** 889–898.
29. GREEN, S. & P. CHAMBON. 1988. Nuclear receptors enhance our understanding of transcription regulation. Trends Genet **4:** 309–315.
30. HÄRD, T., E. KELLENBACH, R. BOELENS, B. A. MALER, K. DAHLMAN, L. P. FREEDMAN, J. CARLSTEDT-DUKE, K. R. YAMAMOTO, J.-Å. GUSTAFSSON & R. KAPTEIN. 1990. Solution structure of the glucocorticoid receptor DNA-binding domain. Science **249:** 157–160.
31. SCHWABE, J. W. R., D. NEUHAUS & D. RHODES. 1990. Solution structure of the DNA-binding domain of the oestrogen receptor. Nature **348:** 458–461.
32. LUISI, B. F., W. X. ZU, Z. OTWINOWSKI, L. P. FREEDMAN, K. R. YAMAMOTO & P. B. SIGLER. 1991. Crystallographic analysis of the interaction of the glucocortoid receptor with DNA. Nature **352:** 497–505.
33. UMESONO, K., K. K. MURAKAMI, C. C. THOMPSON & R. M. EVANS. 1991. Direct repeats as selective response elements for the thyroid hormone, retinoic acid, and vitamin D_3 receptors. Cell **65:** 1255–1266.
34. SCHWABE, J. W. R., L. CHAPMAN, J. T. FINCH & D. RHODES. 1993. The crystal structure of the complex between the oestrogen receptor DNA-binding domain and DNA at 2.4Å: How receptors discriminate between their response elements. Cell **75:** 567–578.
35. MARMORSTEIN, R., M. CAREY, M. PTASHNE & S. C. HARRISON. 1992. DNA recognition by GAL4: Structure of a protein-DNA complex. Nature **356:** 408–414.
36. PAN, T. & J. E. COLEMAN. 1990. GAL4 Transcription factor is not a "zinc finger" but forms a $Zn(II)_2Cys_6$ binuclear cluster. Proc. Natl. Acad. Sci. USA **87:** 2077–2081.
37. KRAULIS, P. J., A. R. C. RAINE, P. L. GADHAVI & E. D. LAUE. 1992. Structure of the DNA-binding domain of zinc GAL4. Nature **356:** 448–450.
38. SUMMERS, M. F., L. E. HENDERSON, M. R. CHANCE, J. W. BESS, JR., T. L. SOUTH, P. R. BLAKE, I. SAGI, G. PERET ALVARADO, R. C. SOWDER, III, D. R. HARE & L. O. ARTHUR. 1992. Nucleocapsid zinc finger detected in retrovi-

ruses: EXAFS studies of intact viruses and the solution-state structure of the nucleocapsid protein from HIV-1. Protein Sci. **1:** 563–574.
39. KLUG, A. & D. RHODES. 1987. "Zinc fingers": A novel protein motif for nucleic acid recognition. Trends Biochem **12:** 464–469.
40. SUZUKI, M. 1993. Common features in DNA recognition helices of eukaryotic transcription factors. EMBO J. **12:** 3221–3226.
41. JORDAN, S. R. & C. O. PABO. 1988. Structure of the λ complex at 2.5Å resolution: Details of the repressor operator interactions. Science **242:** 893–897.
42. KISSINGER, C. R., B. LIU, E. MARTIN-BLANCO, T. B. KORNBERG & C. O. PABO. 1990. Crystal structure of an engrailed homeodomain-DNA complex at 2.8Å resolution: A framework for understanding homeodomain-DNA interactions. Cell **63:** 579–590.
43. NEUHAUS, D., Y. NAKASEKO, J. W. R. SCHWABE & A. KLUG. 1992. Solution structures of two zinc finger domains from SW15, obtained using two-dimensional ^{1}H NMR spectroscopy: A zinc finger structure with a third strand of β-sheet. J. Mol. Biol. **228:** 637–651.
44. SCHWABE, J. W. R. & A. KLUG. 1994. Zinc mining for protein domains. Nature Struct. Biol. **1:** 345–349.

Genetics of Retroviruses[a]

HOWARD M. TEMIN[b]

McArdle Laboratory
1400 University Avenue
University of Wisconsin–Madison
Madison, Wisconsin 53706

We clearly live in a DNA world, as evidenced by this meeting and by its participants–all DNA-based. However, most DNA-based organisms, including people, contain in their genomes representatives of parasitic sequences coding for reverse transcriptase.[1] The most successful of these sequences seem to be those that also code for retroviruses, as seen in the AIDS epidemic and in the interest in retrovirus-mediated somatic gene therapy for genetic diseases, cancer, and AIDS.

These successes of retroviruses result from their having the advantages of RNA viruses (rapid genetic change) and the advantages of some DNA viruses (integration and latency). The usefulness of retroviruses as vectors for somatic gene therapy rests on their ability to integrate and the malleability of their genomes for genetic engineering. This ease results from the organization of a retrovirus DNA genome with control sequences for encapsidation, reverse transcription, integration, transcription initiation, and 3' end formation grouped at the ends of the virus genome[2,3] (FIG. 1). This organization gives the virus autonomy from cellular control sequences and allows easy laboratory manipulation to make any desired retrovirus vector.

Reverse transcriptase is a multifunctional molecule which carries out, in turn, during retrovirus replication, RNA-directed DNA synthesis, RNaseH digestion, strand transfer, and DNA-directed DNA synthesis.[4] Probably as a result of the necessity of being able to carry out all of these functions, retrovirus reverse transcription is very error-prone. The errors include frameshifts, deletions, deletions with insertions, homologous recombination, heterologous recombination, base-pair substitution, and hypermutation (TABLE 1).[5,6] We have proposed that the majority of these errors are connected to the ability of reverse transcriptase to bring about strand transfer and that the high error frequency is advantageous to retroviruses because it allows more genetic variation.[7] The high rate of base-pair substitution occurs as a result of the lack

---

[a] This work was supported by U. S. Public Health Service Grants No. CA-22443 and No. CA-07175 from the National Cancer Institute. H. M. T. was an American Cancer Society Research Professor.

[b] Howard M. Temin died on February 9, 1994. This was his last public address.

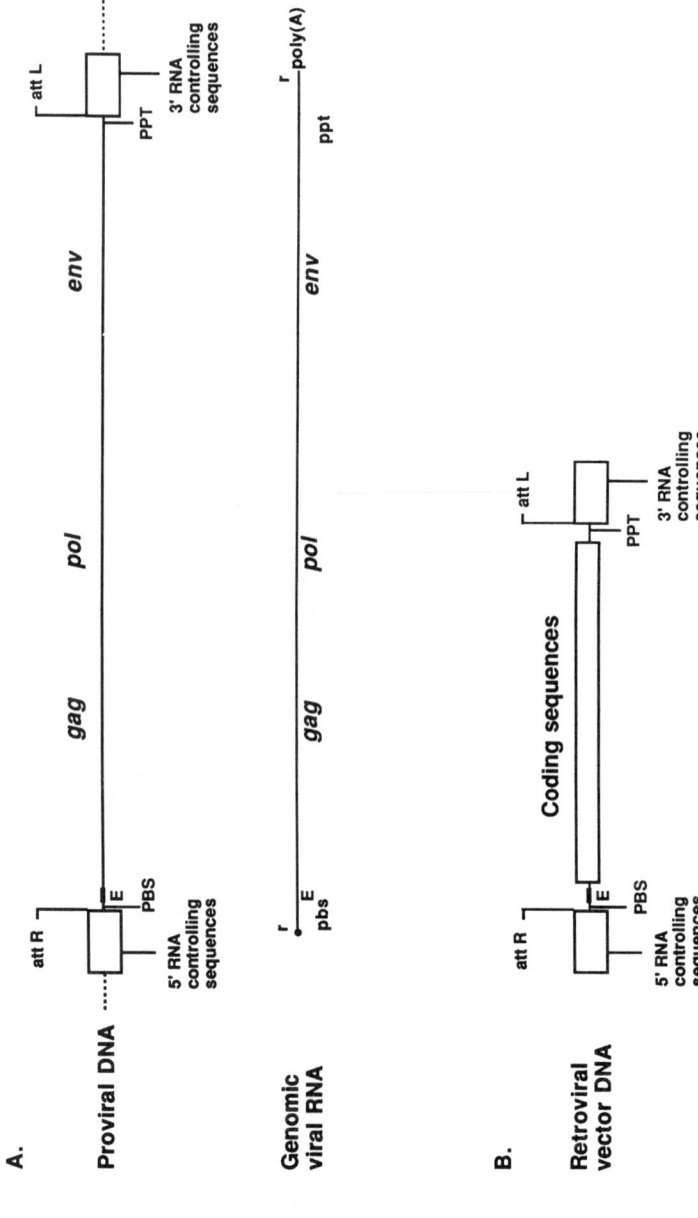

FIGURE 1. Retroviral genomes. (A) DNA and RNA genomes of a simpler retrovirus. Boxes are long terminal repeats; att R, attachment site right; PBS, primer binding site for minus-strand DNA synthesis binding; E, encapsidation site; *gag, pol, env*, viral coding genes; PPT, polypurine tract for primer for plus-strand DNA synthesis; att L, attachment site left; r, repeated sequence in viral RNA. (B) DNA genome of a defective retrovirus vector. Abbreviations as in (A).

TABLE 1. Rates of Errors during a Single Round of Retrovirus Reverse Transcription

Error	Rate per Genome (10^4b)
Frameshift	Up to 50% (depends on sequence)
Base-pair substitution	10%
Deletion and deletion with insertion	10%
Homologous crossover	50%
Nonhomologous crossover	0.01%

of an error-correcting mechanism in reverse transcriptase, the ability of reverse transcriptase to read through mismatches, and a high propensity for the reverse transcriptase to misalign during polymerization, that is, to strand transfer on the template.[8]

Analysis of error rates with retroviruses from two different species—the simpler spleen necrosis virus and the more complex bovine leukemia virus—shows that, although the rate of genetic change in one round of replication is a factor of 2.5 lower for bovine leukemia virus, the types of mutations are similar.[9] This similarity suggests that the same error-forming mechanism is operative in the two viruses, but that errors happen at a lower frequency with bovine leukemia virus. It is easy to see how frameshifting, deletion, and deletion with insertion use mechanisms similar to strand transfer. The readthrough after misincorporation and the misincorporation as a result of misalignment could involve the same characteristic of reverse transcriptase.

Retroviruses have two other special features: the DNA is longer than the RNA (FIG. 1A), and each virion contains two molecules of genomic viral RNA. Strand transfer is needed to synthesize the larger DNA, and the two molecules of RNA are needed for recombination between the two RNAs, a process like strand transfer of the primer molecules.

Our studies of reverse transcription during retrovirus replication used retrovirus vectors and helper cells. Retrovirus vectors are retroviruses modified to carry foreign sequences (FIG. 1B).[10] Most are replication-defective because of deletion of viral coding sequences. Retrovirus helper cells express the missing viral proteins from expression vectors. When retrovirus vector DNA is introduced into retrovirus helper cells by transfection or infection, infectious retrovirus vectors are formed by complementation of the vector RNA with the helper cell proteins. These vector viruses are only competent for one cycle of replication because of the absence of virion proteins in the target cells.

Using this system, we established that only one molecule of virion RNA is sufficient for retrovirus replication.[11] This conclusion is based on the fact that in the absence of recombination the normal primer strand transfers involve only one of the two virion RNA molecules. With the same vector system, we established that dimer RNA is necessary for recombination, the retrovirus version of sex.[12-14] In order to get recombination both parental viruses must be in the same helper cell, so that a heterodimer virion can be formed.

Modified retrovirus vectors were used to show that strand transfer on one template occurs very frequently at certain homopolymer runs. For example, we found a 50% rate of frameshifting at certain homopolymer runs (TABLE 1).[15]

Strand transfer can also be between the two virion RNA molecules in a retrovirus virion, resulting in recombination. Further, when foreign RNA molecules are packaged in the virion, the strand transfer can be to the foreign molecule, a process called nonhomologous recombination. Homologous recombination occurs at a rate of 60% per genome per cycle, one thousand-fold higher than the rate of nonhomologous replication (TABLE 1).[16,17]

Naturally occurring highly oncogenic retroviruses, like reticuloendotheliosis virus strain-T, are natural retrovirus vectors for oncogenes and were formed using nonhomologous recombination. This process was preceded by problems in normal 3' end formation resulting in readthrough transcription and formation of a chimeric RNA.[18]

We modeled this interaction with retrovirus vectors. Analysis of transfers in recombinant proviruses with these vectors showed that the transfers are to short regions of sequence identity in the midst of otherwise nonidentical sequences. This conclusion was drawn from examining the junctions of natural and experimental recombinants.[17,18] It also can be drawn from experiments placing a small amount of sequence identity in the midst of a region of sequence nonidentity and determining how frequently there is recombination in the identical sequence compared to the surrounding sequence. Such experiments involving 20 to 60 base-pairs of sequence identity showed increased recombination by the 60 bp sequence. PCR and sequencing determined that, even with 20 bp of sequence identity in the midst of otherwise nonidentical sequence, there was increased recombination so that 7 of 18 recombinants were in the 20 base sequence identity.[19]

These recombinants at a defined location provided a test of whether or not retrovirus recombination is error-prone.[20] Thirty defined recombination junctions were sequenced. No mutations were found.[19] Therefore, normal retrovirus recombination is not error-prone. (However, forced copy-choice recombination may be different.)

Thus, retroviruses have used a necessity, strand transfer, to increase greatly their genetic variability. We propose that this genetic variability is so high that mutation-driven evolution is possible for some retroviruses; human immunodeficiency virus may well be an example of this process.

ACKNOWLEDGMENTS

I thank G. Pulsinelli for useful comments on the manuscript and S. Yang for the figure.

REFERENCES

1. TEMIN, H. M. 1992. Origin and general nature of retroviruses. *In* The Retroviridae, Vol. 1. J. A. Levy, Ed.: 1–18. Plenum Press. New York.
2. COFFIN, J. M. 1992. Structure and classification of retroviruses. *In* The Retroviridae, Vol. 1. J. A. Levy, Ed.: 19–50. Plenum Press. New York.
3. COFFIN, J. M. 1990. Retroviridae and their replication. *In* Virology, Vol. 1, 2nd ed. B. N. Fields, D. M. Knipe, *et al.*, Eds.: 1437–1500. Raven Press. New York.
4. SKALKA, A. M. & S. P. GOFF, EDS. 1993. Reverse Transcriptase. Cold Spring Harbor Laboratory. Cold Spring Harbor, NY.
5. PATHAK, V. K. & H. M. TEMIN. 1990. Broad spectrum of *in vivo* forward mutations, hypermutations, and mutational hotspots in a retroviral shuttle vector after a single replication cycle: Substitutions, frameshifts, and hypermutations. Proc. Natl. Acad. Sci. USA **87**: 6019–6023.
6. PATHAK, V. K. & H. M. TEMIN. 1990. Broad spectrum of *in vivo* forward mutations, hypermutations, and mutational hotspots in a retroviral shuttle vector after a single replication cycle: Deletions and deletions with insertions. Proc. Natl. Acad. Sci. USA **87**: 6024–6028.
7. TEMIN, H. M. 1993. Retrovirus variation and reverse transcription: Abnormal strand transfers result in retrovirus genetic variation. Proc. Natl. Acad. Sci. USA **90**: 6900–6903.
8. BEBENEK, K. & T. A. KUNKEL. 1993. The fidelity of retroviral reverse transcriptases. *In* Reverse Transcriptase. A. M. Skalka & S. P. Goff, Eds.: 85–102. Cold Spring Harbor Laboratory Press. Cold Spring Harbor, NY.
9. MANSKY, L. M. & H. M. TEMIN. 1994. Lower mutation rate of bovine leukemia virus relative to that of spleen necrosis virus. J. Virol. **68**: 494–499.
10. BORIS-LAWRIE, K. A. & H. M. TEMIN. 1994. Recent advances in retrovirus vector technology. Curr. Opinion Genetics Devel. **3**: 102–109.
11. JONES, J. S., R. W. ALLAN & H. M. TEMIN. 1994. One retroviral RNA is sufficient for synthesis of viral DNA. J. Virol. **68**: 207–216.
12. HU, W.-S. & H. M. TEMIN. 1990. Genetic consequences of packaging two RNA genomes in one retroviral particle: Pseudodiploidy and high rate of genetic recombination. Proc. Natl. Acad. Sci. USA **87**: 1556–1560.
13. TEMIN, H. M. 1991. Sex and recombination in retroviruses. Trends in Genetics **7**: 71–74.
14. STUHLMANN, H. & P. BERG. 1992. Homologous recombination of copackaged retrovirus RNAs during reverse transcription. J. Virol. **66**: 2378–2388.
15. BURNS, D. P. W. & H. M. TEMIN. 1994. High rates of frameshift mutations within homo-oligomeric runs during a single cycle of retroviral replication. J. Virol. **68**: 4196–4203.
16. ZHANG, J. & H. M. TEMIN. 1993. Nonhomologous recombination in retrovirus replication. Science **259**: 234–238.
17. ZHANG, J. & H. M. TEMIN. 1993. 3' Junctions of oncogene-virus sequences and the mechanisms for formation of highly oncogenic retroviruses. J. Virol. **67**: 1747–1751.
18. ZHANG, J. & H. M. TEMIN. 1994. Retrovirus recombination depends on the length of sequence identity and is not error prone. J. Virol. **68**: 2409–2414.
19. PELISKA, J. A. & S. J. BENKOVIC. 1992. Mechanism of DNA strand transfer reactions catalyzed by HIV-1 reverse transcriptase. Science **258**: 1112–1118.

In Memoriam:
Howard Temin, the Fierce Scholar[a]

Howard Temin, who died on February 9, 1994, was driven by the genetic preoccupations of the "phage group" to an insight that was fundamental to the development of contemporary cellular biology. Howard went to Caltech in 1955 to begin graduate studies, and there developed a unique scientific style, blending the influences of Max Delbrück and Renato Dulbecco. His work was marked by a devotion to understanding the genetic issues posed by cancer-inducing viruses. This focus on genetics put him firmly in the traditions of American science dating back to the beginning of the century, and his concern with virus-induced cancer also built on a rich past. But Howard's fierce belief in himself, his deep scholarship, and his remarkable insight allowed him to realize a synthesis that made him one of the most creative scientists of the twentieth century.

Working on his thesis with Harry Rubin in Dulbecco's laboratory, Howard in 1958 developed a quantitative assay for cell transformation by Rous sarcoma virus (RSV). This set the stage for dissecting the components in that complex mixture of viruses inherited from almost 50 years of passage of RSV in animals and cells. But it also allowed cancer induction to be studied as a quantitative variable *in vitro*, a perspective that still illuminates cancer research. Howard's scholarship led him to understand, even as a graduate student, that the history of work on phage and animal viruses showed that a quantitative assay can change the trajectory of a field.

Howard had an even deeper preoccupation at Caltech. His work had led him to the belief that RSV, although an RNA virus, must develop a lysogenic relationship to the transformed cell. Apparently fascinated by the then-developing model of λ phage, he saw that RSV might make a DNA copy of itself. Dulbecco remembered that at Howard's thesis exam, Delbrück was very impressed with everything except Howard's speculation on a possible DNA intermediate in the virus's growth.

Howard left Caltech in 1960, going to the University of Wisconsin, where he spent the rest of his life. In his first 10 years there, he concentrated much of his attention on understanding the differences between normal and transformed cells, ultimately isolating critical serum factors. But throughout that

---

[a] This tribute is reproduced by permission from *Cell* (Vol. 76, pp. 967–968) © 1994 by Cell Press and by the courtesy of Dr. David Baltimore.

period, he was continually trying to test the notion of a DNA intermediate in the viral growth cycle.

His work on the biology of the transformation process showed him that RSV could permanently alter the growth properties of cells, and the permanence of the changes drove his belief that the virus must exert a genetic influence on the cell. He also recognized that mutants of RSV yielded transformants with distinct morphologies, implying that the genotype of the virus impressed itself on the cell. For RSV to do that, he fervently believed, it must transfer its information from the RNA form it has in the virion to a DNA form in the cell.

RNA-dependent DNA synthesis was a heretical notion at the time because the "central dogma" that information flows from DNA to RNA to protein was so firmly established. But driven by the model of phage lysogeny, Howard found it obvious that RNA must be copied into DNA, and he labored on through the 1960s to provide the evidence that would convince others. His was a lonely quest because the tools of those days were so blunt. He tried inhibitors and showed that RSV replication was blocked by actinomycin D, but the scientific community merely yawned and said, rightly, that inhibitors can be misleading. He tried DNA hybridization, but was unwittingly confounded by the endogenous RSV-related DNA in normal cells so that the small differences he could find between uninfected and infected cells were unconvincing.

It took until 1970 for biochemistry, ironically, to come to this geneticist's rescue and bring him from obscurity to center stage. He, with his colleague Satoshi Mizutani, found that RSV virions contain an RNA-dependent DNA polymerase, providing the machinery for the reversed flow of information that Howard had long postulated. Simultaneously, but coming from the biochemical tradition pioneered by Severo Ochoa and Arthur Kornberg, I too found the reverse transcriptase, as *Nature* called it. When we jointly announced the discovery at the 1970 Cold Spring Harbor annual meeting, it was rapidly confirmed by others and was immediately accepted by all but a very few skeptics as verifying what Howard had thought and said for so long. In 1975, Howard and I shared the Nobel Prize with Dulbecco for having set the stage for understanding how viruses can add genes to normal cells and cause them to grow as cancer cells. But the discoveries that flowed from Howard's insight and tenacity have continued to reverberate. The reverse transcriptase became a key tool of the biotechnology industry, which uses it just as viruses use it, to create genes from RNA molecules. HIV was found to be a retrovirus by assay of its reverse transcriptase, and the 13 years of intensive research on retroviruses that followed our 1970 discovery became the base for understanding HIV's role in AIDS. Other viruses that reverse transcribe as well as retrotransposons were found, and by now it is believed that the large repetitive fraction of the mammalian genome mainly arose by reverse transcription. In fact, the growing consensus that an RNA world preceded our DNA-based world suggests that evolution of a reverse transcriptase may have been a key moment in the transition. Howard's insight thus both reached back deeply into evolution and illuminated so much of the science and even industry that followed.

I first knew Howard when, as a high school student, I went for a summer to Jackson Laboratory in Bar Harbor, Maine. Howard returned that summer of 1955 as a recent college graduate to be the guru to the high school group. We venerated him for his wide knowledge and deep commitment to science. There, young scientists were exposed to the power of genetic research—an important influence on Howard. It was perhaps also important that we were exposed to the ideas of Francisco Duran-Reynals, who preached the then-unpopular view that viruses were an important cause of cancer.

When I enrolled at Swarthmore College, from which Howard had already graduated, the stories of his academic feats were central myths of the biology department. It was said that he had read every book in the biology library and they had had to buy new ones to satisfy his voracious appetite for knowledge. They told of Howard's honors exam, a great Swarthmore tradition in which professors come from other schools and administer an oral exam in each subject taken as an honors seminar. Most students quiver, but Howard knew that examining his understanding of biological science was superfluous, so he sat astride a table and took the opportunity to engage the examiners in a dialogue about their work. In a school where the newly developing knowledge of molecular biology could only be seen as a pale outline through a fog, one can imagine that Howard was eager to take advantage of a clearer vantage point. When Howard left Swarthmore, the Watson–Crick DNA structure was only 2 years' old, but Howard was nevertheless ready to help move biology from where it was to where it would go.

Howard lived with a rare intensity of focus on the genetics of cancer-inducing viruses in chickens. From the first, he was interested in both the phenomenon of cell transformation and the genetics of retroviruses, but in recent years the genetic issues dominated his output. His laboratory discovered and did important work on the *rel* oncogene, the prototype of the family that includes NF-κB and *Drosophila dorsal*. It made many contributions to the characterization of retrovirus recombination mechanisms and mutation rates. He was one of the first to use recombinant DNA methods to engineer retroviruses for expression of specific genes, ushering in the present era of hope for gene therapies. Genetic concerns drove his research program right up to his death at age 59.

Howard trained many people who went on to key positions in international science. People from his laboratory have been important in the study of HIV, of other viruses, of *rel* proteins, and of many issues in carcinogenesis and differentiation. John Coffin has emphasized to me Howard's continuing dedication to the careers of his 75 trainees, and I know personally their dedication to and affection for Howard.

Although he traveled relatively little for such a widely honored modern biologist, he often went to Washington in the last 10 years of his life to offer his knowledgeable advice about virology, cancer research, and AIDS. One of his most influential positions was as a member of the National Cancer Advisory Board. He was an untiring advocate for investigator-initiated research because Howard, perhaps mirroring his own history, believed fiercely in the pre-eminence of the individual scientist's contributions. He brought to his advisory roles an uncompromising criticality, a quantitative focus, and a genetic perspective. Although Howard could be harsh in his criticisms, his insistence on rigor brought to national programs a very important perspective that has served this country well and will be sorely missed.

In living his professional life in Madison, Wisconsin, Howard made the conscious choice to stay apart from the bustle of the coasts. It was as if his intensity was so great that it needed the calmer surroundings of the Midwest to cool its ardor. It was not isolation, however, because Howard not only stayed abreast of the details in a wide swath of biological science but was also a student of politics and of general affairs. No one who knew him will forget the amused but penetrating comments he made on the foibles of politicians and the irrationality of political affairs. Much as he wished for more rational approaches, he adopted a sardonic view of fatherly sympathy for the lack of objectivity in political decision-making.

Howard's reaction to the Nobel ceremony was particularly revealing. This is a very regal affair and most recipients of the prize let it flow over them — it is like being a participant in a fairy tale. Not for Howard! He had to be both participant and viewer at once and carried camera and tape recorder through the events. Things didn't flow over Howard; his analytic sense, his apartness, never left him free to become a pure object of attention.

What most people in retrovirology probably remember best about Howard was his active participation in the annual Cold Spring Harbor tumor virus meetings. Howard sat in the front row, attentive to every word, and followed almost every talk with a penetrating question. His desire to understand, his passion to connect with the next generation of virologists was evident to all.

It is the ultimate irony that Howard should have died of lung cancer because he was a vehement crusader against cigarette smoking. He had devoted his life to studying cancer, and he certainly hoped that his insights would lead to new ways of fighting the disease. But he also realized that in today's world, the most effective and cheapest way to cut the incidence of cancer would be for people to stop smoking. That would save hundreds of thousands of lives, and no preventive measure from the cancer research laboratory yet matches that possibility. Howard even chose the venue of the winner's 2-minute response at the Nobel dinner to rail against smoking. It was part of his superrationality — even when being celebrated for his brilliance, he had to remind the world that a simple act of self-control can sometimes do more than that of all of the great minds.

Howard's death at such a young age removes from us one of the treasures of modern science. His uncompromising integrity is a rare commodity in these times of increasing hype and fashion. His unique focus on the genetics of retroviruses made him a special contributor to contemporary science. As biological science revels in the fruits of its successes, growing more powerful and becoming more commercial, it is important to remember that he who toils alone may become the revolutionary, that insight goes further than logic, and that a life lived in uncompromising reach for personal goals is the greatest legacy an individual can leave to the world.

DAVID BALTIMORE
The Rockefeller University
New York, New York 10021

PART IV. BANQUET PROGRAM

In Honor of James D. Watson, Francis Crick, and Maurice Wilkins

On the night of October 15, more than 600 conferees attended the conference banquet held to honor the achievements of Watson, Crick and Wilkins. During that session, after greetings by Rodney Nichols, Chief Executive Officer of the New York Academy of Sciences, Dr. James Watson was given certificates of appreciation by Dr. Stanley Ikenberry, President of the University of Illinois, Sir Crispin Tickell, Warden of Green College, Oxford, and by Dr. Joshua Lederberg, President of the New York Academy of Sciences. In opening the conference that morning, Dr. James Stukel, Chancellor of the University of Illinois at Chicago, welcomed the conferees and brought greetings from the Governor of Illinois, James Edgar. In appreciation of the conference celebrating the advances made by these biomedical scientists and in recognition of the fact that Dr. Watson is a native of Chicago, the Governor proclaimed the week of October 11-18, 1993, Biomedical Science Appreciation Week in the State of Illinois, and Dr. Stukel, on behalf of Governor Edgar, read the proclamation.

Included in this section are photographs taken at the banquet as well as the address given to the conference by Dr. Watson and the introductory remarks of Rodney Nichols, Dr. Lederberg and Sir Crispin Tickell. In addition, Dr. Francis Crick and Dr. Maurice Wilkins, who shared the Nobel Prize with Dr. Watson and were also honored by the conference, sent their prepared remarks, which are contained here as well. Dr. Crick was unable to be present because of illness and Dr. Wilkins, because of a conflicting celebration at University College, London. This section concludes with a poem, "Merry Crickmas," by Dr. Rollin Hotchkiss, initially written to celebrate Dr. Crick's birthday during a Cold Spring Harbor symposium. Dr. Hotchkiss read this poem in tribute to Dr. Crick during the meeting.

In introducing Dr. Lederberg, I made mention of two coincidences. The first was that at the time the conference was conceived, we did not know that Dr. Lederberg would be elected President of the New York Academy of Sciences in the conference year and how extraordinary it was that this was the case. The second was that both Dr. Lederberg and I had the good fortune to come under the influence of Professor Francis Ryan at Columbia College. In fact, Dr. Ryan was Dr. Lederberg's first mentor. Several generations of Columbia students were introduced to the rigors of vertebrate morphology by Professor Ryan, who was, at the same time, one of the pioneers of molecular

STATE OF ILLINOIS
EXECUTIVE DEPARTMENT
Proclamation

WHEREAS, 1993 represents the 40th anniversary of the discovery of the structure of DNA by James D. Watson, Francis Crick, and Maurice Wilkins, representing a momentous event in the biomedical revolution of the 20th century; and

WHEREAS, the New York Academy of Sciences, the University of Illinois at Chicago, and Green College of the University of Oxford are sponsoring an international landmark meeting in Chicago from October 13 through 16, 1993, to celebrate this event; and

WHEREAS, James Dewey Watson, the co-discoverer of the structure of DNA and nobel laureate was born in Chicago on April 6, 1928, and is thereby a citizen of the City of Chicago and the State of Illinois;

THEREFORE, I, Jim Edgar, Governor of the State of Illinois, proclaim October 10-17, 1993, as BIOMEDICAL SCIENCES APPRECIATION WEEK in Illinois in recognition of this historic meeting and the impact of the biomedical sciences on our citizens.

In Witness Whereof, I have hereunto set my hand and caused the Great Seal of the State of Illinois to be affixed.

Done at the Capitol, in the City of Springfield, this THIRTIETH day of SEPTEMBER, in the Year of Our Lord one thousand nine hundred and NINETY-THREE, and of the State of Illinois the one hundred and SEVENTY-FIFTH.

George H. Ryan
SECRETARY OF STATE

Jim Edgar
GOVERNOR

FRANCIS RYAN
(Photo courtesy of Mrs. Elizabeth Ryan.)

genetics. Francis Ryan died prematurely in 1963 at the age of 47. In recognition of his contributions to molecular biology and to the development of the Cold Spring Harbor Laboratory, the 1963 volume of the *Cold Spring Harbor Symposium* was dedicated to him, as is this section.

—DONALD A. CHAMBERS, *Conference Chair*

Greetings

RODNEY W. NICHOLS
Chief Executive Officer
The New York Academy of Sciences
2 East 63rd Street
New York, New York 10021

As we celebrate the fortieth anniversary of the discovery of the double helical structure of DNA, I'd like to express my delight in joining Maclyn McCarty and Rollin Hotchkiss to do so—they were old friends and mentors during the twenty years I spent at The Rockefeller University, a place where much of the groundwork preceding the discovery was done. Drs. McCarty and Hotchkiss embody both the human and the intellectual qualities needed in a good scientist. Their humane commitment to medical progress and their integrity moves me to hold them up as an example to the relatively younger audience we have here. They've helped to build the leadership of the next generation. Perhaps this is what Donald Chambers had in mind when he called a "birthday party." We all understand that we're not only rejoicing in past accomplishment, but also celebrating the potential in all of these fields.

Before the celebration commences, I would like to say a few words about the New York Academy of Sciences, which, as the third oldest scientific society in the United States, continues to reflect its contemporary milieu as it comes to grips with the pressing scientific questions of today. At its inception in 1817, and throughout the nineteenth century, its concerns were primarily ones of classification and exploration into a universe that was becoming more and more susceptible of being known. In our century, especially the latter half of it, the analytical powers have been deepened so much that what we are hearing at this convocation would have been scarcely imaginable a half-century ago.

Today, the Academy, with its broad reach of about 45,000 members in 150 countries, seeks to address questions that can only be answered by harnessing the insights of scientists, engineers, physicians, and social scientists from every part of the academic, industrial, and governmental community. The Academy further seeks to aid society by not only offering reliable information—through its publishing program, its conferences, and its education programs, among others—but also participating more actively in the analysis of the choices for crafting public policy. In so doing, the Academy is a leader among several scientific organizations advancing the case for rock-solid science in conquering disease, husbanding the planet's resources, and fostering wise technological and economic development. As Pasteur said, "There is only one science and the application of science, and they are bound as the fruit is to the tree." The modern New York Academy serves both science and society.

To return to today's conference: so far we've had a series of *tour de force* presentations during which I was continually struck by the repeated intersections of physics, chemistry, biology, and many instruments of modern technology in tackling the questions of understanding the very core mechanisms of life. In a certain sense all of the sciences relate to any of the sciences. Alex Rich's reminiscences of Linus Pauling, showing how Pauling integrated physical and chemical insights into his thinking on biology, demonstrates this essential unity. This search for unifying principles keeps science going and teases us with a dance of interrelatedness. We thank all of you for coming—young and seasoned scientists alike—to join in a retrospective and a prospective view of what has been the greatest biological discovery of the century.

This conference, as one of the many the Academy sponsors each year, draws not only on the intellectual resources of its organizers and participants, but also on the practical and financial help of many sponsors and contributors. We are very grateful to them; without their help, mounting a meeting of this magnitude would have been impossible and we rely on these "silent partners" to help us in many activities of mutual interest. I'd also like to thank those persons from our conference department at the Academy and their counterparts at the University of Illinois at Chicago: their interstate and interdepartmental collaboration over several months has resulted in this splendid meeting. And, of course, the labor of Dr. Donald Chambers and Dr. Rhonna Cohen in undertaking the most challenging of conceptual and the most arduous of administrative tasks must be acknowledged with the deepest gratitude. Their enthusiasm for science and their belief in its benefits for humankind fueled the making of this convocation. Let the celebration begin!

Greetings

JOSHUA LEDERBERG
President
The New York Academy of Sciences
2 East 63rd Street
New York, New York 10021

I'm pleased that Donald Chambers has evoked Francis Ryan's memory: he was so important in the lives of all of his students, and I am proud to have been one of them.

I was going to start by arguing with Gunther Stent as to whether the Renaissance started at the fall of Constantinople in the fifteenth century or the recapture of Toledo and its library in the twelfth . . . but I'll spare you that. For one thing I wasn't there, so I can't speak at first hand for either of these events.

Gunther's remarks about "premature discovery" and its reception do have a lot of bearing on how we do science, so I would like to say something about that. He was right about Mendel, of course. As to the reception of "AMM44,"[1] to the contrary, I *was* there, and have a more complicated view. I just do not believe that those can be compared. This meeting offers too nearly unique a setting and audience to let that go by without comment.

Does *premature* mean:

- "the data do not exist to explain all of the paradoxes and challenges of a new discovery"?
- "the audience is incapable of understanding the challenge"?

Many of Gunther Stent's readers have interpreted premature to mean the latter; else almost every provocative discovery is in some respects premature.

I suggest that the touchstone is the reaction, not just verbally, but in the laboratory. You might see controversy and active inquiry, diligent effort to accumulate experimental data that will resolve the question, which is just what did happen from 1944 on. (The reaction to McClintock's "jumping genes" in maize, also decried as premature, was in fact similar.) Or the claim might be relegated to oblivion. Mendel's case was close to that, happily a rare event in the history of science. To pursue the receptor metaphor: the ligand can be irrelevant, or it can be an agonist, an antagonist, or even both.

My inclinations go closer to Rollin Hotchkiss's remarks that give credit to the importance of organized critical skepticism[2] in maintaining the integrity of science. Even when hindsight tells otherwise, we should be skeptical of instant genuflection to new revelations. For novel claims to be challenged is a necessary and healthy aspect of scientific progress.

I had a lot of sympathy for Rollin Hotchkiss when he talked, both 40 years ago and today, about that last 1% or 0.1% of margin for protein contamination of DNA. That was fighting Avogrado's number: a formidable opponent for the exclusion of protein. As late as 1953, and after his celebrated experiment with Martha Chase, even Al Hershey had some reservations: "My own guess is that DNA will not prove to be a unique determiner of genetic specificity, but that contributions to the question will be made in the near future only by persons willing to entertain the contrary view."[3] Well, his statement was half-right! Too easily forgotten today was Wendell Stanley's error in 1935, claiming that crystalline TMV was a pure protein, only to be corrected by Pirie's finding of a few percent of RNA: a mistake no one wanted to repeat in a hurry. So we should welcome debate and the search for critical and corroborative evidence and applaud Mac McCarty for that extra, arduous repetition to seal the argument. To be ignored is only slightly worse than to be swallowed whole. And there is a lot of revision ahead even for our well established dogmas.[4]

Let me now turn to my proper role, to be the stagehand pulling up the curtain for our star event. To *introduce Jim Watson* is the oxymoron of all time. I have never seen him in a modest mood—you can puzzle whether that refers to his mood, or Francis's, or my own. He does have so much not to be modest about! Since Albert Einstein, no scientist's name has received so much media attention; unlike Einstein he has laid himself bare, displaying, if not exaggerating, every flaw in his true confessions,[5] in the genre of, and by now more widely read than, Rousseau or Augustine. Whether he has reached the age of repentance we have still to see.

Let us turn from the man to the discovery and ask:

- first, what did the double helix do for us scientifically? and
- second, how would it have mattered if that hadn't happened in 1953? (beyond that Matt Meselson would have had another year or two to finish his rotifer paper).

For my first question, let me quickly separate duplicity from helicity. The duplex and its tacit lemma, complementarity, have dominated DNA research for the last 40 years, informing every branch of biology and medicine, and I see neither the need nor have the capability to repeat this information here.[6] The duplex is at the root of DNA as an informative molecule, and of every experiment involving sequence specificity, enzymatic reactivity, biological specificity and so forth: in a word it is inseparably connected with any laboratory or vital test of its primary structure: *pace* the protein interactions, it takes one nucleotide sequence to recognize another: via duplex formation.

Helicity has been more difficult to study, and has taken second place—it is, after all, secondary structure.[7,8] But as Alex Rich began to remark, it is very much involved in the dynamics of DNA in the cell, and obviously in such spheres as regulated gene expression, transcription, and the packaging of DNA into chromatin and chromosomes. It is becoming increasingly important in the study of recombination, mutagenesis, and cancer.

With regard to my second question, it is on a different track from Gunther's—not *who* else would have discovered the structure, but *whether* it would have been discovered. What if X-ray diffraction just didn't exist as a workable method of structural analysis? We could note that only a tiny fraction of contemporary research—an important one!—actually uses those tools, nor depends on the precise molecular coordinates of DNA. Of course, that's an unfair test: who would have thought of the simpler experiments without the structural precedent?

Rollin Hotchkiss has commented on the growing consensus about the centrality of DNA in the late 1940s. Modelled on the study of protein secondary structure, and its denaturation, were the beginnings of hyperchromicity assays and their bearing on DNA secondary structure. Levene's tetranucleotides, never more than a casual heuristic, had been refuted. But the chief obstacle in my view was, in contrast to proteins, a lack of a conveniently available, homogeneous biological specimen on which to conduct biochemical and biophysical analysis. It would eventually come in the form of small bacteriophages. The virtue of X-ray analysis of DNA fibers was its insensitivity to heterogeneity of primary structure: it leapfrogged to the next level of generalization and that surely saved us ten years or more of false starts and stumbling in the dark. Nor can anyone who has actually tried to build models from scratch belittle the ingenuity and insight that went into the construction Watson and Crick made.

I also join Gunther in celebrating the 1953 papers for having generated one of the most far-reaching icons of the twentieth century—the image has even reached perfume bottles; it is redolent of—I was going to say the Caduceus, but that's single-stranded—Hermes' staff. As iconography, it is a creative artistic production that has captured the visual imagination at a symbolic level as well as an intellectual one, for the public as well as the scientific community.

For flash of scientific insight, beacon, and icon, we are all in your debt; and you deserve a little something for having endured us. On behalf of the New York Academy of Sciences, I am privileged to offer this certificate of appreciation to you, Jim Watson.

REFERENCES

1. AVERY, O. T., C. M. MACLEOD & M. MCCARTY. 1944. Studies on the chemical nature of the substance inducing transformation of pneumococcal types. J. Exp. Med. **79:** 137–158.
2. MERTON, R. K. 1973. The Sociology of Science. Theoretical and Empirical Investigations. The University of Chicago Press. Chicago, IL.

3. HERSHEY, A. D. 1953. Functional differentiation within particles of bacteriophage T2. Cold Spring Harbor Symp. Quant. Biol. **18:** 135–139.
4. LEDERBERG, J. 1993. The anti-expert system: Hypotheses an AI program should have seen through. *In* Artificial Intelligence and Molecular Biology. L. Hunter, Ed.: 459–463. AAAI Press. Menlo Park, CA.
5. WATSON, J. D. 1968. The Double Helix. A Personal Account of the Discovery of the Structure of DNA. Atheneum. New York.
6. LEDERBERG, J. 1993. What the double helix (1953) has meant for basic biomedical science. A personal commentary. JAMA **269**(15): 1981–1985.
7. WELLS, R. D. & S. C. HARVEY, EDS. 1988. Unusual DNA Structures. Springer-Verlag. New York.
8. FRANK-KAMENETSKII, M. D. 1993. Unravelling DNA. VCH Publishers. New York.

Intellectual Dawns

SIR CRISPIN TICKELL

Warden, Green College
University of Oxford
Oxford, United Kingdom

It is a disproportionate honor for Green College to be one of the three sponsors of this conference. But as a small community of scholars and teachers, many in the field of genetics, we are more than glad to join the University of Illinois at Chicago and the New York Academy of Sciences. Our representation here today, which includes the three successive Wardens of the youngest Oxford College, well demonstrates our interest and enthusiasm. We are deeply indebted to Professor Donald Chambers, Visiting Fellow of Green College, and to his wife, Dr. Rhonna Cohen, who together made this conference possible.

It would be the height of unwisdom for me to venture into the field of genetics. Instead I thought I would speak on the background to the shock which always follows the introduction of a new idea. A prime example is, of course, the elucidation of DNA forty years ago. We all have our mental models to make sense of the world around us. Indeed the human mind has almost boundless capacity to invent, to justify and to rationalize. We can get locked into a way of looking at things like the prisoners in Plato's cave, who could only see a puppet show of passing images within the cave rather than the reality behind them in the light of day outside.

It is astonishingly difficult for us to think our way back into the mental models of other generations or societies: I give three examples:

The Aztecs: It needs a leap of the imagination to enter the mind of a people who believed that the sun would not rise without the nourishment of blood from human sacrifice. The queues of victims participated with feelings of worship as well as dread. As has been well written, the population seems to have been spellbound by the drama, the beauty, and the terror of the event. For them this was normality.

The intellectual landscape of the Bible: The pre-Darwinian world of fixed species—the world of our great-great-grandparents—is a strange place. I once glimpsed it through the person of a monk on Mount Athos. For him the earth had been created four thousand years ago, and the hands of God and the Devil could be seen manipulating events of daily life. After a few moments, our talk on all but trivialities became meaningless. The gulf was too wide for comprehension.

The fixity of the earth's surface: It should have been obvious to anyone looking at a map, as well as to observant geologists, that chunks of the earth must once have fitted into each other. Yet it needed knowledge of the mechanics—first available in the 1960s—before the idea was generally accepted and its liberating implications could be understood.

New ideas are not always welcome. Plato's prisoners were first dazzled by the glitter of light when they emerged from the darkness. No doubt the Aztecs feared that the sun might not rise and that the earth would dissolve into night. Natural selection sent shudders through accepted religion and truth. The notion of a moving surface to the earth confounded current thinking. Yet after the shock comes the pleasure. Things half understood, or misunderstood, before, fall gracefully into place. Once in place they tend to harden. New structures of orthodoxy are built. Dogma returns. In some cases the thought police in their many guises move in. We then have to await the next revelation or, rather, revolution.

At the basis of our current society is an amalgam of past tradition and recent discovery. Old and new are like geological formations unconformably mixed. As Lord Keynes once said: the difficulty is not in new ideas, but in escaping from old ones. There has been an industrial revolution, which has transformed the production processes, and thereby society itself; a stretching out of time to lengths that we can read about but scarcely comprehend; a stretching out of distance in the same fashion; a revolution in information technology and transfer, which is only now beginning; and the new understanding of ourselves and the mechanisms of inheritance which we are here to celebrate today.

In terms of belief and action the main results have been

1. an extraordinary population explosion of our species, and damage to other species and forms of life on the planet;
2. a breakdown of unified world views in earlier models, or more specifically religions and philosophies;
3. a culture that is not appropriate for management of the problems which confront us. We have to ask ourselves whether the models in our minds will lead us to be exploiters, engineers, stewards, or gardeners of the natural estate.

But today we should rejoice in one of the most liberating intellectual revolutions of our time. The prisoners are out of the cave and becoming used to the light. The shock is past. The pleasure lingers. The dogmas are undergoing change. Our eyes are on the next horizons. It is a good time to be alive.

What the Double Helix Has Meant for Basic Biomedical Science[a]
A Personal Commentary

JOSHUA LEDERBERG

The Rockefeller University
1230 York Avenue
New York, New York 10021-6399

INTRODUCTION

The article published by Watson and Crick in 1953[1] was the landmark pointer to our contemporary model of DNA as a macromolecular structure. This lay on a well-worn path of biophysical analysis, reducing microscopic anatomy to the molecular level. It also helped inspire an enormous body of biochemical research that has defined DNA as *the* informational molecule, a discontinuity that has been labeled the Biological Revolution of the twentieth century. As a piece of structural analysis, the idea of the double helix includes the concepts (1) that DNA is a duplex structure, comprising two paired complementary strands, associated by secondary, noncovalent bonds; (2) that the strand pairs are coiled, forming a double helix; and (3) that these are antiparallel – the orientation of one strand being in the opposite polarity from the other.

The most novel features of DNA are associated with its duplicity, rather than its helicity. Linear polymers rarely form stiff straight rods; folding into coils is the norm. The genetic functions of DNA are inextricably associated with its duplex structure, and hardly at all with its helical shape; this is reflected in the preoccupation of DNA research with its role as an informational molecule. However, we shall see a recent concentration of interest in supercoiling. Inevitably, the biochemical interactions of DNA with other molecules, be they regulatory proteins or chemotherapeutic inhibitors, will often be intimately wound up with the precise three-dimensional conformation of the helix. This is also proxy for higher orders of coiling, interactions

---

[a] This article has been reprinted by permission from the *Journal of the American Medical Association*, Volume 269, number 15 (April 21, 1993), pp. 1981–1985. © 1993, American Medical Association.

with histones and other DNA-binding proteins, and the organization of DNA into chromosomes.

DNA can be built in either an antiparallel or a parallel format, although the former adds a note of symmetry that may account for the prevalence of the antiparallel in nature. For parallel DNA a different enzyme would be needed to recognize and replicate the 3' versus the 5' end. Recognizing this asymmetry, Watson and Crick[1] speculated that DNA was anti-parallel prior to concrete observational evidence for this conformation.

Rarely has a structural determination been coupled so promptly with functional implications. Watson and Crick[1] immediately inferred that DNA duplexes were formed automatically when each strand was replicated, and that this involved the assembly of nucleotides, one by one, complementary to the existing structure.[2] They overreached the mark by suggesting that this might be possible even without the intervention of specific anabolic enzymes, the discovery of which we owe to the prodigious labors of Arthur Kornberg and his school in the 1960s. But in imputing autocatalytic powers to the DNA double helix, Watson and Crick[1] might lay claim to having anticipated the enzymatic functions of RNA (if not DNA), an iconoclasm that earned the Nobel Prize in 1989 for Sidney Altman and Thomas Cech.

Despite the intellectual revolution initiated by Watson and Crick,[1] we might still ask the question: At what point was the welfare of any patient altered by specific knowledge of the double helix? This is a question I agonized over during the 1970s, and its first answer was perhaps the work of Y. W. Kan on the prenatal diagnosis of hemoglobin disorders, using DNA hybridization (1978). How rapidly we have moved in the interval is recounted by Caskey[3] in the companion article. Why did that take 25 years? One may simply point to the enormous edifice of contributory knowledge that now bridges the most reductionist aspects of DNA structure to pathological manifestations.

HISTORICAL BACKGROUND OF WATSON AND CRICK

The biological role of DNA was still enmeshed in controversy in 1953. Nucleic acids had been extracted from pus cells by Miescher in 1869, and from the beginning were associated with cell nuclei. These substances are now known to be macromolecules composed of a linear array of nucleotides joined by phosphodiester bonds. Cytologists writing in the early 1900s remarked on the association of nucleic acids with chromosomes and speculated that this basophilic material in chromatin might be the substance of genetic continuity. This brilliant anticipation was, however, submerged by a misleading observation, namely the apparent loss of basophilia in the chromosomes of oocytes, leading E. B. Wilson (1925) to remark that "the continued presence of 'chromatin' [i.e., basi-chromatin] is essential to the genetic continuity of the chromosome has, however, become an antiquated notion." We now know that these chromosomes become remarkably unraveled in keeping with their massive involvement in transcription, associated proteins then overshadowing the continuity of the DNA.

This skepticism was reinforced by the apparent monotony of DNA structure embedded in Phoebus Levene's first analyses of DNA. They contained only four constituent nucleotides—each comprising a phosphate group, a sugar, and one of the four bases: cytosine (C), thymine (T), adenine (A), or guanine (G). Within the limited analytical precision available in the 1920s, these appeared to be present in exact stoichiometric equivalence. Hence the provisional hypothesis of DNA as a tetranucleotide, although it was well-recognized that its molecular weight and other key parameters had yet to be ascertained. Nor was there any biological system or array of sources to tell that one DNA preparation was in any way different from any other. Such a simple molecule seemed a poor candidate for the miraculous capabilities of the gene. On the other hand, proteins contained an abundant variety of constituent amino acids (eventually 20). More important, dozens, even hundreds of proteins were isolated with vastly different biological, physical, and chemical properties, including wide disparities in composition. The 1920's saw the most exciting developments in protein chemistry, even the crystallization of urease and of pepsin and the demonstration that enzymes were pure proteins (Sumner, 1926; Northrop, 1930). The cap seemed to be a similar characterization of the tobacco mosaic virus, claimed to be pure protein by Wendell Stanley in 1935. This was, however, soon to be corrected by Bawden and Pirie in 1937, who found phosphorus and carbohydrate in infectious concentrates of tobacco mosaic virus and inferred the presence of RNA. Stanley, nevertheless, received the Nobel Prize in chemistry in 1946, together with Sumner and Northrop. By that time, Stanley acknowledged "that the nucleic acid could not be removed without causing loss of virus activity and there was general agreement that the virus was a nucleoprotein." Thus, this prize was a noble reinforcement of the primacy of proteins as the seat of biological specificity.

The breakthrough challenge to that dogma was thrust forth in 1944 by Oswald T. Avery, Colin MacLeod, and Maclyn McCarty. They had studied the diverse serological types of the pneumococcus and followed up Griffith's report (1928) that these could be altered or transformed by extracts of other strains. The gist of the 1944 study was that the transforming substance was DNA! This was contrary to expectations that the carbohydrate antigen or some associated protein would be the transforming substance. Avery, a member of the same Rockefeller Institute as Wendell Stanley, was intimately familiar and impressed with the difficulties of characterizing biopolymers. Though fully cognizant of the biological implications of the discovery, he was even more hesitant to dwell upon them—but did include a remark that "The inducing substance has been likened to a gene. . . ."

Their claims, of course, aroused intense critical controversy, largely around the obvious question whether their DNA preparations were still contaminated with traces of biologically active protein. Avogadro's number, 6×10^{23} per mole, would allow a residuum of 10^7 protein molecules per microgram of a preparation that was 99.99% protein-free, at the limit of analytical detectability. The sensitivity of the active materials to deoxyribonuclease might be ascribed to a protective rather than informational function of the DNA. Likewise, the insensitivity to proteases might be an attribute of a nucleoprotein complex.

My own role in the debate was a willingness, even desire, to believe—but a sense of responsibility that the issue was too important to be regarded as closed until there was no escape. It was not clear what feasible experiments (short of *ab initio* synthesis of DNA) could ultimately seal all these infinitesimal loopholes. One might go along with "DNA" as a working hypothesis, and some did. Most biologists blurred their judgments by talking about nucleoproteins—not necessarily informed by the distinction they were implying. Some might have meant something like "protein" or "nucleic acid" or a combination thereof, but please do not ask the role of the constituents. A rare few gambled on the DNA—as in some sense did Watson and Crick,[1] although they would have enjoyed working out its structure regardless of its biological implications. In the event, the final elucidation of DNA structure was a horse race. By Watson's own account, only a few weeks would have separated their priority from the looming insights of Maurice Wilkins and Rosalind Franklin (who had provided the critical experimental data) or of Linus Pauling.

The biological significance of the pneumococcus transformation was also problematical. It looked like a transfer of genetic information; but until 1951, the only markers tested were the serotype antigens. Could one extrapolate from those to genes in general, particularly given that the very idea of a bacterial genetics was in its infancy?

After the 1944 bombshell, more chemical attention was given to the tetranucleotide model, and signs of greater chemical complexity emerged. Of particular import were the deviations of the four bases from the simplistic 1:1:1:1 ratio, found by Erwin Chargaff. Furthermore, DNA from different sources exhibited different base composition. So perhaps DNA could be more complex, more diversified than previously thought—could be rehabilitated as a candidate for the gene. During the 1940s, the Feulgen cytochemical test for DNA and analyses indicating constancy of DNA per genome in somatic cells and a halving in germ cells also added to DNA's respectability. But these findings did not necessarily prove more than a structural or scaffolding role for the DNA. The pneumococcus transformation remained the only biological assay for a genetic role for DNA—in contrast to the innumerable enzyme and immunological assays available for candidate proteins.

This impasse was overcome by the broadening of phage research, sternly governed by Max Delbrück's genius, to embrace a wider range of chemical studies of phage infection. A critical one was the 1952 double-labeling experiment of Hershey and Chase. Most of the S-35 label (capsid protein) was excluded from infected cells; most of the P-32 (DNA) entered and was transmitted to the phage progeny. This experiment is often cited as the crowning blow on behalf of the "DNA-only" model. But Hershey himself did not go so far—well aware that "most" is not "all," he was still referring to "nucleoprotein" in 1953—and this at the same Cold Spring Harbor Symposium that sponsored a critical discussion of the paper by Watson and Crick.[1]

The article by Watson and Crick[1] did not, of course, bear directly on the loopholes in Avery's claims. It did add a further note of plausibility to a DNA-only concept of the gene. In the absence of any serious contradiction, this gradually hardened from working hypothesis to central dogma. The most serious challenge today is the prion hypothesis: that some "infectious" agents may be devoid of nucleic acid. This is still contentious at an experimental level: the hypothesis least in conflict with nucleic doctrine is that the infectious prion is a sort of epitaxial primer of aggregation of a host-determined protein. This still leaves obscure how and whether different prions could maintain and propagate their identity in a genetically defined host.

Long after many other lines of evidence converged to support an informational role of DNA—e.g., the colinearity of DNA sequences with protein products (Yanofsky), genetically active DNA was eventually synthesized in the chemical laboratory (Khorana) and replicated enzymologically (Kornberg), fully vindicating Avery *et al.* and those who gave their faith to these propositions.

THE FLOWERING OF MOLECULAR GENETICS

Since the rediscovery in 1900 of Mendel's 1865 work, genetics has had an extraordinary development, even without the benefit of tangible physical and chemical models of the genetic material. The biological phenomena of mutation and of sexual crossing (genetic recombination) opened the door to experiments in which existing organisms were the reagents. Genomes could be mixed by crossing, and new combinations of factors segregated into the offspring. Likewise, fruit flies could be subjected to radiation, and variant or mutant forms discovered. Genetic information is organized into linear chromosomes, and the processes of meiosis in gametogenesis: precise synapsis of homologues and crossing-over or segmental exchange of chromosome parts allowed powerful dissection of fine structure on a scale that rivals that of microchemical analysis. These methods continue to play an indispensable role in the denomination and mapping of mutant genes. By 1941, through the work of Beadle and Tatum, the groundwork of biochemical genetics had been laid—the role of genes in the prescription of protein products, and the use of mutations in the dissection of metabolic pathways. Indeed, many of these ideas had been anticipated by Archibald Garrod's studies of human biochemical defects at the very dawn of genetics.

Since 1953, we have had a new language for the description of genes: they are now segments of DNA that can be defined and manipulated as chemical entities. The linguistic transition has been conceptually smooth, though marked by occasional generational quarrels. Understandably, very few individuals can combine erudite knowledge of the life histories of a wide range of organisms in their natural habitats with focused and specialized knowledge of biochemical manipulations in the laboratory. Nor have many radical revisions of genetic doctrine issued from the molecular perspective. We have had to acknowledge that genes, as bits of DNA, are subject to a wider range of chemical and biological interactions than was previously thought—especially with other DNA. The icon of stability of genomes has been shaken by the discovery of transposable elements, first noted in maize by McClintock in 1951; these remained inexplicable until they could be studied as DNA molecules. And concentrating on DNA now allows us to inject genes with viruses, needles, even "shotguns," into a range of cellular targets including the germ line, providing a technical revolution in the construction of new genotypes in all kinds of organisms—bacteria, plants and mammals.

Meanwhile, other advances, notably the extension of recombination analysis to somatic cells in culture by cell fusion, have extended the technical power of genetic analysis in ways compatible with, but not dependent on, the double helix. It is paradoxical that the human chromosome number, $2n = 46$, was not correctly understood until 1956 (Tjio and Levan), and that for about 20 years thereafter this was at least as important in the development of human genetics as was the structure of DNA.

The adumbration of DNA-based research, molecular genetics, since 1953 would embrace a substantial fraction of world science. Many encyclopedic monographs struggle to record the details and promptly become obsolete. We can hardly do more herein than summarize the major headings, following an imprecise dichotomy distinguishing topological DNA—an informational duplex—from mechanical DNA—a three-dimensional geometric object.

DNA AS AN INFORMATIONAL DUPLEX

Denaturation and Hybridization

The most elementary aspect of the duplex is the separability of its strands, using temperature or chemical denaturants. A-T base pairs melt (separate from one another) at a lower temperature than G-C pairs, so melting curves can distinguish DNA of different base composition. Single strands once separated can also be reannealed, allowed to rejoin, the kinetics allowing the discovery that much DNA (in eukaryotes) has a repetitive or a redundant sequence. Radioactively labeled probes can be used to ferret out target homologous DNA with high precision.

Homology and Evolution; Polymorphism Within the Species

These and related methods can be used as quantitative indices of the genetic relatedness of divers species, supplanting the subjectively evaluated morphological criteria used in systematics heretofore. Within the species, genetic polymorphism can now be described at the DNA level—one astonishing finding is that humans are typically heterozygous with a prevalence of two or three per thousand, i.e., almost once in every gene. As most of these base substitutions have no perceptible phenotypic effect, random drift (rather than selectible or adaptive change) may predominate in evolutionary change (Kimura).

Mutagenesis and DNA Repair

The vulnerability of genes to mutational change in response to X-rays was known empirically since 1927 (Muller), and to chemicals since 1944 (Auerbach). Early hopes that chemical mutagenesis would be a direct path to the chemistry of the gene were not substantiated. Most chemical mutagens react with amino acids as well as DNA bases. The exceptions are nuclein base analogues which may be misincorporated into DNA; but these were discovered much later. Above all, we now understand that the initial lesions in DNA would usually be lethal, and that eventual mutations are the result of intricate repair metabolism that occasionally misfires.

Transcription; Genetic Code

The "central dogma" of information flow has emerged, that DNA → (transcription) RNA → (translation) protein. The base sequence of DNA is transcribed faithfully into a messenger RNA copy. This in turn governs the assembly of a polypeptide sequence, each three-base frame of RNA encoding one particular amino acid. The polypeptide then folds (perhaps with the guidance of a chaperone) into a pre-ordained protein three-dimensional shape, which can then function as an enzyme, antibody, hormone, structural unit, and so forth. This folding process is not yet fully computable. There may even be circumstances where a given polypeptide might have alternative foldings, but this is not accepted dogma.

The details of messenger RNA synthesis have become much more intricate. Primary transcripts are usually processed, only some of the RNA tracts being spliced together to form the final message. The other "intervening sequences," or introns, may be the major part of the RNA – their functions remain obscure. As with repeated sequences, they may reflect "selfish DNA," whose presence in the genome has little to do with their adaptive value to the overall organism. In other examples, RNA may be edited in other ways before translation is completed.

Enzymology: Nucleases, Ligase, Replication; Reverse Transcriptase

For a legion of brilliant and tireless investigators, the DNA structural model has been the platform for isolating a host of enzymes involved in every aspect of DNA metabolism. Besides giving us that metabolic map, explaining how DNA is replicated, sliced, stitched, spliced and repaired, these enzymes are the vital technical tools for further study of DNA and for the engineering of new constructs.

Some viruses, notoriously the retroviruses (including human immunodeficiency virus), exhibit a reverse transcriptase, whereby RNA→DNA. This knowledge is indispensable to the virologist. It has also given some of the most valuable tools for studying RNA, e.g., messenger, by allowing the production of DNA copies for input into other technology.

Tools for Engineering: DNA Splicing; PCR

These sempstering tools have founded the multi-billion dollar biotechnology industry. DNA tailored *in vitro*, with inserts from human or a variety of other sources, can be patched into convenient host garments (from bacteria to cows) for the easier exhibition of a variety of products – growth factors, enzymes, immunizing antigens, replacement therapeutics (like clotting factors) – in unlimited variety. Related technology is used to target specific host genes, to elucidate their functions in physiology and development.

The PCR (polymerase chain reaction) has been the instrument of the "democratization of molecular biology." With it a single DNA molecule in a messy mixture can be fished out and amplified *ad libitum*, most importantly at low cost and with simple instruments. High-school students do experiments today that would have been doctoral dissertations 15 years ago. The applications range widely, from forensics and diagnosis of genetic disease to the hunt for new viruses and the revival of fossil DNA. At its heart, a synthetic DNA probe is a rational, linear, digital signature to locate any counterpart in the analysand. Its core of combinatorial specificity can be contrasted with that of antibodies, which is founded on three-dimensional shapes of the immunoglobulin and its targets.

Drug Discovery

DNA combinatorics has reached a new peak in a paradigm for drug discovery that mimics natural evolution.[4] Randomized DNA sequences are expressed on host cells (or phages), and these are then selectively screened for specificities of binding to specific reagents—usually receptors for which agonists or antagonists are sought. The cell expressing the desired epitope can then be grown out for larger-scale production and testing. In one application, the mammalian antibody-forming mechanism can be emulated, and mutant immunoglobulin polypetides selected for the desired specificity. RNA can fold into stereospecific objects; hence, randomized RNA molecules can be directly selected and replicated with reverse transcriptase.

Human Genome Project

With the availability of all of these tools, the image has firmed of establishing the complete DNA sequence of the human genome. As a scientific objective, this is uncontroversial. The controversy pertains to the primacy given to the staging of the effort. Should it be a once and for all technological production, mindless of the ancillary interest in some genes or DNA tracts compared to others? Does it need to be a centralized project, administered top-down with the trappings (and political appeal) of other Big Science? Or can it be left to the cumulative efforts of hundreds or thousands of laboratories, each digging more deeply at some features of the terrain, and intent on going much further than establishing a sequence of bases? In fact, we are seeing the emergence of constructive compromise among these visions; and at the same time the technologies of mapping and sequencing are advancing to where the costs of a unified project need no longer prejudice more individualized efforts.

In any case, sequence information is but the beginning of more intensive inquiry into the polymorphisms, regulatory factors and gene functions associated with any DNA segment.

DNA AS A HELIX

Higher Orders of Organization

The visible chromosome is a packaging of DNA, histones, and accessory proteins three or four orders of coiling beyond the double helix. Cytological observation leaves no doubt that the morphological expression of the chromosome reflects functional allocation of different genes; but we are at the mere beginning of understanding.

Gene Regulation and Morphogenesis

The basic outlines of the central dogma now consensually agreed upon, the core challenge of molecular biology has been the path from the gene to the organism. Given that, to some approximation, each somatic cell has the identical genotype, (1) how is gene expression differentially modulated, and (2) how is this transmitted in cell lineages?

A multitude of DNA-binding proteins have been found which do modulate gene expression: transcriptional regulators. As a three-dimensional interaction, protein binding is fully sensitive to three-dimensional shape and the major and minor grooves of the double helix, as well as the base sequences contained therein. In addition, if not in consequence of bound proteins, some tracts of DNA are methylated shortly after DNA replication, in ways correlated with gene activation.

How these properties are locally transmitted remains a matter of speculation, but may well be bound up with local methylation.

DNA Supercoiling; Topoisomerases; Other Conformations

The standard double helix exhibits a pitch of about 10 base pairs per complete turn. If nothing else, the processes of replication and transcription would entail the unraveling and rewinding of the helices: this is the task of enzymes generically called topoisomerases. These can transiently cut single strands to permit the relief of torsional stress, then rejoin them. In its natural habitat, DNA is often found in states of positive or negative supercoiling, often correlated with maintained gene expression. In addition, many cytotoxic and cancer chemotherapeutic agents seem to be topoisomerase inhibitors, and most owe some of their specificity to the momentary DNA-supercoil status of a given cell. It is particularly intriguing that environmental signals can modulate that status, often by regulating the production of the various topoisomerases.

At least *in vitro*, DNA can undergo a spontaneous transition to a totally different, kinked and left-handed conformation called Z-DNA. Tracts rich in G-C pairs are especially prone to this shift. The importance of Z-DNA *in vivo* is hotly contested.

DNA conformations plainly confer different chemical reactivity on the bases, a principle exploited by the footprinting methods used to study conformation. This must have some implications for localized chemical mutagenesis—a matter not yet systematically studied.

TRIUMPH OF MECHANISM

The dominion of the DNA paradigm has been the triumph of mechanistic interpretation in twentieth-century biology. It is sometimes remarked that human personality is nothing but the individual's 3 billion base pairs—an assertion that fascinates some, terrifies others, and has much to do with the debate about the Human Genome Project. If we could believe that existing genotypes had achieved more than a tiny fraction of the human potential—in culture, in intellect, in compassion, in a sane ordering of affairs—we could elevate the genome to that pedestal of nemesis. On the other hand, we do know that many, probably most, individuals labor under some potentially remediable burden of hereditary origin. As much to understand the better nurturing of human development, a *euphenics*, as to intervene in genetic constitution, *eugenics*, it does behoove us to learn all we can about genetic polymorphisms and their impact on human health and capability. It is particularly important to distinguish interventions in germ cells from those in the somatic cells, and to communicate that it is only the latter that are intended to be the targets of the new gene therapies.

SELECTED READINGS

It would be a precious exercise to provide specific documentation for every detail of this commentary; it would be both arduous and redundant—many single points would deserve a library. The up-to-date detail can be found in standard texts of molecular biology (a few are listed) and in the volumes of *Annual Review of Biochemistry*. The leading historical monographs on DNA are also listed.

Molecular Biology

ALBERTS, B., D. BRAY, J. LEWIS, M. RAFF, K. ROBERTS & J. D. WATSON. 1989. Molecular Biology of the Cell. Garland Publishing. New York.
BERG, P. & M. SINGER. 1992. Dealing With Genes: The Language of Heredity. University Science Books. Mill Valley, CA.
DARNELL, J. E., H. F. LODISH & D. BALTIMORE. 1990. Molecular Cell Biology. Scientific American Books. New York.
KORNBERG, A. & T. A. BAKER. 1992. DNA Replication, 2nd ed. W. H. Freeman. New York.
STRYER, L. 1988. Biochemistry. W. H. Freeman. New York.

WATSON, J. D., N. H. HOPKINS, J. W. ROBERTS, J. A. STEITZ, A. M. WEINER. 1987. Molecular Biology of the Gene. Benjamin/Cummings. Menlo Park, CA.
WELLS, R. D. & S. C. HARVEY, Eds. 1988. Unusual DNA Structures. Springer-Verlag. New York.

History of DNA

CRICK, F. H. C. 1988. What Mad Pursuit: A Personal View of Scientific Discovery. Basic Books, New York.
DUBOS, R. J. 1976. The Professor, the Institute, and DNA. Rockefeller University Press. New York.
FRUTON, J. S. 1972. Molecules and Life. Wiley-Interscience. New York.
JUDSON, H. F. 1979. The Eighth Day of Creation. Simon & Schuster. New York.
LEDERBERG, J. 1987. Genetic recombination in bacteria: A discovery account. Annu. Rev. Genet. **21**: 23–46.
MCCARTY, M. 1985. The Transforming Principle. W.W. Norton. New York.
OLBY, R. C. 1974. The Path to the Double Helix. University of Washington Press. Seattle.
PORTUGAL, F. H. & J. S. COHEN. 1977. A Century of DNA. MIT Press. Cambridge, MA.
SAYRE, A. 1975. Rosalind Franklin and DNA. W.W. Norton. New York.
WATSON, J. D. 1968. The Double Helix. Atheneum. New York.
WATSON, J. D. & J. TOOZE. 1981. The DNA Story. W.H. Freeman. San Francisco.

REFERENCES

1. J. D. WATSON & F. H. C. CRICK. 1953. Molecular structure of nucleic acids: A structure for deoxyribose nucleic acid. Nature **171**: 737–738.
2. J. D. WATSON & F. H. C. CRICK. 1953. Genetic implications of the structure of deoxyribonucleic acid. Nature **171**: 964–967.
3. CASKEY, C. T. 1993. Molecular medicine: A spin-off from the helix. JAMA **269**: 1986–1992.
4. PEI, D., H. D. ULRICH & P. G. SCHULTZ. 1991. A combinatorial approach toward DNA recognition. Science **253**: 1408–1411.

Values from a Chicago Upbringing[a]

JAMES D. WATSON

Cold Spring Harbor Laboratory
Bungtown Road
Cold Spring Harbor, New York 11724

When this conference marking forty years of the double helix was proposed to be held in Chicago, I knew I should accept. Emotionally and intellectually I was more formed by Chicago than by experiences anywhere else. Outsiders often sneer at Chicago, but I know otherwise. There, in 1928, I was born into a family with three paramount values. One was the importance of books and the belief that knowledge would liberate mankind from superstition, which for my father, brought up to be an Episcopalian, meant religion. The second value was birds. From his adolescence, my father was addicted to observing birds, and later I happily joined him, knowing that it liberated me from the Sunday services of my mother's church. By the time I entered South Shore High School, I also was obsessive about trying to find rare birds in Jackson Park, around Wolf Lake, or out in the Indiana sand dunes at Tremont. It gave romance to my life and more than compensated for being even shorter than my finally 5'2" sister, Betty, two years my junior. Our third family value was the nobility of the Democratic Party, led by my first real hero, Franklin Delano Roosevelt. Then if you had a family car, you could afford to be a Republican, but if you had been knocked down by the Depression, common sense made you a Democrat.

Beginning when I was about 12, my father and I every Friday evening made the mile-long walk to the library on 73rd Street to browse among its stacks and invariably bringing several home to digest during the following week. Our house was also filled with books, the more recent of which came from the Book of the Month Club, but most of which came from the used book shops in the Loop or Hyde Park. Dad worshiped persons of reason and took particular pleasure in reading the thoughts of the great philosophers. His library also had the occasional book on science, and it was those books, not the ones on philosophy, which I pored over when the weather was too unpleasant for bird watching. Learning about evolution particularly caught my fancy, with Darwin's theory of natural selection providing a rational way to think about the diverse forms of life that first excited me through trips to the Field Museum.

[a] Copyright © James D. Watson, 1994.

Even before I entered the University of Chicago in 1943, I had begun to daydream about being a scientist, although I had to wonder whether I was bright enough to enter this world filled with geniuses. All too clearly I had not entered the University of Chicago at 15 because of my high IQ. I was far from the child wonder that, say, Wally Gilbert, several years later, grew up to be in Washington. My premature departure from South Shore High School instead reflected the fact that Robert Hutchins, still the almost boy president of the University of Chicago, considered American high schools disasters that never could be reformed. And instead of wasting money failing to improve them, he had the simple solution of getting kids into college two years earlier. That I was one of the first entrants into Hutchins' "four-year college" owed much to my Southside-Irish-raised mother, who had gone to the nearby University of Chicago. It was she who saw that I filled out the application form for a tuition scholarship which later let me attend college, initially needing only from my family the two 3-cent fares for the daily streetcar rides of some 30 minutes.

My first two years at the U of C were superficially not very successful, with my largely B grades continuing to expose my non-genius qualities. But they prepared me for the future by instilling upon me three new values. The first was to focus on original sources instead of textbooks—read the great books themselves, not the interpretations of others. The second value was the importance of theory. Of course, you have to know some facts, but much more important is how to put them together in some rational scheme. And thirdly, you had to concentrate on learning how to think as opposed to improving memorization skills. Initially, to my annoyance, the big comprehensive exams that gave us our grades for the entire year often seemed to bear no relation to what you learned in your lectures. With time I realized that I did not have to take notes, but instead could concentrate on whether the lecturer's words actually made sense. In retrospect, I now realize I was acquiring the mental habits which later made me acceptable first to Luria and Delbrück and later to Francis Crick.

And Chicago being then the Second City with its *Chicago Tribune* and the University of Chicago not as old as Harvard, I saw no reason to treat authority with much reverence. You were never held back by manners, and crap was best called crap. Offending somebody was always preferable to avoiding the truth, though such bluntness did not make me a social success with most of my classmates. Thus it was lucky that for much of my college life, I was still too short to see the need to effectively move outside the security of my family home. But being honest about what is bad and false leads nowhere, unless you hold equally strong values about what is good. To escape from the false leads of the present, you have to have good advice as to where science should go. Happily, my first scientific hero was the famed U of C geneticist, Sewell Wright, whose course on physiological genetics I audited in the spring of 1946.

The gene had suddenly come to the forefront of my attention several months before through reading *What Is Life* by the Austrian theoretical physicist, Erwin Schrödinger. Soon after its publication in the States, I spotted this slim book in the biology library, and upon reading it was never the same. The gene's being the essence of life was clearly a more important objective of study than how birds migrate, the scientific topic that previously I could not learn enough about. So I considered myself very fortunate that, as a Zoology major, I had the opportunity three times a week to listen to one of the world's best geneticists, whose interests ranged from how genes work to the mathematical foundations of evolutionary genetics. Even though Sewell Wright was not an inspired teacher, using notes written on $3'' \times 5''$ file cards, I never wanted his lectures to end. By the term's end, I had made the decision to have the study of the gene as my life's principal work.

Initially I very much wanted to move on to Caltech as a graduate student since it, more than any other major American university, had a biology department dominated by geneticists. In addition, I was aware that its leading chemist, the already very celebrated Linus Pauling, was interested in the structure of the macromolecules found in cells. But they rejected me, and I went to my second choice, Indiana University, which then stood out because the year before it had added to its faculty Herman J. Muller, perhaps the only American geneticist with intellectual credentials superior to those of Sewell Wright. In addition, I was told that two of its younger biologists, Tracy Sonneborn and Salvador Luria, were both very clever and I might want to do my thesis work with them. Very soon after my arrival, I began to realize how lucky I was that Caltech had turned me down. There I would have felt myself intellectually inferior to many of my peers with much better backgrounds in physics, math, and chemistry. But at Indiana, I was the only incoming Ph.D. student who already had started to think about genes. Moreover, it very quickly appeared that the rigor of my Chicago background made it easier for me than for my classmates to focus on the big picture as opposed to details that go nowhere.

So I now feel very warm toward the city of Chicago and its great university on the Midway that gave me not only the intellectual goal that has so dominated the remainder of my life, but also the intellectual obstacle course that its great books demanded I run. Of course, I would never have come to the double helix without the subsequent testing of many other intellectual waters. But it was Chicago that first taught me that I had to be different than others if later I was to succeed.

Already, by 20 years of age, I knew that if I did what should be done to promote knowledge, I was bound to upset those with different values. Furthermore, there would exist many future situations where compromise between divergent views is not the way to proceed. Some ways to move ahead are clearly bad and little is gained by compromising on important issues with those whom you have no reason to respect. Needless to say, there is no point in being obstinate on minor issues, nor does it make sense to have more than one real fight at a time. But on the plus side, I have generally found that at least in the areas of pure science, it has not been hard to distinguish between the good and bad guys. So I have passed my almost 50 years in science more dominated by either elation or boredom than from the anxiety that comes when you are not sure that you are following the right path into the future.

The time ahead, however, may prove much harder to control to our liking as genetics more and more enters into the public arena. Our growing ability to unscramble human genetic destinies will increasingly have an impact on how humans view themselves and justify their behavior toward others. Our children will more be seen not as expressions of God's will, but as results of the uncontrollable throw of genetic dice that do not always give us the results we want. At the same time, we will increasingly have the power, through prenatal diagnosis, to spot the good throws and to consider discarding through abortion the bad ones. But to so proceed flies in the face of the long-cherished idea that all human life is sacred and intrinsically worthwhile. So there is bound to be deep conflict between those persons who want to maintain revered values of the past and those individuals who wish to have their moral values reflect the world as now revealed by observations and experiments of modern science. In particular, we are increasingly going to be accused of unwisely "playing God" when we use genetics to improve the quality of either current or future human life. Partly these accusations reflect the objections of individuals who don't think we have the right to do "God's" work. But I also sense that sometimes the uneasiness comes from the fear that we might someday use genetic procedures in Hitler-like ways, using our scientific powers to further discriminate against unpopular political and racial groups.

But diabolical as Hitler was, and I don't want to minimize the evil he perpetuated using false genetic arguments, we should not be held in hostage to his awful past. For the genetic dice will continue to inflict cruel fates on all too many individuals and their families who do not deserve this damnation. Decency demands that someone must rescue them from genetic hells. If we don't play God, who will?

DNA: A Cooperative Discovery[a]

FRANCIS H. C. CRICK
The Salk Institute
10010 N. Torrey Pines Road
La Jolla, California 92037

It is surely obvious that it is the DNA molecule (and its RNA sister) that should be honored rather than the midwives who delivered it to the light of day. But DNA, wonderful as it is, cannot by itself make an after-dinner speech, so Jim and I must do our best on its behalf.

I need hardly tell you that it is the base-pairing which is its key feature, not merely for its biological importance, but also because it is at the root of the enormously powerful set of methods we group together under the headings "recombinant DNA" and "DNA sequencing." It is the invention of these techniques—the second DNA revolution—that has powered both genetic engineering and the Human Genome Project. But rather than sing the praises of the nucleic acids, I shall instead mention briefly some of those whose ideas or experimental work laid the essential foundations for the discovery of the double helix in 1953.

First and foremost, I should remind you of Rosalind Franklin, whose contributions have not been sufficiently acknowledged in these meetings on the fortieth anniversary of the discovery. It was Rosalind who clearly showed the existence of the two forms of DNA—the A and the B form. It was Rosalind who painstakingly determined the density, the exact cell dimensions, and the symmetry of the A form, evidence that suggested very strongly that the structure had two chains (not just one), running in opposite directions. And we must also salute Maurice Wilkins, not only for starting experiments on the structure of DNA at King's College, London, but also for the many years of painstaking work, after Rosalind left King's, which showed that the X-ray diffraction pictures of DNA fibers did indeed fit variants of the double-helical model. Whether Maurice would have embarked on DNA without the pioneer work of Bill Astbury at Leeds, I don't know. Astbury was an adventurous but not a meticulous experimenter, as well as being a sloppy model builder, but he was right in attributing the 3.4Å meridional reflection to the stacking of the bases.

On the chemical side it would not have been possible to build correct models without the general chemical formula for DNA, established largely by

---

[a] Owing to Dr. Crick's illness at the time of the banquet this speech was not delivered there.

the work of Lord Todd and his colleagues. But perhaps the most vital information was provided by the careful work of Erwin Chargaff, which led him to his rule for the relative amounts of the four bases. Equally fundamental was the paper by Avery, MacLeod, and McCarty showing that the transforming factor of pneumococcus was almost certainly made of DNA. Another important clue was provided by Gulland, whose work on the titration curve of DNA suggested, but did not quite prove, that DNA in solution was held together by hydrogen bonds.

The whole approach that Jim and I used – combining rather inadequate X-ray data with the restrictions to model building imposed by the chemistry – was largely inspired by Linus Pauling, who not only gave us the very pertinent example of the α-helix, but also whose group had established many of the key distances and angles needed to build accurate models.

All these workers, among many others, provided the many pertinent bits of information that Jim and I needed to guess the structure. We merely put the spark of an idea to the confusing assembly of facts and speculations. So when we salute the DNA molecule – and how much more we know about it today – we can see that the discovery of the double helix was possible because so many good scientists had provided vital information about different aspects of it. It is this cooperative aspect of science that I would like to stress, even though, inevitably, there is bound to be some degree of competition between groups working in exactly the same scientific niche.

Fortunately the DNA molecule knows nothing of all this. It has just been sitting there, doing its stuff, for almost 4 billion years, waiting to be discovered. May it long prosper!

DNA at King's College, London[a]

MAURICE H. F. WILKINS

Professor Emeritus
King's College, Strand
London WC2R 2LS, England

In the *physics* department at King's College, London, we contributed to one of the most important *biological* advances of the century. Let me explain how this came about. The Department had been involved in very many outstanding advances in science—first with Wheatstone and Clerk Maxwell, and then with three Nobel Laureates this century.

Physics had helped greatly in special areas of biology, such as the study of nerves and muscle; and chemistry (which is a physical science) led to the very important subject of *bio*chemistry. Also, biologists created genetics, which is a quantitative science that deals with the ultimate units of inheritance and is like physics, which is a quantitative science that deals with the ultimate units of matter.

In the 1930s the great Niels Bohr recognized the growing links between physics and biology and organized meetings to explore those links. Special attention was given to genetics and X-ray diffraction.

In 1946 John Randall became Head of Physics at King's College, London. He was famous for inventing (with Boot) the cavity magnetron, which enormously increased the power of radar and, it has often been claimed, was the most decisive scientific contribution to winning the War. Randall had been influenced by scientists who had been at Bohr's meetings, such as the X-ray diffractionists Bernal and Astbury and the geneticist C. D. Darlington, who had a special interest in DNA and proteins. Randall decided to turn over a large part of his physics department to the study of biological problems. The result was that Randall's laboratory—together with the Cambridge protein X-ray lab of Perutz and Kendrew—became one of the main routes to the DNA double helix.

The other main route also connected with Bohr. Max Delbrück, a theoretical physicist, was much influenced by Bohr, and after the war Delbrück became the leader of an important new group of geneticists in the United States who stimulated Jim Watson to take up the challenge of gene structure.

Randall was a leader of much vision and energy. He planned to study

---

[a] Talk like that at the King's 1993 DNA Commemoration when A. R. Stokes, H. R. Wilson, and R. G. Gosling (Rosalind Franklin's coworker) also spoke.

biology with all kinds of physical methods—such as electron microscopes, ultraviolet and infrared, spectroscopy, and ultrasound—and the Medical Research Council and Rockefeller Foundation were farsighted in giving him the cash to employ biologists and biochemists to work alongside the physicists at King's College.

As early as 1941 one of Darlington's geneticists had suggested that Randall should use X-rays to study DNA in sperm heads, but the MRC wanted us to study the structure of *cells* and not the structure of molecules (such as DNA). But when we had built new ultraviolet and polarized light microscopes to study cells, I began to use these microscopes to study the structure of DNA. Actually, most scientists then believed that genes were made of protein and *not* DNA, but we were lucky and knew Dermott Taylor in the King's College biochemistry department; he had worked in New York close to Oswald Avery, who had shown by genetic transformation that genes were made of DNA and *not* protein. Taylor told us to trust Avery.

In early 1950, Signer, the Swiss expert on extracting DNA from living cells, came to London with some of his very high-quality DNA and distributed it to research workers who wanted to study it. That was a generous act in the best tradition of science! I found his DNA was unique because, after being wetted, it could be pulled into thin spider's-web-like fibers of remarkable uniformity with the molecules beautifully aligned. I felt that these fibers might have a very regular structure and would give sharp X-ray diffraction patterns. Raymond Gosling was then a research student and the only person in King's using X-ray diffraction. He had begun to study DNA in sperm heads, which Randall was investigating in the electron microscope. Gosling agreed we should try to get diffraction patterns from the DNA fibers, even though they were extremely thin. I found it possible to mount fibers all parallel on a wire frame so that they formed a bundle to put in the X-ray beam.

Randall had special knowledge of weak diffraction and told us that X-rays scattered by the air in the camera might obscure the DNA diffraction so we passed hydrogen through the camera. Sealing it up properly was not possible, but a condom helped in an awkward place. We could find no better camera and just pushed on. But we did take special care of the DNA and treated it like a living substance, keeping the hydrogen moist so the DNA would not dry out. The result was a much sharper and crystalline diffraction pattern than anyone had obtained before. Astbury's patterns had given strong indications of crystallinity, but now it was really obvious—genes had a crystalline structure!

Gosling measured the positions of all the spots on the pattern. There was a repeat every 2.8 nm along the length of the molecule. The molecules packed together like cylinders of 2.0-nm diameter and there was strong diffraction 0.3–0.4 nm along the length of the molecule.

Alex Stokes noticed that apart from the 0.3–0.4-nm diffraction there was nothing else along the length, and he pointed out that that was a sign that

DNA was a helix. That conclusion fitted in well with the molecules' packing together like cylinders. This was encouraging, but no one could tell us how to find the structure of something as big and complicated as DNA.

However, one thing was clear: we must get sharper patterns and obviously we should use X-ray equipment specially designed for fine fibers. The physics department in 1950 was strong on instrumentation and of course it has continued to develop that tradition. We thought carefully about what kind of equipment would be needed. We were then very lucky to find that Ehrenburg at Bernal's lab at Birkbeck College had designed a new micro-focus X-ray tube, which was just what we wanted and he very kindly gave us one! When we fitted a micro-camera on it we hoped to get patterns from a single DNA fiber ($\sim 1/10$ mm in diameter).

Before we had got far with setting up the new equipment, Rosalind Franklin arrived and, since she was an experienced experimental X-ray worker, it was agreed she and Gosling would complete and try out the new equipment. I continued studying with microscopes how DNA fibers absorbed water and swelled and lengthened and how the flat bases tilted and slid over each other as the length of the molecule altered. I also made some rough tests with the old equipment to see whether DNA had the same structure in all kinds of living things. That certainly seemed to be so, supporting the idea that DNA was the universal gene material. Since only the Signer DNA gave fibers, I used smeared lumps of DNA gel. At the center of the patterns was a diffuse X and Stokes pointed out that that would be given by helical DNA with an angle of ascent (like a spiral staircase) of about 40°. We were encouraged and thought the helical hypothesis might greatly help us to find the structure.

Fairly soon Rosalind Franklin and Raymond Gosling finished setting up the fine-focus X-ray equipment and linked it to a micro-camera. We expected the equipment would give better results, but what they achieved exceeded our hopes. There were no leaks with the camera, the water content of the fibers rose higher than before, and they found an exciting new diffraction pattern. They called it B and they sharpened the crystalline pattern too (and called it A).

Meanwhile I had been reading (like Jim Watson) Linus Pauling's new publication on helical proteins. Pauling led the world in applying quantum mechanics to chemical bonding and he knew more than anyone else what the exact lengths and angles were in chemical bonds. He built a very precise molecular model of what he called the α-helix. But I was puzzled that this very brilliant man had no proper way of calculating the X-ray diffraction from his model and therefore no proper way of testing it.

I thought that there must be some simple way of calculating diffraction from helices and if anyone could do it, it was Alex Stokes – he was brilliant at linking maths and physics and he claimed that he worked it all out on the train between London and Welwyn Garden City where he lived. (Later on, or maybe I should say about the same time, Francis Crick worked it out too.)

Stokes used Bessel functions and called his diagram "Waves at Bessel-on-Sea" and I stuck it on the laboratory notice board.

The marvellous new B pattern of DNA had a central helical cross (like the one on my rough patterns), but now the cross was sharply divided into streaks corresponding to the pitch of the helix. We were very excited by the agreement between the calculated diffraction from a helix and the new B pattern, but Rosalind Franklin was more cautious about the helical idea.

Several of us at King's thought (as Jim Watson and Francis Crick did) that we should assume that DNA was a helix and build Pauling-type models of DNA. Rosalind Franklin agreed that Pauling was a great genius, but she thought, like many professional X-ray workers, that most scientists were not geniuses like Pauling and would make more progress by sticking to sound, well-established methods. In line with that, she had begun with Raymond a systematic study of DNA diffraction by measuring all the many individual diffraction beams in the very detailed A pattern. They developed new methods and did a very good job, which was essential in the longer term for arriving at the exact model, but it did not lead quickly to the structure. To do that we needed *brilliant short cuts* and *faith* in the helical hypothesis, which was, of course, the way Watson and Crick worked.

But late in 1951 we at King's were moving very much in the right direction. Rosalind Franklin thought about how DNA absorbed water and swelled and argued that the phosphate groups must be on the *outside* of the molecule. In contrast, Pauling in 1953 and Watson and Crick in 1951, built helical models with phosphate groups *inside* the molecule. That was going in quite the wrong direction.

At King's very late in 1951 (but a year and a half before the double helix discovery), Bruce Fraser, a physics research student in Bill Price's spectroscopy group, built a DNA model that showed very well the general way we were thinking in our laboratory. Bill Price knew a great deal about chemical bonds and energies. That knowledge helped Fraser to build his model. It was helical, with phosphates on the outside and the flat bases stacked inside. The bases on one chain were hydrogen-bonded to those on the other chains. The model had most of the features of the double helix, except, of course, that it had *three* chains instead of *two* and, because of that, could not have the very special and unique system of basepairs, which was such an impressive feature of Watson and Crick's model.

The reason why the Fraser model had three chains was that we at King's thought that density and water content data on DNA meant that there were three chains. We barely considered the possibility there were only *two* chains. We were not alone in making that mistake, both Pauling and Astbury falling into the same physical chemistry trap.

The Fraser model looked very promising, but because it had three chains, it could not be adjusted to fit the X-ray data. That was very disappointing.

There was also the difficulty that we had a shortage of good DNA. I went to Switzerland to see Signer, but he had stopped making DNA, and material from other workers failed to give good patterns.

Even so I obtained a good helical B pattern from sperm heads of cuttle fish (very strange creatures!). I sent a sketch of the helical features in a letter to Francis Crick. (Jim Watson, writing to Delbrück, put three exclamation marks after that news item!)

But, on the whole, King's did not move much closer to the double helix during 1952. I would like to mention, however, that Rosalind Franklin's notebooks show that on February 23, 1953 she began seriously to interpret the B pattern in terms of helices and was beginning to consider a *two*-chain helix. Also, in January 1953 I had begun to think that the bases in DNA must be arranged in planar pairs hydrogen-bonded together, but I had not worked out precise systems of bonds. Altogether our laboratory had, I think, a fairly long way to go before it could reach the Watson-Crick double helix. Their wonderful model was completed by March 12, 1953. Its ability to explain how genes can copy themselves shook the scientific world.

After Watson and Crick had finished their model, the important task remained of proving that the structure was correct. To do that we had to improve our patterns considerably and adjust the model exactly so that it fitted precise X-ray data. Professors Fuller and Arnott and Drs. Marvin, Langridge, and Spencer took leading parts in that considerable task.

"The Night Before Crickmas": A Poem and Deliverance

ROLLIN D. HOTCHKISS

Professor Emeritus
The Rockefeller University
New York, New York 10021

We have had the pleasure of celebrating the insights of forty and fifty years ago, which set us on the route to the rich findings that have been presented and to some exciting visions of the future. In thanking Dr. Donald Chambers, his colleagues and helpers, and the New York Academy of Sciences' staff, for the brilliant design and implementation of this most successful symposium, I'd like to recall to you that the organizers' original plans had included other pioneers of those early days. The participation of Francis Crick, Linus Pauling, Maurice Wilkins, and Erwin Chargaff had been sought, and only unforeseen difficulties—some of them arising only in the last few weeks—prevented their being with us. I know that they would have had a great deal to share with us, and would also have enjoyed our recognition of their contributions. Perhaps too, Maurice Wilkins would have brought some remembrance of his colleague, the late Rosalind Franklin, whose work and spirit are part of what we honor at this celebration.

There are other anniversaries than those of discoveries—and on one marking a birthday of Francis Crick, sometime late in the '60s, at Cold Spring Harbor, Jim Watson and friends celebrated by gathering there a lot of his colleagues. On that sunny June day, Crick, co-inventor of the double helix model, was challenged with an alternative model. After graciously and enthusiastically welcoming it, he signaled that he did not feel obliged in the least to alter his basic philosophy.

On that same occasion I presented Francis and the group with a short reflection of my own on the DNA helix development, which I now propose to share with you. I must warn biophysicists that these reflections do not follow the usual pattern of making equal angles with the perpendicular. In fact, it can perhaps be said that my reflections came at an angle rather oblique to the normal. Nevertheless, you will probably perceive, and I assure you, that they were an affectionate tribute to several of these colleagues that we miss here today.

Like all scientists, I based my contribution on earlier ones; I'll not quote here the obvious sources like *Nature*—known to every schoolchild—but duty obliges me to cite a more obscure source, a bit of American folklore entitled "A Visit from Saint Nicholas," which begins, "'Twas the night before Christmas and all through the house. . . ."

Merry Crickmas

T'was the night before Crickmas, and all through the colleges
    Not a scientist was thinking–they were only adding to knowledges
Save Monod in his think-cap, explaining induction for the ages
    Without a mutation in the whole forty pages–
When out of X-ray there arose such a clatter:
    A new explanation of the arrangement of matter.

It explained to Franklin the sense of her data
    Which she would have seen too, sooner or later,
And aroused Maurice Wilkins to earnest debates
    On the need to have more data updates.

–Well, in seeking to balance the base pair budget
    And predict some sure facts without having to fudge it,
They almost did pull an *il faut ne pas* gaffe[a]
    By rifling the pockets of Erwin Chargaff,
Who promptly responded in jest and in choler
    As on the fence, off-the-cuff, writer and scholar.

Also sifting these fragments of jetsam and flotsam
    Was that wraith-like young drifter, J. Dewey Watson–
Who translated Crick's lore of X-ray diffraction
    Into everyday language to our great satisfaction.

So–the theory caught on, among those interested,
    Who forgot for some time it had still to be tested.
By the time it had come, more or less, to be proved,
    Our young crystallographer was theoretically grooved

---

[a] I am sure someone with a tidy mind will bristle at the apparent infraction of school book French usage: the "*il faut ne pas.* . . ." (when *il ne faut pas* is satisfactory for "one mustn't . . ."!) Ring it up to poetic license–or "all gall, but not at all Gallic"–or my willful occupation of a French island colony in a dream world–BUT I ask the privilege: my own hearing of the line insists on a broad "il faut. . . .", a pause, then a shift to a savage, eyes-narrowed "*ne pas*!", squeezed out through tensely clenched teeth! [This *gaffe* is . . . something that one must . . . just *never* do!]
(Neither my poetic, nor my dramatic, license has been renewed, by the way.)

R.D.H. May 1994

So he read and he traveled to pick up fresh news
  And so solve the code: choose and pick, pick and choose,
Selecting the best, sometimes hard, sometimes lenient,
  And rejecting the worst—or the plain inconvenient.

So theories, ideas and dogmas were fathered,
  Dividing his colleagues into the hot and the bothered.
But the point of it all, I am glad to relate,
  He told us all how—and it really was great:

Kiss and tell, kiss and tell, the double strands move,
  Then along comes an enzyme in linear groove:
Spread the word, spread the word, to proline or glycine,
  But hold on to the end, and avoid puromycin.
Go ahead, have some fun, shift your frame, kiss and tell:
  You can get it all back maybe, after a spell.

How good that we had someone who could read it
  When all of the rest of us quite failed to heed it!
Merry Crickmas, then!—all of us, we've been lucky, it's true,
  And congratulate all—you and me, me and you,
And toast high to heaven for sending this boon—
  —and let's celebrate Crickmas each forthcoming June!

Photos from the Banquet

Stanley Ikenberry, Joshua Lederberg, Donald Chambers, James Watson, Sir Crispin Tickell, Rodney Nichols.

Rollin Hotchkiss, Walter Gilbert, Eric Kandel.

PHOTOS FROM THE BANQUET

Left: James Watson cutting a DNA "birthday cake" at a party held at the University of Illinois at Chicago.
Above: James Watson signing autographs.

Gunther Stent and François Jacob.

James Watson, Sir Crispin Tickell, Joshua Lederberg, and Rodney Nichols.

PART V. MOLECULAR, CELLULAR, AND INTEGRATIVE BIOLOGY

Introduction

R. JOHN SOLARO[a]

Department of Physiology and Biophysics
College of Medicine
University of Illinois-Chicago
Chicago, Illinois 60612-7342

Obviously, it is a great honor to be asked to introduce such distinguished scientists. The contributions of Drs. Tonegawa, Jacob, and Nirenberg that led to their Nobel prizes need not be restated here. What is more interesting to today's scientists are the following questions: What are they thinking about now? What are they doing now? Fortunately for us, they have responded to these questions in their presentations as part of this celebration of the contributions of Watson and Crick.

As a physiologist working at the interface between molecular, cellular, and integrative biology, I try to understand and make sense of molecular signalling processes in terms of functional behavior of the heart. It is particularly gratifying to see that each of these major figures in the history of DNA and molecular biology is now also thinking in terms of integrated actions of molecules, cells, and organisms. I think there's a lesson here for young scientists trying to peek into what will be "hot" in the next decade.

Dr. Tonegawa provides a clear demonstration of the power of the use of the new biology in understanding how the parts work at the organismic level. The use of the "knockout" mouse to study memory takes us from the realm of the detail of the kinase-substrate-phosphatase paradigm to learning and memory in the context of the whole mouse swimming in a water maze.

Dr. Nirenberg is now also working on the central nervous system. As he alludes to in his presentation, an attraction to neurobiology and a move from the deciphering of genetic code stems from the parallel between the information transfer in both realms of investigation. His work is aimed at understanding the rules by which the central nervous system is constructed. Again, tools of molecular biology are used to approach a fundamental physiological question. Importantly, interpretation of the actions of the gene and its regulators are viewed in the physiological context of dynamics and concentration gradients.

---

[a] Address for correspondence: R. John Solaro, Ph.D., Department of Physiology and Biophysics, College of Medicine (M/C 901), University of Illinois-Chicago, 901 South Wolcott, Chicago, Illinois 60612-7342.

The presentation by Dr. Jacob offers valuable lessons not only in the conduct of one's day-to-day science, but also in facing the task of understanding integrated control mechanisms in eukaryotes. He reminds us of the successes of two colleagues working at the blackboard. The synergism of this thinking in pairs is evident, and emergent ideas are certainly more than the sum of the thinking of the individuals. An analogy is at the level of *trans-* and *cis-*activating factors in gene regulation. Here, integrated activity is certainly more than the sum of each of the actions of the regulators and offers a rich diversity of control mechanisms. This process involves emergent properties of regulation owing to aggregation. The complexes of factors are dubbed by Professor Jacob with the delightful term *aggregulates*.

A recent monograph,[1] *The Logic of Life*, is a series of essays dealing explicitly with the challenge of integrative physiology. I think the book demonstrates provocative ideas for those of us facing the future in biological sciences. The presentations by Drs. Tonegawa, Nirenberg, and Jacob provide excellent examples of how the best thinkers are facing these challenges.

REFERENCE

1. BOYD, C. A. R. & D. NOBLE. 1993. The Logic of Life. Oxford University Press. Oxford, UK.

Mammalian Learning and Memory Studied by Gene Targeting

SUSUMU TONEGAWA
Center for Cancer Research
Massachusetts Institute of Technology
Building E17-353
77 Massachusetts Avenue
Cambridge, Massachusetts 02139-4307

This is a unique and interesting conference, and I wish to extend my congratulations to Jim Watson, Frances Crick, and Maurice Wilkins on the 40th anniversary of the discovery of the double helix. In this brief presentation I would like to focus on recent studies in which we employed a genetic approach to study mammalian learning and memory.

The gene-targeting technology to study learning and memory in mammals was pioneered by Mario Capecchi and others and is revolutionizing many subfields of mammalian biology. Neurobiology or neuroscience should be no exception. In this paper, I will focus on our recent attempt to apply this technique to dissect the cellular and molecular mechanisms underlying mammalian learning and memory. The key phenomenon is the long-term potentiation, or LTP, which is an electrophysiological manifestation of a long-lasting increase in the strength of a neuronal synapse that has been stimulated in an appropriate fashion. The initial phase of LTP induction is relatively clear; LTP is induced by the binding of a neurotransmitter, glutamate, combined with depolarization of the postsynaptic membrane, which leads to an influx of calcium ion. What follows the rise of the calcium concentration is less clear at this time, but various studies suggest that the increased calcium concentration triggers a biochemical cascade that involves calcium/calmodulin–dependent kinase-II, calcium-dependent protein kinase, protein kinase C, and phosphotases, etc. Eventually, the signal is thought to be transmitted to the nucleus of the postsynaptic cell to establish the LTP; that is, new protein synthesis may be required. A role of so-called retrograde messenger which would increase the neurotransmitter release from the presynaptic terminal has also been suggested. Although LTP has been widely studied as a candidate information-storage mechanism, at least for some forms of learning and memory, the actual evidence supporting this notion is not very extensive. The main support for LTP as a memory mechanism is the observation that the pharmacological agents such as AP5 known to block hippocampal NMDA receptors prevent the induction of LTP as well as spatial learning in rodents. However, the problem with this evidence is that blocking NMDA receptors disrupts the synaptic function and potentially interferes with the *in vivo* computational ability of hippocampal circuits. Perhaps the impairment of learning that was observed results not from the deficit in LTP *per se*, but simply from incorrect operation of hippocampal circuits that lack NMDA receptor function.

In order to re-examine the relation between LTP and learning, we focused on kinases that are presumably in the downstream of NMDA receptors in the biochemical cascade triggered by the calcium ion influx. We produced a strain of mouse mutants by knocking out the corresponding gene by the gene-targeting procedure. The first mutant mouse of this program in my laboratory was made by Alcino Silva, who is now at the Cold Spring Harbor Laboratory. In this strain of mutant mouse, the α subtype of CaM kinase II is deleted, and I will mention only the summary of the data obtained with the α CaMKII mutant mice because it is now more than a year since these data were published. The first important conclusion drawn was that LTP in the CA1 region of the hippocampus of the mutant mouse was severely impaired, but no impairment occurred in the ordinary postsynaptic mechanisms such as the NMDA receptor function. The second important message we got was that the capability of hippocampus-dependent spatial learning was severely impaired. So these results supported the notion which had been held by some neuroscientists on the basis of earlier pharmacological studies, namely, that LTP is the cellular basis for some types of learning.

We have continued our program of investigating the molecular mechanism of LTP and LTP's relation to learning and memory by producing a second mutant strain—mice lacking the γ isoform of the calcium and phospholipid-dependent protein kinase C (PKC). PKC γ is one of the nine known PKC isoforms and is expressed specifically in the brain and spinal cord, being particularly abundant in the hippocampus. As far as ordinary synaptic transmission is concerned, we could not distinguish the slices from the hippocampus of the PKC γ mutants and those from the wild-type littermates. However, when we looked at the LTP in the CA1 region of the hippocampus, we saw a greatly diminished LTP in the mutant compared to the wild-type littermates.

We then subjected the PKC γ mutant mice to several different learning tasks. The one I will focus on here is called the Morris water maze, which consists of a circular pool 1.2 meters in diameter and tests the animal's ability to acquire spatial learning. The pool is filled with opaque water and a small platform is submerged just below the surface of the water. The platform location is kept constant throughout the training, but a mouse to be trained is allowed to start swimming from a random location in multiple training sessions. Mice are born swimmers, but they don't like to be in water so they try to find a way to escape from it. However, since the water is opaque they cannot see the platform, so the mouse will initially swim more or less randomly trying to find an exit and run into the platform only accidentally. However, as the training is repeated, a wild-type mouse learns the location of the platform fairly precisely and eventually swims directly to it regardless of the starting site. It has previously been shown that the strategy employed by the mouse is to map the location of the hidden platform using the multiple objects surrounding the pool as landmarks. This learning is termed spatial learning and has been shown to be hippocampus-dependent. If the mouse receives a lesion in the hippocampus it cannot learn this task well. So, what one does is to repeat the trials, record the time required for the mouse to find the platform, and plot it against the number of trials. The PKC γ mutant mice improved their performance as well as the wild-type mice did.

In order to confirm that the PKC γ mutant mice indeed use the spatial strategy to improve the performance we must examine the trained mice by appropriate tests. One test is called the probe trial. In this test we first attempt to train a mouse for a particular location of the platform so that the mouse has a memory of a specific location of the platform that could easily last for at least a few weeks. We then put the mouse back in the cage and remove the platform from the pool. We return the mouse to the pool and let it swim for a fixed period of time, like 60 seconds, and then record the fraction of the 60 seconds spent in each of the four equivalent imaginary quadrants (namely, the training, left, right, and opposite quadrants).

As expected, wild-type mice spend more time in the training quadrant compared to any of the other three quadrants, indicating that they have the spatial memory. The mutant mice also have the spatial memory; they showed clear selectivity to the training platform.

We then subjected the trained mice to a second test, called a platform crossing test. In this test trained mice are again tested in a pool without a platform, but in this case one counts the number of times a trained mouse crosses the site at which the platform was situated during the training session. We then compared this number with the number of times the mouse crossed the equivalent site in each of the three non-training quadrants. Again, the wild-type mice showed clear site-selectivity. Mutant mice also showed site-selectivity, but when we analyzed it statistically, we noticed that there was a slight deficiency in the site-selectivity with mutants. Nevertheless, mutant mice clearly indicated the acquisition of spatial learning.

So we are left to conclude that the PKC γ mutant mice can acquire spatial learning despite the severe deficit in the LTP in the hippocampal CA1 region. This is an apparent contradiction of the findings made with α CaMKII mutant mice that I summarized above. So we then asked what hippocampal synaptic plasticity could be the basis for the spatial learning in the PKC γ mutants. We went back to electrophysiology and found that the second major form of synaptic plasticity, called long-term depression (LTD), is intact in the CA1 re-

TABLE 1. Summary of Data from Mutant and AP5-Treated Mice

Analysis	PKCγ (−/−)	αCaMKII (−/−)	AP5-Treated
LTP (conventionally induced)	−	−	−
LTP (primed)	+	−	Not done
LTD	+	−	−
Learning (spatial/contextual)	+[a]	−	−

[a] Partial impairment.

NOTE: Data are from the following sources: LTP and LTD in PKCγ-mutant mice are from Abeliovich et al.[3,4] LTP in αCaMKII-mutant mice is from Silva et al.[1]; LTD in αCaMKII mice is from Stevens et al.[6] Learning in αCaMKII mice is from Silva et al.[2] AP5 LTP is from Collingridge and Singer[7]; AP5 LTD is from Dudek and Bear[8]; and AP5 learning is from Morris et al.[9]

gion of the PKC γ mutant. LTD is similar to LTP except that it is a long-lasting reduction of synaptic efficacy. Like LTP, LTD depends on NMDA receptors and calcium ion influx and can potentially serve as a synaptic mechanism for learning and memory. However, because of the delay in the discovery, LTD has not been studied as a memory mechanism as much as LTP. In any case, our data indicated that the hippocampal LTD could be the basis for the observed spatial learning by PKC γ mutant mice. However, the matter was more complex.

We also found that if a low-frequency stimulation is given prior to a tetanus, LTP can be induced in PKC γ mutant mice. So this type of LTP, which we call primed LTP, can also serve as the mechanism for spatial learning.

We also went back to the α CaMKII mutant mice in light of these new findings with the PKC γ mutant mice. We found that LTP cannot be induced with α CaMKII mutant slices even when a low-frequency stimulation was given prior to the high-frequency stimulation (i.e., no primed LTP). We also found that LTD is impaired in the α CaMKII mutant slices. TABLE 1 lists three different types of mice (AP5-treated α CaMKII mutant, PKC γ mutant, and normal mice) as well as absence or presence of the three different types of synaptic plasticity (LTP, LTD, and primed LTP) and spatial learning capability. This table shows that LTD and primed LTP correlate better with the spatial learning capability than does the conventional LTP, which is not essential for spatial learning. However, the moderate learning deficit observed in PKC γ mutant mice suggests that conventional LTP contributes to learning. Contextual learning is a broader term which includes spatial learning. We have carried out another set of behavioral experiments in which contextual learning capability was tested using a different paradigm (fear conditioning). The results were very similar to those of the Morris water maze.

We are encouraged by the progress made to date with this new genetic approach. Our goal is to uncover the relationship between a function of a specific gene, that for synaptic plasticity, and learning. We are aware of some shortcomings of this approach, however. For instance, one would like to be able to improve the gene-targeting technology such that a disruption of a specific gene can be induced in an adult animal in response to an appropriate agent (inducible knockout). Another desired improvement would be to restrict the knockout to a particular region of the brain (tissue-specific knockout). These technical improvements are currently actively pursued not only in my laboratory, but also in several other laboratories throughout the world. Considering the number of highly qualified applicants I receive for postdoctoral and graduate studies in this field, I would predict that this new genetic strategy for the analysis of mammalian behaviors will really flourish in the coming years.

REFERENCES

1. SILVA, A., R. PAYLOR, J. M. WEHNER & S. TONEGAWA. 1992. Impaired spatial-learning in α calcium/calmodulin kinase II mutant mice. Science **257**: 206–211.
2. SILVA, A., C. STEVENS, J. WEHNER, S. TONEGAWA & Y. WANG. 1992. Deficient hippocampal long term potentiation in α calcium/calmodulin kinase II mutant mice. Science **257**: 201–206.
3. ABELIOVICH, A., C. CHEN, Y. GODA, A. J. SILVA, C. F. STEVENS & S. TONEGAWA. 1993. Modified hippocampal long term potentiation in PKCγ mutant mice. Cell **75**: 1253–1262.
4. ABELIOVICH, A., R. PAYLOR, C. CHEN, J. J. KIM, J. M. WEHNER & S. TONEGAWA. 1993. PKCγ mutant mice exhibit moderate deficits in contextual learning. Cell **75**: 1263–1271.
5. GRANT, S. G., T. J. O'DELL, K. A. KARL, P. L. STEIN, P. SORIANO & E. R. KANDEL. 1992. Impaired long-term potentiation, spatial learning, and hippocampal development in *fyn* mutant mice. Science **258**: 1903–1910.
6. STEVENS, C. F., S. TONEGAWA & Y. WANG. 1994. The role of calcium-calmodulin kinase II in three forms of synaptic plasticity. Curr. Biol. **4**(8): 687–693.
7. COLLINGRIDGE, G. L. & W. SINGER. 1990. Excitatory amino acids and synaptic plasticity. Trends Pharmacol. Sci. **11**: 290–296.
8. DUDEK, S. M. & M. F. BEAR. 1992. Homosynaptic long-term depression in area CA1 of hippocampus and effects of N-methyl-D-aspartate receptor blockade. Proc. Natl. Acad. Sci. USA **89**: 4363–4367.
9. MORRIS, R. G. M., E. ANDERSON, G. S. LYNCH & M. BAUDRY. 1986. Selective impairment of learning and blockade of long-term potentiation by an *N*-methyl-D-aspartate receptor antagonist AP5. Nature **319**: 774–776.

Circuits

FRANÇOIS JACOB
Institut Pasteur
25, rue du Dr. Roux
75724-Paris Cedex 15, France

PAIRS ARE MORE THAN 1+1

At the beginning of molecular biology, much of the work was done by pairs, or couples, or duets of scientists—suffice it to mention George Beadle and Edward Tatum, Salvatore Luria and Max Delbrück, Max Perutz and John Kendrew, James Watson and Francis Crick, François Jacob and Jacques Monod, Matthew Meselson and Franklin Stahl, among others. The papers signed by each of these pairs corresponded to some specific steps in the development of molecular biology. Why was the role of pairs so important then? Why was this time particularly favorable to the coupling of scientists? It is probably not only due to the interdisciplinary character of molecular biology; to the fact that, in many instances, experiments involved the use of various techniques issued from different fields; nor to the complexity of experiments that were easier to perform in cooperation than in isolation.

It seems to me that it was on the theoretical rather than the experimental side of biology that pairs could display their talents and efficiency. At the beginning of a science, when the landscape is still fuzzy and open, more opportunities are offered to theory-making, to models. And theories or models are less easy to concoct from internal monologues of isolated scientists than from dialogues between two different minds accustomed to cooperate, to argue, to criticize one another, to confront two different ways of looking at the world; in short to work in common or against each other. Not to mention the obvious fact that it is much more fun to do the work between the two rather than alone. Ideas spring faster. They bounce on the partner. They develop by branching out like trees. Phantasms are more easily nipped in the bud.

Actually, the work takes a completely different turn when performed by a pair. Very soon, some rules specific for that pair and that game become established. New words are coined. A particular vocabulary comes into use. Quite often, during some especially hot discussion, when the dialogue speeds up like a table tennis party, the excitement increases to such a point that each of the protagonists begins to answer before the other finishes his sentences so that often other people who happen to attend such sessions can no longer follow the argument.

PAIRS AT THE INSTITUT PASTEUR

At the Institut Pasteur, I became a member of two successive pairs. One, with Elie Wollman, to investigate lysogeny and bacterial conjugation; and the other with Jacques Monod, to analyze the lactose (Lac) system of *Escherichia coli*, on which Jacques had been working for many years. When looking back at this period, the aspects of the research which immediately stand up are the results that came as surprises. The world is geared in such a way that human beings cannot know what they are the most interested in, that is: what is going to happen tomorrow. This, of course, applies to science. At any time, scientists can predict what will happen in the next five years, for example. This prognostication, however, is the least interesting part of science. The really interesting aspect is precisely what we cannot predict, what will come as a surprise, most often from some unknown people, in some unexpected place. To be exciting, science has to contain some dose of uncertainty. In most instances, experiments are planned and their outcome is predicted on a basis of probability. It is certainly very pleasant to see the results coming out as expected. Yet in many instances such a result does not tell much. It is mainly when the outcome of an experiment turns out as a surprise that the world becomes really exciting and provocative. As nicely stated by Lewis Thomas,[1] the interest of the work is in a way measured by the intensity of the astonishment it brings.

At the Institut Pasteur, we thus had some good series of surprises. To mention only a few: with Elie Wollman, the so-called zygotic, or erotic, induction of prophage development; the coitus interruptus experiment by placing mating bacteria in a Waring blender; the circularity of the bacterial chromosome. With Jacques Monod: the results of the PaJaMo experiment performed with Arthur Pardee; and also the isolation of completely unexpected mutants such as the first dominant negative clones in the regulation of Lac and Lambda; the first operator, *cis*-acting constitutive for Lac. Actually, there is also another type of surprise that arose during all this work: after having produced a model that one hardly believes oneself, the amazement of finding that it contains some element of truth; that the world, or a piece of the world, does ultimately come out as one had imagined it to be! At least for a while! This occurred when I succeeded in producing a strain diploid for the Lac region, carrying on each chromosome a dominant negative regulator i^s gene and selecting for Lac-permease activity. Among the mutants growing under these conditions, several had a deletion fusing the Lac genes with a purine operon. Instead of being induced by β-galactosides, the Lac genes had become repressible by purines.[2] That the result could correspond so precisely to what the model predicted was indeed a surprise!

THE CONTROL UNIT IN BACTERIA

For several years, Jacques Monod and I spent hours everyday in his office, in front of the blackboard, half talking, half sketching models. Progressively, from the combined work on lysogeny and on the lactose system, after days of discussions, of drawing arrows and arrows, there emerged the so-called "operon model," a representation of protein synthesis and its control through regulatory circuits.[3] This model introduced the idea of the control unit for gene expression: a regulatory protein and its DNA target. The regulatory protein can react with one or several effectors; the target controls the initiation step for the transcription of sequences linked to it. In bacteria, a target can control one or several different genes.

Of course, when we proposed this model for the regulation of gene expression in bacteria, our hope was that similar control units, with the necessary addition of complexity, would also be at work to regulate gene action in higher organisms, in particular in those phenomena that underlie embryonic development and cellular differentiation. These control units could be considered as basic elements of circuitry. They could be combined and geared in an infinite number of ways to achieve a variety of regulatory functions. In the conclusions of the 1961 *Cold Spring Harbor Symposium for Quantitative Biology*,[4] we thus produced a number of hypothetical circuits that could account for various situations, in particular for alternative switches or for permanent lock-up of certain genetic pathways.

After the operon model was published, a number of alternative models were proposed for the control of gene expression in bacteria, some of these suggestions being radically different from our model. They turned out to be wrong. A few years later, the isolation of the Lac repressor by Gilbert and Müller-Hill[5] and that of the phage lambda repressor by Ptashne[6] allowed a molecular analysis of the control unit. These studies supported, with minor modifications, the main lines of the operon model and resolved the molecular details of the interactions between protein and operator.

CIRCUITS IN HIGHER ORGANISMS

For a long time, the impossibility of performing fine genetic analysis on higher organisms precluded the study of their regulatory circuits, in particular of those involved in embryonic development. Again, a variety of models was proposed without much experimental basis (see, for instance, Britten and Davidson[7]). The rapid development of DNA recombinant technology, on the one hand, and the remarkable possibilities offered by yeast and *Drosophila* genetics, on the other, have given access to the control mechanisms of gene expression in eukaryotic systems. And, as in bacteria, the bulk of gene control turned out to operate on transcription by a variety of switching devices. These devices are based on control units made of regulatory proteins and their DNA targets, which appear to follow, with an added complexity, principles similar to those found in bacteria. I cannot resist the pleasure of quoting extracts of an article entitled "Glimpses of Allostery in the Control of Eukaryotic Gene Expression" by De La Brousse and McKnight[8]:

> Thirty years ago, Jacob and Monod described how genes are selectively switched on and off in bacterial cells. According to their paradigm for gene regulation, specialized DNA sequences, later termed *cis*-regulatory sequences, are interpreted and mobilized by *trans*-acting regulatory proteins . . .
>
> Genes of eukaryotic organisms work the same way; the names and faces have been changed to protect those of us too proud to realize that we have been busily rediscovering the wheel. *cis*-regulatory DNA sequences are fashionably called enhancers and silencers; *trans*-acting regulatory proteins are collectively termed transcription factors. In the end, however, in eukaryotes as in bacteria, transcription factors interact with regulatory DNA sequences to activate or repress genes.
>
> Why, then, all the hoopla over eukaryotic transcription factors? Because modern biologists simply can't escape them. Molecular oncologists were insufficiently nimble to avoid transcription factors: they are encoded by many oncogenes and tumor suppressor genes misregulated by various genetic aberrations. Developmental biologists, despite valiant efforts, couldn't flank the wave: transcription factors control even the most ephemeral aspects of ontogeny. Physiologists, immunologists, endocrinologists, neurobiologists, virologists, none have found refuge from the blight.

Not only do regulatory proteins work in eukaryotes as in prokaryotes to activate or repress genes, but also eukaryotic proteins are often structurally and evolutionary related to the prokaryotic ones. The realization that the products of homeotic genes in insects are transcription factors came from observed similarities between their amino acid sequences and those of bacterial repressor proteins.[9,10] In the same way, the repetition of a same motif in a frog transcription factor suggested the mechanism of action of nuclear hormone receptors.[11] There exists a striking similarity between the structure of the lambda repressor with its DNA operator and that of *Antennapedia* homeobox protein of *Drosophila* with its DNA target.[12]

What is the basis of the increase in complexity found in the regulatory circuits of higher organisms as compared with those of bacteria? Complexity did not arise only from an increasing intricacy in the circuitry. It came also from two other mechanisms, two combinatorial systems that allow, at two different levels, the construction of a large variety of circuits with a limited number of basic elements. The first of these systems deals with the structure of regulatory proteins. Like many proteins, transcription factors are most generally mosaic structures. They are composed of modules, of domains, each one carrying a particular recognition site: for a DNA sequence, or for a growth factor, a hormone, or some other protein, etc. The 3-D shape of these modules, their electrostatic charge and their capacity of hydrophobic reactivity determine their possibilities of recognition and interactions. Regulatory genes are the products of some kind of Meccano linking together relatively short DNA fragments in which discrete polypeptidic modules are encoded. With the additional variability provided by mutations, it is thus possible to produce an important repertoire of transcription factors by combining a limited number of short DNA sequences.

The second mechanism providing complexity is found in the protein aggregates involved in the control circuits of higher forms. The expression of many genes is precisely controlled, in time and space, by a series of specific transcription factors which bind near the promoter or on some enhancer. Thus, in the vicinity of the initiation site for transcription, a group of proteins assemble to form a specific aggregate that activates or inhibits gene transcription. It is the reunion of these proteins, their interactions, their reactivity towards certain compounds such as hormones or growth factors or metabolites, that insure the specific regulation of that gene. Such specific, often transient, complexes, made of nonspecific transcription factors, might be called *aggregulates*, a recombinant word between aggregate and regulation.

A protein molecule can harbor only a limited number of specific sites, whether catalytic or interactive. Hence the importance of complexes, of aggregates, in which various proteins carrying different recognition sites, can interact and build specific molecular constructions playing a structural role—as in cytoskeleton, viruses, ribosomes, etc.—or a regulatory role—as in aggregulates. Such multimolecular constructions make it possible to increase, to diversify, and to refine interactions between cellular components and reactions which, otherwise, would ignore each other.

In prokaryotes, a single protein with several sites—one for a DNA sequence, others for some metabolites—is probably sufficient to ensure, for a large part, the control of gene expression. With eukaryotic cells and pluricellular organisms, these proteins—activators and repressors—are no longer sufficient. To arrange complex regulatory circuits, such as those operating in development and cell differentiation, evolution has made use of aggregulates. These have several advantages. First, they allow an almost infinite number of interactions between a variety of molecular species. Second, aggregulates make it easy to see how the control of a given gene can be precisely built during embryonic development. Some components are added according to a given order in time and space, as they are synthesized at certain stages of development and in certain cells. If a single activator is missing, the gene will not be expressed. Only in those tissues where, and at the time when, the missing element is produced, will gene expression be activated. Finally, with aggregulates, cells can build a large repertoire of regulatory circuits with a limited set of transcription factors. A gene can be activated by the combination of protein x, y and z, while another is activated by a different combination x, y and a. This combinatorial system frees organisms from the constraint of making a particular transcription factor for each of the many genes active in cells.

REFERENCES

1. THOMAS, L. 1974. The Lives of a Cell. Viking Press. New York.
2. JACOB, F., A. ULLMANN & J. MONOD. 1965. Délétions fusionnant l'opéron lactose et un opéron purine chez *Escherichia coli*. J. Mol. Biol. **13**: 704–719.
3. JACOB, F. & J. MONOD. 1961. Genetic regulatory mechanisms in the synthesis of proteins. J. Mol. Biol. **3**: 318–356.
4. MONOD, J. & F. JACOB. 1961. Teleonomic mechanisms in cellular metabolism, growth and differentiation. Cold Spring Harbor Symp. Quant. Biol. **26**: 389–401.
5. GILBERT, W. & B. MULLER-HILL. 1966. Isolation of the Lac Repressor. Proc. Natl. Acad. Sci. USA **56**: 1891–1898.
6. PTASHNE, M. 1967. Isolation of the λ phage repressor. Proc. Natl. Acad. Sci. USA **57**: 306–312.
7. BRITTEN, R. J. & E. H. DAVIDSON. 1969. Gene regulation for higher cells: A theory. Science **165**: 349–357.
8. DE LA BROUSSE, F. C. & S. L. MCKNIGHT. 1993. Glimpses of allostery in the control of eukaryotic gene expression. Trends Genetics **9**: 151–154.
9. SCOTT, M. P. & A. J. WEINER. 1984. Structural relationships among genes that control development: Sequence homology between the Antennapedia, Ultrabithorax, and fushi tarazu loci of *Drosophila*. Proc. Natl. Acad. Sci. USA **81**: 4115–4119.
10. SHEPHERD, J. C. W., W. MCGINNIS, A. E. CARRASCO, E. M. DE ROBERTIS & W. J. GEHRING. 1984. Fly and frog homeo domains show homologies with yeast mating type regulatory proteins. Nature **310**: 70–71.
11. WEINBERGER, C., S. M. HOLLENBERG, M. G. ROSENFELD & R. M. EVANS. 1985. Domain structure of human glucocorticoid receptor and its relationship to the v-erb-A oncogene product. Nature **318**: 670–672.
12. QIAN, Y. Q., M. BILLETER, G. OTTING, M. MULLER, W. J. GEHRING & K. WUTHRICH. 1989. The structure of the *Antennapedia* homeodomains determined by NMR spectroscopy in solution: Comparison with prokaryotic repressors. Cell **59**: 573–580.

The NK-2 Homeobox Gene and the Early Development of the Central Nervous System of *Drosophila*

MARSHALL NIRENBERG,[a]
KOHZO NAKAYAMA,[b] NORIKO NAKAYAMA,[b]
YONGSOK KIM,[c] DERVLA MELLERICK,[a]
LAN-HSIANG WANG,[a]
KEITH O. WEBBER,[d] AND RAJNIKANT LAD[e]

[a] *Laboratory of Biochemical Genetics*
National Heart, Lung, and Blood Institute
National Institutes of Health
Bethesda, Maryland 20892

[b] *Department of Immunobiology*
Cancer Research Institute
Kanazawa University
Takaramachi 13-1
Kanazawa, Ishikawa 920, Japan

[c] *Laboratory of Molecular Cardiology*
National Heart, Lung, and Blood Institute
National Institutes of Health
Bethesda, Maryland 20892

[d] *Laboratory of Molecular Biology*
National Cancer Institute
National Institutes of Health
Bethesda, Maryland 20892

[e] *Department of Psychiatry*
Hospital of the University of Pennsylvania
Philadelphia, Pennsylvania

My colleagues and I deciphered the genetic code gradually, over a period of five years, between 1961 and 1965. I then stopped working on the code and began working in the field of neurobiology. The logic that connects the genetic code to neurobiology is that information is processed in both genetic and neural systems.

My interest in the NK-2 homeobox gene[1] of *Drosophila* stems from the observation that NK-2 is the earliest predominantly neural gene regulator that has been found thus far that is expressed in the ventrolateral neurogenic an-

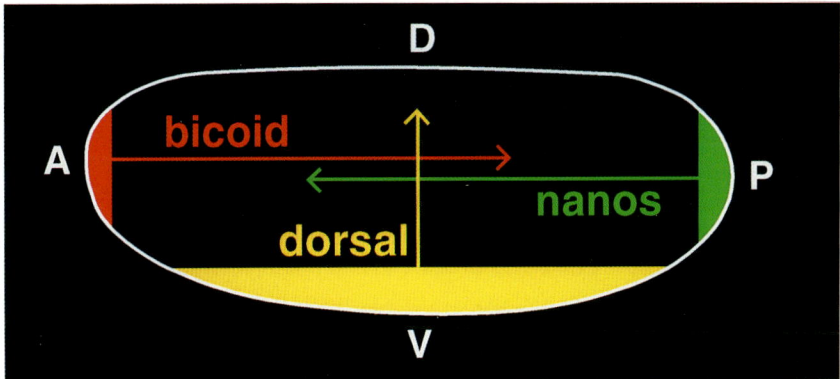

FIGURE 1. Side view of stage 5 *Drosophila* embryo illustrates concentration gradients of proteins that regulate gene expression. The concentration of bicoid homeobox protein is high in the anterior (A) region and low in the posterior (P) region of the embryo. A concentration gradient of nanos protein is established in the posterior to anterior direction. A concentration gradient of dorsal protein is established in nuclei in the ventral (V) towards the dorsal (D) region of the embryo.

lage, which gives rise to part of the central nervous system of the embryo. I will tell you what we know about the NK-2 gene and relate these findings to the early development of the *Drosophila* embryo and the central problem of understanding the principles that are used initially to construct part of the central nervous system of the embryo. The studies on NK-2 were performed by my colleagues, Yongsok Kim, Kohzo Nakayama, Noriko Nakayama, Dervla Mellerick, Lan-Hsiang Wang, Keith Webber, and Rajnikant Lad.

The nucleus of a fertilized *Drosophila* embryo undergoes 13 rounds of nuclear division in the first 130 minutes of embryonic development, resulting in an embryo that consists of a single cell with approximately 5,000 nuclei. Most of the nuclei move to the periphery during stage 4 (80–130 minutes after fertilization, nuclear divisions 10–13[2,3]) and cell membranes form around each nucleus between 130 and 170 minutes after fertilization (stage 5). The anterior-posterior and ventral-dorsal axes of the embryo are established and different cell types are generated during stages 4 and 5 by the formation of concentration gradients of proteins that regulate gene expression (FIG. 1). A concentration gradient of bicoid homeobox protein is established in the anterior-posterior direction (high concentration in the anterior portion of the embryo, low concentration in the posterior portion[4–6]); and a concentration gradient of nanos protein is established in the posterior to anterior direction.[7–10] Concomitantly, a concentration gradient of dorsal protein is established in nuclei, with the highest concentration of dorsal protein in nuclei in the ventral portion of the embryo and the lowest concentration in nuclei in the dorsal part of the embryo.[11–14] These gene regulators and terminal gene regulators initiate the induction or repression of other genes that encode proteins that regulate gene expression and result in dynamically changing patterns of gene expression in different parts of the embryo, depending upon the concentrations of gene regulators that the nuclei were exposed to. The anterior-posterior gradients of gene regulators result in the formation of vertical stripes of equivalent nuclei that were exposed to the same concentrations of gene regulators, whereas the ventral to dorsal gradient of dorsal protein results in the formation of horizontal stripes of equivalent nuclei (for reviews see Refs. 15–18). In effect, the embryo is divided into a bilaterally symmetric checkerboard of clusters of nuclei that express different combinations of genes for proteins that regulate genes. The position of a nucleus in the embryo therefore determines the initial developmental fate of the nucleus.

In FIGURE 2 is shown the composite nucleotide sequence of NK-2 cDNA and genomic DNA and the deduced amino acid sequence of NK-2 protein.[19] The NK-2 gene contains 3 exons and 2 introns; introns 1 and 2 are approximately 1.6[81] and 3.1[19] kb in length, respectively. 3 NK-2 protein contains two regions near the N-terminus that consist almost entirely of alternating acidic and basic amino acid residues or pairs of acidic and basic amino acid residues. The protein contains multiple Ala repeats, an Asn repeat, an acidic domain, followed by a homeodomain, which is not closely related to any other *Drosophila* homeodomain. The C-terminal region of NK-2 protein contains a 17–amino acid residue sequence of unknown function termed the NK-2 box (amino acid residues 631–647) that has been highly conserved during evolution and a His-Ala repeat. The NK-2 gene was shown to reside on the sex chromosome at 1C1-5.[1]

The NK-2 homeodomain has been conserved during evolution. FIGURE 3A compares the amino acid sequence of the *Drosophila* NK-2 homeodomain[1,19] and NK-2-like homeodomains from *Xenopus*,[20] mouse,[21,22] planaria,[23,24] leech,[25] and tapeworm.[26] The similarity in amino acid sequence ranges from 95 to 67 percent. The mouse genome contains six copies of the NK-2 gene, presumably formed by gene duplication.

The amino acid residues that comprise NK-2 homeodomain α-helices I, II, and III were determined by NMR.[27] Binding of a 77–amino acid residue protein that contains the NK-2 homeodomain to a high-affinity NK-2 binding site in DNA results in an increase in the length of α-helix III from 11 to 19 amino acid residues (from NK-2 homeodomain residues 42–52 to 42–60) and also increases the stability of the secondary structure of the homeodomain.[27]

Xenopus and mouse proteins with NK-2-like homeodomains also contain the highly conserved 17–amino acid residue NK-2 box after the homeodomain (94–77% homology with the NK-2 box sequence of *Drosophila* NK-2 protein) (FIG. 3B). The *Drosophila* NK-3[1,29] (*bagpipe*)[28] homeodomain protein also contains a sequence related to the NK-2 box (47% homology); how-

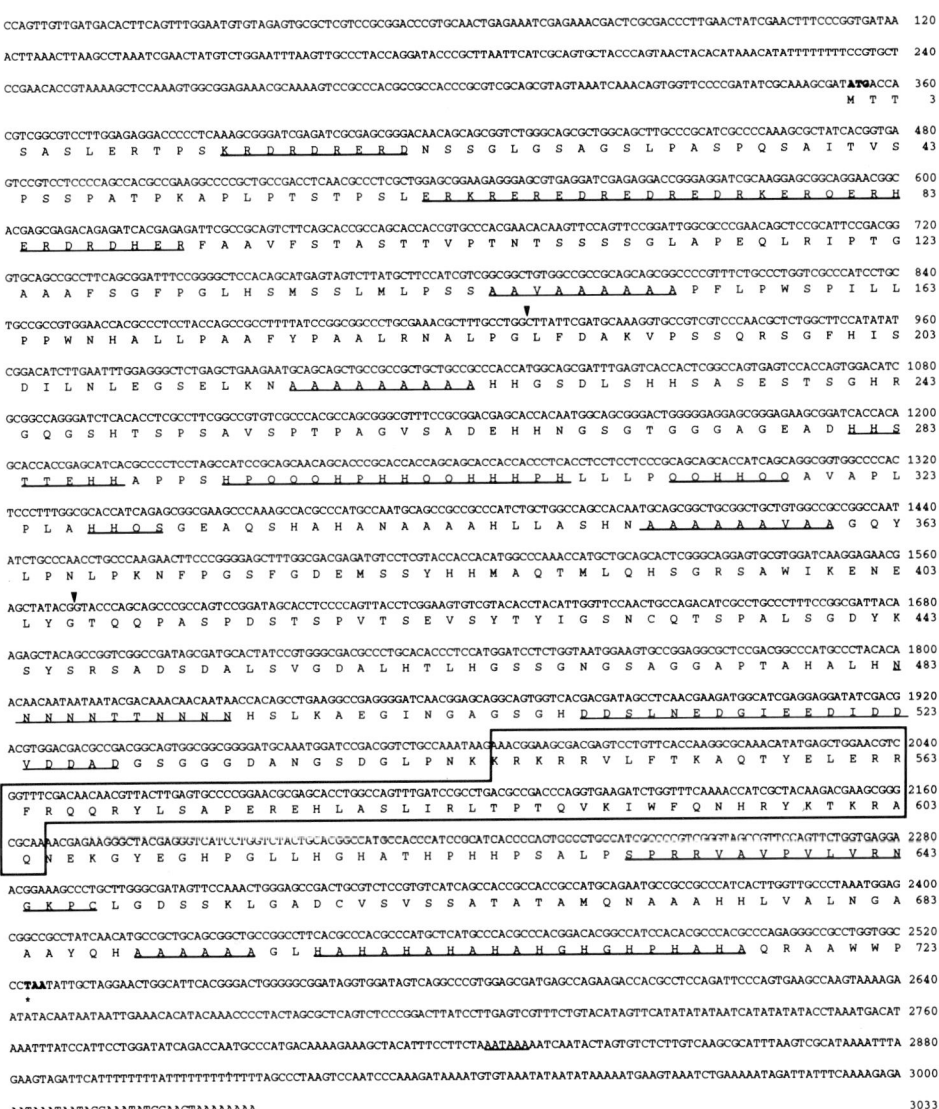

FIGURE 2. The nucleotide sequence of NK-2 cDNA and the deduced amino acid sequence of NK-2 protein (from Ref. 19). The homeodomain is enclosed within a box. Other interesting amino acid sequences are underlined, such as an acidic domain before the homeodomain and the highly conserved NK-2 box sequence after the homeodomain (amino acid residues 631–647). An inverted triangle shows the position of an intron.

A

		α-HELIX 1	α-HELIX 2	α-HELIX 3 α-HELIX 3		% HOMOLOGY	REF.
		1 11 22	28 38	42 52	60		
NK-2	d	KRKRRVLFTKAQTYELERRFRQQRYLSAPEREHLASLIRLTPTQVKIWFQNHRYKTKRAQ				100	1, 19
XeNK-2	x	-------S---		-----M---R		95	20
Nkx-2.2	m	-------S---		-----M---R		95	21
Nkx-2.4	m	-----SQ--V-------K--K------------M-H------------------M--QA				83	21
Nkx-2.1	m	R------SQ--V-------K--K------------M-H------------------M--QA				82	21
Nkx-2.3	m	R--P----SQ--VF------K---------------SLK--S---------R---C--QR				75	21, 22
Nkx-2.5	m	R--P----SQ--V-------K----------DQ---VLK--S---------R---C--QR				73	22
Nkx-2.6	m	Q--S----SQ--VLA-----K-----T---------ALQ--S---------R---S-SQR				70	22
Dth-1	p	--------S-K-IL----H---KK-----------N--G-S--------------M---H				80	23, 24
Dth-2	p	R----I--SQ--I-------K--K-----------N--N-----------------C--S-				82	23, 24
Lox 10	l	R----I--SQ--I----------K----------TF-G-------------------KSK				80	25
EgHbx-3	t	QS------N-F-ISQ--K---K----T-Q--QE--HT-G-------------A--M--LF				67	26

B

SPECIES		ΔAA	% HOMOLOGY	REF.
NK-2	d	26	SPRRVAVPVLVR·NGKPC	100	1, 19
XeNK-2	x	11	------------·D----	94	20
Nkx-2.2	m	11	------------·D----	94	21
Nkx-2.3	m	14	P-----------·D----	88	21
Nkx-2.4	m	23	-----------K·D----	88	21
Nkx-2.1	m	34	-----------K·D----	88	21
Nkx-2.5	m	13	PA--I-------·D----	77	22
Nkx-2.6	m	13	PA---------L·D----	77	22
NK-3(bagpipe)	d	10	ASK--P-Q----ED-STT	47	1, 28-29

FIGURE 3. (A) Comparison of the amino acid sequence of the (d) *Drosophila* NK-2 homeodomain with similar homeodomains from (x) *Xenopus*, (m) mouse, (p) planaria, (l) leech, and (t) tapeworm (from Ref. 19). The symbol (−) represents the same amino acid residue as NK-2. The amino acid residues of *Drosophila* NK-2 homeodomain α-helix 1, α-helix 2, and α-helix 3 were determined in NMR.[27] In the absence of DNA α-helix 3 extends from amino acid residue 42 through 52. However, binding of the NK-2 homeodomain to a high-affinity NK-2 site in DNA increases the length of α-helix 3 (residues 42–60).[27] (B) The NK-2 box is a highly conserved 17–amino acid residue sequence that is found after the homeodomain in proteins related to NK-2. ΔAA represents the number of amino acid residues between the end of the homeodomain and the beginning of the conserved NK-2 box sequence. The symbol (-) represents the same amino acid residue; (·) represents the absence of an amino acid residue.

ever, a highly conserved NK-2 box was not detected in the planarian homeodomain proteins, Dth-1 or Dth-2.

Northern analysis of poly A⁺ RNA from *Drosophila* at various stages of development showed that NK-2 mRNA is present in highest concentration in 3–6-hr *Drosophila* embryos and then progressively decreases during further embryonic development.[19] No NK-2 mRNA was detected in 0–3-hr *Drosophila* embryos; therefore, no maternal NK-2 mRNA was found. Larvae and pupae contain greatly reduced levels of NK-2 mRNA compared with that of 3–6-hr embryos; however, an increase in NK-2 mRNA was found in adult flies.

FIGURE 4 shows the distribution of NK-2 mRNA in *Drosophila* embryos as a function of developmental age, determined by *in situ* hybridization.[19]

NK-2 gene expression is initiated during stage 4 in bilaterally symmetrical longitudinal stripes, one stripe on each side, in the ventral (i.e., medial) half of the ventrolateral neurogenic anlage (FIG. 4A). By stage 5, when the first cell membranes are forming around the nuclei, the stripes of nuclei that express NK-2 extend from 0 to 90% of the embryo length and each stripe is 6 or 7 nuclei in width (FIG. 4B). NK-2 is also expressed in part of the procephalic neuroectodermal anlage, the endodermal anterior and posterior midgut anlagen, and the hindgut anlage. The ventral mesodermal primordium invaginates during gastrulation, bringing the longitudinal NK-2 positive bands of neuroectodermal cells closer to the ventral midline, separated only by ventral midline mesectodermal cells, which do not contain NK-2 mRNA (FIGS. 4C–F). At first the level of NK-2 mRNA in the band of NK-2–positive cells is fairly homogeneous; however, during early gastrulation clusters of cells with high levels of NK-2 mRNA appear that are separated by vertical stripes of cells, 1 or 2 cells in width, that contain lower levels of NK-2 mRNA, apparently due to repression of the NK-2 gene (FIGS. 4E–F). Initially one cluster of cells with a high level of NK-2 mRNA is formed per hemisegment; later two clusters of NK-2–positive cells appear per hemisegment (FIGS. 4G–J). Germ band extension results in an increase in the length of the band of cells that synthesize NK-2 mRNA and a concomitant decrease in the width of the band to 2 to 3 cells per side (FIG. 4I). Hence, the neuroectodermal cells that synthesize NK-2 mRNA give rise to medial and paramedial neuroblasts that continue to synthesize NK-2 mRNA. Ganglion mother cells and neurons were found that express the NK-2 gene that perhaps are the progeny of neuroblasts that express the NK-2 gene. However, during later embryonic development, the abundance of NK-2 mRNA decreases in some neurons and is extinguished in others. Some neurons that express the NK-2 gene form commissures and others contribute to longitudinal connectives.

FIGURE 5 shows schematic diagrams of cross-sections of embryos at the cellular blastoderm stage (stage 5) before nuclei have been enclosed by cell membranes, and at the end of gastrulation (end of stage 7) after ventral mesodermal primordium cells have invaginated. These cross-sections of embryos illustrate ventral-dorsal patterning during early development of the *Drosophila* embryo. The ventral to dorsal concentration gradient of dorsal protein[11–14] activates the *twist*[30–34] and *snail*[34–36] genes in the most ventral nuclei, which correspond to the mesodermal anlage, and the NK-2 gene is activated in the ventral (medial) half of the ventrolateral neuroectodermal anlage. A gene regulator specific for dorsal (lateral) neuroectoderm has not been identified thus far. Dorsal protein represses the *decapentaplegic (dpp)* gene,[37–40] which encodes a protein that is a homologue of TGF-β,[37,41] and the *zen-1* and *zen-2* homeobox genes[42–46] in nuclei in the ventral and lateral parts of the embryo, but not in nuclei that become dorsolateral epidermoblasts or dorsal amnioserosa,

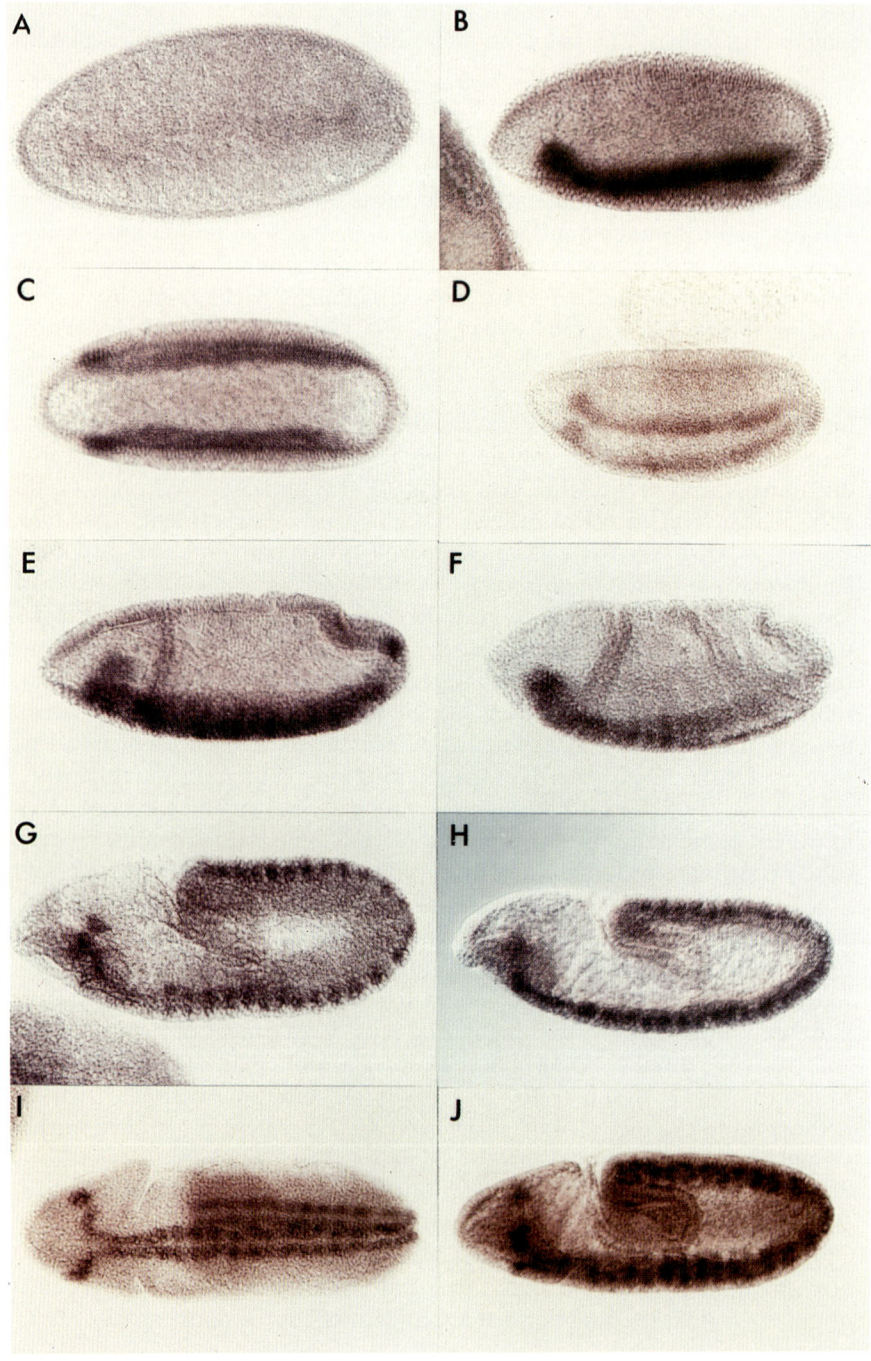

respectively. Hence, the concentration gradient of dorsal protein establishes the ventral-dorsal axis of the embryo and divides the embryo into longitudinal bands of nuclei that have different developmental fates. After cell membranes form, the most ventral cells, the mesodermal anlage, invaginate, which brings the mesectodermal cells to the ventral midline.

Neuroectodermal cells gradually segregate as neuroblasts between about 3.5 and 7.3 hours after fertilization[47-49] (FIG. 6). Eventually a monolayer of neuroblasts and glioblasts separated by ventral midline mesectodermal cells is formed above the epidermal cells. Doe[49] has shown that thirty-one neuroblasts or glioblasts delaminate per hemisegment, that each is a unique cell type, and that the relative position of each neuroblast or glioblast in the set is determined.

A *Drosophila* embryo contains 14 parasegments and additional segments in the head region. Approximately 800 ventrolateral neuroblasts or glioblasts segregate from the medial and lateral neuroectoderm per embryo and additional neuroblasts and glioblasts are formed from mesectodermal cells.[49] The pattern of neuroblasts is repeated in different segments possibly with some variation. However, some genes that encode proteins that regulate gene expression are known to be expressed only by cells in a single segment, or a few segments. Although most of the proof is lacking, it is likely that many neuroblasts also express segment-specific gene regulators and that most of the neuroblasts per side eventually will be found to be unique cell types.

Three longitudinal stripes of neuroblasts or glial precursors can be distinguished on each side that are the precursors of neurons and glia of the ventral nerve cord: ventral midline mesectodermal cells that separate the right and left halves of the ventral nerve cord, medial neuroblasts or glial precursors that express the NK-2 homeobox gene, and lateral neuroblasts or glial precursors that have little or no NK-2 mRNA.

The monolayer of neuroblasts that give rise to the ventral nerve cord is also divided along the anterior-posterior axis of the embryo into 14 parasegments; most parasegments consist of posterior compartment neuroblasts that

FIGURE 4. Distribution of NK-2 mRNA in *Drosophila* embryos as a function of developmental age (from Ref. 19). The RNA probe used for *in situ* hybridization was from the 3'-untranslated region of NK-2 cDNA; the probe did not contain the homeobox. (**A**) Expression of the NK-2 gene is initiated during stage 4, the syncytial blastoderm stage. (**B**) Stage 5–6, side view. (**C**) Ventral view, stage 6, early gastrulation. (**D**) Ventrolateral view in late stage 6 embryo. (**E**) Side view of embryo; gastrulation is almost completed. Late stage 7, about 185 minutes after fertilization. Notice the apparent segmentation of the NK-2–positive region. (**F**) Late stage 7 illustrating apparent segmentation of NK-2–positive region. (**G**) Side view stage 9 embryo 3.7–4.3 hours after fertilization. Two clusters of neuroectodermal cells and/or neuroblasts that contain NK-2 mRNA can be seen per hemisegment. (**H**) Side view of stage 9–10 embryo. (**I**) Ventral view of stage 10 embryo. Two clusters of medial neuroectodermal cells and/or neuroblasts that contain NK-2 mRNA are present per hemisegment. (**J**) Stage 9–10 embryo, side view.

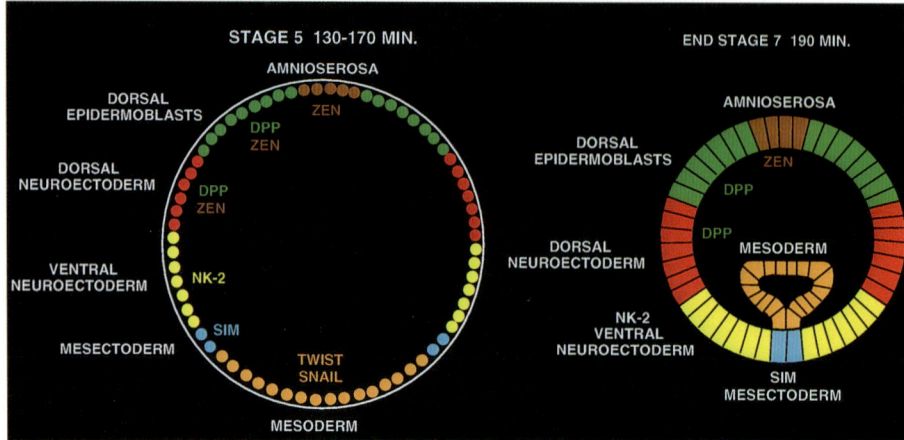

FIGURE 5. Schematic drawings of cross-sections of embryos before gastrulation (stage 5) and after gastrulation (end of stage 7) to show the ventral neuroectoderm nuclei or cells (*yellow*) that express the NK-2 gene.

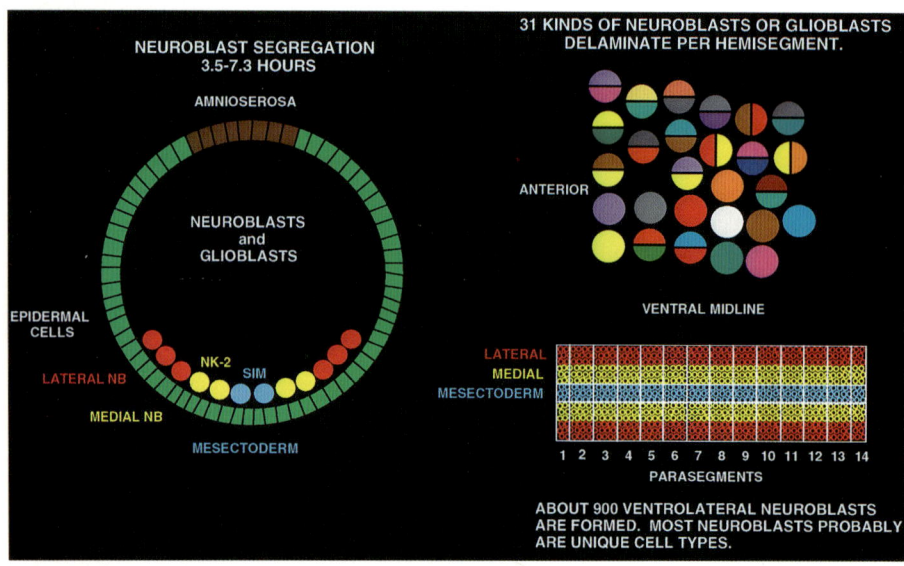

FIGURE 6. Some of the neuroectodermal cells delaminate and form a monolayer of neuroblasts and/or glioblasts immediately above the epidermal cell layer. The figure on the *left* is a schematic illustration of a cross-section of an embryo to show the monolayer of neuroblasts or glioblasts that delaminate from the neuroectodermal cell layer. *Blue*: ventral midline mesectodermal cells that express the *sim* gene. *Yellow*: medial neuroblasts that express the NK-2 gene. *Red*: lateral neuroblasts and/or glioblasts. *Upper right panel* shows 31 kinds of neuroblasts or glioblasts that delaminate per average hemisegment. Four of the delaminated cells have migrated to other positions; hence, only 27 delaminated cells are shown. *Lower right panel*: Ventral view of the monolayer of neuroblasts and/or glioblasts to illustrate ventral-dorsal and anterior-posterior patterns. The *yellow* medial neuroblasts and/or glioblasts express the NK-2 gene.

express the engrailed (en) homeobox protein[49,50] and anterior compartment cells that do not express en. Hence the 31 neuroblasts or glial precursor cells that segregate per hemisegment are divided into four groups of cells, depending on the position of the cells and the expression of the NK-2 and *en* genes: medial anterior compartment cells that have high levels of NK-2 mRNA but no en mRNA, lateral anterior compartment cells that lack NK-2 and en mRNA, medial posterior compartment cells that have high levels of both NK-2 and en mRNA, and lateral posterior compartment cells that have en mRNA and low levels of NK-2 mRNA.

Neuroblasts start to divide soon after they segregate from the neuroectodermal cell layer. Each neuroblast division gives rise to a small ganglion mother cell and a large neuroblast, which becomes smaller with each division (FIG. 7). Each ganglion mother cell divides only once and gives rise to two neurons. The first neuroblasts to originate divide about eight times, whereas the last neuroblasts divide five times.[47-48] Therefore, a single neuroblast may be the precursor of 10 to 16 neurons.

The neuroblasts that express the NK-2 gene[51] in a thoracic segment at the end of neuroblast segregation (late stage 11) are shown schematically in FIGURE 8. It should be emphasized that the abundance of NK-2 mRNA changes dynamically during development. NK-2 is expressed by two longitudinal columns of medial neuroblasts on each side; however, the abundance of NK-2 mRNA usually is higher in the column of neuroblasts adjacent to the mesectodermal ventral midline cells than in the second column of neuroblasts. All neuroblasts in the posterior compartment express the NK-2 gene; however, the lateral neuroblasts contain much less NK-2 mRNA than do the medial neuroblasts. Hence, a medial to lateral gradient of NK-2 mRNA is established in both the anterior and posterior compartments. The amount of NK-2 mRNA in neuroblasts 2-1, 2-3, 5-1, and 5-2 (that is, immediately after or before the posterior compartment) decreases during development, resulting in the formation of two clusters of neuroblasts that express the NK-2 gene per hemisegment, one cluster in the anterior compartment and the second consisting of posterior compartment neuroblasts. Some ganglion mother cells and neurons also express the NK-2 gene; however, the levels of NK-2 mRNA decrease markedly in some cells during later stages of embryonic development. These results show that about half of the ventrolateral neuroblasts express the NK-2 gene and that medial neuroblasts contain higher levels of NK-2 mRNA than do intermediate or lateral neuroblasts.

The pattern of expression of the NK-2 gene also was determined in various mutant lines of flies as a function of developmental age.[51] The NK-2 gene was found to be activated initially by dorsal in the ventral half of the embryo. However, the NK-2 gene normally is not expressed in the mesodermal anlage because of repression by snail, in the mesectodermal anlage because of repression by sim, or in part of the lateral neuroectodermal anlage or dorsal epi-

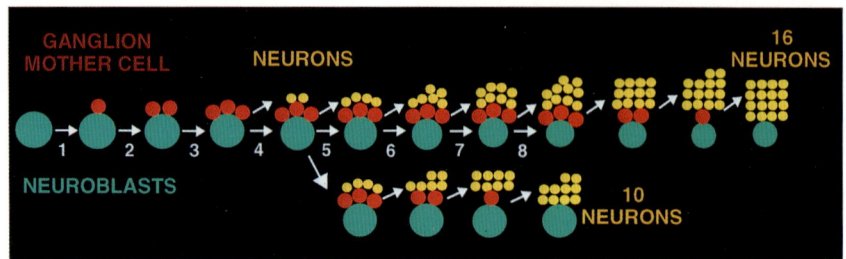

FIGURE 7.

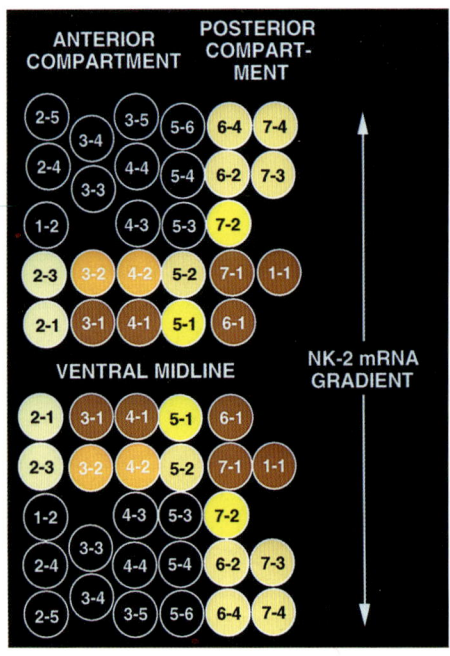

FIGURE 8.

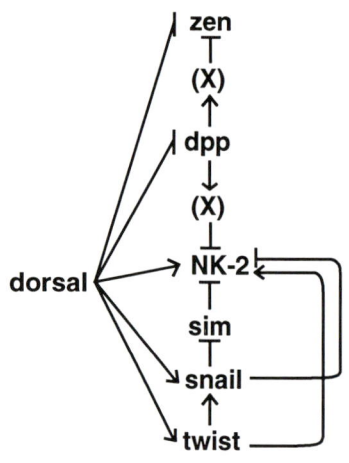

FIGURE 9.

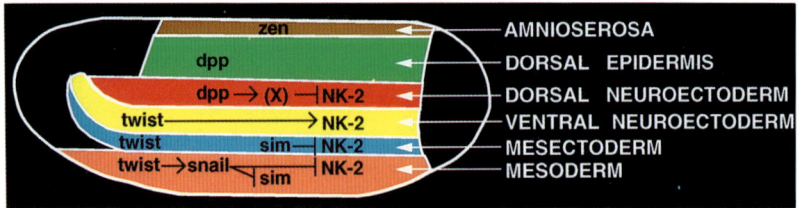

FIGURE 10.

dermal anlage because of repression mediated by dpp. Both dorsal and twist were found to be required to activate the NK-2 gene in the hindgut primordium and the posterior midgut primordium.

During stage 4 dorsal activates the *twist, snail,* and NK-2[51] genes (FIG. 9). Dorsal represses *dpp* and the *zen* genes in the ventral and lateral portions of the embryo; however, the concentration of dorsal is too low for effective repression of *dpp* or *zen* genes in the dorsolateral and dorsal portions of the embryo, respectively (for reviews see Refs. 52–54). *snail* is expressed only in the most ventral nuclei, which comprise the mesodermal anlage.[55–56] *sim* is activated in both the mesectodermal and mesodermal anlage, but *sim* is expressed only in the mesectodermal anlage because of repression by snail.[34,57–60] The NK-2 gene is activated in the mesodermal, mesectodermal, ventral neuroectodermal, and part of the dorsal neuroectodermal anlagen, but is repressed by *snail* in the mesodermal anlage, by *sim* in the mesectodermal anlage, and by a gene regulator that has not thus far been identified whose repression is mediated by dpp in the dorsal neuroectodermal anlage.[51] *twist* is expressed in the mesodermal, mesectodermal, and the ventral portion of the ventral neuroectodermal anlagen.[61,62,35] *twist* protein activates the *snail* gene in the mesodermal primordium[36] and the NK-2 gene in the ventral portion of the neuroectoderm.[51] *snail* represses the *sim* and NK-2 genes in the mesodermal anlage,[51,59,60] while *sim* represses the NK-2 gene in the mesectodermal anlage.[51] The hierarchical organization of gene regulation results in the appear-

---

FIGURE 7. Neuroblast division. A neuroblast (*green*) divides 5–8 times. Each neuroblast division is unequal and gives rise to a slightly smaller neuroblast and a much smaller ganglion mother cell (*red*). Each ganglion mother cell divides once, giving rise to two neurons (*yellow*).

FIGURE 8. A ventral view of neuroblasts in a thoracic segment at the end of stage 11 is shown using the neuroblast nomenclature of Doe.[49] Neuroblasts shown in color express the NK-2 gene (from Ref. 51). The varying darkness of the neuroblast color from *brown* to *tan* represents the relative abundance of NK-2 mRNA; for example, in order of decreasing abundance of NK-2 mRNA, we see *brown* (neuroblast 4-1), *orange* (4-2), *yellow* (5-1), *pale yellow* (5-2), and *tan* (2-1). No NK-2 mRNA was detected in black neuroblasts. Medial neuroblasts closest to the ventral midline usually contain more NK-2 mRNA than do neuroblasts in a more lateral position in that vertical row. All posterior compartment neuroblasts express the NK-2 gene. A medial to lateral NK-2 mRNA gradient is present.

FIGURE 9. Regulation of NK-2 gene expression deduced from the patterns of expression of the NK-2 gene in various mutant lines of flies (from Ref. 51). An *arrowhead* represents gene activation, while a terminal *bar* represents gene repression. (X) corresponds to an unidentified repressor mediated by *dpp*.

FIGURE 10. Side view of an embryo showing the ventral-dorsal pattern of the anlagen indicated and the hierarchically organized regulation of NK-2 gene expression.[51] An *arrowhead* corresponds to gene activation, while a terminal *bar* represents repression. dpp indirectly mediates repression of the NK-2 gene in dorsal neuroectoderm, via (X), an unidentified repressor. The NK-2 gene is repressed by sim in the mesectodermal anlage and by snail in the mesodermal anlage.

ance of six horizontal stripes which, from ventral to dorsal, comprise the mesodermal, mesectodermal, ventral neuroectodermal, dorsal neuroectodermal, dorsoepidermal, and amnioserosa anlagen, as shown in FIGURE 10.

The ventral border of the horizontal stripe of nuclei that synthesize NK-2 mRNA is created by repression of the NK-2 gene by snail initially and then by sim, while the dorsal border of the stripe is created by a different, unidentified species of repressor mediated by dpp.[51] Thus, the ventral and dorsal borders of the NK-2–positive stripe of nuclei are created independently by different species of repressors. The width of the NK-2–positive stripe of nuclei and the position of the stripe on the ventral-dorsal axis of the embryo are not fixed, but can be shifted by the combined effects of proteins that induce and repress the NK-2 gene. Rao, Vaessin, Jan, and Jan[63] have shown previously that the position of the neuroectoderm in the *Drosophila* embryo is shifted in appropriate mutants. Levine and his colleagues[64–66] have shown that the leading and trailing edges of a vertical stripe of nuclei that express the *eve* gene (eve stripe 2) are formed independently by giant and Krüpple, respectively, which repress the *eve* gene. Although the gradient of inducer and the repressors are different, the formation of a horizontal stripe of nuclei that express the NK-2 gene resembles the formation of a vertical stripe of nuclei that express the *eve* gene.

Neuroectodermal cells develop at different rates and segregate as neuroblasts at different times, depending upon their position in the embryo (for review see Ref. 67). Therefore, it is likely that the expression of a proneural gene is the rate-limiting step in the development of ventral neuroectodermal cells and segregation of medial neuroblasts. Our working hypothesis is that NK-2 is a proneural gene required for the formation of medial neuroblasts. Deletion of the NK-2 gene and some neighboring genes is a homozygous lethal deficiency and results in embryos with grossly defective ventral nerve cords that lack many neurons compared to wild-type embryos.[68] At the present time we are trying to obtain specific mutations of the NK-2 gene.

Initially all nuclei in the ventrolateral neurogenic anlage are committed to the neuroblast pathway of development. However, only about 25% of the neuroectodermal cells segregate as neuroblasts; most of the remaining neuroectodermal cells become ventrolateral epidermoblasts. Campos-Ortega and others have identified a set of neurogenic genes, *Notch*, *Delta*, *almondex*, *big brain*, *master mind*, *neuralized*, and the *Enhancer of split* [*E(spl)*] complex of genes, whose expression is required to turn off the neuroblast pathway of development and/or turn on the epidermoblast pathway of development (for reviews see Refs. 67, 69, and 70). *Delta* and *Notch* encode cell membrane proteins that interact with one another and contain multiple EGF repeats[71–75] (Delta and Notch proteins are thought to function as a ligand and corresponding receptor). The *E(spl)* complex of genes contains a cluster of related genes that encode similar basic helix-loop-helix DNA binding proteins (HLH-m3, HLH-m5, HLH-m7, HLH-m8 [E(spl)], HLH-mβ, HLH-mγ, and HLH-mδ, which are thought to be required for epidermoblast development,[76-79] although some redundancy of HLH proteins is likely. The available information suggests that direct contact between a segregated neuroblast and neighboring neuroectodermal cells turns off the neural pathway of development and activates the epidermoblast pathway of development in the neuroectodermal cells, a process termed lateral inhibition. The switch from neuroectodermal to epidermoblast pathway of development is blocked by mutation of a neurogenic gene (or by deletion of the *E(spl)* complex of genes resulting in overproduction of neuroblasts and underproduction of epidermoblasts). Mellerick and Nirenberg[80] have shown that a null mutation of the *Delta* gene or deletion of the *E(spl)* gene complex results in overproduction of neuroblasts that express the NK-2 gene. These results show that neuroectodermal cells that express the NK-2 gene are sensitive to lateral inhibition and suggest that one or more of the E(spl) HLH proteins repress the NK-2 gene. Delta probably represses the NK-2 gene indirectly by signalling activation of *E(spl)* HLH genes. One possibility that remains to be explored is that repression of the NK-2 gene by E(spl) HLH proteins may extinguish an NK-2-dependent pathway for medial neuroblast development and activate the epidermal pathway of development.

The NK-2 homeodomain was expressed in *E. coli* and purified to essential homogeneity. Binding studies to oligodeoxynucleotides showed that the consensus nucleotide sequence for NK-2 homeodomain binding is TNAAGTGG, and that the K_D is approximately 2×10^{-10} M.[81] Twenty high-affinity and additional lower-affinity NK-2 homeodomain binding sites were found in 2.2 kb of the 5'-upstream region of the NK-2 gene,[81] which suggests that NK-2 protein may be required to maintain NK-2 gene expression.

Little is known about the functions of the NK-2 homeodomain protein. Our working hypothesis is that NK-2 is a proneural gene that may be required for the development of a subset of neuroblasts.

REFERENCES

1. KIM, Y. & M. NIRENBERG. 1989. *Drosophila* NK-homeobox genes. Proc. Natl. Acad. Sci. USA **86:** 7716–7720.
2. FOE, V. E. & B. M. ALBERTS. 1983. Studies of nuclear and cytoplasmic behavior during the five mitotic cycles that precede gastrulation in *Drosophila* embryogenesis. J. Cell Sci. **61:** 31–70.
3. CAMPOS-ORTEGA, J. A. & V. HARTENSTEIN. 1985. The embryonic development of *Drosophila melanogaster*. Springer-Verlag. Berlin-Heidelberg-New York.
4. FRONHÖFER, H. G. & C. NÜSSLEIN-VOLHARD. 1986. Organization of anterior pattern in the *Drosophila* embryo by the maternal gene *bicoid*. Nature **324:** 120–125.
5. DRIEVER, W. & C. NÜSSLEIN-VOLHARD. 1988. The bicoid protein determines

position in the *Drosophila* embryo in a concentration-dependent manner. Cell **54:** 95–104.
6. DRIEVER, W. & C. NUSSLEIN-VOLHARD. 1988. A gradient of bicoid protein in *Drosophila* embryos. Cell **54:** 83–93.
7. IRISH, V., R. LEHMANN & M. AKAM. 1989. The *Drosophila* posterior-group gene *nanos* functions by repressing hunchback activity. Nature **338:** 646–648.
8. WANG, C. & R. LEHMANN. 1991. Nanos is the localized posterior determinant in *Drosophila*. Cell **66:** 637–647.
9. WHARTON, R. P. & G. STRUHL. 1991. RNA regulatory elements mediate control of *Drosophila* body pattern by the posterior morphogen nanos. Cell **67:** 955–967.
10. LEHMANN, R. & C. NUSSLEIN-VOLHARD. 1991. The maternal gene *nanos* has a central role in posterior pattern formation of the *Drosophila* embryo. Development **112:** 679–691.
11. STEWARD, R., S. B. ZUSMAN, L. H. HUANG & P. SCHEDL. 1988. The dorsal protein is distributed in a gradient in early *Drosophila* embryos. Cell **55:** 487–495.
12. STEWARD, R. 1989. Relocalization of the dorsal protein from the cytoplasm to the nucleus correlates with its function. Cell **59:** 1179–1188.
13. ROTH, S., D. STEIN & C. NUSSLEIN-VOLHARD. 1989. A gradient of nuclear localization of the dorsal protein determines dorsoventral pattern in the *Drosophila* embryos. Cell **59:** 1189–1202.
14. RUSHLOW, C. A., K. HAN, J. L. MANLEY & M. LEVINE. 1989. The graded distribution of the dorsal morphogen is initiated by selective nuclear transport in *Drosophila*. Cell **59:** 1165–1177.
15. INGHAM, P. W. 1987. The molecular genetics of embryonic pattern formation in *Drosophila*. Nature **335:** 25–34.
16. ST. JOHNSTON, D. & C. NÜSSLEIN-VOLHARD. 1992. The origin of pattern and polarity in the *Drosophila* embryo. Cell **68:** 201–219.
17. HOCH, M. & H. JÄCKLE. 1993. Transcriptional regulation and spatial patterning in *Drosophila*. Curr. Opin. Genet. Dev. **3:** 566–573.
18. KORNBERG, T. B. & T. TABATA. 1993. Segmentation of the *Drosophila* embryo. Curr. Opin. Genet. Dev. **3:** 585–595.
19. NAKAYAMA, K., N. NAKAYAMA, Y. S. KIM, K. O. WEBBER, R. LAD & M. NIRENBERG. The NK-2 Homeobox Gene of *Drosophila*. Proc. Natl. Acad. Sci. USA. In press.
20. SAHA, M. S., R. B. MICHEL, K. M. GULDING & R. M. GRAINGER. 1993. A *Xenopus* homeobox gene defines dorsal-ventral domains in the developing brain. Development **118:** 193–202.
21. PRICE, M., D. LAZZARO, T. POHL, M-G. MATTEI, U. RÜTHER, J. C. OLIVO, D. DUBOULE & R. DI LAURO. 1992. Regional expression of the homeobox gene *Nkx-2.2* in the developing mammalian forebrain. Neuron **8:** 241–255.
22. LINTS, T. J., L. M. PARSONS, L. HARTLEY, I. LYONS & R. P. HARVEY. 1993. *Nkx-2.5*: A novel murine homeobox gene expressed in early heart progenitor cells and their myogenic descendants. Development **119:** 419–431.
23. GARCIA-FERNANDEZ, J., J. BAGUNA & E. SALO. 1993. Genomic organization and expression of the planarian homeobox genes *Dth-1* and *Dth-2*. Development **118:** 241–253.
24. GARCIA-FERNANDEZ, J., J. BAGUNA & E. SALO. 1991. Planarian homeobox genes: Cloning, sequence analysis, and expression. Proc. Natl. Acad. Sci. USA **88:** 7338–7342.
25. NARDELLI-HAEFLIGER, D. & M. SHANKLAND. 1993. *Lox 10*, a member of the

NK-2 homeobox gene class, is expressed in a segmental pattern in the endoderm and in the cephalic nervous system of the leech *Helobdella*. Development **118**: 877–892.
26. OLIVER, G., M. VISPO, A. MAILHOS, C. MARTINEZ, B. SOSA-PINEDA, W. FIELITZ & R. EHRLICH. 1992. Homeoboxes in flatworms. Gene **121**: 337–342.
27. TSAO, D. H. H., J. M. GRUSCHUS, L.-H. WANG, M. NIRENBERG & J. A. FERRETTI. 1994. Elongation of helix III of the NK-2 homeodomain upon binding to DNA: A secondary structure study by NMR. Biochemistry **33**: 15053–15060.
28. AZPIAZU, N. & M. FRASCH. 1993. *tinman* and *bagpipe*: two homeobox genes that determine cell fates in the dorsal mesoderm of *Drosophila*. Genes Dev. **7**: 1325–1340.
29. WEBBER, K. O., Y. KIM, R. LAD, K. NAKAYAMA & M. NIRENBERG. Unpublished manuscript.
30. THISSE, C., F. PERRIN-SCHMITT, C. STOEZEL & B. THISSE. 1991. Sequence specific transactivation of the *Drosophila twist* gene by the dorsal gene product. Cell **65**: 1191–1201.
31. JIANG, J., D. KOSMAN, Y. T. IP & M. LEVINE. 1991. The dorsal morphogen gradient regulates the mesoderm determinant *twist* in early *Drosophila* embryos. Genes Dev. **5**: 1881–1891.
32. PAN, D., J. D. HUANG & A. J. COUREY. 1991. Functional analysis of the *Drosophila* twist promoter reveals a dorsal-binding ventral activator region. Genes Dev. **5**: 1892–1901.
33. PAN, D. & A. J. COUREY. 1992. The same dorsal binding sites mediate both activation and repression in a context dependent manner. EMBO J. **11**: 1837–1842.
34. LEPTIN, M. 1991. twist and snail as positive and negative regulators during *Drosophila* mesoderm development. Genes Dev. **5**: 1568–1576.
35. KOSMAN, D., Y. T. IP, M. LEVINE & K. ARORA. 1991. Establishment of the mesoderm-neuroectoderm boundary in the *Drosophila* embryo. Science **254**: 118–122.
36. IP, Y. T., R. E. PARK, D. KOSMAN, K. YAZDANBAKHSH & M. LEVINE. 1992. dorsal-twist interactions establish *snail* expression in the presumptive mesoderm of the *Drosophila* embryo. Genes Dev. **6**: 1518–1530.
37. ST. JOHNSTON, R. D. & W. M. GELBART. 1987. *decapentaplegic* transcripts are localized along the dorsal-ventral axis of the *Drosophila* embryo. EMBO J. **6**: 2785–2791.
38. IRISH, V. F. & W. M. GELBART. 1987. The *decapentaplegic* gene is required for dorsal-ventral patterning of the *Drosophila* embryo. Genes Dev. **1**: 868–879.
39. RAY, R. P., K. AROR, C. NUSSLEIN-VOLHARD & W. M. GELBART. 1991. The control of cell fate along the dorsal-ventral axis of the *Drosophila* embryo. Development **113**: 35–54.
40. HUANG, J-D., D. H. SCHWYTER, J. M. SHIROKAWA & A. J. COUREY. 1993. The interplay between multiple enhancer and silencer elements defines the pattern of *decapentaplegic* expression. Genes Dev. **7**: 694–704.
41. PADGETT, R. W., R. D. ST. JOHNSTON & W. M. GELBART. 1987. A transcript from a *Drosophila* pattern gene predicts a protein homologous to the transforming growth factor β family. Nature **325**: 81–84.
42. DOYLE, H., K. HARDING, T. HOEY & M. LEVINE. 1986. Transcripts encoded by a homeobox gene are restricted to dorsal tissues of *Drosophila* embryos. Nature **323**: 76–79.

43. DOYLE, H. J., R. KRAUT & M. LEVINE. 1989. Spatial regulation of *zerknullt*: A dorsal-ventral patterning gene in *Drosophila*. Genes Dev. **3**: 1518-1533.
44. RUSHLOW, C., H. DOYLE, T. HOEY & M. LEVINE. 1987. Molecular characterization of the *zerknullt* region of the *Antennapedia* gene complex in *Drosophila*. Genes Dev. **1**: 1268-1279.
45. IP, Y. T., R. KRAUT, M. LEVINE & C. A. RUSHLOW. 1991. The dorsal morphogen is a sequence-specific DNA-binding protein that interacts with a long range repression element in *Drosophila*. Cell **64**: 439-446.
46. JIANG, J., C. A. RUSLOW, Q. ZHOU, S. SMALL & M. LEVINE. 1992. Individual dorsal morphogen binding sites mediate activation and repression in the *Drosophila* embryo. EMBO J. **11**: 3147-3154.
47. HARTENSTEIN, V. & J. A. CAMPOS-ORTEGA. 1987. Early neurogenesis in wild-type *Drosophila melanogaster*. Roux's Arch. Dev. Biol. **196**: 473-485.
48. HARTENSTEIN, V., E. RUDLOFF & J. A. CAMPOS-ORTEGA. 1984. The pattern of proliferation of the neuroblasts in the wild-type embryo of *Drosophila melanogaster*. Roux's Arch. Dev. Biol. **193**: 308-325.
49. DOE, C. Q. 1992. Molecular markers for identified neuroblasts and ganglion mother cells in the *Drosophila* central nervous system. Development **116**: 855-863.
50. PATEL, N. H., B. SCHAFER, C. S. GOODMAN & R. HOLMGREM. 1989. The role of segment polarity genes during *Drosophila* neurogenesis. Genes Dev. **3**: 890-904.
51. MELLERICK, D. M. & M. NIRENBERG. Dorsal-ventral patterning genes restrict NK-2 homeobox gene expression to the ventral half of the central nervous system of *Drosophila* embryos. Dev. Biol. In press.
52. FERGUSON, E. L. & K. V. ANDERSON. 1991. Dorsal-ventral pattern formation in the *Drosophila* embryo: The role of zygotically active genes. Curr. Top. Dev. Biol. **25**: 17-43.
53. IP, Y. T. & M. LEVINE. 1992. The role of the *dorsal* morphogen gradient in *Drosophila* embryogenesis. Seminars Dev. Biol. **3**: 15-23.
54. STEWARD, R. & S. GOVIND. 1993. Dorsal-ventral polarity in the *Drosophila* embryo. Curr. Opin. Genet. Dev. **3**: 556-561.
55. BOULAY, J. L., C. DENNEFELD & A. ALBERGA. 1987. The *Drosophila* developmental gene *snail* encodes a protein with nucleic acid binding fingers. Nature **330**: 395-398.
56. ALBERGA, A., J. L. BOULAY, E. KEMPE, C. DENNEFELD & M. HAENLIN. 1991. The *snail* gene required for mesoderm formation in *Drosophila* is expressed dynamically in derivatives of all three germ layers. Development **111**: 983-992.
57. THOMAS, J. B., S. T. CREWS & C. S. GOODMAN. 1988. Molecular genetics of the *single-minded* locus: A gene involved in the development of the *Drosophila* nervous system. Cell **52**: 133-141.
58. CREWS, S. T., J. B. THOMAS & C. S. GOODMAN. 1988. The *Drosophila single-minded* gene encodes a nuclear protein with sequence similarity to the *per* gene product. Cell **52**: 143-151.
59. NAMBU, J. R., R. G. FRANKS, S. HU & S. T. CREWS. 1990. The *single-minded* gene of *Drosophila* is required for the expression of genes important for the development of CNS midline cells. Cell **63**: 63-75.
60. KASAI, Y., J. R. NAMBU, P. M. LIEBERMAN & S. T. CREWS. 1992. Dorsal-ventral patterning in *Drosophila*: DNA binding of snail protein to the *single-minded* gene. Proc. Natl. Acad. Sci. USA **89**: 3414-3418.
61. THISSE, B., C. STOETZEL, M. EL MESSAL & F. PERRIN-SCHMITT. 1987. Genes of

the *Drosophila* maternal dorsal group control the specific expression of the zygotic gene *twist* in the presumptive mesodermal cells. Genes Dev. **1:** 709–715.
62. THISSE, B., C. STOETZEL, C. GOROSTIZA-THISSE & F. PERRIN-SCHMITT. 1988. Sequence of the *twist* gene and nuclear localization of its protein in endomesodermal cells of early *Drosophila* embryos. EMBO J. **7:** 2175–2183.
63. RAO, Y., H. VAESSIN, L. Y. JAN & Y. N. JAN. 1991. Neuroectoderm in *Drosophila* embryos is dependent on the mesoderm for positioning but not for formation. Genes Dev. **5:** 1577–1588.
64. SMALL, S., R. KRAUT, T. HOEY, R. WARRIOR & M. LEVINE. 1991. Transcriptional regulation of a pair-rule stripe in *Drosophila*. Genes Dev. **5:** 827–839.
65. STANOJEVIC, D., S. SMALL & M. LEVINE. 1991. Regulation of a segmentation stripe by overlapping activators and repressors in the *Drosophila* embryo. Science **254:** 1385–1387.
66. SMALL, S., A. BLAIR & M. LEVINE. 1992. Regulation of even-skipped stripe 2 in the *Drosophila* embryo. EMBO J. **11:** 4047–4057.
67. CAMPOS-ORTEGA, J. A., also GOODMAN, C. S. 1993. *In* The Development of *Drosophila melanogaster*. M. Bate & A. Martinez-Arias, Eds. Cold Spring Harbor Laboratory Press. Cold Spring Harbor, New York.
68. MELLERICK, D. & M. NIRENBERG. Unpublished material.
69. CABRERA, C. V. 1992. The generation of cell diversity during early neurogenesis in *Drosophila*. Development **115:** 893–901.
70. FORTINI, M. E. & S. ARTAVANIS-TSAKONAS. 1993. *Notch*: Neurogenesis is only part of the picture. Cell **75:** 1245–1247.
71. VAESSIN, H., K. A. BREMER, E. KNUST & J. A. CAMPOS-ORTEGA. 1987. The neurogenic gene *Delta* of *Drosophila melanogaster* is expressed in neurogenic territories and encodes a putative transmembrane protein with EGF-like repeats. EMBO J. **6:** 3433–3440.
72. WHARTON, K. A., K. M. JOHANSEN, T. XU & S. ARTAVANIS-TSAKONAS. 1985. Nucleotide sequence from the neurogenic locus *Notch* implies a gene product that shares homology with proteins containing EGF-like repeats. Cell **43:** 567–581.
73. REBAY, I., R. J. FLEMING, R. G. FEHON, L. CHERBAS, P. CHERBAS & S. ARTAVANIS-TSAKONAS. 1991. Specific EGF repeats of *Notch* mediate interactions with *Delta* and *Serrate*: Implications for *Notch* as a multifunctional receptor. Cell **67:** 687–699.
74. FEHON, R. G., P. J. KOOH, I. REBAY, C. L. REGAN, T. XU, M. A. T. MUSKAVITCH & S. ARTAVANIS-TSAKONAS. 1990. Molecular interactions between the protein products of the neurogenic loci *Notch* and *Delta*, two EGF-homologous genes in *Drosophila*. Cell **61:** 523–534.
75. BRAND, M. & J. A. CAMPOS-ORTEGA. 1990. Second-site modifiers of the split mutation of *Notch* define genes involved in neurogenesis in *Drosophila melanogaster*. Roux's Arch. Dev. Biol. **198:** 275–285.
76. KLAMBT, C., E. KNUST, K. TIETZE & J. A. CAMPOS-ORTEGA. 1989. Closely related transcripts encoded by the neurogenic gene complex *Enhancer of split* of *Drosophila melanogaster*. EMBO J. **8:** 203–210.
77. DELIGAKIS, C. & S. ARTAVANIS-TSAKONAS. 1992. The *Enhancer of split* [E (spl)] locus of *Drosophila* encodes seven independent helix-loop-helix proteins. Proc. Natl. Acad. Sci. USA **89:** 8731–8735.
78. KNUST, E., H. SCHRONS, F. GRAW & J. A. CAMPOS-ORTEGA. 1992. Seven genes of the *Enhancer of split* complex of *Drosophila melanogaster* encode helix-loop-helix proteins. Genetics **132:** 505–518.

79. SCHRONS, H., E. KNUST & J. A. CAMPOS-ORTEGA. 1992. The *Enhancer of split* complex and adjacent genes in the 96F region of *Drosophila melanogaster* are required for segregation of neural and epidermal progenitor cells. Genetics **132:** 481–503.
80. MELLERICK, D. & M. NIRENBERG. Manuscript in preparation.
81. WANG, L.-H., R. CHMELIK & M. NIRENBERG. Manuscript in preparation.

PART VI. DNA AND MOLECULAR MEDICINE

Introduction

LORD WALTON OF DETCHANT

Green College
University of Oxford
Oxford, OX2 6HG United Kingdom

I am delighted and honored to have been invited to chair this session, first because I was Warden of Green College between 1983 and 1989, following Sir Richard Doll, and being in turn succeeded by Sir Crispin Tickell. Thus all three Wardens of Green College in Oxford are participating in this notable and exciting meeting. And it was during my Wardenship that our distinguished Chairman, Donald Chambers, was a Visiting Scholar of that College, upon which he left a lasting impression because of his lively contributions to College life in a scientific, intellectual and social sense. It is a pleasure to be associated with the New York Academy of Sciences as I have had the privilege of contributing to several meetings organized by that notable body in the past, and I have also been privileged to visit the University of Illinois, another co-sponsor, as a visiting lecturer on more than one occasion, the first time some 40 years ago when I was warmly greeted by the great Percival Bailey.

It is, of course, our purpose at this meeting to pay homage to the seminal discoveries of Francis Crick, James Watson, and Maurice Wilkins, who it can truthfully be said laid the foundations of modern molecular biology. In this session we shall be dealing with the way in which DNA recombinant technology and all of its scientific progeny have led to burgeoning developments in molecular medicine, a burgeoning that has already transformed, and will soon transform even more fundamentally, our knowledge and management of inherited disease. As President of the World Federation of Neurology, and as one with a life-long interest in research into the muscular dystrophies and other neuromuscular diseases, I have watched with growing delight, even at times amazement, the way in which the fundamental discoveries of basic laboratory science have led, and are indeed continuing to lead, to practical developments in disease management. When in the early 1950s I reported the first case of Duchenne dystrophy in a morphological female with Turner's syndrome and an XO chromosome constitution, thus giving final confirmation of X-linked inheritance of this disease, I firmly believed that localization and characterization of the causal gene, bringing with it the prospects of effective treatment, might come during my professional lifetime. As we worked on indirect techniques of carrier detection in the female relatives of boys suffering from the condition, and as knowledge grew, I was able to say that even though detection of the gene seemed rather like finding a single misprint in a book the size of the Bible, we eventually found ourselves in the correct building, later on the right floor and subsequently in the appropriate room. Finally, through the superb work of Lou Kunkel and his colleagues, building, as he freely acknowledged, upon the findings of many others throughout the world, in 1987 the gene was localized, characterized, and sequenced and soon the missing or abnormal gene product, namely dystrophin, was identified. Now, in that disease precise carrier detection and antenatal diagnosis through chorionic biopsy have become a reality, prevention via embryo biopsy and pre-implantation diagnosis is becoming so, and work on naturally occurring models in the mouse and the dog and cat leads us to believe that gene therapy for this tragic progressive disease may just lie round the corner. That discovery alone, in which I confess to having a close personal interest, has brought hope to thousands across the world, but there are innumerable other examples about which we shall be hearing at this meeting which impinge not just upon the clinical neurosciences, but upon medicine as a whole.

In the history of human disease, 40 years may seem a professional lifetime to some, but in science it is a relatively short timespan, and what once seemed like science fiction has become, through the fundamental contributions of those whom we honor at this symposium, a shining success story with unlimited potential for the improvement of human health. I therefore have much pleasure in opening this exciting session.

The Molecular Basis for Phenotypic Diversity of Genetic Disease

DAVID J. WEATHERALL

Institute of Molecular Medicine
University of Oxford
John Radcliffe Hospital
Oxford OX2 9DU, England

INTRODUCTION

The phenotypic effects of what appear to be mutations at the same gene locus may be extremely variable, a phenomenon which is reflected in terms like "penetrance" and "expresstivity." It has important implications for predictive genetics, particularly now that thoughts are turning to the widespread prenatal detection of monogenic disease, or gene therapy in early life.[1]

Once single-gene disorders became amenable to study at the molecular level, it was assumed that the reasons for their phenotypic variability would become apparent. However, progress in this important aspect of medical genetics has been slow. Indeed, except in the case of the hemoglobin disorders there is still very little information about the inherited and acquired factors that modify the expression of these conditions.

The problem of phenotypic heterogeneity of inherited disease has implications which stretch beyond a better understanding of monogenic disorders. One of the major aims of modern medical genetics is to try to dissect the complex interactions between the genotype and environment which underlie common intractable diseases such as heart disease, cancer, major psychoses, and many other chronic disorders.[1] Before tackling these complex problems it would be extremely useful to have some knowledge of the way in which the phenotypes of more simple monogenic diseases can be modified. We need to understand how much heterogeneity can be generated at a single locus, how many other loci may be involved, and to what extent acquired and environmental factors are implicated.

In this short review we shall summarize the molecular basis for the heterogeneity of the human globin disorders and what little is known about the causes for the clinical variability of other monogenic diseases. We shall then consider the implications of this information for the broader problems of the analysis of human polygenic disease.

MOLECULAR BASIS FOR THE PHENOTYPIC VARIABILITY OF THE HUMAN GLOBIN DISORDERS

The genetic disorders of human hemoglobin are the commonest single-gene disorders in man. They were among the first to be analyzed by the methods of recombinant DNA technology and a great deal is known about the molecular basis for their heterogeneity, though many questions still remain.

Human Hemoglobin and Its Disorders

Human adult hemoglobin is a mixture of proteins, consisting of a major component, hemoglobin A (Hb A), with a minor component (Hb A_2), which constitutes about 2.5 percent of the total. During early development there are several embryonic hemoglobins, after which the main hemoglobin during intrauterine life is Hb F.

The structure of these hemoglobins is similar.[2] Each consists of two separate pairs of identical globin chains. Except for some of the embryonic hemoglobins, all the normal human hemoglobins have one pair of α chains: in Hb A they are combined with β chains ($\alpha_2\beta_2$), in Hb A_2 with δ chains ($\alpha_2\delta_2$), and in Hb F with γ-chains ($\alpha_2\gamma_2$). Hemoglobin F is a mixture of two molecular forms with the formulas $\alpha_2\gamma_2^{136Gly}$ and $\alpha_2\gamma_2^{136Ala}$; the γ-chains containing glycine at position 136 are designated $^G\gamma$ chains and those that contain alanine in this position are called $^A\gamma$ chains. Before the eighth week of intrauterine life there are three embryonic hemoglobins, Gower 1 ($\zeta_2\epsilon_2$), Gower 2 ($\alpha_2\epsilon_2$), and Portland ($\zeta_2\gamma_2$).

The β-like globin genes are controlled by a gene cluster on chromosome 11 in which the different genes are arranged in the order 5'-ϵ-$^G\gamma$-$^A\gamma$-$\psi\beta$-δ-β-3'. The α-like gene cluster is on the chromosome 16, p13.3, and the genes are arranged in the order 5'-ζ-$\psi\zeta$-$\psi\alpha$-$\alpha2$-$\alpha1$-θ-3'. The arrangement of these clusters, together with the sites of their major regulatory elements (locus control regions), is shown in FIGURE 1.

There are several distinct classes of inherited disorders of hemoglobin. First there are the structural variants of the α, β, δ and γ genes. Most of these result from single amino acid substitutions in the particular globin genes involved, although there are some more subtle structural variants due to frameshifts, internal deletions, or chain termination mutations. The second group comprises the thalassemias, a heterogeneous group of disorders characterized by the reduced output of one or more of the globin chains. The important forms are the α-, β-, and $\delta\beta$-thalassemias. They are further subclassified according to whether the particular globin chain is produced ineffectively, or not at all; β^0-thalassemia describes a condition in which no β chains are produced, whereas β^+- or β^{++}-thalassemia is used to designate a wide spectrum of dis-

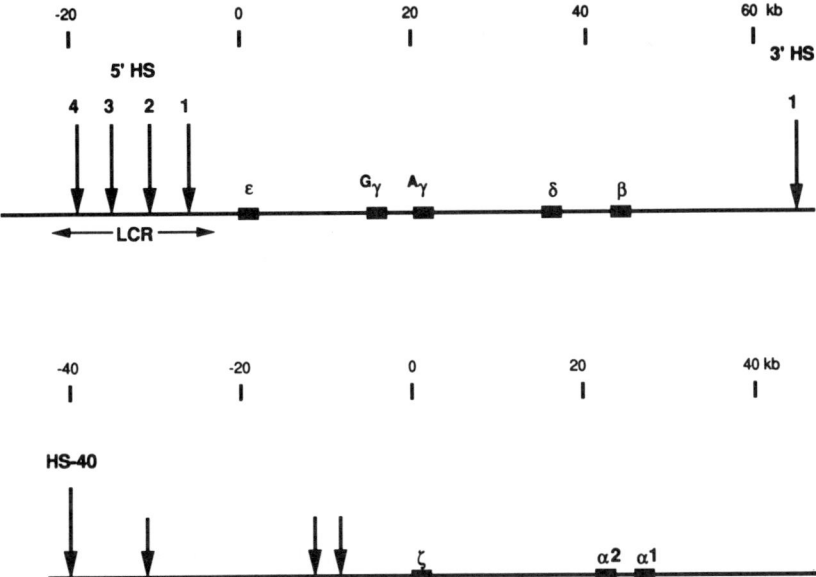

FIGURE 1. The human globin gene clusters. The hypersensitive sites (HS) and major control regions (locus control region–LCR and HS40) are shown.

orders in which the amount of β chain production is reduced by a variable amount. The α-thalassemias are similarly subclassified. Finally, there is a heterogeneous collection of conditions that go under the general title "hereditary persistence of fetal hemoglobin," in which Hb F is produced in variable levels in adult life.

Overall, the genotype/phenotype relationships for the structural hemoglobin disorders depend on the site of the particular amino acid substitution and its effect on the stability and function of the hemoglobin molecule.[2] Different clinical syndromes may be produced because of the synthesis of high or low oxygen affinity variants or unstable hemoglobins. The commonest structural variant, Hb S, is responsible for sickle cell anemia. Although this was the first human disease to be defined at the molecular level, progress in understanding its remarkable clinical heterogeneity has been disappointing. It is known that the disease may be modified by factors such as the coinheritance of α thalassemia or different genes that cause elevated fetal hemoglobin levels in adult life. But these factors only seem to account for a modest amount of the remarkable clinical heterogeneity of this condition. Currently, we seem a long way from being able to predict the clinical course and outlook for an individual patient with sickle cell anemia or its variants.

Much more progress has been made in determining the molecular basis for the remarkable heterogeneity of the thalassemias. We shall illustrate this by describing some of the factors which modify the phenotype of the two commonest classes, the β- and α-thalassemias.

The Molecular Basis for Phenotypic Variability of β-Thalassemia

The clinical manifestations of the β-thalassemias are remarkably heterogeneous.[3] The picture associated with the homozygous or compound heterozygous states range from life-threatening anemia and complete transfusion-dependence, through milder forms of anemia which require infrequent or no transfusion, to a completely symptomless condition in which the hemoglobin level may be only a little below normal. Even more surprisingly, a similar degree of heterogeneity characterizes the heterozygous states for β-thalassemia. In most cases they are characterized by a mild anemia with small, underhemoglobinized red cells. But it is now clear that some forms of the β-thalassemia trait are completely silent, with no hematologic abnormalities whatever. At the other end of the spectrum, there are much more severe forms, which are characterized by a dominant pattern of inheritance, with moderate to severe anemia in those affected with only a single abnormal β-globin gene.

The major mechanisms underlying the phenotypic heterogeneity of the β-thalassemias are summarized in FIGURE 2. They are most easily appreciated from an understanding of the pathophysiology of the disease.[3] The main reason for the defective red cell production that characterizes all the thalassemias is imbalanced globin chain synthesis. In β-thalassemia the defective output of β-globin chains leads to an excess of α chains, which precipitate in the red cell precursors and cause their destruction, either within the bone marrow or after the red cells are liberated into the circulation. It is now clear that most of the factors which modify the phenotype of β-thalassemia are involved with determining the degree of imbalanced globin chain production; the fewer excess α chains that are free to precipitate, the less severe the anemia, and hence the clinical phenotype.

Some, but not all of the heterogeneity of β-thalassemia can be ascribed to the many different mutations which involve the β-globin genes.[4,5] The hundred or more varieties of β-thalassemia are composed of nonsense and frameshift mutations which completely inactivate the β-globin genes, mutations involving the splice junctions or the consensus sequences close to them, mutations which activate cryptic splice sites in the exons or introns, and mutations which involve the upstream promoter elements or the β-globin gene termination codon. Rarely, the β-globin genes are partially or completely deleted. Usually, nonsense and frameshift mutations are associated with a severe transfusion-dependent phenotype. This is also the case for some, but not all, of the splice mutations. Mutations of the invariant GT or AG dinucleotides at the donor (5') and recipient (3') sites at exon/intron boundaries always

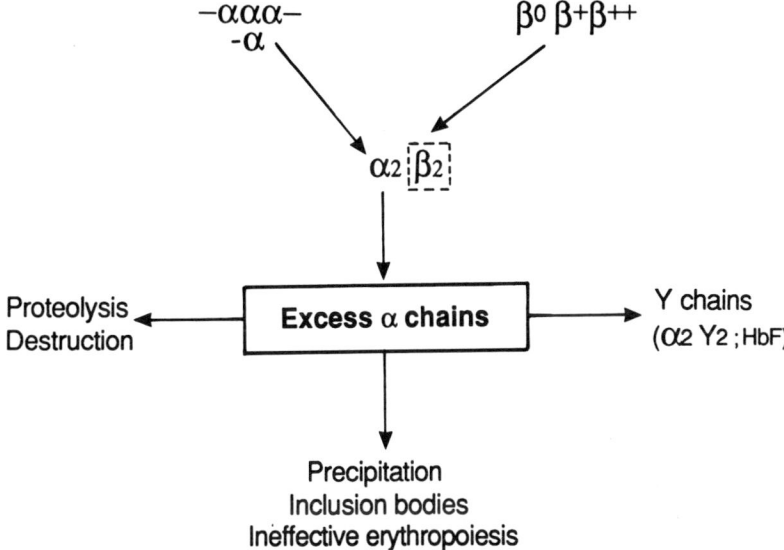

FIGURE 2. Summary of the main mechanisms for modifying globin chain imbalance in β-thalassemia.

cause β^0 thalassemia. On the other hand, some mutations involving consensus sequences in introns, or cryptic splice sites, may result in an extremely mild defect in β chain production. The same applies to many of those which involve the upstream promoter elements. Homozygosity for particular splice or promoter site mutations may be associated with an extremely mild phenotype.

Largely as the result of compound heterozygosity for different β-thalassemia mutations, almost every gradation of clinical severity may occur, ranging from a transfusion-dependent to a completely asymptomatic disorder.[4-6] For example, the co-inheritance of chromosomes with a β^0-thalassemia (codon 39 C→T) mutation and an extremely mild β^+-thalassemia (IVS 1-6 C→T) mutation results in a fairly severe anemia with occasional transfusional requirement. On the other hand, the inheritance of a moderately severe β^+-thalassemia allele together with a mild β^+-thalassemia allele results in only a moderate degree of anemia which is often associated with normal development without transfusion. Thus, simply because of heterogeneity of mutations involving the β globin genes alone, a wide spectrum of clinical variability can be observed in the homozygous or compound heterozygous states.

But the situation is much more complex than this. For example, in Mediterranean populations many patients with the common splice mutation (IVS 1-110 G→G) are transfusion-dependent, but there is a small subgroup who are not. It turns out that they have inherited two α^+-thalassemia genes (see the next section). The result of this remarkable experiment of nature is that, although these individuals have a severe deficit of β-globin, the co-existent deficit of α-globin results in more balanced globin chain synthesis. Thus, although these patients have poorly hemoglobinized red cells, the degree of α-globin precipitation is less and, hence, because they have more effective erythropoiesis, their anemia is not so severe. Because there is a broad spectrum of β-thalassemias, and patients may inherit one or more α-thalassemia genes, this simple interaction also accounts for a wide degree of heterogeneity of the phenotypes of β-thalassemics.[5,6]

There is an additional factor that modifies the expression of β-thalassemia. Some patients who are homozygous for severe forms of β^0-thalassemia are not transfusion-dependent, have relatively high hemoglobin levels, and grow and develop normally. There is increasing evidence that this occurs because they have an inherited ability to synthesize larger levels of fetal hemoglobin than normal in response to the β-thalassemia. As shown in FIGURE 2, persistent γ chain production is extremely valuable in β-thalassemia, because γ chains combine with α chains to form Hb F and hence there is less α chain precipitation and ineffective erythropoiesis.

The mechanisms for the variable persistence of Hb F production in β-thalassemia and sickle cell anemia is not yet understood.[5] In a few cases there are point mutations in the promoter regions of the γ globin genes, or major deletions involving the β-globin gene cluster, but in the majority the γ-globin genes are intact. There is some evidence that a polymorphism at position -158 in the $^G\gamma$ genes[7] may be directly involved in increased Hb F production in association with the hematologic stress of the β chain hemoglobinopathies.[8-11] Furthermore, it is becoming apparent that some of the determinants for increased Hb F production that behave in this way are unlinked to the β-globin gene cluster.[12] In short, there is also considerable heterogeneity of the factors that determine the degree to which γ chain synthesis may persist into adult life.

What is the basis for the heterogeneity of the heterozygous states for β-thalassemia? In most cases this condition is characterized by mild anemia and small red cells. Occasionally, however, it is clinically and hematologically silent. The presence of a "silent" form of β-thalassemia of this type can only be ascertained by the finding of a moderately severe form of β-thalassemia in a patient in whom one parent has typical β-thalassemia trait and the other appears to be normal; the added mild reduction of β chain production due to the silent allele transforms β-thalassemia trait to a more severe disease in the compound heterozygote. At least two molecular mechanisms for these silent forms of β-thalassemia have been discovered.[5] One involves an upstream promoter mutation at position -101 from the β-globin gene[13]; the other is due to a mutation in a cryptic splice site in exon 1 of the β-globin gene.[14]

In some families, particularly of northern European origin, there is a form of moderately severe β-thalassemia which is inherited as a mendelian domi-

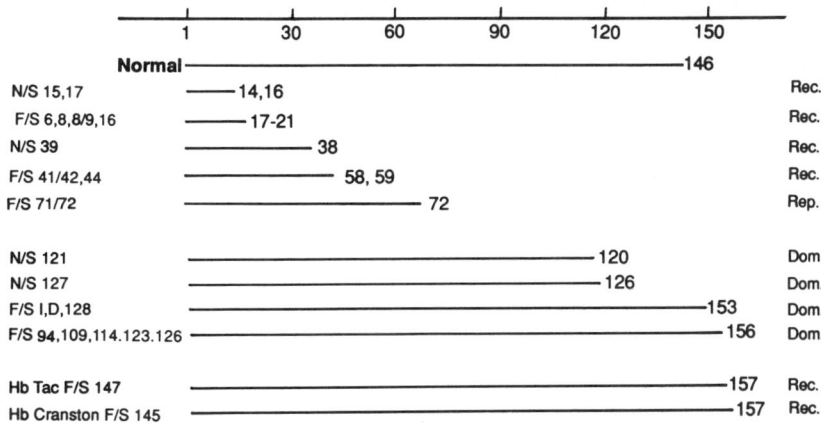

FIGURE 3. A comparison of the lengths of the abnormal gene products resulting from mutations of the β-globin genes. Dom = a dominant phenotype in heterozygotes; Rec = recessive inheritance.

nant. At first sight it was difficult to understand how the phenotype of heterozygous β-thalassemia could be more severe than that encountered with a mutation that completely inactivates the β-globin gene. However, recent studies have thrown some light on this conundrum.[15,16] It turns out that many of the mutations associated with dominant β-thalassemia involve exon 3 of the β-globin gene. They include frameshifts, premature chain termination mutations, and complex rearrangements which lead to the synthesis of truncated or elongated and highly unstable β-globin gene products.

A comparison of the lengths of abnormal gene products due to nonsense or frameshift mutations in the β-globin gene (FIG. 3) has suggested a mechanism to explain why host heterozygous forms of β-thalassemia are mild while those due to mutations in exon 3 are more severe.[15] Non-sense or frameshift mutations that produce truncated β chains up to about 72 residues in length are usually associated with the mild phenotype of typical β-thalassemia trait. Presumably these short β chain fragments are degraded and the small excess of α chains are removed in the same way. However, many exon 3 mutations produce longer truncated products or elongated but unstable β chains. It is likely that the severe phenotypes associated with these products reflect their heme binding properties and instability. Those with only 72 residues or less cannot bind heme, whereas those truncated to residue 120 or longer should bind heme since only helix H of the β chain is missing. Furthermore, such heme binding products should have secondary structure and hence be less susceptible to proteolytic degradation. It is believed, therefore, that the large inclusions that are found in the red cell progenitors of these heterozygous patients consist of aggregates of precipitated β chain products together with

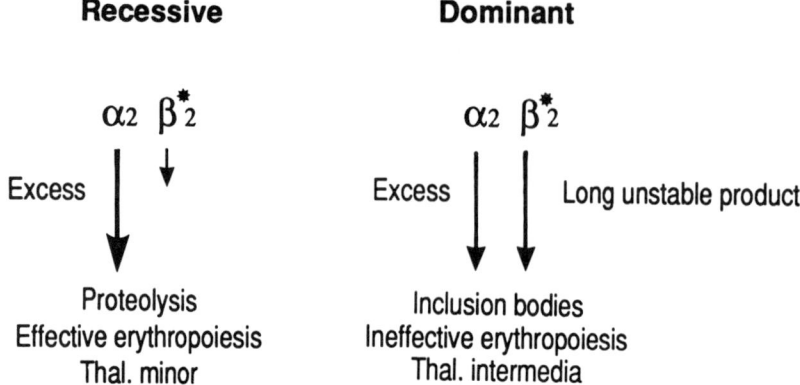

FIGURE 4. A summary of the mechanisms for recessive and dominant inheritance of β-thalassemia.

excess α chains (FIG. 4). This would explain the marked degree of ineffective erythropoiesis that is observed in this interesting condition.

Finally, heterozygous or homozygous β-thalassemia may be further modified by the inheritance of larger numbers of α globin genes than usual.[5,17] For example, the coinheritance of β-thalassemia trait with five α genes (that is, a triplicated α gene complex on one chromosome) may be associated with more severe anemia than is usually the case in the trait alone. Because there is a wide heterogeneity of the number of α globin genes in some populations, this offers another relatively common source of heterogeneity of the β-thalassemias. Again, it is mediated through variability of the imbalance of globin chain production.

In summary, much of the heterogeneity of the homozygous and heterozygous states for β-thalassemia can be explained by the wide range of different mutations which involve the β-globin gene, and by variability at two other related sets of genes, the α and γ gene families. Whether, as has been suggested, heterogeneity of the systems that are involved in the proteolytic degradation of the α globin chains may also be implicated is uncertain. But what is now clear is that much of the heterogeneity of this common disease can be ascribed to the various interactions at three gene loci. There are also, of course, a number of acquired and environmental factors that also modify the β-thalassemia phenotype.

The Heterogeneity of α-Thalassemia

Since there are two α-globin genes per haploid genome (that is, four in normal individuals), a good deal of the phenotypic variability of α-thalassemia can be ascribed to gene dosage effects consequent on deletions of one or more of these genes. The α^0-thalassemias usually result from partial or complete deletion of both of the linked α-globin genes, or from deletions upstream from the α gene cluster which remove the α-globin locus control region (LCR). α^+-thalassemia results from the deletion or inactivation by mutation of one of the linked pairs of α-globin genes.[17]

On the basis of simple dosage effects, several phenotypes can be observed. The homozygous state for α^0-thalassemia results in an absence of α-globin chain synthesis and death *in utero* or in the early neonatal period, the hemoglobin Bart's hydrops syndrome. Loss of three of the four α genes, usually by the co-inheritance of α^0- and α^+-thalassemia of the deletion or non-deletion type, results in hemoglobin H disease, that is, a condition associated with a variable degree of anemia and the presence of β^4 tetramers, or Hb H, in the peripheral blood. Because excess γ or β chains produce relatively stable homotetramers, there is less ineffective red cell production and the anemia is more hemolytic in character than that in β-thalassemia; Hb H precipitates in older red cells in the peripheral blood, causing their rigidity. But it is a useless oxygen carrier and therefore the phenotype may be more severe than the level of anemia suggests.

Alpha thalassemia trait may reflect the heterozygous state for α^0-thalassemia (-/$\alpha\alpha$) or the homozygous state for α^+-thalassemia (-α/-α). The phenotypic consequences of the loss of two α-globin genes seems to be the same whether they are on the same or opposite pairs of homologous chromosomes 16, although there are minor differences, depending on which of the pair of linked α-globin genes is deleted.

But there are some more subtle phenotype/genotype relationships in the α-thalassemias.[17,18] While the homozygous states for the deletion forms of α^+-thalassemia (-α/-α) are associated with very mild hematological changes characteristic of the thalassemia trait, the homozygous state for non-deletion forms of α^+-thalassemia due to mutations involving the upstream (α2) globin gene are associated with a more severe deficit of α-globin production, and the phenotype of Hb H disease. It appears that if the upstream α-globin gene is intact, but inactivated by a mutation, its downstream partner is unable to increase its output (FIG. 5). Whether this is simply a positional effect depending on the proximity of α genes to the α-globin LCR, or whether it reflects competition for activation factors which can bind to the intact promoters in the case of the mutated, intact upstream α gene, is not known. But clearly the position of the α-globin genes relative to each other is of considerable importance in determining their compensatory increase in response to an inactivated partner.

These subtle phenotypic changes relating to the presence of deletion or non-deletion forms of α-thalassemia are also reflected in the clinical heterogeneity of Hb H disease in many parts of the world. Overall, those that result

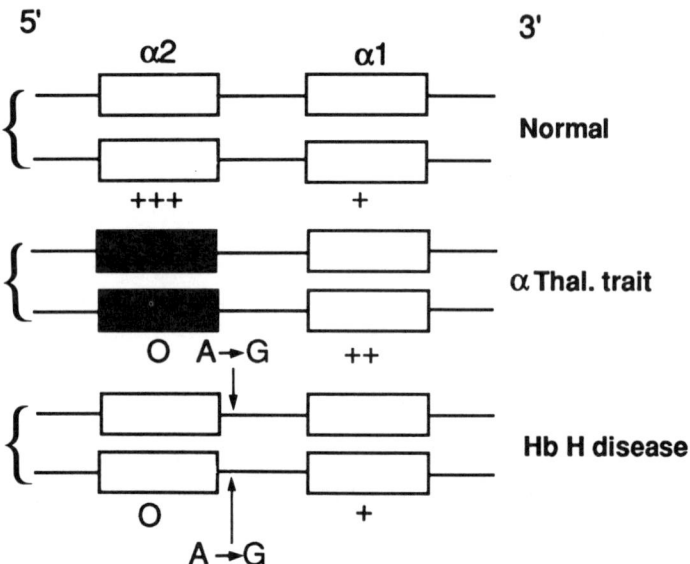

FIGURE 5. The differential effects on the output of the α1 gene when the α2 gene is deleted (*solid box*) or inactivated by a mutation (*open box*; the mutation shown is an A→G change in the polyA adenylation site).

from the inheritance of α^0-thalassemia together with deletion forms of α-thalassemia (--/-α) seem to be less severe than those in which there is the co-inheritance of α^0- and α^+-thalassemia due to non-deletion forms (--/$^T\alpha$).[17,18]

More recently it has become apparent that α-thalassemia may show further heterogeneity due to completely different mechanisms. In 1981 three patients were described who had a varying degree of mental retardation associated with a phenotype of α-thalassemia.[19] From studies of this interesting condition over recent years it is now clear that there are two different varieties of the α-thalassemia mental retardation syndrome.[20]

First, there is a group of patients who have relatively mild mental handicap and a variable constellation of facial and skeletal dysmorphisms. These individuals have long deletions involving the α-globin gene cluster, removing at least one megabase. It appears that this condition can arise in several ways, including unbalanced translocation involving chromosome 16, truncation of the tip of chromosome 16, or the loss of the α-globin gene cluster and parts of its flanking regions by other mechanisms.

The second group is characterized by defective α-globin chain synthesis associated with severe mental retardation and a relatively homogeneous pattern of dysmorphology. These patients have a particularly mild form of α-thalassemia and so far have all had a male karyotype. There are no abnormalities of the α-globin genes, the activity of which appears to be reduced in both *cis* and *trans*. Since these chromosomes direct the synthesis of normal amounts of α-globin gene in mouse erythroleukemia cells, it appears that the α-thalassemia is due to a deficiency in the *trans*-acting factor involved in the regulation of the α-globin genes. It has now been confirmed that the genetic determinant that is responsible for this unusual form of α-thalassemia and mental retardation is located on the long arm of the X chromosome. It seems likely, therefore, that there is an X chromosome encoded *trans*-acting factor that regulates α-globin gene synthesis and that is deficient in these children. Whether this factor is also involved in development or, as seems more likely, the condition represents a contiguous gene syndrome that involves the locus for the α-globin gene regulation factor together with other genes remains to be determined.

The Overall Heterogeneity of the Thalassemias

The molecular mechanisms for the phenotypic variability of the thalassemias are well-established. Overall, a great deal of the clinical variability of these syndromes can be explained by the heterogeneity of mutations at the respective loci involved, variable expression at one or two related loci, and more subtle gene dosage or position effects. It is also becoming clear that heterogeneity of loci unlinked to the α- or β-globin gene cluster may have important effects in modifying their activities. There is potential for further heterogeneity due to mutations of the major LCRs which control the α and β gene clusters. Similarly, there may be inherited variability of the modification of the secondary effects of defective global chain production. These include the effectiveness of proteolysis of excess α chains, the stability of the homotetramers in α-thalassemia, and the pathways which limit the rate of oxidant damage to the red cells consequent on imbalanced globin chain synthesis.

THE PHENOTYPIC HETEROGENEITY OF OTHER MONOGENIC DISEASES

Despite the extraordinary phenotypic diversity of the thalassemias, the underlying molecular mechanisms that are responsible for this diversity are remarkably straightforward and, given an understanding of the molecular pathology of the disease, predictable. It is surprising, therefore, that more has not been learnt about the reasons for the clinical diversity of other monogenic diseases in which there is a genuine understanding of the pathophysiology. For example, in the case of the many different mutations that have been found in the LDL receptor gene as the basis for monogenic hypercholesterolemia, there is, so far, very little evidence as to why the disorder has such a variable clinical course. In one family a second locus is involved, which seems to have caused a reduction in the cholesterol level in some family members, but otherwise, despite our increasing knowledge about the genetic control of cholesterol metabolism, the reason for phenotypic diversity of this disease is not clear.[21] The same applies to cystic fibrosis.[22] There is some evidence that certain internal deletions of the dystrophin gene may give rise to a particularly mild phenotype of Duchenne muscular dystrophy,[23] and that individuals who have large deletions of the factor VIII or IX genes may be less prone to develop antibodies when treated with replacement therapy for hemophilia or Christmas disease. But for the most part the reasons for the heterogeneity of many genetic diseases are still not clear.

There are, however, some novel molecular mechanisms emerging that were not seen in the globin field and that have the potential to modify the phenotype of monogenic disease. A number of important genetic neurological disorders, including Huntington's disease, the fragile X syndrome, myotonic dystrophy and spinobulbar muscular atrophy appear to result from the gradual expansion in the numbers of trinucleotide repeats within or adjacent to the particular genes involved.[24] These potentially unstable regions may gradually increase in size because of augmented repeat number until, when a critical length is reached, a clinical phenotype results. In the case of the fragile X syndrome, increasing repeat numbers give rise to the phenomena of premutation; with 50–200 copies, although there is no phenotypic change, babies of affected mothers are more likely to be affected. When the numbers reach 200–2000, the syndrome is expressed. In Huntington's disease, the number of repeats is related to the sex of the parent and the age of onset.[25,26] And it appears that, by gene conversion or other mechanisms, the repeat numbers may be reduced, with reversion to a normal phenotype.

Some single-gene disorders may also be modified by epigenetic effects. A well-known example is the variable expression of X-linked disorders in female carriers, presumably due to skewed inactivation of the X chromosome. Parental imprinting may also modify the expression of some single-gene disorders and contiguous gene syndromes.[27]

DISCUSSION

Although there is much to be learnt, it is clear that the potential for phenotypic diversity among the single gene disorders is quite remarkable. In the case of the thalassemias and some other single-gene disorders some of this clinical variability can be explained by molecular diversity of the underlying mutations, combined with heterogeneity at a few other loci. However, even in the case of the globin disorders many questions remain, and there is still the possibility that further genetic heterogeneity of systems a long way removed from the basic defects in globin synthesis may be involved in determining the phenotype. Furthermore, the richly varied genotypes that underlie these diseases may all be modified by secondary pathology and environmental factors. This is likely to be the case for many monogenic diseases.

The recent demonstration of the expansion of trinucleotide repeats as the basis for the fragile X syndrome, Huntington's disease, and other monogenic disorders provides another important mechanism for generating phenotypic variability. Although in the case of the fragile X syndrome it appears that a critical repeat number is required before the disease is manifest, the premutation phenomenon indicates the potentials of systems like this for generating heterogeneity. In the case of Huntington's disease there is already some evidence that the number of repeats may be related to the age of onset of the disease. In effect, this mechanism may help to explain puzzling phenomena such as increased penetrance in successive generations, or anticipation, and parental sex bias in the transmission of the most severe forms of some genetic diseases, features that were explained previously on the basis of imprinting and variable DNA methylation.

These observations have important implications with respect to the genetic analysis of the common polygenic diseases of Western society. It is becoming increasingly clear that heart disease, cancer, rheumatic disease, the major psychoses, and our other intractable disorders, reflect a complex interaction between our genetic make-up and the environment with, in many cases, the added complication of the diverse effects of aging. One of the major goals of medical genetics is to try to determine the major genes involved in these polygenic systems, both for predictive purposes and for a better understanding of the pathophysiology.

Some of the problems that we may encounter as we explore these complex polygenic systems can be predicted by considering them in their evolutionary context.[28] There is very good evidence that the thalassemias and other common hemoglobin disorders reached their current high frequency through selection of heterozygotes against *P. falciparum* malaria. In the few thousand years to which man has been exposed to malaria, literally hundreds of different mutations have been selected in this way, involving not just the hemoglobin genes but also those for many other red cell proteins and the HLA-DR system. Thus this single environmental agent, acting over a relatively short evolutionary period, has produced enormous genetic and phenotypic diversity.

It is quite possible that many of the genes that make us more or less susceptible to diseases like atheroma and diabetes reflect much more ancient and complex polymorphisms.[28] For example, the worldwide epidemic of insulin-resistant diabetes, which has a particularly strong genetic component, seems to reflect the action of a Western diet and lifestyle on populations with a particular genetic susceptibility. As pointed out by Neel, this phenomenon could reflect a "thrifty genotype" which was selected during periods of dietary deprivation.[29] From its distribution it is possible that it evolved during the original dispersion of the main streams of different races. If this is true, and because there are so many genes involved in glucose and insulin metabolism, it seems likely that there will have been many different polymorphisms involved in producing the thrifty genotype, some of which are extremely ancient.

Given the considerable complexity of the genotype/phenotype relationships for monogenic disease, and the likelihood that the important genes in the polygenic systems that underlie some of our most important diseases may reflect extremely ancient and diverse polymorphisms, it may be very difficult to define some of the key loci that are involved in susceptibility or resistance to common disease. Recent work using the mouse model of type 1 diabetes suggests that this will be the case; at least five genes, and possibly twice that many, have already been identified as important factors in the generation of this autoimmune disease.[30] Sorting out the relative roles of these different genes, and how they interact with each other and with the environment, will be a major challenge.

REFERENCES

1. WEATHERALL, D. J. 1991. The New Genetics and Clinical Practice: 3. Oxford University Press. Oxford.
2. HUISMAN, T. H. J. 1993. The structure and function of normal and abnormal haemoglobins. Baillière's Clin. Haematol. **6**: 1–30.
3. WEATHERALL, D. J. & J. B. CLEGG. 1981. The Thalassaemia Syndromes. Blackwell Scientific Publications. Oxford.
4. KAZAZIAN, H. H. 1990. The thalassemia syndromes: molecular basis and prenatal diagnosis in 1990. Sem. Hematol. **27**: 209–228.
5. WEATHERALL, D. J. 1994. The Thalassemias. *In* The Molecular Basis of Blood Diseases: 2. G. Stamatoyannopoulos, A. W. Nienhuis, P. W. Majerus & H. Varmus, Eds.: 157–205. W. B. Saunders.
6. WAINSCOAT, J. S., J. M. OLD, D. J. WEATHERALL & S. H. ORKIN. 1983. The molecular basis for the clinical diversity of β thalassaemia in Cypriots. Lancet **i**: 1235–1237.
7. GILMAN, J. G. & T. H. J. HUISMAN. 1985. DNA sequence variation associated with elevated fetal $^G\gamma$ globin production. Blood **66**: 783–787.
8. THEIN, S. L., M. SAMPIETRO, J. M. OLD, M. D. CAPPELLINI, G. FIORELLI, B. MODELL & D. J. WEATHERALL. 1987. Association of thalassaemia intermedia with a beta-globin gene haplotype. Brit. J. Haemat. **65**: 370–373.
9. NAGEL, R. L., M. E. FABRY, J. PAGNIER, I. ZOHOUN, H. WAJCMAN, V. BAUDIN & D. LABIE. 1985. Hematologically and genetically distinct forms of sickle cell anemia in Africa. N. Eng. J. Med. **312**: 880–884.
10. LABIE, D., O. DUNDA-BELKHODJA, F. ROUABHI, J. PAGNIER, A. RAGUSA & R. L. NAGEL. 1987. The –158 site 5' to the $^G\gamma$ gene and $^G\gamma$ expression. Blood **66**: 1463–1465.
11. KULOZIK, A. E., B. C. KAR, R. K. SATAPATHY, B. E. SERJEANT, G. R. SERJEANT & D. J. WEATHERALL. 1987. Fetal hemoglobin levels and β^S globin haplotypes in an Indian population with sickle cell disease. Blood **69**: 1742–1746.

12. JEFFREYS, A. J., V. WILSON, S. L. THEIN, D. J. WEATHERALL & B. A. J. PONDER. 1986. DNA "fingerprints" and segregation analysis of multiple markers in human pedigrees. Am. J. Hum. Genet. **39:** 11.
13. GONZALEZ-REDONDO, J. H., T. A. STOMING, A. KUTLAR, F. KUTLAR, K. D. LANCLOS, E. F. HOWARD, Y. J. FEI, M. AKSOY, C. ALTAY, A. GURGEY, A. N. BASAK, G. D. EFREMOV, G. PETKOV & T. H. J. HUISMAN. 1989. A C→T substitution at nt −101 in a conserved DNA sequence of the promoter region of the β-globin gene is associated with "silent" β-thalassemia. Blood **73:** 1705–1711.
14. ORKIN, S. H., S. E. ANTONARAKIS & D. LOUKOPOULOS. 1984. Abnormal processing of β Knossos RNA. Blood **64:** 311.
15. THEIN, S. L., C. HESKETH, P. TAYLOR, P. TEMPERLEY, R. M. HUTCHISON, J. M. OLD, W. G. WOOD, J. B. CLEGG & D. J. WEATHERALL. 1990. Molecular basis for dominantly inherited inclusion body β thalassemia. Proc. Natl. Acad. Sci. USA **87:** 3924–3928.
16. KAZAZIAN, H. H., C. E. DOWLING, R. L. HURWITZ, M. COLEMAN & J. G. I. ADAMS. 1989. Thalassemia mutations in exon 3 of the β-globin gene often cause a dominant form of thalassemia and show no predilection for malarial-endemic regions of the world. Am. J. Hum. Genet. **45:** A242.
17. HIGGS, D. R., M. A. VICKERS, A. O. M. WILKIE, I.-M. PRETORIUS, A. P. JARMAN & D. J. WEATHERALL. 1989. A review of the molecular genetics of the human α-globin gene cluster. Blood **73:** 1081–1104.
18. HIGGS, D. R. 1993. α-thalassaemia. Baillière's Clin. Haematol. **6:** 117–150.
19. WEATHERALL, D. J., D. R. HIGGS, C. BUNCH, J. M. OLD, D. M. HUNT, L. PRESSLEY, J. B. CLEGG, N. C. BETHLENFALVAY, S. SJOLIN, R. D. KOLER, E. MAGENIC, J. L. FRANCIS & D. BEBBINGTON. 1981. Hemoglobin H disease and mental retardation. A new syndrome or a remarkable coincidence? N. Eng. J. Med. **305:** 607.
20. GIBBONS, R. J., A. O. WILKIE, D. J. WEATHERALL & D. R. HIGGS. 1991. A newly defined X linked mental retardation syndrome associated with α thalassaemia. J. Med. Genet. **28:** 729–733.
21. HOBBS, H. H., M. S. BROWN & J. L. GOLDSTEIN. 1992. Molecular genetics of the LDL receptor gene in familial hypercholesterolemia. Hum. Mutat. **1:** 445–466.
22. TSUI, L.-C. 1991. Probing the basic defect in cystic fibrosis. Curr. Biol. **1:** 4–10.
23. HOFFMAN, E. P. & L. M. KUNKEL. 1989. Dystrophin abnormalities in Duchenne/Becker muscular dystrophy. Neuron **2:** 1019–1229.
24. MANDEL, J.-L. 1993. Question of expansion. Nature Genet. **4:** 8–9.
25. TELENIUS, H., H. P. H. KREMER, J. THEILMANN, S. E. ANDREW, E. ALMQVIST, M. ANVRET, C. GREENBERG, J. GREENBERG, G. LUCOTTE, F. SQUITIERI, E. STARR, Y. P. GOLDBERG & M. R. HAYDEN. 1993. Molecular analysis of juvenile Huntington disease: the major influence of $(CAG)_n$ repeat length is the sex of the affected parent. Hum. Mol. Genet. **2:** 1534–1540.
26. STINE, O. C., N. PLEASANT, M. L. FRANZ, M. H. ABBOTT, S. E. FOLSTEIN & C. A. ROSS. 1993. Correlation between the onset age of Huntington's disease and length of the trinucleotide repeat in IT-15. Hum. Mol. Genet. **2:** 1547–1549.
27. HALL, J. G. 1990. Genomic imprinting: review and relevance to human diseases. Am. J. Hum. Genet. **46:** 857–873.
28. WEATHERALL, D. J. 1992. The Role of Nature and Nurture in Common Diseases. The Harveian Oration of the Royal College of Physicians, London.
29. FLINT, J., R. M. HARDING, A. J. BOYCE & J. B. CLEGG. 1993. The population genetics of the haemoglobinopathies. Baillière's Clin. Haematol. **6:** 215–262.

30. TODD, J. A., T. J. AITMAN, R. J. CORNALL, S. GHOSH, J. R. HALL, C. M. HEARNE, A. M. KNIGHT, J. M. LOVE, M. A. MCALEER, J-B. PRINS, N. RODRIGUEZ, M. LATHROP, A. PRESSEY, N. H. DELARTO, L. B. PETERSON & L. S. WICKER. 1991. Genetic analysis of autoimmune type I diabetes mellitus in mice. Nature **351:** 542–546.

A Molecular Switch for the Consolidation of Long-Term Memory: cAMP-Inducible Gene Expression

CRISTINA M. ALBERINI, MIRELLA GHIRARDI,
YAN-YOU HUANG, PETER V. NGUYEN,
AND ERIC R. KANDEL

*Center for Neurobiology and Behavior
College of Physicians & Surgeons of Columbia University; and
Howard Hughes Medical Institute
722 West 168th Street
New York, New York 10032*

As this symposium illustrates, molecular genetics has brought about a dramatic unification within the biological sciences. The ability to sequence genes, and to infer the amino acid sequence of the proteins they encode, has revealed unanticipated relationships between proteins encountered in different contexts. As a result, there is now a general blueprint for cell function that provides a common conceptual framework for several, previously unrelated, disciplines: genetics, biochemistry, immunology, development, cell biology, and neurobiology. A parallel and potentially equally profound unification is occurring between neural science, the science of the brain, and cognitive psychology, the science of the mind. The ability to study the neuronal basis of mental function is providing a new impetus for examining cognitive processes such as perception, action, language, learning, and memory. To what degree can these two independent and disparate strands be brought together? Can molecular biology enlighten the study of mental processes? In this brief review we outline the possibility of a *molecular biology of cognition*, using as examples several simple forms of memory and learning in invertebrates and vertebrates.

LEARNING IS NOT A UNITARY MENTAL FACULTY BUT HAS AT LEAST TWO MAJOR FORMS

One of the major conceptual advances of recent cognitive psychological studies is the finding that learning is not a unitary faculty of the mind, but consists of at least two distinct mental processes (for review, see Ref. 1): learning about people, places, and things (explicit or declarative forms of learning), and learning motor skills and perceptual strategies (implicit or procedural forms of learning). These two major forms of learning have been localized to different neural systems within the brain.[2,3] Explicit learning importantly involves regions within the temporal lobe of the cerebral cortex, including the *hippocampus*, whereas implicit learning involves only the specific sensory and motor systems recruited for the particular perceptual or motor skills utilized during the learning process.[4] As a result, implicit learning can be studied in a variety of reflex systems, including those of invertebrates such as *Aplysia*, *Limax*, *Hermissenda*, and *Drosophila*.[5,6] By contrast, explicit forms of learning are best studied in mammals.[7-9]

The finding of two phenotypically different forms of learning raises the question: To what degree do they share common molecular steps? One clue to shared mechanisms comes from the study of memory storage, that is, the retention of information acquired through learning. Memory for both implicit and explicit forms of learning has stages, and is commonly divided into at least two temporally distinct components: short-term memory that lasts minutes to hours, and long-term memory that can last days, weeks, or even years. Studies of memory storage for both implicit as well as explicit learning indicate that both involve a switch or *consolidation mechanism* that appears to have common features. In both cases the consolidation of memory from a labile short-term form to a stable, self-maintained long-term form requires the induction of genes and proteins. With both implicit and explicit memory a transient application of inhibitors of mRNA and protein synthesis selectively blocks the induction of long-term memory without affecting short-term memory.[10-13] A similar application of inhibitor has no effect on the maintenance of long-term memory once it is established.

What genes and proteins contribute to the consolidation switch that turns on the long-term process? To what degree are they conserved in the two major forms of learning? Here, we first describe the insights that have been gained from studies of elementary forms of implicit learning in *Aplysia* and *Drosophila*. We then briefly consider long-term potentiation (LTP) in the hippocampus, a type of synaptic plasticity that is thought to contribute to explicit forms of memory storage in mammals.

SENSITIZATION OF THE GILL-WITHDRAWAL REFLEX IN *APLYSIA*: AN IMPLICIT FORM OF LEARNING

The molecular mechanisms contributing to implicit memory storage have been most extensively studied for the gill-withdrawal reflex of the marine snail, *Aplysia californica*. As is true for other defensive reflexes, the gill-withdrawal reflex can be modified by several different forms of learning. We will focus here on only one: sensitization.

Sensitization is a form of learning in which an animal learns to strengthen its reflex responses to previously neutral stimuli following the presentation of an aversive stimulus. When *Aplysia* is presented with a noxious stimulus to the tail, the animal recognizes the stimulus as aversive and learns to enhance its reflex response to an innocuous stimulus applied to the siphon. The duration of the consequent memory is a function of the number of sensitizing stimuli applied to the tail.[14] A single noxious stimulus to the tail produces short-term memory. The resulting enhancement of the withdrawal reflex lasts for minutes and does not require new protein synthesis.[11,15] By contrast, four or five noxious stimuli to the tail produce long-term memory lasting one to two days, which requires new protein synthesis. Further training leads to an even more enduring memory reflected as a reflex enhancement lasting several weeks.

A number of key components of the neural circuit for the gill-withdrawal reflex have been identified. Sensitization leads to modification of several of these (FIG. 1). One site that has been studied extensively is the monosynaptic connection between the sensory and motor neurons. This connection carries a representation of both short- and long-term memory, which is expressed as an enhanced release of transmitter from the synaptic terminals of the sensory neurons (presynaptic facilitation) (FIG. 1). The molecular steps leading to enhanced transmitter release activated by short-term memory involve a phosphorylation cascade mediated by cyclic AMP and protein kinase A[16,17] as well as protein kinase C.[17-19] Long-term memory leads to enhanced transmitter release by means of cAMP-mediated gene expression.

The monosynaptic component between the sensory and motor neurons of the gill-withdrawal reflex can be reconstituted in dissociated cell culture. This has allowed a more detailed analysis of the mechanisms involved in presynaptic facilitation of transmitter release. The reconstituted circuit in culture undergoes presynaptic facilitation in response to serotonin (5-HT), a neuromodulator released *in vivo* during sensitizing stimulation applied to the tail.[14,15,20-22] As with behavioral sensitization,[11] the amplitude and duration of the synaptic facilitation *in vitro* is a function of the number of applications of 5-HT. A brief pulse of 5-HT or cAMP produces a short-term facilitation lasting only minutes; 4 or 5 pulses of 5-HT or cAMP separated by 20-min intervals elicit a long-term facilitation lasting more than 24 hours. As with behavioral sensitization, long-term facilitation requires RNA and protein synthesis during a brief critical time window that corresponds to the period of 5-HT application.[15] One hour after the last 5-HT application, long-term facilitation is no longer susceptible to disruption by inhibitors of RNA or protein synthesis.[15] This consolidation period evident on the cellular level, during which long-term facilitation is capable of being blocked by inhibitors of protein and RNA synthesis, corresponds to the consolidation phase of long-term memory. This finding of an elementary cellular representation of consolidation allows us to ask: What is the molecular nature of memory consolidation?

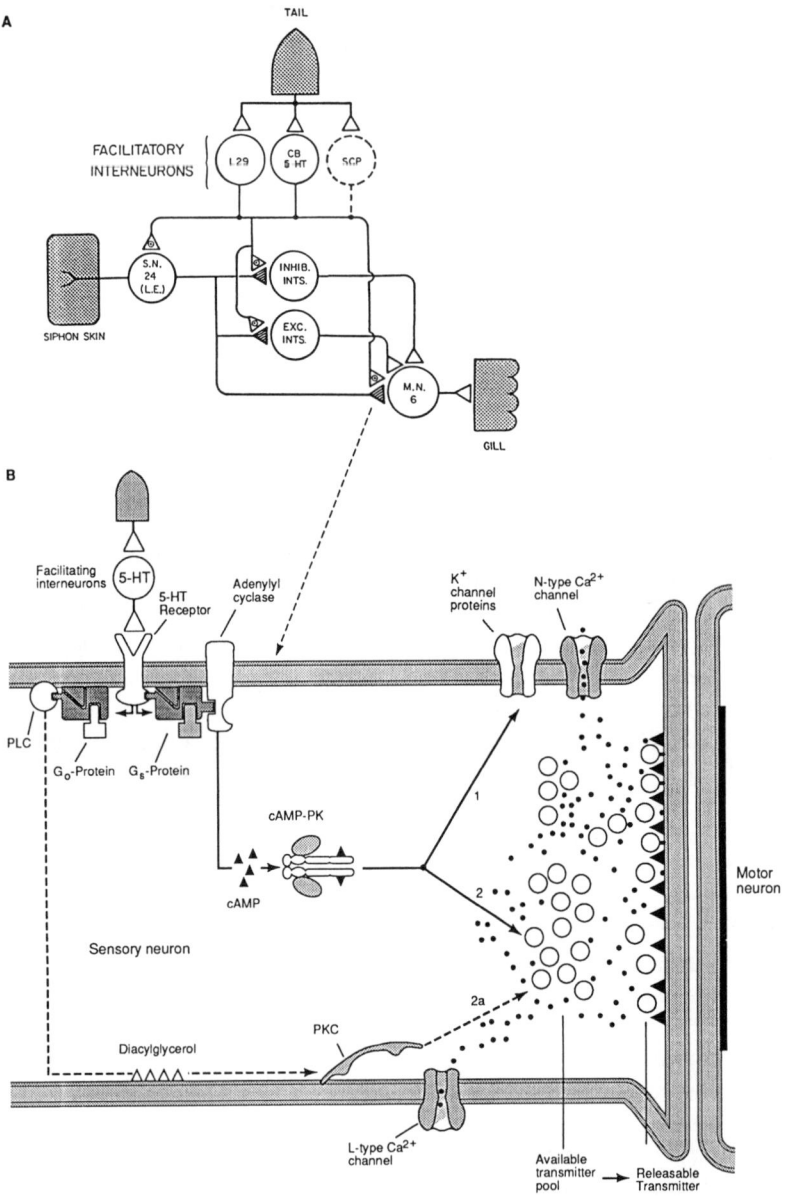

LONG-TERM FACILITATION IS ASSOCIATED WITH THE SYNTHESIS OF "EARLY" AND "LATE" PROTEINS

Since long-term synaptic plasticity, like long-term memory, requires protein synthesis during a consolidation period, Barzilai et al.[23] first analyzed the proteins synthesized in *Aplysia* sensory neurons after exposure to sensitizing training. They found that both behavioral training and application of 5-HT or cAMP induce a sequence of early and late changes in protein synthesis, consistent with the consolidation period's being a time during which a cascade of gene expression is activated, allowing early regulatory proteins to control the expression of late effector genes (FIG. 2). The early proteins have the features of immediate-early gene products. Their expression is rapid (within 15 to 30 min), transient, and dependent on transcription. The immediate-early genes encode not only effector proteins, but also regulatory proteins such as transcription factors that act on later effector genes.

cAMP-DEPENDENT GENE EXPRESSION IS REQUIRED FOR LONG-TERM FACILITATION

The data from Barzilai et al. suggested that repeated application of facilitatory neurotransmitter activates not only cytoplasmic second messengers, but also a cascade of early and late genes whose products are required for long-term synaptic plasticity.[24] How does cAMP activate transcription? What immediate-early genes are activated by 5-HT and cAMP? What are the late effector genes? Studies by Bernier et al.[25] and Bacskai et al.[26] showed that serotonin, acting on the sensory neurons, stimulates the synthesis of cAMP, which activates the protein kinase A (PKA) catalytic subunit by binding to the regulatory subunit. By imaging the free catalytic and regulatory subunits

---

FIGURE 1. Presynaptic facilitation underlies sensitization of the gill-withdrawal reflex in *Aplysia*. **(A)** Sensitization is produced by a noxious stimulus to the tail of *Aplysia*. Sensory neurons innervating the tail excite facilitating interneurons, some of which use serotonin (5-HT) as their transmitter. Inhibitory and excitatory interneurons make synapses with the axonal terminals of sensory neurons from the siphon skin, where they enhance transmitter release by presynaptic facilitation. **(B)** Postulated biochemical steps during presynaptic facilitation in the sensory neuron. 5-HT enhances transmitter release by binding to a receptor that engages a G-protein, which increases the activity of adenylyl cyclase, thereby increasing cAMP levels in the presynaptic terminal. cAMP activates the cAMP-dependent protein kinase (PKA) by releasing the enzyme's catalytic subunit from its complex with the regulatory subunit. The catalytic subunit phosphorylates K$^+$ channels, thereby decreasing the K$^+$ current, prolonging the action potential, increasing Ca^{2+} influx through N-type Ca^{2+} channels, and augmenting transmitter release. In addition, 5-HT also increases the availability of transmitter by mobilizing vesicles to a releasable pool near the active zones where release occurs (pathway 2). This second pathway involves both PKA and protein kinase C (PKC), which is activated by 5-HT via a G-protein that activates a phospholipase (PLC) that produces diacylglycerol in the membrane. Diacylglycerol then stimulates PKC.

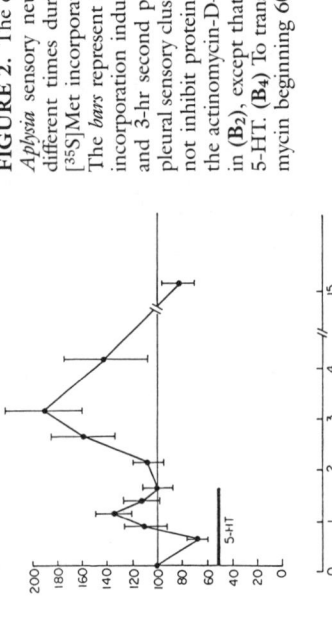
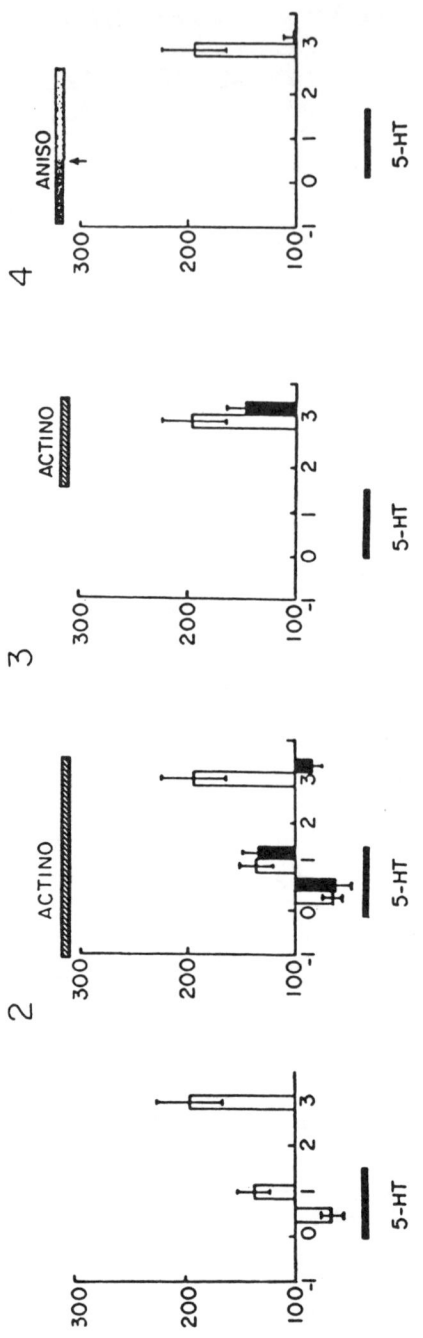

FIGURE 2. The effect of a 1.5-hr application of 5-HT on the rate of overall protein synthesis in *Aplysia* sensory neurons. (**A**) Incorporation of [^{35}S]Met into pleural sensory cluster proteins at different times during and after treatment with 5-HT. The results are expressed as the ratio of [^{35}S]Met incorporated into 5-HT–treated cluster over unstimulated cluster from the same animal. The *bars* represent SEM. (**B**) The effect of actinomycin and anisomycin on the peaks of [^{35}S]Met incorporation induced by 5-HT. (**B$_1$**) The *open histograms* represent the 0.5-hr trough, 1-hr peak, and 3-hr second peak. The data are the same as those of (**A**). (**B$_2$**) To block transcription, the pleural sensory clusters were preincubated with 100 μg/ml actinomycin-D (a concentration that does not inhibit protein synthesis) 1 hour prior to treatment with 5-HT. The *solid histograms* represent the actinomycin-D–treated clusters. (**B$_3$**) The experimental protocol was identical to that described in (**B$_2$**), except that actinomycin-D was applied 90 min after the sensory neurons were treated with 5-HT. (**B$_4$**) To transiently block translation, the sensory cells were preincubated with 20 μM anisomycin beginning 60 min before and lasting until 30 min after the application of 5-HT.

of PKA, Bacskai et al. found that a single pulse of 5-HT increases the free catalytic subunit concentration in the cytoplasm of the sensory neuron and especially in the presynaptic terminals (FIG. 3A).[26] With repeated pulses of 5-HT, the catalytic subunit translocates to the nucleus of the sensory neurons (FIG. 3A),[26] where it appears to phosphorylate one or more CREB-related transcription factors that activate cAMP-inducible genes.[27,28] In fact, Dash et al. and Kaang et al. found data suggesting that one of the substrates of protein kinase A is a CREB-like protein that binds to the cyclic AMP response element (CRE) (FIGS. 3B and 3C). By injecting oligonucleotides containing somatostatin CRE into sensory neurons, Dash et al. blocked long-term facilitation without affecting short-term facilitation (FIG. 3B), showing that CRE-binding proteins are critical for expression of long-term facilitation. Kaang et al.[28] expressed in the sensory neurons of *Aplysia* a chimeric transacting factor consisting of the mammalian CREB activation domain fused to a GAL4 DNA-binding domain. This fusion protein was able to transactivate the reporter gene (chloramphenicol acetyltransferase) in response to repeated 5-HT application (FIG. 3C). They then tested whether endogenous phosphorylation by PKA was required for the induction of this transacting activity. They compared the activity of the wild-type chimera CREB-GAL4 to a mutant chimera (CREB-GAL4 SA 119) in which the residue serine 119 (essential for the activity of mammalian CREB), was substituted with an alanine, and found that this substitution abolished the ability of 5-HT to induce the transacting activity of CREB-GAL4. In addition, they found that the kinase essential for this activity is likely PKA, since a mutation that inactivates only the PKA phosphorylation site (the substitution of arginine 117 with an alanine) but leaves the CaM kinase consensus site intact blocked the 5-HT-dependent transactivation (FIG. 3C). Taken together, these data imply that CREB-related transcription factors are required for long-term facilitation and are activated by the PKA-dependent phosphorylation induced by 5-HT.

ApC/EBP IS AN IMMEDIATE-EARLY GENE INDUCED DURING THE CONSOLIDATION PHASE OF LONG-TERM FACILITATION IN *APLYSIA*

To understand which cAMP-dependent transcriptional events are activated during long-term facilitation, we next focused on cAMP-regulated transcription factors. Some of the transcription factors known to be activated by cAMP belong to a family known as CCAAT enhancer-binding protein (C/EBP). A member of this family, C/EBPβ, is expressed in the rat pheochromocytoma PC12 cell line, where it has been shown to be activated by cAMP and to regulate the expression of the *c-fos* gene by binding to the sequence ATTAGGACAT (enhancer response element, ERE) in the *c-fos* promoter.[29] Since nerve cells

A1

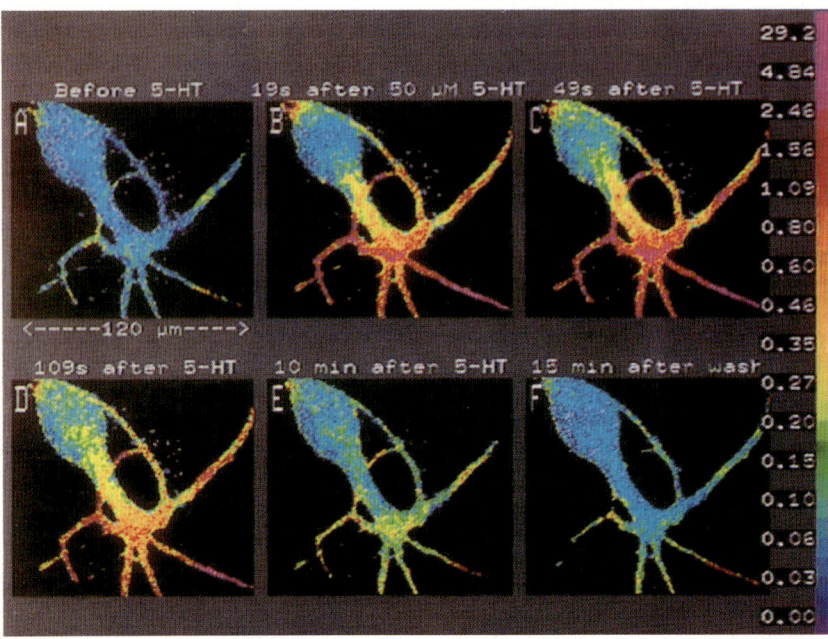

A2

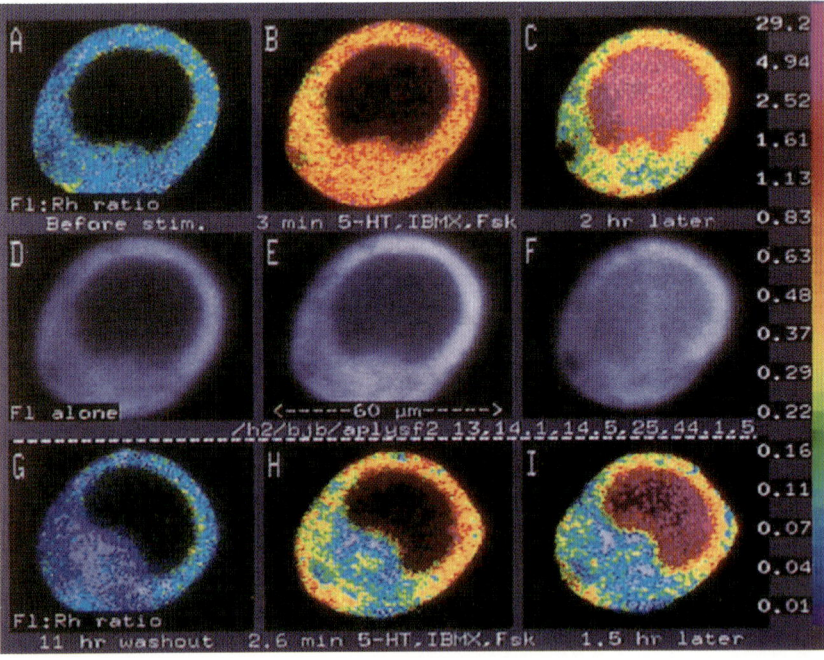

FIGURE 3A1. Gradient in [cAMP] in *Aplysia* sensory neurons after uniform bath application of serotonin (5-HT). A single, cultured *Aplysia* sensory neuron was microinjected with FICRhR. Digital fluorescence images are the result of simultaneous acquisition of two confocal single-wavelength emission images (500 to 530 nm and < 560 nm) at a plane just above the glass substream. After subtraction of background, a ratio was calculated (short over long wavelength), and the image was pseudocolored from blues to reds, which correspond to low to high ratios and low to high concentrations of free cAMP (scale on *right* in μM cAMP). **(A)** Before treatment. **(B)** After uniform bath application of 50 μM 5-HT, a striking gradient of [cAMP] develops between the cell body and the distal processes. The increase in [cAMP] persisted with some habituation while 5-HT remained in the bath **(C** through **E)**, but after 5-HT was washed away **(F)**, [cAMP] returned to near control levels **(A)**.

FIGURE 3A2. Translocation of C subunit of cAMP-dependent protein kinase into and out of the nucleus. A single neuron was microinjected with FICRhR and imaged at a plane 20 to 30 μm above the cover slip **(A)**. The nucleus excluded the labeled protein and was therefore not fluorescent; the cytoplasm, however, was very bright and exhibited negligible concentrations of cAMP. **(B)** Soon after application of 25 μM forskolin, 0.1 mM isobutylmethylxanthine (IBMX), and 20 μM 5-HT to raise intracellular [cAMP], the fluorescence ratio increased in the cytoplasm, although the nucleus remained dim. **(C)** The nucleus became brightly fluorescent 2 hours later. This apparent high [cAMP] in the nucleus is an artifact generated by entry of C subunits and exclusion of R subunits, causing a fluorescein/rhodamine ratio higher than that caused by dissociated holoenzyme. **(D** through **F)** Single-wavelength images of fluorescein fluorescence (distribution of C subunit) from the corresponding images of **(A)** through **(C)**. Confocal sections at different planes of focus (not shown) confirmed that the C subunit entered the nucleus and was not just concentrated around its periphery. The R subunits stayed mostly or completely within the cytoplasm, as determined by excitation of rhodamine alone (not shown) or by red pseudocolor of nucleus in C. **(G)** The same cell was washed and allowed to recover overnight. The emission in the cytoplasm fell to low levels, indicating reconstitution of holoenzyme, and the nucleus was again relatively devoid of fluorescence. Blue patches probably represent degradation of protein still labeled with rhodamine. **(H)** After treatment with 25 μM forskolin and 0.5 mM IBMX, the fluorescence ratio increased, showing that, after more than 12 hours in a cell, much of the FICRhR was still responsive to changes in [cAMP]. **(I)** Nuclear translocation could be observed again after an additional 1.5 hours of stimulation.

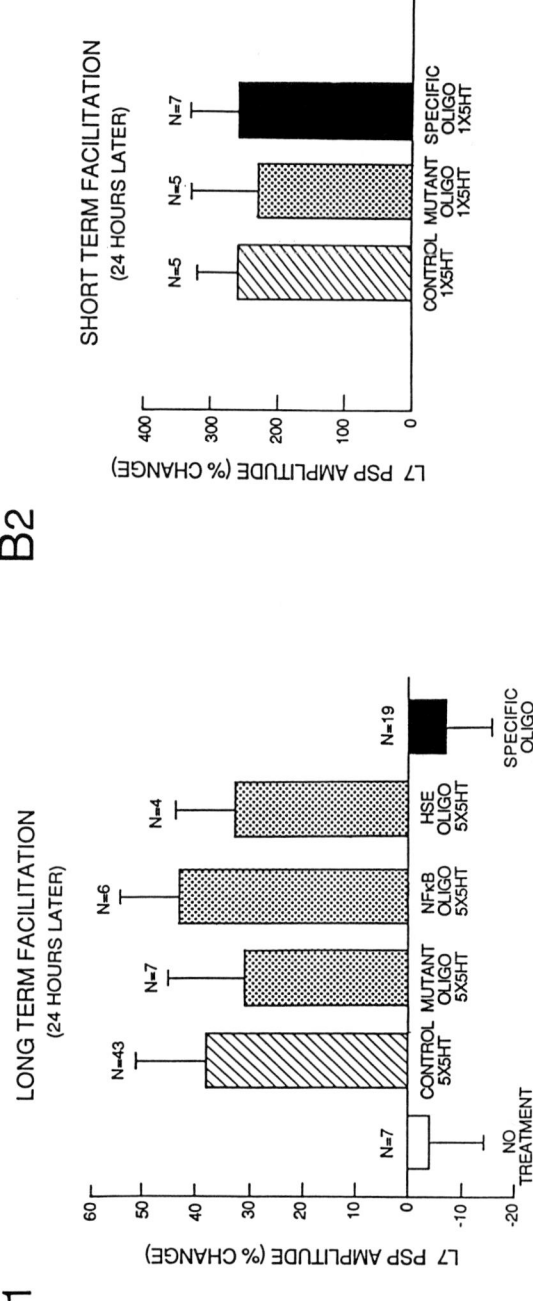

FIGURE 3B1. Injection of CRE oligonucleotides blocks 5-HT–induced long-term facilitation. Summary of the blockade of the 5-HT–induced increase in EPSP amplitude by CRE injection. The height of each *bar* is the percentage change in the EPSP amplitude ± SEM retested 24 hours after treatment. (A two-tailed *t*-test comparison of means indicated that the decrease in EPSP in cultures injected with CRE oligonucleotide is significantly different ($p < 0.05$) from the increase in the EPSP in the cells injected with either the mutant or NF-^{k}B oligonucleotides). (B2) Summary of the pooled data for short-term facilitation 24 hours after injection. In contrast to long-term facilitation, the 5-HT (5 µM) was applied after the EPSP was first depressed. Five stimuli were given with an interstimulus interval of 30 sec, and this resulted in 70–80 percent depression in EPSP amplitude. Application of 5-HT after the fifth stimulus produced an increase in EPSP amplitude by the seventh stimulus. The increase in short-term facilitation was measured by calculating the percentage increase in the seventh EPSP amplitude as compared with the fifth EPSP amplitude. Because the facilitation here was of a depressed EPSP, the percentage facilitation is larger than the long-term, where only the nondepressed EPSP was examined.

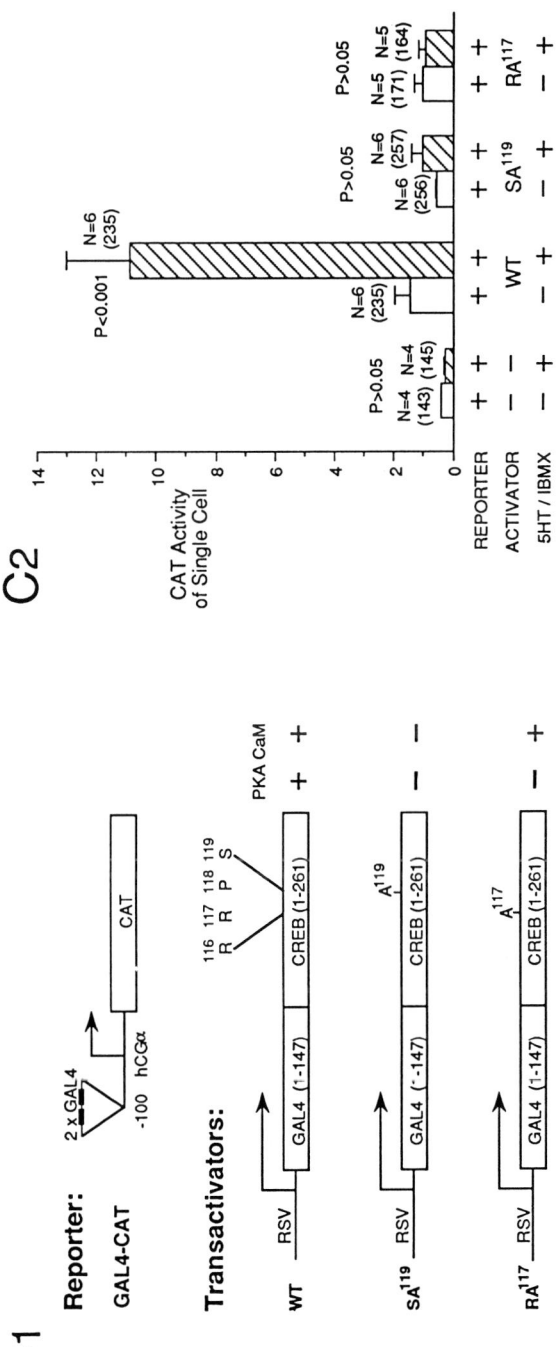

FIGURE 3C. 5-HT/IBMX lead to activation of CREB through the phosphorylation of PKA. (C1) DNA constructs used for transactivation experiments. **Reporter:** DNA-binding domain of yeast GAL4 transcription factor (amino acids 1–147) fused to wild-type or mutated forms of mammalian CREB transactivation domains (amino acids 1–261). Wild-type and mutated phosphorylation consensus sequences are indicated in single-letter amino acid code above constructs. The wild-type sequence (RRPS) is a phosphorylation substrate for both protein kinase A (PKA) and calcium/calmodulin-dependent kinase (CaM) at Ser^{119}. SA^{119} contains a substitution of Ser^{119} with Ala^{119}, preventing phosphorylation by either kinase. RA^{117} contains a substitution of Arg^{117} with Ala^{117}, abolishing the PKA site, but leaving intact the CaMK consensus phosphorylation sequence. GAL4-CREB fusion proteins were constitutively expressed in pNEXδ vector and injected with the reporter construct (GAL4-CAT) containing two copies of the GAL4 site upstream of human chorionic gonadotropin α-subunit gene driving the CAT gene. (C2) Quantitative analysis of 5-HT–regulated CREB transactivation. The wild-type (WT) GAL4-CREB transactivator enhances 5-HT/IBMX-mediated expression of CAT, whereas mutant (SA^{119} and RA^{117}) GAL4-CREB show no enhancement. Pairs of sensory clusters were injected and treated in parallel, each lane represents CAT activity of one sensory cluster. CAT activity is expressed as the percentage of substrate acetylated ($\times 10^3$); histograms show pooled data, N = number of animals and value in parenthesis is number of injected neurons. Mean, SEM and p value from two-tailed paired test are shown.

of *Aplysia* contain specific binding activity toward ERE, Alberini et al.[30] used the ERE to isolate a clone that displayed specific binding activity toward C/EBP DNA binding elements. The protein sequence inferred from the DNA sequence corresponds to a 286–amino acid polypeptide with the characteristic basic region-leucine zipper (b-zip) domain at the C-terminus that is highly homologous to the b-zip domains of the C/EBP family members. The sequence of the genomic ApC/EBP showed that in the 5′ region of the gene a nonpalindromic CRE site is present (ApCRE) 19 bp upstream from the putative TATA box. This site may represent a regulatory element recognized by CREB-like DNA binding proteins.

ApC/EBP IS REQUIRED FOR LONG-TERM FACILITATION

Where is ApC/EBP expressed? How is it regulated? Does it have a role in learning-related synaptic facilitation? To address these questions, we first investigated the expression of ApC/EBP by determining its mRNA concentration in untreated or 5-HT-treated animals. The level of ApC/EBP expression was undetectable in untreated CNS, but it increased significantly after a 2-hr exposure to 5-HT, 8-Bromo-cAMP, or forskolin (FIG. 4A).

The induction of mRNA ApC/EBP by 5-HT is rapid. It is detectable within 15 minutes after exposure to either treatment (FIG. 4B). To determine whether this rapid induction is caused by the direct action of a constitutively expressed factor, we examined the action of 5-HT on ApC/EBP mRNA in the presence of protein synthesis inhibitors (anisomycin or emetine). In the presence of a protein synthesis inhibitor, 5-HT caused a superinduction of the ApC/EBP mRNA (FIG. 4A), indicating that ApC/EBP transcription is induced by 5-HT and cAMP as an immediate-early gene via a constitutively expressed transcription factor. We therefore next investigated the effects of blocking the activity of ApC/EBP on short- and long-term facilitation of sensory-motor synapses in reconstituted monosynaptic circuits using three different approaches. First, we interfered with the binding of the transcription factor to its DNA binding element by injecting ERE oligonucleotides that compete for the binding activity of the endogenous ApC/EBP to its target sequence in the nucleus of the sensory cells. Second, we blocked specifically the synthesis of endogenous ApC/EBP by injecting ApC/EBP antisense RNA into the sensory cells. Third, we blocked the binding activity of ApC/EBP to its DNA target sites by injecting into the sensory cells a specific antibody able to disrupt the binding of ApC/EBP and its target sequences. We found that all of these conditions blocked long-term facilitation (FIG. 5). By contrast, the injection of the same molecules had no effect on short-term facilitation (FIGS. 5A3, B3, C3). These data show that the binding of ApC/EBP to its target DNA sequences is required in order to induce long-term facilitation lasting 24 hours.

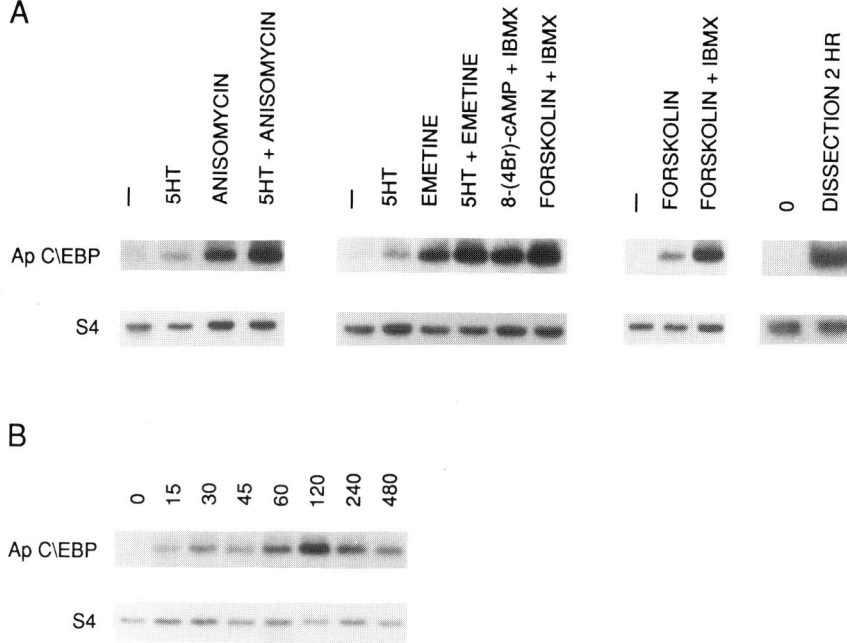

FIGURE 4. Induction of ApC/EBP mRNA. (**A**) ApC/EBP mRNA expression in CNS of untreated *Aplysia*, of *Aplysia* treated *in vivo* with the indicated drugs for 2 hr at 18°C, or dissected without treatment and kept at 18°C in culture medium. Four independent experiments are shown, in which 10 μg of total RNA extracted from CNSs of untreated (−) or treated *Aplysia*, as indicated, were electrophoresed, blotted, and hybridized with ^{32}P-labeled ApC/EBP (*top*) or S4 (*bottom*) probes. The latter encodes the *Aplysia* homologue of S4 ribosomal protein,[59] which is constitutively expressed and used as a loading control. 0 indicates RNA extracted immediately after dissection of *Aplysia* CNS. Two-hour dissection represents RNA extracted from *Aplysia* CNS dissected and incubated in culture medium for 2 hr at 18°C. (**B**) Time course of ApC/EBP mRNA induction following 5-HT treatment. Times of treatment are indicated. Five μg of total RNA extracted from total CNS of *in vivo* treated *Aplysia* were analyzed as described in (**B**).

How long does the transcription factor need to be active? One possibility is that the binding of ApC/EBP to its target sequences is required throughout the maintenance period of the facilitation. Alternatively, the long-term facilitation may become self-perpetuating as a result of subsequent expression of stable, effector genes. To distinguish between these hypotheses, we injected ERE oligonucleotides into sensory cells at various times from 1 to 12 hours after giving five pulses of 5-HT. We found that the blocking effect of the specific oligonucleotide was progressively reduced when the injection was performed at longer intervals after the training, with facilitation no longer affected by the injection at 12 hours after the training (FIG. 6). Therefore, the induction of ApC/EBP during the 5-HT treatment leads to the activation of

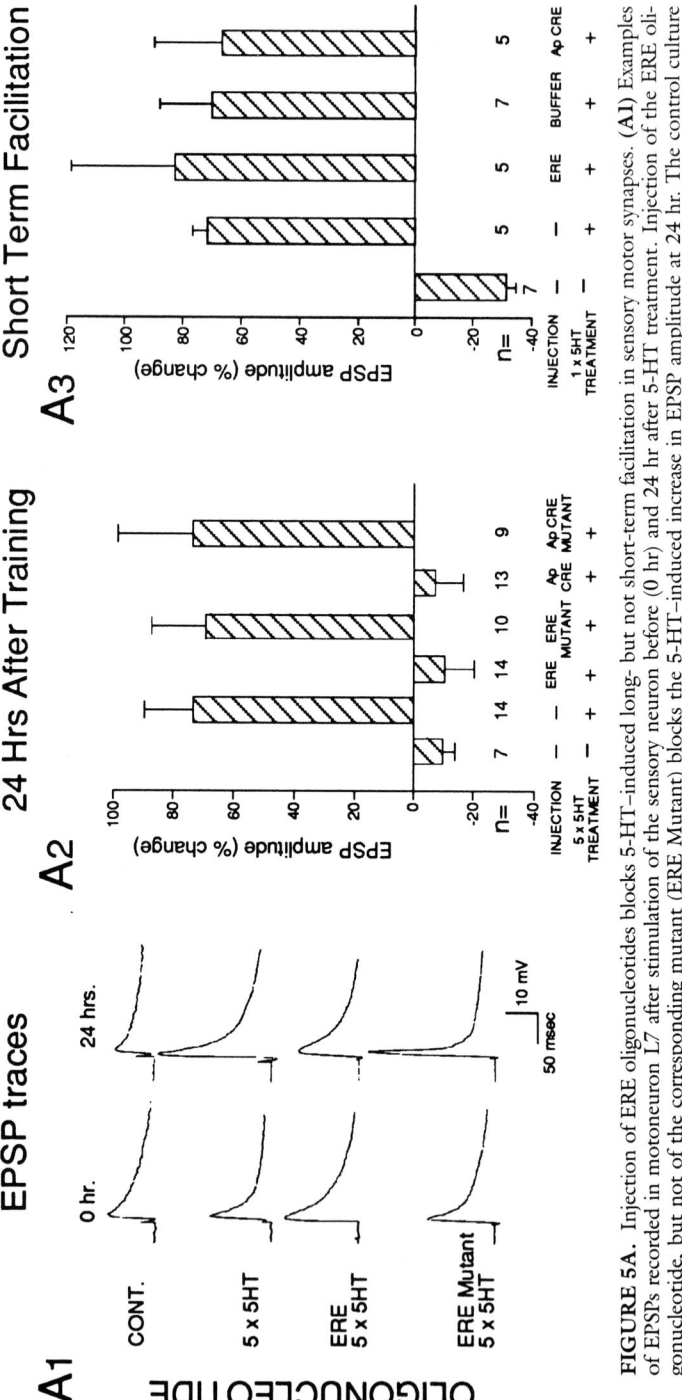

FIGURE 5A. Injection of ERE oligonucleotides blocks 5-HT–induced long- but not short-term facilitation in sensory motor synapses. (A1) Examples of EPSPs recorded in motoneuron L7 after stimulation of the sensory neuron before (0 hr) and 24 hr after 5-HT treatment. Injection of the ERE oligonucleotide, but not of the corresponding mutant (ERE Mutant) blocks the 5-HT–induced increase in EPSP amplitude at 24 hr. The control culture did not receive 5-HT applications or oligonucleotide injections. (A2) *Bar graph* representing the effects of oligonucleotide injections in long-term facilitation. The height of each bar corresponds to the mean percentage change ± SEM in EPSP amplitude tested 24 hr after 5-HT treatment. A one-way analysis of variance and Newman Keuls' multiple-range test indicate that 5-HT treatment significantly increases the EPSP amplitude in noninjected cells, as well as in ERE Mutant and ApCRE Mutant injected cells, relative to the control (not 5-HT–treated and noninjected cells) ($p < 0.01$). On the contrary, the EPSP amplitude change in ERE or ApCRE injected cells was not significantly different from that of control cells that were neither injected nor treated. (A3) Bar graph representing the mean EPSP amplitude percentage change ± SEM of short-term facilitated cells injected with ERE oligonucleotides or with ApCRE, or with buffer. A one-way analysis of variance and a comparison of the means show a significant effect of 5-HT in increasing EPSP amplitude in noninjected, in ERE, in ApCRE, or in buffer-injected cultures compared to the control ($p < 0.01$).

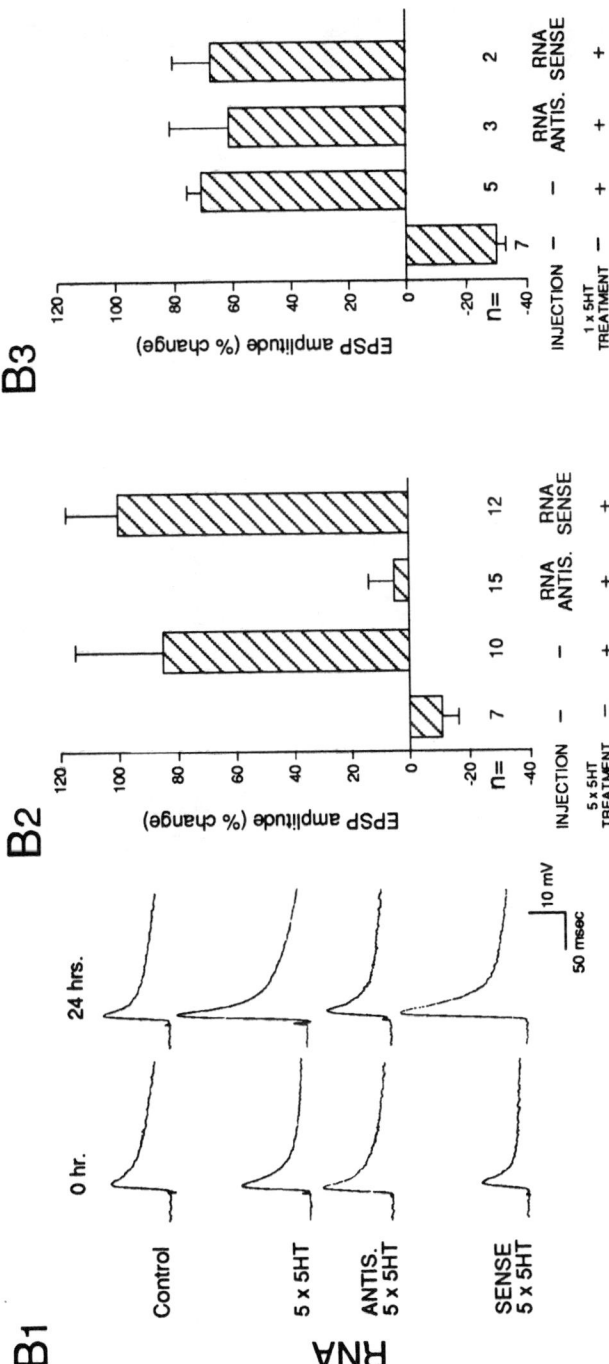

FIGURE 5B. Injection of ApC/EBP antisense RNA blocks 5-HT–induced long- but not short-term facilitation in the sensory motor synapses. **(B1)** Examples of EPSPs recorded in motoneuron L7 after stimulation of the sensory neuron before (0 hr) and 24 hr after the 5-HT treatment. Injection of the ApC/EBP antisense RNA but not of the ApC/EBP sense RNA prevents the 5-HT–induced increase in EPSP amplitude at 24 hr. The control culture did not receive 5-HT applications or RNA injections. **(B2)** *Bar graph* representing the effects of RNA injections in long-term facilitation. The height of each bar corresponds to the mean percentage change ±SEM in EPSP amplitude tested 24 hr after 5-HT treatment. A one-way analysis of variance and Newman Keuls' multiple-range test indicate that 5-HT treatment significantly increases the EPSP amplitude in noninjected cells as well as in ApC/EBP sense-RNA–injected cells relative to the control (not 5-HT–treated and noninjected cells) ($p < 0.01$). On the contrary, the EPSP amplitude change in ApC/EBP antisense-RNA–injected and 5-HT–treated cells was not significantly different from the control, nontreated, noninjected cultures. **(B3)** *Bar graph* representing short-term facilitation of cells injected with ApC/EBP antisense or sense RNA. A one-way analysis of variance and a comparison of the means show a significant effect of 5-HT in increasing EPSP amplitude in noninjected, in ApC/EBP antisense, or in ApC/EBP-sense RNA–injected cells compared to the control cultures that were neither injected nor 5-HT–treated ($p < 0.01$).

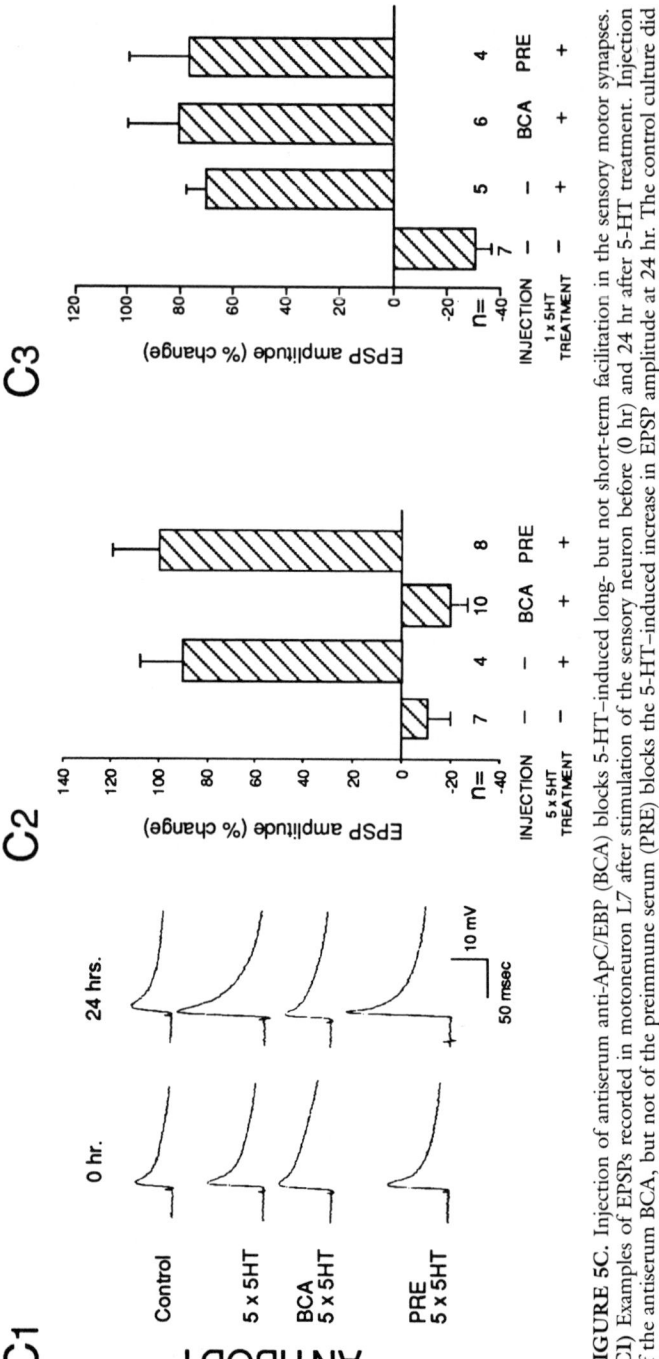

FIGURE 5C. Injection of antiserum anti-ApC/EBP (BCA) blocks 5-HT–induced long- but not short-term facilitation in the sensory motor synapses. (C1) Examples of EPSPs recorded in motoneuron L7 after stimulation of the sensory neuron before (0 hr) and 24 hr after 5-HT treatment. Injection of the antiserum BCA, but not of the preimmune serum (PRE) blocks the 5-HT–induced increase in EPSP amplitude at 24 hr. The control culture did not receive 5-HT applications or any injections. (C2) *Bar graph* representing the effects of injection of the antiserum BCA or the preimmune serum on long-term facilitation. The height of each bar corresponds to the mean percentage change ±SEM in EPSP amplitude tested 24 hr after 5-HT treatment. A one-way analysis of variance and Newman Keuls' multiple-range test indicate that 5-HT treatment significantly increases the EPSP amplitude in non-injected cells as well as in cells injected with preimmune serum relative to the control (not 5-HT–treated and noninjected cells) ($p < 0.01$). In contrast, the EPSP amplitude change in BCA injected cells was not significantly different from that observed in control cultures. (C3) *Bar graph* representing short-term facilitation of cells injected with BCA antiserum or preimmune serum. A one-way analysis of variance and a comparison of the means shows a significant effect of 5-HT in increasing EPSP amplitude in noninjected, in BCA-, or in PRE-injected cells compared to the controls that were neither injected nor 5-HT–treated ($p < 0.01$).

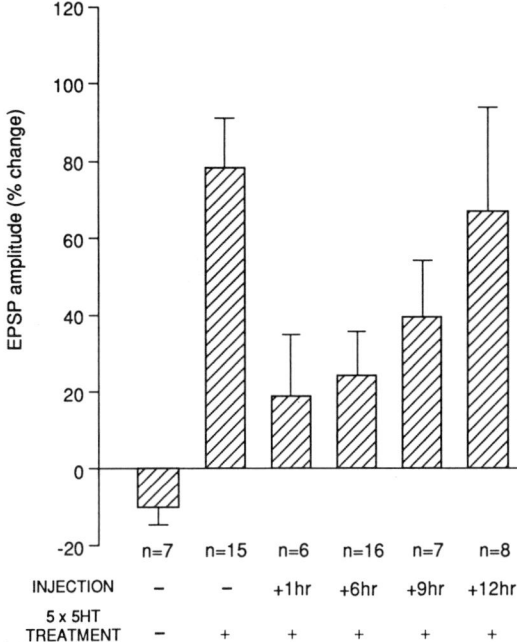

FIGURE 6. Time course of ERE effect following 5-HT treatment. *Bar graph* represents the percentage of change ±SEM in EPSP amplitude recorded 24 hr after five pulses of 5-HT from cocultures injected with ERE oligonucleotide at the indicated times after the end of 5-HT applications.

a cascade of self-perpetuating transcriptional events essential for the late phase of long-term facilitation.

cAMP-DEPENDENT GENE EXPRESSION IS A MOLECULAR SWITCH REQUIRED FOR CONSOLIDATION OF LONG-TERM FACILITATION IN *APLYSIA*

A schematic summary of the molecular events contributing to short- and long-term presynaptic facilitation is shown in FIGURE 7. The binding of serotonin to its surface receptors activates adenylyl cyclase, which catalyzes the synthesis of cAMP. cAMP binds to the regulatory subunit of the cAMP-dependent protein kinase (PKA), leading to the activation of its catalytic subunit. PKA acts on at least two classes of substrates to produce facilitation of the transmitter release. First, it phosphorylates K^+ channels or associated proteins, which leads to a reduction of the outward K^+ current and results in a broadening of the action potential and increased Ca^{2+} influx into the presynaptic neuron. Second, PKA also seems to act directly on the machinery in-

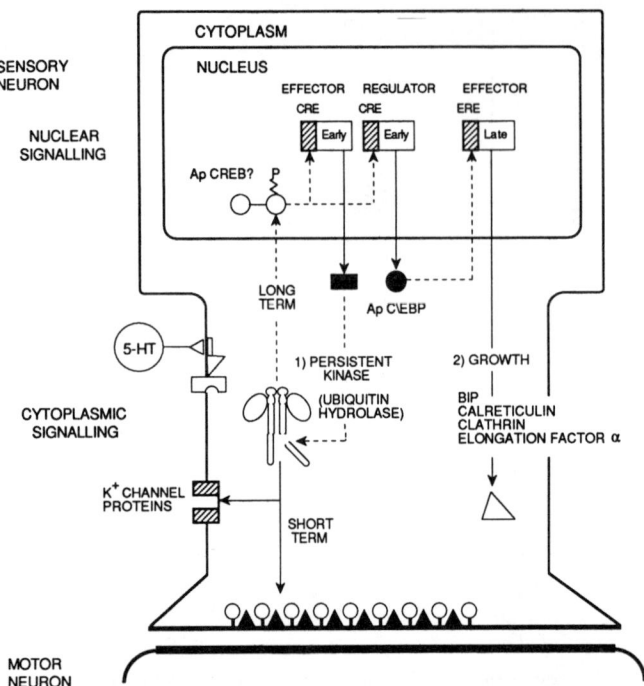

FIGURE 7. Schematic model of the activation pathways involved in *Aplysia* short- and long-term sensitization. See text for discussion.

volved in the exocytotic release of transmitter. These modifications take place in the presynaptic terminals and lead to short-term facilitation.

With repeated or prolonged application of 5-HT, the PKA catalytic subunit translocates to the nucleus, where it acts on nuclear substrates which appear to include transcription factors of the CREB/ATF family. Thus, the activity of CRE-binding proteins is necessary for long-term facilitation.

Transcription is the key event leading to long-term facilitation and the accompanying structural changes. This transcription-dependent phase includes the rapid induction of the transcription factor ApC/EBP, an immediate-early gene. This suggests that the induction of long-term facilitation requires the activation of a cascade of genes with constitutively active proteins regulating the expression of immediate-early genes, one of which is the transcription factor ApC/EBP. In turn, the early regulatory genes will likely lead to the expression of late target genes. The switching on of a self-maintaining mechanism by immediate-early genes explains why the characteristic protein synthesis-dependent phase is brief: the induction of regulatory factors is the limiting step that allows the expression of late phase events. In addition to regulatory factors, early effectors also are synthesized during the consolidation phase.

Among these effectors is the C-terminal ubiquitin hydrolase that seems to participate in the proteolytic cleavage of the PKA regulatory subunit, maintaining the enzymatic activity of the catalytic subunit in the absence of a cAMP increase.[31,32]

Although PKA enzymatic activity is necessary for the first 10 hours following repeated 5-HT application, it is not maintained.[33] The late phase is characterized by morphological changes. Morphological studies have shown that structural changes appear within 1 hr after 5-HT or tail-shock training[34] and persist for days or weeks. Moreover, the decay of these structural changes seems to parallel the decay of behavioral memory.[35,36]

What molecular mechanisms underlie the formation of new synaptic connections? The synaptic growth is associated with a downregulation of NCAM-related apCAMs on the surface membrane of the sensory neuron.[37] Downregulation is achieved by activation of the endosomal pathway leading to internalization and apparent degradation of apCAM.[38] *Aplysia* expresses two types of isoforms, a transmembrane form and a phosphoinositol-linked form. Which of the two types of apCAM isoforms is internalized? To address this question, Craig Bailey, Bong-Kiun Kaang, and their colleagues selectively expressed epitope-tagged constructs of the two isoforms in cultured sensory neurons. By combining thin-section electron microscopy with gold-conjugated antibodies they have found that serotonin elicits a 68% decrease in the density of gold-labeled complexes bound to the transmembrane form of apCAM at the surface membrane, and a 24-fold increase in their internalization. By contrast, serotonin has no effect on either the surface distribution or internalization of the phosphatidylinositol-linked isoform of apCAM. The selective internalization of the transmembrane form highlights the potential regulatory significance of its intracellular domain, which contains a PEST sequence (thought to mediate protein degradation) and has two consensus sites for MAP kinase phosphorylation. Deletions of, or mutations in, the cytoplasmic tail should allow determination of which part of this molecule triggers internalization and which part targets degradation.

IMPLICIT FORMS OF LEARNING IN *DROSOPHILA* ALSO USE CREB AND THE cAMP CASCADE

Drosophila show classical conditioning to olfactory cues paired with shock. Several single gene mutants have been isolated that cannot learn the task although their behavior is otherwise normal. Two mutations have been analyzed in particular detail and each involves a step in the cAMP cascade.[39] *Dunce* involves a defect in the cAMP phosphodiesterase, whereas *rutabaga* is defective in the calcium-calmodulin–dependent adenylyl cyclase. Expression of an inhibitor of PKA using a heat-shock promoter also blocks the learning.

Recently, Tully *et al.*[13] have shown that spaced training gives rise to a long-term memory that lasts at least seven days and is blocked by inhibitors of protein synthesis. This long-term memory is selectively blocked by the heat-shock-induced expression of a dominant negative inhibitor of CREB, a CREM-like transcription repressor. Thus, several forms of long-term memory for implicit forms of learning require CREB- and cAMP-induced gene expression.

IMPLICIT AND EXPLICIT FORMS OF LEARNING SHARE SOME COMMON MECHANISMS

The studies in *Aplysia* suggest that the switch from short-term to long-term memory for simple reflexive forms of learning involves the induction by cAMP and CREB of a set of immediate early genes that participate in the growth of new synaptic connections that underlie the long-term process. Is there a similar molecular switch for memory consolidation in the mammalian brain that might contribute to the establishment of long-term memory storage for more complex explicit forms of learning?

Studies in humans and experimental animals have indicated that structures within the temporal lobe, such as the hippocampus, are particularly critical for explicit memory storage.[4] Are there cellular mechanisms within the hippocampus for storing explicit forms of memory? In 1973 Timothy Bliss and Terry Lømo first demonstrated that neurons in the hippocampus have remarkable plastic capabilities of the kind that would be required for learning.[40] A brief, high-frequency train of action potentials in any one of three neural pathways within the hippocampus produces an increase in synaptic strength in that pathway that can last for hours or days. This strengthening is called *long-term potentiation* or LTP.

LTP in the mammalian hippocampus shares some of the mechanisms used for synaptic facilitation in *Aplysia*. One form of LTP, called mossy fiber LTP, occurs at synapses between the dentate gyrus granule cells and CA3 pyramidal cells, and primarily involves a cAMP-dependent enhancement of transmitter release from the presynaptic terminals (FIG. 8). By contrast, Schaffer collateral LTP in CA1 is much more complex. Its induction involves calcium influx into the postsynaptic cell through the NMDA receptor channel and the recruitment in the postsynaptic cell of several second-messenger pathways involving tyrosine kinases, protein kinase C, and calcium/calmodulin kinase II. In addition to these inductive steps in the postsynaptic cell, Schaffer collateral LTP also involves an enhancement of transmitter release from the presynaptic neuron.[41-43] This enhanced release is thought to be mediated by one or more retrograde messenger signals (perhaps nitric oxide or carbon monoxide) that diffuse from the postsynaptic cell to the terminals of the presynaptic neuron.[44-46]

Similar to the presynaptic facilitation in *Aplysia*, both mossy fiber and Schaffer collateral LTP have distinct temporal phases, indicating that both short- and long-term memory have a cellular representation. There is an early phase produced by a single tetanic stimulation that lasts 1–3 hours and requires only covalent modification of preexisting proteins. In addition there is a late phase induced by repeated tetanic stimulation that persists for many hours, and is dependent on protein and RNA synthesis, and on PKA activity (FIG. 8).[47–49] As is the case with behavioral memory and presynaptic facilitation in *Aplysia*, there is a consolidation switch on the cellular level, and this requirement for transcription in LTP also has a critical time window.[50] Since the late transcription-dependent phase of LTP is blocked by inhibitors of PKA,[47,48] this suggests that in mammalian LTP as in *Aplysia* presynaptic facilitation cAMP-inducible genes need to be expressed (FIG. 8). Consistent with these findings, Bourtchuladze *et al.*[51] have found that mice that have a knockout of several critical CREB isoforms have a defect in LTP in the CA1 region that affects the late phase of LTP preferentially. These mice have normal learning and short-term memory of context conditioning, a hippocampal-based learning task, but they selectively lack long-term memory.

AN OVERALL VIEW

One of the key unifying findings emerging from these molecular studies is that the genetic switch for hippocampal LTP, required for explicit forms of learning, seems to share important similarities with that utilized in *Aplysia* and in *Drosophila*. Thus, molecular studies of cognition are revealing, on a mechanistic level, previously unsuspected similarities between different classes of learning, and suggest the interesting possibility that a common set of genetic mechanisms may be involved in a variety of learning-related long-term enhancements of synaptic transmission.

The data in *Aplysia* imply that these mechanisms include activation of immediate-early genes. Analysis of immediate-early gene expression in vertebrate brains has shown that many immediate-early genes are strongly induced in hippocampus and certain regions of the neocortex by treatments that lead to long-term potentiation.[52–55] It is now clear that CREB participates in hippocampal-based plasticity and long-term memory for explicit tasks. It will be of further interest to investigate whether cAMP-dependent immediate-early genes, perhaps of the C/EBP family, are also required for long-term neuronal plasticity in mammals.

The apparent conservation of some steps in the molecular mechanisms for long-term synaptic plasticity may reflect the fact that long-term memory storage commonly involves structural changes. Thus, in *Aplysia* the self-sustaining long-term process is expressed in the growth of new connec-

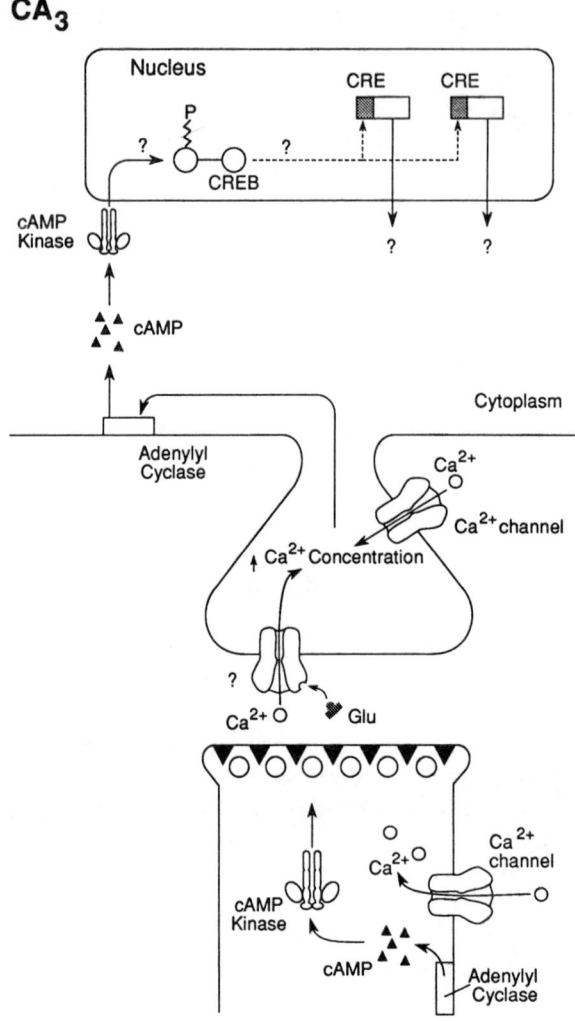

FIGURE 8. Schematic models of the early and late phases of LTP in hippocampal region CA3. See text for discussion.

tions.[36] This process is switched on by the cAMP-mediated induction of a cascade of immediate-early genes.

That the late phase of the two forms of LTP in the hippocampus also involves cAMP raises the possibility that, in the hippocampus as well, cAMP and protein kinase A are recruited because they may be able to access immediate-early genes that control the molecular machinery for the growth and terminal differentiation of new synaptic connections. By delineating the genes and proteins recruited by the different forms of NMDA-dependent and -independent forms of LTP, this possibility can now be tested.

Finally, the finding that the cAMP pathway is necessary for neuronal changes related to long-term memory storage does not exclude the participation of other second-messenger pathways. Indeed, the CREB proteins, the nuclear targets of PKA, are not only targets for the cAMP pathway. CREB is a multifunctional transcription factor that can also be activated by CaM kinase II as well as by various neurotrophins such as NGF and BDNF.[56-58] It therefore will be interesting to see whether these neurotrophins can modulate the late phase of LTP and whether this modulation contributes to the growth of new synaptic connections following learning.

REFERENCES

1. POLSTER, M. R., L. NADEL & D. L. SCHACHTER. 1991. Cognitive neuroscience. Analysis of memory: A historical perspective. J. Cogn. Neurosci. **3**: 95–116.
2. MILNER, B. 1985. Memory and the human brain. *In* How We Know. M. Shafto, Ed. Harper & Rowe. San Francisco.
3. ZOLA-MORGAN, S., L. R. SQUIRE & D. G. AMARAL. 1986. Human amnesia and the medial temporal region: Enduring memory impairment following a bilateral lesion limited to field CA1 of the hippocampus. J. Neurosci. **6**: 2950–2967.
4. SQUIRE, L. R. 1992. Memory and the hippocampus: A synthesis from findings with rats, monkeys and humans. Psychol. Rev. **99**: 195–231.
5. HAWKINS, R. D., G. A. CLARK & E. R. KANDEL. 1987. Cell biological studies of learning in simple vertebrate and invertebrate systems. *In* Handbook of Physiology, Section 1: The Nervous System. Vol. V: Higher Functions of the Nervous System, Part 1. V. B. Mountcastle, F. Plum & S. R. Geiger, Eds.: 25–83. American Physiological Society. Bethesda, MD.
6. DRAIN, P., E. FOLKERS & W. G. QUINN. 1991. cAMP-dependent protein kinase and the disruption of learning in transgenic flies. Neuron **6**: 71–82.
7. MORRIS, R. G. M., F. SHENK, F. TWEEDIE & L. E. JARRAD. 1990. Ibotenate lesions of hippocampus and or subiculum: Dissociating components of allocentric spatial learning. Eur. J. Neurosci. **2**: 1016–1028.
8. GRANT, S. G. N., T. J. O'DELL, K. A. KARL, P. L. STEIN, P. SORIANO & E. R. KANDEL. 1992. Impaired long-term potentiation, spatial learning, and hippocampal development in *fyn* mutant mice. Science **258**: 1903–1910.
9. SILVA, A. J., R. PAYLOR, J. M. WEHNER & S. TONEGAWA. 1992. Impaired spatial learning in α-calcium-calmodulin kinase II mutant mice. Science **257**: 206–211.
10. DAVIS, H. P. & L. R. SQUIRE. 1984. Protein synthesis and memory. A review. Psychol. Bull. **96**: 518–559.
11. CASTELLUCCI, V. F., H. BLUMENFELD, P. GOELET & E. R. KANDEL. 1989. Inhibitor of protein synthesis blocks long-term behavioral sensitization in the isolated gill-withdrawal reflex of *Aplysia*. J. Neurobiol. **20**: 1–9.
12. CROW, T. & J. FORRESTER. 1990. Inhibition of protein synthesis blocks long-term enhancement of generator potentials produced by one-trial in vivo conditioning in *Hermissenda*. Proc. Natl. Acad. Sci. USA **87**: 4490–4494.

13. TULLY, T., T. PREAT, S. C. BOYNTON & M. DEL VECCHIO. 1994. Genetic dissecting consolidated memory in *Drosophila melanogaster.* Cell. **79:** 35–47.
14. FROST, W. N., V. F. CASTELLUCCI, R. D. HAWKINS & E. R. KANDEL. 1985. Monosynaptic connections from the sensory neurons of the gill- and siphon-withdrawal reflex in *Aplysia* participate in the storage of long-term memory for sensitization. Proc. Natl. Acad. Sci. USA **82:** 8266–8269.
15. MONTAROLO, P. G., P. GOELET, V. F. CASTELLUCCI, J. MORGAN, E. R. KANDEL & S. SCHACHER. 1986. A critical period for macromolecular synthesis in long-term heterosynaptic facilitation in *Aplysia.* Science **234:** 1249–1254.
16. BRUNELLI, M., V. CASTELLUCCI & E. R. KANDEL. 1976. Synaptic facilitation and behavioral sensitization in *Aplysia:* Possible role of serotonin and cyclic AMP. Science **194:** 1178–1181.
17. GHIRARDI, M., O. BRAHA, B. HOCHNER, P. G. MONTAROLO, E. R. KANDEL & N. DALE. 1992. Roles of PKA and PKC in facilitation of evoked and spontaneous transmitter release at depressed and nondepressed synapses in *Aplysia* sensory neurons. Neuron **9:** 479–489.
18. BRAHA, O., N. DALE, B. HOCHNER, M. KLEIN, T. W. ABRAMS & E. R. KANDEL. 1990. Second messengers involved in the two processes of presynaptic facilitation that contribute to sensitization and dishabituation in *Aplysia* sensory neurons. Proc. Natl. Acad. Sci. USA **87:** 2040–2044.
19. SACKTOR, T. C. & J. H. SCHWARTZ. 1990. Sensitizing stimuli cause translocation of protein kinase C. in *Aplysia* sensory neurons. Proc. Natl. Acad. Sci. USA **87:** 2036–2039.
20. CLARK, G. A. & E. R. KANDEL. 1984. Branch-specific heterosynaptic facilitation in *Aplysia* siphon sensory cells. Proc. Natl. Acad. Sci. USA **81:** 2577–2581.
21. RAYPORT, S. G. & S. SCHACHER. 1986. Synaptic plasticity *in vitro*: Cell culture of identified *Aplysia* neurons mediating short-term habituation and sensitization. J. Neurosci. **6:** 759–763.
22. GLANZMAN, D. L., S. L. MACKEY, R. D. HAWKINS, A. DYKE, P. E. LLOYD & E. R. KANDEL. 1989. Depletion of serotonin in the nervous system of *Aplysia* reduces the behavioral enhancement of gill withdrawal as well as the heterosynaptic facilitation produced by tail shock. J. Neurosci. **9:** 4200–4213.
23. BARZILAI, A., T. E. KENNEDY, J. D. SWEATT & E. R. KANDEL. 1989. 5-HT modulates protein synthesis and the expression of specific proteins during long-term facilitation in *Aplysia* sensory neurons. Neuron **2:** 1577–1586.
24. GOELET, P., V. F. CASTELLUCCI, S. SCHACHER & E. R. KANDEL. 1986. The long and the short of long-term memory—a molecular framework. Nature **322:** 419–422.
25. BERNIER, L., V. F. CASTELLUCCI, E. R. KANDEL & J. H. SCHWARTZ. 1982. Facilitatory transmitter causes a selective and prolonged increase in adenosine 3': 5'-monophosphate in sensory neurons mediating the gill and siphon withdrawal reflex in *Aplysia.* J. Neurosci. **2:** 1682–1691.
26. BACSKAI, B. J., B. HOCHNER, M. MAHAUT-SMITH, S. R. ADAMS, B.-K. KAANG, E. R. KANDEL & R. Y. TSIEN. 1993. Spatially resolved dynamics of cAMP and protein kinase A subunits in *Aplysia* sensory neurons. Science **260:** 222–226.
27. DASH, P. K., B. HOCHNER & E. R. KANDEL. 1990. Injection of the cAMP-responsive element into the nucleus of *Aplysia* sensory neurons blocks long-term facilitation. Nature **345:** 718–721.
28. KAANG, B. K., E. R. KANDEL & S. G. N. GRANT. 1993. Activation of cAMP-responsive genes by stimuli that produce long-term facilitation in *Aplysia* sensory neurons. Neuron **10:** 427–435.
29. METZ, R. & E. ZIFF. 1991. cAMP stimulates the C/EBP-related transcription

factor RNFIL-6 to translocate to the nucleus and induce *c-fos* transcription gene. Genes Dev. **5:** 1754–1766.
30. ALBERINI, C. M., M. GHIRARDI, R. METZ & E. R. KANDEL. 1994. C/EBP is an immediate-early gene required for the consolidation of long-term facilitation in *Aplysia*. Cell **76:** 1099–1114.
31. BERGOLD, P. J., J. D. SWEATT, I. WINICOV, K. R. WEISS, E. R. KANDEL & J. H. SCHWARTZ. 1990. Protein synthesis during acquisition of long-term facilitation is needed for the persistent loss of regulatory subunits of the *Aplysia* cAMP-dependent protein kinase. Proc. Natl. Acad. Sci. USA **87:** 3788–3791.
32. HEGDE, A. N., A. L. GOLDBERG & J. H. SCHWARTZ. 1993. Regulatory subunits of cAMP-dependent protein kinases are degraded after conjugation to ubiquitin: A molecular mechanism underlying long-term synaptic plasticity. Proc. Natl. Acad. Sci. USA **90:** 7436–7440.
33. MONTAROLO, P. G., M. GHIRARDI & E. R. KANDEL. 1992. Contribution of persistent PKA activity to the serotonin-induced long-term facilitation of *Aplysia* sensory motor synapses in culture. Soc. Neurosci. Abstr. **18:** 712.
34. BAILEY, C. H., M. CHEN, E. R. KANDEL & S. SCHACHER. 1993. Early structural changes associated with long-term presynaptic facilitation in *Aplysia* sensory neurons. Soc. Neurosci. Abstr. **19:** 16.
35. BAILEY, C. H. & M. CHEN. 1989. Time course of structural changes at identified sensory neuron synapses during long-term sensitization in *Aplysia*. J. Neurosci. **9:** 1774–1780.
36. BAILEY, C. H. & E. R. KANDEL. 1993. Structural changes accompanying memory storage. Annu. Rev. Physiol. **55:** 397–426.
37. MAYFORD, M., A. BARZILAI, F. KELLER, S. SCHACHER & E. R. KANDEL. 1992. Modulation of an NCAM-related adhesion molecule with long-term synaptic plasticity in *Aplysia*. Science **256:** 638–644.
38. BAILEY, C. H., P. MONTAROLO, M. CHEN, E. R. KANDEL & S. SCHACHER. 1992. Inhibitors of protein and RNA synthesis block structural changes that accompany long-term heterosynaptic plasticity in *Aplysia*. Neuron **9:** 749–758.
39. YIN, J. C. P., J. S. WALLACH, M. DEL VECCHIO, E. L. WILDER, H. ZHUO, W. G. QUINN & T. TULLY. 1994. Induction of a dominant-negative CREB transgene specifically blocks long-term memory in *Drosophila*. Cell. **79:** 49–58.
40. BLISS, T. V. P. & T. LØMO. 1973. Long-lasting potentiation of synaptic transmission in the dentate area of the anesthetized rabbit following stimulation of the perforant path. J. Physiol. (London) **232:** 331–356.
41. BEKKERS, J. M. & C. F. STEVENS. 1990. Presynaptic mechanism for LTP in the hippocampus. Nature **346:** 724–729.
42. MALINOW, R. & R. W. TSIEN. 1990. Presynaptic enhancement shown by whole-cell recordings of LTP in hippocampus slices. Nature **346:** 177–180.
43. MALGAROLI, A. & R. W. TSIEN. 1992. Glutamate induced LTP of the frequency of miniature synaptic currents in cultured hippocampal neurons. Nature **357:** 134–139.
44. O'DELL, T. J., L. HUANG, T. M. DAWSON, J. L. DINERMAN, S. H. SNYDER, E. R. KANDEL & M. C. FISHMAN. 1994. Endothelial NOS and blockade of LTP by NOS inhibitors in mice lacking neuronal NOS. Science **265:** 542–546.
45. SCHUMAN, E. M. & D. V. MADISON. 1991. A requirement for the intercellular messenger nitric oxide in LTP. Science **254:** 1503–1506.
46. O'DELL, T. J., R. D. HAWKINS, E. R. KANDEL & O. ARANCIO. 1991. Tests of the roles of two diffusible substances in long-term potentiation: Evidence for nitric oxide as a possible early retrograde messenger. Proc. Natl. Acad. Sci. USA **88:** 11285–11289.

47. FREY, U., Y.-Y. HUANG & E. R. KANDEL. 1993. Effects of cAMP simulate a late stage of LTP in hippocampal CA1 neurons. Science **260:** 1661–1664.
48. HUANG, Y.-Y. & E. R. KANDEL. 1994. Recruitment of long-lasting and protein kinase A-dependent long-term potentiation in the CA1 region of hippocampus requires repeated tetanization. Learning & Memory **1:** 74–82.
49. HUANG, Y.-Y., X.-C. LI & E. R. KANDEL. 1994. cAMP contributes to mossy fiber LTP by initiating both a covalently-mediated early phase and a macromolecular synthesis dependent late phase. Cell. **79:** 69–79.
50. NGUYEN, P. V., T. ABEL & E. R. KANDEL. 1994. Requirement of a critical period of transcription for induction of a late phase of LTP. Science **265:** 1104–1107.
51. BOUTCHULADZE, R., B. FRENGUELLI, J. BLENDY, D. CIOFFI, G. SCHUTZ & A. J. SILVA. 1994. Deficient long-term memory in mice with a targeted mutation of the cAMP-response element-binding protein. Cell **79:** 59–68.
52. COLE, A. J., D. W. SAFFEN, J. M. BARABAN & P. F. WORLEY. 1989. Rapid increase of an immediate-early gene messenger RNA in hippocampal neurons by synaptic NMDA receptor activation. Nature **340:** 474–476.
53. DRAGUNOW, M., R. W. CURRIE, R. L. M. FAULL, H. A. ROBERTSON & K. JANSEN. 1989. Immediate-early genes, kindling and long-term potentiation. Neurosci. Behav. Rev. **13:** 301–313.
54. SONNENBERG, J. L., F. J. RAUSCHER III, J. I. MORGAN & T. CURRAN. 1989. Regulation of proenkephalin by proto-oncogenes *fos* and *jun*. Science **246:** 1622–1625.
55. MORGAN, J. I. & T. CURRAN. 1991. Stimulus-transcription coupling in the nervous system: Involvement of the inducible proto-oncogenes *fos* and *jun*. Annu. Rev. Neurosci. **14:** 421–451.
56. DASH, P. K., K. A. KARL, M. A. COLICOS, R. PRYWES & E. R. KANDEL. 1991. cAMP response element-binding protein is activated by Ca^{2+}/calmodulin- as well as cAMP-dependent protein kinase. Proc. Natl. Acad. Sci. USA **88:** 5061–5065.
57. SHENG, M., M. A. THOMPSON & M. E. GREENBERG. 1991. A Ca^{2+} regulated transcription factor phosphorylated by calmodulin-dependent kinase. Science **252:** 1427–1430.
58. GINTY, D., A. BONNI & M. E. GREENBERG. 1994. Nerve growth factor activates a Ras-dependent protein kinase that stimulates *c-fos* transcription via phosphorylation of CREB. Cell **77:** 713–728.
59. THOMAS, M., D. M. BEDWELL & M. NOMURA. 1987. Regulation of an operon gene expression in *Escherichia coli*. A novel form of translation coupling. J. Mol. Biol. **196:** 333–345.

Molecular Analysis of Duchenne Muscular Dystrophy: Past, Present, and Future[a]

KAY E. DAVIES, JONATHON M. TINSLEY, AND DEREK J. BLAKE

Molecular Genetics Group
Institute of Molecular Medicine
John Radcliffe Hospital
Oxford OX3 9DU, United Kingdom

INTRODUCTION

At the end of the 1970s, it was realized that human genetic disease mutations could be traced through families using restriction fragment length polymorphisms (RFLPs).[1] Unlike protein polymorphisms, RFLPs are present at frequent intervals along the DNA and are distributed throughout the genome. More recently, a complete genetic map of the human genome has been constructed based on DNA sequence variation. It is now possible to map any human genetic disorder where affected families are available.[2] This has led to the chromosomal localization and thence the isolation of the genes responsible for the most common single gene disorders. Our understanding of the molecular basis of Duchenne muscular dystrophy (DMD) has depended very much on the development of this "new genetics" technology.

DMD is a severe muscle-wasting disease affecting approximately 1 in 3,000 boys.[3] Affected boys are usually confined to a wheelchair before the age of twelve and die in their late teens or early twenties. Becker muscular dystrophy (BMD) is an allelic disorder with a later onset and a much longer survival rate. Before the advent of DNA recombinant technology, the carrier detection tests for DMD were not totally reliable and prenatal diagnosis was not possible.

LINKAGE ANALYSIS

The DMD gene was first localized to Xp21 by cytogenetic analysis of rare affected females with balanced X;autosome translocations.[4] The normal X

[a] This work was financially supported by the Medical Research Council, the Muscular Dystrophy Group of Great Britain and Northern Ireland, and the Muscular Dystrophy Association.

chromosome was inactivated in these patients and the break on the translocated X chromosome was in Xp21, disrupting the function of the gene. Genetic analysis using RFLPs confirmed the localization of the gene in Xp21 and also showed that the BMD gene was probably allelic.[5] The RFLPs provided markers for reliable prenatal diagnosis and carrier detection for the disease (see Refs. 6 and 7 for review).

THE DYSTROPHIN GENE

The DMD gene was subsequently identified by taking advantage of a patient who had a deletion in Xp21[8] and by cloning one of the translocation breakpoints in an affected female.[9] Analysis of the genomic organization of the gene shows that it is the largest gene so far identified covering over 2.5 megabases and containing at least 79 exons.[10-12] The unusually high mutation rate at the DMD locus may in part be accounted for by the large size of the gene. The product of the DMD gene is known as dystrophin since its absence leads to muscular dystrophy.

The majority of DMD patients have out-of-frame deletions of the gene which results in the loss of dystrophin.[13,14] In BMD, the reading frame is maintained, and the patients produce partially functional truncated dystrophin molecules.[15]

The 14kb dystrophin mRNA is expressed predominantly in skeletal, cardiac and smooth muscle with lower levels in brain.[8,16,17] The muscle and brain transcripts are identical, apart from a differing first exon, and they are regulated from their own promoters;[18,19] the brain promoter is only active in neuronal cells, whereas expression from the muscle promoter occurs in cultures of differentiated myogenic cells, and primary glial cells.[20,21] A third promoter, which is expressed in cerebellar Purkinje neurons, lies between the muscle promoter and the second exon of dystrophin.[22] Alternatively spliced isoforms derived from the C-terminal coding region of dystrophin have also been described.[19,23] The significance of these isoforms at the RNA and protein level has yet to be elucidated.

THE DYSTROPHIN PROTEIN

Dystrophin is a 427kDa protein localized to the cytoplasmic face of the sarcolemma, enriched at myotendinous junctions and the postsynaptic membrane of the neuromuscular junction.[24-28] Dystrophin co-localizes with β-spectrin and vinculin in three distinct domains at the sarcolemma overlaying both I bands and M lines.[29] High-resolution immunofluorescence shows that dystrophin lies in an array of thick bands localizing at the sites of attachment of the sarcomeres to the muscle plasma membrane.[30,31] Electron microscopic analysis indicates that the carboxy-terminal domains are bound to the protoplasmic half of the plasmalemma.[32] Thus, dystrophin forms an intricate part of the muscle cytoskeleton and probably provides an important link between the normal contractile apparatus and the sarcolemma. Dystrophin is also localized at the postsynaptic regions of neurons.[33]

The 5′ end of dystrophin contains an actin binding site.[34,35] The central rod domain consists of a number of repeats which show similarity to spectrin and probably give the molecule a flexible rod-shaped structure.[36] Deletions of the 5′ end and central rod domain of the protein can be associated with mild as well as severe phenotypes. However, mutations in the cysteine-rich domain and first half of the carboxy-terminal domain of dystrophin almost always result in severe phenotypes, indicating their importance for the normal function of dystrophin.[13] Loss of the last 200 amino acids of dystrophin produces a DMD/BMD-like phenotype, whereas further loss in this domain results in severe DMD.[37] The carboxy-terminal has no homology with any other identified sequences apart from the related protein, utrophin, which is discussed below.

DYSTROPHIN AND THE DYSTROPHIN-ASSOCIATED GLYCOPROTEINS

The carboxy-terminal region of dystrophin binds a glycoprotein complex.[38,39] Detailed analysis shows that the cysteine-rich domain and the first half of the carboxy-terminal domain of dystrophin contains the glycoprotein binding sites.[40] This dystrophin-associated glycoprotein (DAG) complex consists of cytoskeletal (59k), transmembrane (50k, 43k, 35k and 25k) and extracellular (156k) components.[41] In subcellular fractionation procedures, the 156k DAG co-purifies with laminin, which is the major component of the extra-cellular matrix.[42] Thus the DAG complex, embedded in the sarcolemma, provides the link between the internal cytoskeleton of the muscle cell and the extracellular matrix. Loss of dystrophin destroys this link and hence leads to muscle degradation.

OTHER GENES TRANSCRIBED FROM THE DMD LOCUS

Studies of the expression of dystrophin isoforms have provided evidence for the occurrence of smaller gene products of 5.8kb, 4.8kb, and 2.2kb, which are transcribed from the 3′ end of the DMD locus. We have called these apo-dystrophin-2, -1, and -3, respectively. The apo-dystrophin-2 mRNA is transcribed from a promoter within the 3′ end of the DMD locus, but no further information regarding its site of expression is yet available.[43] The 4.8kb apo-dystrophin-1 mRNA and 2.2kb apo-dystrophin-3 mRNA are expressed in all tissues examined including muscle from the dystrophin-deficient *mdx* mouse.[43-45] Apo-dystrophin-1 (and probably apo-dystrophin-3) is transcribed from a promoter lying between exons 62 and 63 of dystrophin. Both transcripts have a 5' end distinct from dystrophin containing an identical untranslated region and first 7 amino acids. The 3' sequences of apo-dystrophin-1 are identical to those of dystrophin except for the loss of two exons and the addition of 31 unique residues at the very carboxy-terminus.[19,46] Apo-dystrophin-3 has a translational stop codon in an identical position to the truncated dystrophin isoform (dystrophin-ΔC) described by Feener *et al.*[19] (Tinsley *et al.*, unpublished data). The function of these small transcripts from the DMD locus remains to be determined.

UTROPHIN

Utrophin (dystrophin-related protein [DRP] or DMD-like [DMDL] protein) is an autosomally encoded related protein of dystrophin.[47] It is localized to human chromosome 6q24 and to the proximal region of mouse chromosome 10.[48] The localization to mouse chromosome 10 is of interest since this is very close to the *dystrophia muscularis* mutation (*dy*). The *dy* mutation is recessive and results in a severe neuromuscular disease. However, apparently normal utrophin is found in the muscle of these mice.[49]

The utrophin gene is multiexonic and approximately 1 megabase in size.[50] The corresponding mRNA is 13kb and encodes a protein of 395 kDa.[51] Comparison with the amino- and carboxy-terminus functional domains of dystrophin predicts utrophin to have very similar protein binding functions and hence a related function to dystrophin. In the rod domain both molecules contain a large central region which is probably extended and predominantly α-helical.

The range of tissues where utrophin is expressed is much wider than that of dystrophin. Utrophin has been demonstrated in normal and *mdx* mouse brain, stomach, kidney, spleen, liver and lung, and mRNA has been detected in human placenta and adult and fetal skeletal muscle, liver, intestine (smooth muscle), kidney, and testis.[52] Utrophin is expressed at higher levels in fetal muscle compared to that of the adult.[49,52]

The localization of utrophin in adult skeletal muscle is almost exclusively at the postsynaptic membranes of neuromuscular junction regions.[49,52,53] High-resolution analysis of rat neuromuscular junctions suggests that utrophin is precisely localized at the crests of the junctional folds along with the acetylcholine receptors whereas dystrophin is more dispersed around these junctions.[54] Interestingly, a weak but consistently detectable sarcolemma staining of utrophin is observed in muscle sections of DMD patients and *mdx* mice.[52-57] Although utrophin can be detected at the sarcolemma at times of muscle regeneration in other disorders,[58] it has yet to be determined whether utrophin is capable, in part at least, of replacing dystrophin in adult tissue.

Localization of utrophin is also seen in vascular and myometrial smooth muscle and glial and Schwann cells. In the brain, utrophin is expressed in neuronal, glial, and vascular cells and is enriched in neurologic cells. Electron microscope studies indicate that utrophin is located at the end feet of perivascular astrocytes, which suggests a function related to the maintenance of the blood–brain barrier.[59] The distribution of utrophin in the brain is conserved to the Elasmobranch fish, spanning more than 400 million years of evolution. This suggests that utrophin must play an important physiological role.[59]

From the amino acid sequence of the carboxy-terminus of utrophin it would not be surprising that utrophin is also capable of binding with the DAGs in muscle. Matsumara et al.[60] showed *in vitro* that utrophin co-purifies with the DAG complex from membranes of the dystrophin-deficient *mdx* mouse. What is potentially even more interesting is that in small-caliber skeletal and cardiac muscle fibers which show little or no altered pathology, the level of 156k DAG is near normal and the level of utrophin is increased approximately four-fold. In *mdx* quadriceps, which show muscle necrosis, there is no obvious change in the levels of utrophin, although there is a drastic reduction of the 156k DAG. It may be that increased utrophin expression can maintain the integrity of the DAG complex and consequently prevent muscle necrosis. If this is the case, then mechanisms to upregulate utrophin expression may be of important therapeutic use in the treatment of DMD.

GENE THERAPY STRATEGIES

The introduction of dystrophin to DMD muscle is going to be a difficult task because the protein is so large and muscle cells are such a large component of the body. Early studies using myoblast transfer were disappointing (see Ref. 61 for review) and other approaches are now being tested. Gene transfer has been achieved using direct injection,[62] retroviral transfer,[63] and adenoviral infection,[64] but the efficiency of all these techniques is low. The treatment of the disease by upregulating utrophin in skeletal muscle presents an exciting possibility and is currently being tested.[65] An improvement in both gene delivery and levels of expression are needed before any of these methods will have any therapeutic value.

CONCLUSION

The last ten years have witnessed enormous advances in our understanding of DMD. Reliable prenatal diagnosis and carrier detection is available. The function of dystrophin is now being unravelled. The challenge now is to find ways of treating the disease before it is time to celebrate 50 years of the double helix!

ACKNOWLEDGMENTS

We are grateful to Helen Blaber for assistance with the references. The text is based on a review by Tinsley *et al.*[66] Readers are also referred to a recent review by Ahn and Kunkel.[67]

REFERENCES

1. BOTSTEIN, D. & R. L. WHITE. 1980. Construction of a genetic linkage map using restriction fragment length polymorphisms. Am. J. Hum. Genet. **32:** 314–331.
2. WEISSENBACH, J., G. GYAPAY, C. DIB, A. VIGNAL, J. MORISSETTE, P. MILLASSEAU, G. VAYSSEIX & M. LATHROP. 1992. A second generation linkage map of the human genome. Nature **356:** 794–801.
3. EMERY, A. E. 1989. Clinical and molecular studies in Duchenne muscular dystrophy. Prog. Clin. Biol. Res. **306:** 15–28.
4. JACOBS, P. A., P. A. HUNT, M. MAYER & R. D. BART. 1981. Duchenne muscular dystrophy (DMD) in a female with an X/autosome translocation: Further evidence that the DMD locus is at Xp21. Am. J. Hum. Genet. **33:** 513–518.
5. DAVIES, K. E., P. L. PEARSON, P. S. HARPER, J. M. MURRAY, T. O'BRIEN, M. SARFARAZI & R. WILLIAMSON. 1983. Linkage analysis of two cloned DNA sequences flanking the Duchenne muscular dystrophy locus on the short arm of the human X chromosome. Nuc. Acid. Res. **11:** 2303–2312.
6. KOENIG, M., E. P. HOFFMAN, C. J. BERTELSON, A. P. MONACO, C. FEENER & L. M. KUNKEL. 1987. Complete cloning of the Duchenne muscular dystrophy (DMD) cDNA and preliminary genomic organisation of the DMD gene in normal and affected individuals. Cell **50:** 509–517.
7. FORREST, S. M., G. S. CROSS, A. SPEER, D. GARDNER-MEDWIN & K. E. DAVIES. 1987. Preferential deletion of exons in Duchenne and Becker muscular dystrophies. Nature **326:** 638–640.
8. MONACO, A. P., R. L. NEVE, C. COLLETTI-FEENER, C. J. BERTELSON, D. M. KURNIT & L. M. KUNKEL. 1986. Isolation of candidate cDNAs for portions of the Duchenne muscular dystrophy gene. Nature **323:** 646–650.
9. BURGHES, A. H. M., C. LOGAN, X. HU, B. BELFALL, R. G. WORTON & P. N. RAY. 1987. A cDNA clone from the Duchenne/Becker muscular dystrophy gene. Nature **328:** 434–437.
10. COFFEY, A. J., R. G. ROBERTS, E. D. GREEN, C. G. COLE, R. ANAND, F. GIANELLI & D. R. BENTLEY. 1992. Construction of a 2.6-Mb contig in yeast artificial chromosomes spanning the human dystrophin gene using an STS-based approach. Genomics **12:** 474–484.
11. MONACO, A. P., A. P. WALKER, I. MILLWOOD, Z. LARIN & H. LAHRACH. 1992. A yeast artificial chromosome contig containing the complete Duchenne muscular dystrophy gene. Genomics **12:** 465–473.
12. ROBERTS, R. G., A. J. COFFEY, M. BOBROW & D. R. BENTLEY. 1992. Determi-

nation of the exon structure of the distal portion of the dystrophin gene by vectorette PCR. Genomics **13**: 942–950.
13. KOENIG, M., A. H. BEGGS, M. MOYER *et al.* 1989. The molecular basis for Duchenne versus Becker muscular dystrophy: Correlation of severity with type of deletion. Am. J. Hum. Genet. **45**: 498–506.
14. FORREST, S. M., T. J. SMITH, G. S. CROSS, A. P. READ, N. S. T. THOMAS, R. C. MOUNTFORD, P. S. HARPER, R. T. GEIRSSON & K. E. DAVIES. 1987. An effective strategy for prenatal prediction of Duchenne and Becker muscular dystrophy. Lancet **2**: 1294–1297.
15. MONACO, A. P., C. J. BERTELSON, S. LIECHTI-GALLATI, H. MOSER & L. M. KUNKEL. 1988. An explanation for the phenotypic differences between patients bearing partial deletions of the DMD locus. Genomics **2**: 90–95.
16. CHAMBERLAIN, J. S., J. A. PEARLMAN, D. M. MUZNU, G. A. GIBBS, J. E. RANIER, A. A. REEVES & C. T. CASKEY. 1988. Expression of the murine Duchenne muscular dystrophy gene in muscle and brain. Science **239**: 1416–1418.
17. NUDEL, U., K. ROBZYK & D. YAFFE. 1988. Expression of the putative Duchenne muscular dystrophy gene in differentiated myogenic cell cultures and in the brain. Nature **331**: 635–638.
18. NUDEL, U., D. ZUK, P. EINAT, E. ZEELON, Z. LEVY, S. NEUMAN & D. YAFFE. 1989. Duchenne muscular dystrophy gene product is not identical in muscle and brain. Nature **337**: 76–78.
19. FEENER, C. A., M. KOENIG & L. M. KUNKEL. 1989. Alternative splicing of human dystrophin mRNA generates isoforms at the carboxy terminus. Nature **338**: 509–511.
20. BARNEA, E., D. ZUK, R. SIMANTOV, U. NUDEL & D. YAFFE. 1990. Specificity of expression of the muscle and brain dystrophin gene promoters in muscle and brain cells. Neuron **5**: 881–888.
21. CHELLY, J., G. HAMARD, A. KOULAKOFF, J. C. KAPLAN, A. KAHN & N. Y. BERWALD. 1990. Dystrophin gene transcribed from different promoters in neuronal and glial cells. Nature **344**: 64–65.
22. GORECKI, D. C., A. P. MONACO, J. M. J. DERRY, A. P. WALKER, E. A. BARNARD & P. J. BARNARD. 1992. Expression of four alternative dystrophin transcripts in brain regions regulated by different promoters. Hum. Mol. Genet. **1**: 505–510.
23. BIES, R. D., S. F. PHELPS, M. D. CORTEZ, R. ROBERTS, C. T. CASKEY & J. S. CHAMBERLAIN. 1992. Human and murine dystrophin mRNA transcripts are differentially expressed during skeletal muscle, heart, and brain development. Nucl. Acid. Res. **20**: 1725–1731.
24. WATKINS, S. C., E. P. HOFFMAN, H. S. SLAYTER & L. M. KUNKEL. 1988. Immunoelectron microscopic localization of dystrophin in myofibres. Nature **333**: 863–866.
25. CULLEN, M. J., J. WALSH, L. V. NICHOLSON & J. B. HARRIS. 1990. Ultrastructural localization of dystrophin in human muscle by using gold immunolabelling. Proc. Roy. Soc. Lond. Biol. **240**: 197–210.
26. SAMITT, C. E. & E. BONILLA. 1990. Immunocytochemical study of dystrophin at the myotendinous junction. Mus. & Ner. **13**: 493–500.
27. BYERS, T. J., L. M. KUNKEL & S. C. WATKINS. 1991. The subcellular distribution of dystrophin in mouse skeletal, cardiac, and smooth muscle. J. Cell. Biol. **115**: 411–421.
28. SEALOCK, R. M., H. BUTLER, N. R. KRAMARCY, K.-X. GAO, A. A. MURANE, K. DOUVILLE & S. C. FROEHNER. 1991. Localization of dystrophin relative to

acetylcholine receptor domains in electric tissue and adult and cultured skeletal muscle. J. Cell. Biol. **113:** 1133–1144.
29. PORTER, G. A., G. M. DMYTRENKO, J. C. WINKELMANN & R. J. BLOCH. 1992. Dystrophin colocalizes with beta-spectrin in distinct subsarcolemmal domains in mammalian skeletal muscle. J. Cell. Biol. **117:** 997–1005.
30. MINETTI, C. H., W. CHANG, R. MEDORI, A. PRELLE, M. MOGGIO, S. D. JOHNSEN & E. BONILLA. 1991. Dystrophin deficiency in young girls with sporadic myopathy and normal karyotype. Neurology **41:** 1288–1292.
31. MASUDA, T., N. FUJIMAKI, E. OZAWA & H. ISHIKAWA. 1992. Confocal laser microscopy of dystrophin localization in guinea pig skeletal muscle fibres. J. Cell. Biol. **119:** 543–548.
32. SQUARZONI, S., P. SABATELLI, M. C. MALTARELLO, A. CATALDI, P. R. DI & N. M. MARALDI. 1992. Localization of dystrophin COOH-terminal domain by the fracture-label technique. J. Cell. Biol. **118:** 1401–1409.
33. LIDOV, H. G., T. J. BYERS, S. C. WATKINS & L. M. KUNKEL. 1990. Localization of dystrophin to postsynaptic regions of central nervous system cortical neurons. Nature **348:** 725–728.
34. HEMMINGS, L., P. A. KUHLMAN & D. R. CRITCHLEY. 1992. Analysis of the actin-binding domain of alpha-actinin by mutagenesis and demonstration that dystrophin contains a functionally homologous domain. J. Cell. Biol. **116:** 1369–1380.
35. WAY, M., B. POPE, R. A. CROSS, J. KENWRICK-JONES & A. G. WEEDS. 1992. Expression of the N-terminal domain of dystrophin in coli and demonstration of binding to F-actin. FEBS Lett. **301:** 243–245.
36. PONS, F., N. AUGIER, R. HEILIG, J. LEGER, D. MORNET & J. J. LEGER. 1990. Isolated dystrophin molecules as seen by electron microscopy. Proc. Natl. Acad. Sci. USA **87:** 7851–7855.
37. ROBERTS, R. G., M. BOBROW & D. R. BENTLEY. 1992. Point mutations in the dystrophin gene. Proc. Natl. Acad. Sci. USA **89:** 2331–2335.
38. CAMPBELL, K. P. & S. D. KAHL. 1989. Association of dystrophin and an integral membrane glycoprotein. Nature **338:** 259–262.
39. ERVASTI, J. M., S. D. KAHL & K. P. CAMPBELL. 1991. Purification of dystrophin from skeletal muscle. J. Biol. Chem. **266:** 9161–9165.
40. SUZUKI, A., M. YOSHIDA, H. YAMAMOTO & E. OZAWA. 1992. Glycoprotein-binding site of dystrophin is confined to the cysteine-rich domain and the first half of the carboxy-terminal domain. FEBS Lett. **308:** 154–160.
41. ERVASTI, J. M. & K. P. CAMPBELL. 1991. Membrane organization of the dystrophin-glycoprotein complex. Cell **66:** 1121–1131.
42. IBRAGHIMOV, B. O., J. M. ERVASTI, C. J. LEVEILLE, C. A. SLAUGHTER, S. W. SERNETT & K. P. CAMPBELL. 1992. Primary structure of dystrophin-associated glycoproteins linking dystrophin to the extracellular matrix. Nature **355:** 696–702.
43. BLAKE, D. J., D. R. LOVE, J. M. TINSLEY, G. E. MORRIS, H. TURLEY, K. GATTER, G. DICKSON, Y. H. EDWARDS & K. E. DAVIES. 1992. Characterisation of a 4.8kb transcript from the Duchenne muscular dystrophy locus expressed in Schwannoma cells. Hum. Mol. Genet. **1:** 103–109.
44. LEDERFEIN, D., Z. LEVY, N. AUGIER, D. MORNET, G. E. MORRIS, O. FUCHS, D. YAFFE & U. NUDEL. 1992. A 71-kilodalton protein is a major product of the Duchenne muscular dystrophy gene in brain and other nonmuscle tissues. Proc. Natl. Acad. Sci. USA **89:** 5346–5350.
45. HUGNOT, J. P., H. GILGENKRANTZ, N. VINCENT, P. CHAFEY, G. E. MORRIS, A. P. MONACO, Y. BERWALD-NETTER, A. KOULAKOFF, J.-C. KAPLAN, A. KHAN & J. CHELLY. 1992. Distal transcript of the dystrophin gene initiated from an

alternative first exon and encoding a 75-kDa protein widely distributed in nonmuscle tissues. Proc. Natl. Acad. Sci. USA **89:** 7506–7510.
46. GENG, Y., P. SICINSKI, D. GORECKI & P. J. BARNARD. 1991. Developmental and tissue-specific regulation of mouse dystrophin: The embryonic isoform in muscular dystrophy. Neuromusc. Disord. **1:** 125–133.
47. LOVE, D. R., D. F. HILL, G. DICKSON, N. K. SPURR, B. C. BYTH, R. F. MARSDEN, F. WALSH, Y. H. EDWARDS & K. E. DAVIES. 1989. An autosomal transcript in skeletal muscle with homology to dystrophin. Nature **339:** 55–58.
48. BUCKLE, V. J., J. L. GUENET, C. D. SIMON, D. R. LOVE & K. E. DAVIES. 1990. Localisation of a dystrophin-related autosomal gene to 6q24 in man, and to mouse chromosome 10 in the region of the dystrophia muscularis (*dy*) locus. Hum. Genet. **85:** 324–326.
49. LOVE, D. R., G. E. MORRIS, J. M. ELLIS, U. FAIRBROTHER, R. F. MARSDEN, J. F. BLOOMFIELD, Y. H. EDWARDS, C. P. SLATER, D. J. PARRY & K. E. DAVIES. 1991. Tissue distribution of the dystrophin-related gene product and expression in the *mdx* and *dy* mouse. Proc. Natl. Acad. Sci. USA **88:** 3243–3247.
50. PEARCE, M., D. J. BLAKE, J. M. TINSLEY, B. C. BYTH, L. CAMPBELL, A. P. MONACO & K. E. DAVIES. 1993. The utrophin and dystrophin genes share similarities in genomic structure. Hum. Mol. Genet. **2:** 1765–1772.
51. TINSLEY, J. M., D. J. BLAKE, A. ROCHE, U. FAIRBROTHER, J. RISS, B. C. BYTH, S. J. KNIGHT, J. KENWRICK-JONES, G. K. SUTHERS, D. R. LOVE, Y. H. EDWARDS & K. E. DAVIES. 1992. Primary structure of dystrophin-related protein. Nature **360:** 591–593.
52. KHURANA, T. S., S. C. WATKINS, P. CHAFEY, J. CHELLY, F. M. S. TOME, M. FARDEAU, J.-C. KAPLAN & L. M. KUNKEL. 1991. Immunolocalization and developmental expression of dystrophin related protein in skeletal muscle. Neuromusc. Disord. **1:** 185–194.
53. NGUYEN, T. M., J. M. ELLIS, D. R. LOVE, K. E. DAVIES, K. C. GATTER & G. E. MORRIS. 1991. Localization of the DMDL gene-encoded dystrophin-related protein using a panel of nineteen monoclonal antibodies: Presence at neuromuscular junctions, in the sarcolemma of dystrophic skeletal muscle, in vascular and other smooth muscles, and in proliferating brain cell lines. J. Cell. Biol. **115:** 1695–1700.
54. BESWICK, G. S., L. V. B. NICHOLSON, C. YOUNG, E. O'DONELL & C. SLATER. 1992. Different distributions of dystrophin and related proteins at nerve-muscle junctions. Neuroreport **3:** 857–860.
55. KHURANA, T. S., E. P. HOFFMAN & L. M. KUNKEL. 1990. Identification of a chromosome 6-encoded dystrophin-related protein. J. Biol. Chem. **265:** 16717–16720.
56. OHLENDIECK, K., J. M. ERVASTI, K. MATSUMURA, S. D. KAHL, C. J. LEVEILLE & K. P. CAMPBELL. 1991. Dystrophin-related protein is localized to neuromuscular junctions of adult skeletal muscle. Neuron **7:** 499–508.
57. TANAKA, H., T. ISHIGURO, C. EGUCHI, K. SAITO & E. OZAWA. 1991. Expression of a dystrophin-related protein associated with the skeletal muscle cell membrane. Histochemistry **96:** 1–5.
58. HELLIWELL, T. R., T. M. NGUYEN, G. E. MORRIS & K. E. DAVIES. 1992. The dystrophin-related protein, utrophin, is expressed on the sarcolemma of regenerating human skeletal muscle fibres in dystrophies and inflammatory myopathies. Neuromusc. Disord. **2:** 177–184.
59. KHURANA, T. S., S. C. WATKINS & L. M. KUNKEL. 1992. The subcellular distribution of chromosome 6-encoded dystrophin-related protein in the brain. J. Cell. Biol. **119:** 357–366.
60. MATSUMURA, K., J. M. ERVASTI, K. OHLENDIECK, S. D. KAHL & K. P. CAMP-

BELL. 1992. Association of the dystrophin-related protein with dystrophin-associated proteins in *mdx* mouse muscle. Nature **360**: 588–591.
61. DUBOWITZ, D. 1992. Transferring myoblasts in Duchenne dystrophy. Clinical results are disappointing. Brit. Med. J. **305**: 844–845.
62. ACSADI, G., G. J. DICKSON, D. R. LOVE, A. JANI, F. WALSH, A. GURUSINGHE, J. WOLFF & K. E. DAVIES. 1991. Human dystrophin expression in mdx mice after intramuscular injection of DNA constructs. Nature **352**: 815–818.
63. DUNCKLEY, M. G., D. R. LOVE, K. E. DAVIES, F. S. WALSH, G. E. MORRIS & G. DICKSON. 1992. Retroviral-mediated transfer of a dystrophin minigene into mdx mouse myoblasts in vitro. FEBS Lett. **296**: 128–134.
64. RAGOT, T., N. VINCENT, P. CHAFEY, E. VIGNE, H. GILGENKRANTZ, D. COUTON, J. CARTAUD, P. BRIAND, J. C. KAPLAN, M. PERRICAUDET *et al*. 1993. Efficient adenovirus-mediated transfer of a human minidystrophin gene to skeletal muscle of mdx mice. Nature **361**: 647–650.
65. TINSLEY, J. M. & K. E. DAVIES. 1994. Utrophin: A potential replacement for dystrophin? Neuromus. Disord. **3**: 537–539.
66. TINSLEY, J. M., D. J. BLAKE, M. PEARCE, S. J. KNIGHT, J. KENWRICK-JONES & K. E. DAVIES. 1993. Dystrophin and related proteins. Curr. Opin. Genet. Develop. **3**: 484–490.
67. AHN, A. H. & L. M. KUNKEL. 1993. The structural and functional diversity of dystrophin. Nature Genet. **3**: 283–290.

Transgenic Mouse Models of Disease: Altering Adipose Tissue Function *in Vivo*

SUSAN R. ROSS,[a,d] REED A. GRAVES,[b]
LISA CHOY,[c] VERONICA SOLEVEVA,[a]
AND BRUCE M. SPIEGELMAN[c]

[a] *Department of Biochemistry (m/c 536)*
University of Illinois School of Medicine
Chicago, Illinois 60612

[b] *Gastroenterology Section*
Department of Medicine
University of Chicago
Chicago, Illinois 60637

[c] *Dana Farber Cancer Institute, and*
Department of Cell Biology
Harvard Medical School
Boston, Massachusetts 01225

More than 40 years after the ground-breaking experiments of Avery, McLeod, McCarty, Hotchkiss, and others, showing that genetic information could be transferred between bacteria and that this information was DNA, similar experiments in mammals have been achieved.[1-5] The ability to introduce cloned DNA into the germline of living organisms to create transgenic animals with new genetic information represented a major advance to a number of areas of study, and more than a decade after the first transgenic mice were created, the technique has become essentially a routine laboratory procedure. Some of the uses for transgenic technology are listed in TABLE 1 and have been reviewed extensively by others.[6-9] In this paper, we would like to focus on the first three aspects listed as we have applied them to a specific tissue, fat: the ability to target expression of genes to particular tissues and the use of this targeted expression to both alter tissue function and to cause tumors in particular tissues, thereby creating mouse models of human diseases.

One of the diseases that can be best approached using transgenic models is obesity. Obesity refers to a heterogenous group of disorders in which there is an excess of either or both adipocyte cell mass and number. Adipocytes play

---

[d] Present address: Department of Microbiology, and Cancer Center, University of Pennsylvania, Philadelphia, Pennsylvania 19104.

TABLE 1. Uses for Transgenic Animals

1. *Study of gene regulation:* Definition of what DNA sequences are necessary to get correct developmental and tissue-specific expression of a gene.
2. *Define or change gene function:* Studies of the physiological or developmental consequences of introducing an "unregulated" gene in the animal.
3. *Oncogenesis:* Studies of consequences of expression of an activated or unregulated oncogene in specific tissue.
4. *Gene therapy:* Models for the correction of genetic defects.
5. *Gene modification:* Creation of animal models for human disease by replacing a "good" gene with a "bad" one.
6. *Identification of new genes:* Insertional mutagenesis.
7. *Factory:* The use of transgenic animals to produce unlimited quantities of proteins.

a central role in maintaining energy balance in animals, storing energy in the form of triglycerides during periods of nutritional abundance, and releasing it in a metabolically useful form at times of nutritional deprivation. In rodents that inherit genes that cause them to be obese, triglyceride storage in adipose tissue is more efficient and catabolism of these stores is slower than in their normal, non-obese counterparts.[10,11] The retention of such genes may be advantageous to animals living in the wild, since an increased ability to store energy would result in an increased survival rate when no food is available. Indeed, it has been shown in the laboratory that animals that have inherited even one copy of an "obesity" gene survive a prolonged fast longer than their otherwise genetically identical siblings that lack such a gene.[10]

There are several animal models in which obesity is inherited; recently, one of the genes that cause obesity, the lethal yellow Ay mutation, was cloned.[12] This mutation causes obesity in a semi-dominant manner and results in the novel expression of a gene not normally produced in fat. It is not yet known how the gene product alters adipocyte metabolism. In contrast to the Ay gene, there are a number of other known mutations that result in obesity only in the homozygous recessive state and the genes for these have not yet been identified.[11-14] Some of the genetic forms are likely due to aberrant signalling from the hypothalamus and not to mutations in the fat cells themselves.[15] It has long been known that physical or chemical lesions of the hypothalamus induce obesity.[15,16] The hypothalamus secretes hormones involved in many different biological functions, including releasing factors necessary to stimulate secretion of pituitary, adrenal, and thyroid hormones. Many of these hormones regulate the metabolism of the adipocyte in addition to other cell types in the animal.

Because the defect that results in obesity is systemic, when adipocytes are removed from an obese animal and either cultured *in vitro* or transplanted into a non-obese animal, they revert to a normal phenotype. Thus, the best ex-

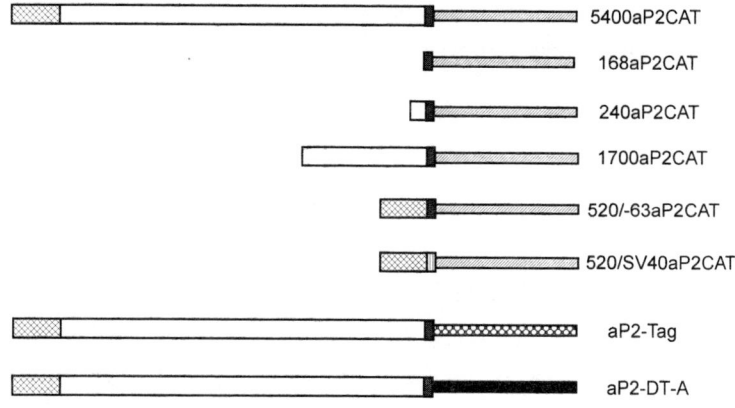

FIGURE 1. Maps of the DNA constructs used to make transgenic mice.

perimental system for studying both the effects of the obesity genes on adipocyte metabolism and the result of altering the function of this cell on overall metabolism is in animals. We have been altering adipocyte function in transgenic mice by directing expression of linked heterologous genes to this tissue. Such animals have provided us with a system for the study of the function of endogenous genes thought to be important in adipocyte metabolism and systemic energy balance. In addition, we can now manipulate the ability of fat cells to store lipid in transgenic animals by directing gene expression to adipose tissue and studying the effect on other metabolic pathways.

IDENTIFICATION OF A FAT CELL ENHANCER

The adipocyte P2 (aP2) gene was originally cloned from differentiating cultured adipocytes and encodes an abundant protein belonging to the intracellular lipid carrier protein family expressed predominantly in adipose tissue.[17] Although aP2 has been shown to bind fatty acids such as retinoic acid *in vitro*, its specific function *in vivo* is unknown.[18] The level of expression of this protein does not vary greatly under catabolic and anabolic conditions[19] or in obese mice relative to their lean siblings.[20] Thus, the regulatory region of the ap2 gene was ideal for achieving expression of linked heterologous genes specifically in adipose tissue.

Previous work in adipocyte tissue culture cells had indicated that short constructs containing as few as 168 base pairs upstream from the transcription start site of the aP2 gene was sufficient to achieve cell type–specific expression.[21-25] Surprisingly, when we made transgenic mice containing varying amounts of the aP2 upstream regulatory region linked to the bacterial chloramphenicol acetyltransferase (CAT) gene (FIG. 1), high levels of CAT activity

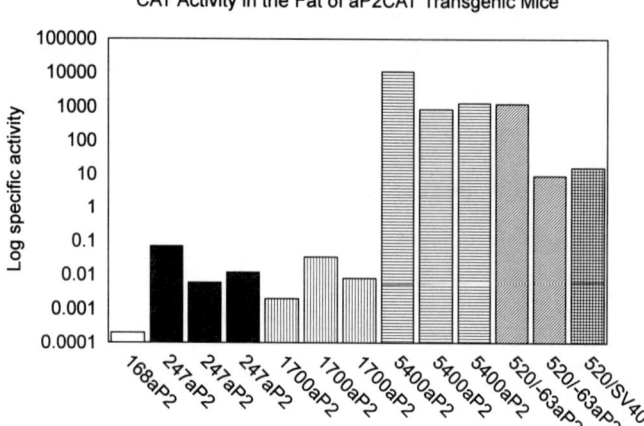

FIGURE 2. CAT-specific activity in extracts made from the WAT of the various aP2-CAT transgenic mice. One strain of the 168aP2CAT and 520/SV40CAT, two strains of the 520/−63aP2CAT, and three strains of the 247aP2CAT, 1700aP2CAT, and 5400aP2CAT were studied. CAT-specific activity is the cpm of acetylated chloramphenicol produced per μg protein per minute of reaction time.

in adipose tissue were not observed until the 5′ flanking region extended to −5.4kb (5400aP2 mice in FIG. 2).[26,27] We further mapped this enhancer to a 520bp fragment between −4.9 and −5.4 that functioned both in tissue culture cells and in transgenic mice (520/−63 in FIG. 2) and showed that this enhancer worked both with the aP2 promoter and with a heterologous SV40 promoter (520/SV40 in FIG. 2).[28] Significant expression of transgenes containing the fat-specific enhancer was not seen in other tissues.[26,27]

Thus, we identified an enhancer that was the major determinant of fat-specific expression of aP2 and that could be used to drive the expression of heterologous genes. This gave us the ability to alter the function of adipose cells in the context of the whole animal.

TRANSGENIC MICE WITH HIBERNOMAS: ISOLATION OF A BROWN FAT CELL TISSUE CULTURE LINE

Mammals contain two types of adipose tissue, termed white (WAT) and brown (BAT), based on the color. WAT primarily stores energy in the form of triglycerides, while BAT functions as an energy-dissipating tissue and is the major organ responsible for non-shivering thermogenesis, especially in newborn and hibernating mammals. In addition, BAT is activated to dissipate energy as a defense against obesity in certain animal models of overfeeding.[29–31] In general, the two types of adipose tissue are found in distinct depots. Histologically, white adipocytes contain unilocular lipid droplets and few mitochondria, while brown adipocytes contain multilocular lipid droplets and a large number of mitochondria. BAT is also more highly innervated than WAT.

In spite of these morphological differences, BAT and WAT are remarkably similar at the level of gene expression. Although there are quantitative differences in the levels of a number of genes that are expressed in WAT and BAT, many if not all of the genes expressed in WAT are also found in BAT. Only one protein, the mitochondrial uncoupling protein (UCP), has been found to be uniquely expressed in BAT. UCP is found on the inner mitochondrial membrane and acts as a proton channel that uncouples respiration from ATP synthesis, thus allowing the release of thermal energy.[32] UCP expression is stimulated by catecholamines released by the sympathetic nervous system.[33,34]

One of the genes abundantly expressed in both WAT and BAT is aP2. To create a model for studying adipocyte transformation *in vivo*, we placed the SV40 large T tumor antigen gene under the control of the regulatory region from the aP2 gene and created transgenic mice (aP2-Tag, FIG. 1).[35] Most of the transgenic mice rapidly developed brown fat tumors (hibernomas) in their interscapular BAT as well as abnormalities in their WAT. We were able to generate three pedigrees in which hibernoma formation was a heritable trait, because these animals developed hibernomas at >2 months after birth. The latency of hibernoma formation was probably related to the level of expression of the transgene in the different founders, since it has been previously shown that the amount of Tag expressed in various transgenic mice influences both the onset and severity of tumor formation.[36-38]

BAT was more severely affected by T antigen expression than was WAT in the aP2-SV40 mice. There are several possible reasons why this occurred. First, BAT develops in mice before WAT and since transformation occurred relatively early in development, BAT might be expected to be more affected in these animals. Second, it is possible that aP2 expression, and hence expression of the transgene, was higher in pre- or neonatal BAT than WAT, which would influence the latency and degree of transformation. Because neonatal WAT depots are so small, it was impossible to determine the level of gene expression in this tissue at early times. Finally, early-stage brown adipocytes might retain a higher potential to reenter the mitotic cycle than early-stage white adipocytes. Whatever the mechanism, the high tumor potential of BAT was also observed in transgenic mice that expressed SV40 under the control of the amylase 1a promoter; these mice developed predominantly hibernomas despite the fact that the T antigen protein was expressed in several tissues in addition to WAT and BAT.[38]

Although the hibernomas maintained differentiated characteristics and expressed UCP mRNA, the major function of this tissue, thermogenesis, was defective. The hibernoma-bearing mice were unable to thermoregulate at

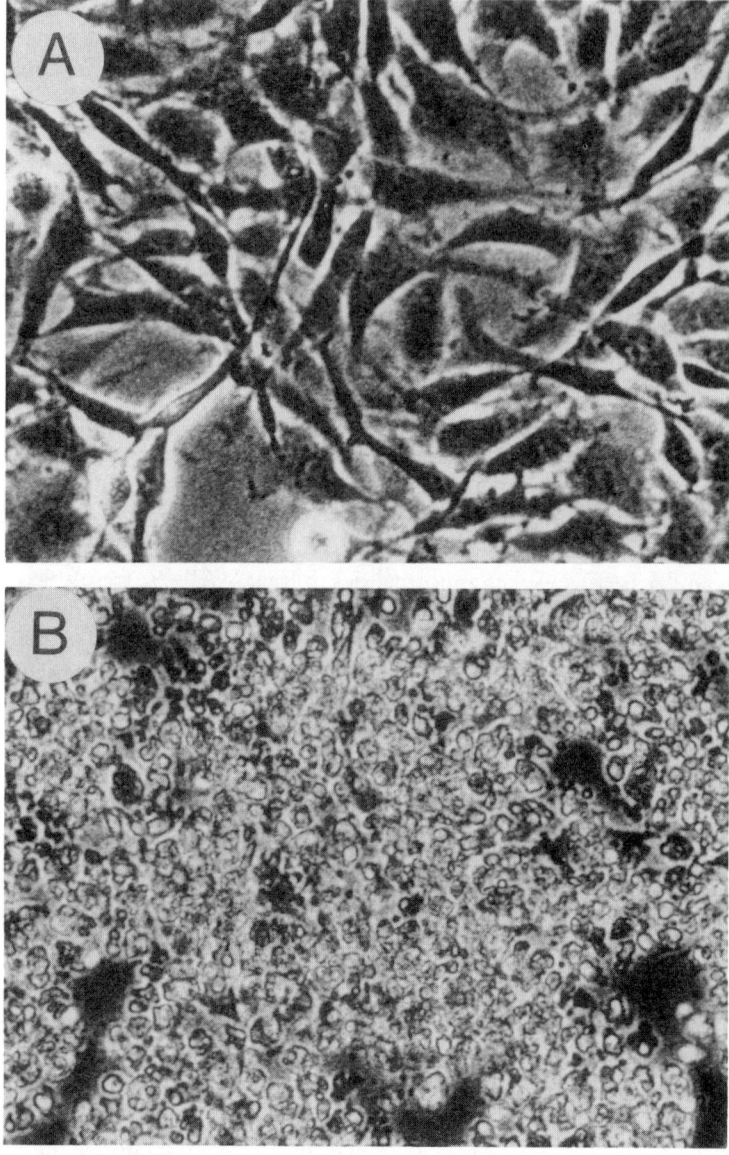

FIGURE 3. Differentiation of HIB 1B cells. (**A**) HIB 1B cells 1 day preconfluence. (**B**) HIB 1B cells 7 days after they were induced to differentiate with hydrocortisone, 3-isobutyl-1-methylxanthine, and indomethacin, as described.[35]

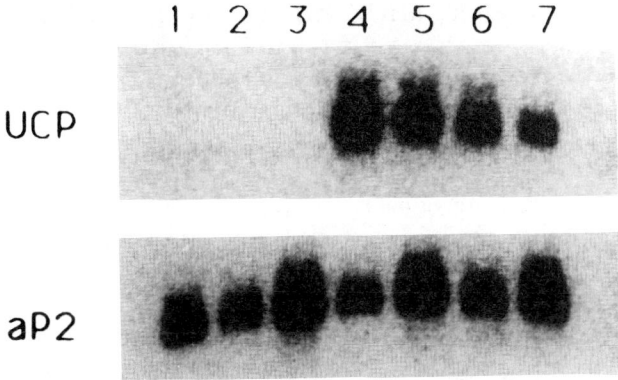

FIGURE 4. Northern blot analysis of preadipose and adipose HIB 1B cells treated with various β agonists and hybridized to UCP or aP2 probes. *Lane 1*: untreated HIB 1B cells; *lane 2*: undifferentiated HIB 1B cells treated with D7114; *lane 3*: HIB 1B adipocytes; *lanes 4–7*: HIB 1B adipocytes treated with N6,O2'-dibutyryladenosine 3',5'-cyclic monophosphate (*lane 4*), norepinephrine (*lane 5*), isoproterenol (*lane 6*), or D7114 (lane 7).

4°C, although their basal body temperatures at room temperature were normal.[35] This was probably due to the poor tissue organization and lack of proper innervation in the tumors that is required for appropriate regulated thermogenesis.

In addition to creating an *in vivo* model of adipocyte transformation, the generation of the aP2-Tag mice allowed us to develop a BAT tissue-culture line. Several tumors were used to establish cultured cell lines. These cell lines had a fibroblastic appearance (FIG. 3A). All of the cell lines expressed the aP2-Tag transgene and endogenous aP2 RNA (FIG. 4), indicating that they were at least partially differentiated along the adipocyte lineage.

At least one of the lines, HIB 1B, could be induced to differentiate into a more adipocyte-like cell that accumulated small lipid droplets (FIG. 3B). These cells expressed UCP RNA upon stimulation with dibutyryl cAMP, norepinephrine, isopreteronol or a β_3 adrenergic agonist, ICI D7114 (FIG. 4),[35,39] all of which have been shown to induce UCP expression in primary BAT cultures.

Much effort has been directed toward finding compounds that will activate BAT thermogenesis as an approach to therapeutics in obesity. Increasing caloric expenditure through increased metabolic rates is an attractive approach to therapies for obesity and makes the brown adipocyte a potential target for drug development. However, the study of brown adipose cell function has been hindered by the lack of established cell lines that differentiate into brown adipocytes expressing the UCP gene. These cells represent the first established cell model that can be used to study both the regulation of this gene and the metabolism of this cell type and should accelerate research on the function and therapeutic potential of BAT.

TRANSGENIC MICE RESISTANT TO OBESITY

Obesity leads to a general perturbation of the general metabolic state of the individual and is a major risk factor for a number of clinical diseases, such as non-insulin-dependent diabetes (NIDDM), cardiovascular disease, hyperlipidemia, and hypertension. Because of these associated diseases, the life expectancy of the obese individual is decreased.[40] A key question in obesity research is the nature of the link between adiposity and its accompanying clinical disorders.

As described above, some genetic forms of obesity are thought to originate from inappropriate hypothalamic function. Moreover, chemical lesions to the hypothalamus, such as neonatal administration of monosodium glutamate (MSG), induce obesity.[16] In obese animals, the levels of many hormones whose secretion is controlled by releasing factors produced in the hypothalamus is altered. For example, MSG-treated mice have foreshortened body axes, which is thought to be due to reduced pituitary function stemming from decreased growth hormone–releasing hormone secretion from the hypothalamus.[15,16] Moreover, inappropriate levels of gonadotropin-releasing hormones are thought to contribute to infertility in obese animals.[16] It is likely, however, that the effect on hormonal levels is also the result of the general perturbation of the physiological state of these animals. It is also probable that the hypothalamic defect(s) that causes obesity affects tissues in addition to fat.

Because we could target gene expression to adipose tissue, we were able to create transgenic mice that could be used to determine which of the clinical correlates of obesity were the direct result of excess adipose tissue or of the systemic defect that causes the obesity. We used the aP2 regulatory region to direct expression of an attenuated form of the diphtheria toxin A chain gene to fat (aP2-DT-A, FIG. 1).[41] The DT-A protein ribosylates eukaryotic elongation factor-2, resulting in its inactivation and the inhibition of protein synthesis; this protein behaves as a cytotoxic agent only when expressed intracellularly and so does not kill surrounding cells. The attenuated 176 gene has been reported to be about 15- to 30-fold less potent than the wild-type DT-A gene[42,43] and therefore has to be expressed at higher levels to completely kill cells in which it is expressed. Because the level of transgene expression varied in the different transgenic mouse pedigrees, owing to copy number and position effects, the use of the attenuated DT-A gene resulted in animals with a range of phenotypes depending on the amount of toxin gene produced in adipose tissue.

Nine SW outbred founder animals that contained the aP2-DT-A construct were produced.[44] Because of the different levels of transgene expression, four different phenotypes were observed (TABLE 2). High transgene expression (strains A, B, 6 and 8) resulted in neonatal lethality accompanied by the presence of a chylomicron- and lipid globule-filled fluid in the peritoneal cavity,

TABLE 2. Phenotype of aP2-DT-A Mice

Strain	Expression	Neonatal Lethality[a]	MSG-Obesity
A	++++	+	NA[b]
B	++++	+	NA
1	+	−	R[c]
2	++++	+	NA
6	++++	+	NA
8	++	−	R
12	−	−	S[d]
13	−	−	S
18	+++	+/−	ND[e]

[a] Neonatal lethality accompanied by chylous ascites.
[b] NA, not applicable.
[c] R, resistant to MSG-induced obesity.
[d] S, sensitive to MSG-induced obesity.
[e] ND, not determined.

termed chylous ascites.[45,46] This phenotype presumably resulted from an inability to metabolize dietary lipid because of fat cell dysfunction. Those newborn transgenic pups that expired within a few days after birth expressed very high levels of the DT-A transgene specifically in adipose tissue (TABLE 2). Expression of the DT-A transgene was not detected in other tissues, such as liver or brain.

The remaining founders fell into three additional classes. Two strains of transgenic animals did not express the transgene in any tissue (#12 and #13, TABLE 2), presumably because of silencing effects of the integration site, as is often seen in transgenic animals. Strains #8 and #18 expressed the transgene specifically in adipose tissue at about 100-fold lower levels than was seen in the newborn BAT of the neonatal lethal strains and we have termed such mice intermediate expressors. Finally, strain #1 expressed barely detectable levels of the transgene (low expressors).

These results indicated that loss of adipose tissue function prevented normal postnatal development; however, only those mice that expressed highs level of the toxin gene were so affected. In contrast, strains #1, #8, and #18 were viable. Young mice of the #8 and #18 strains (<4 months) had amounts of adipose tissue similar to that of their non-transgenic littermates (TABLE 3). However, as these mice aged, they developed progressively smaller adipose tissue depots with histological abnormalities. Strain #1, which had the lowest level of transgene expression, had no apparent alteration in adipose tissue depots until they were >1 year of age. The loss of adipose tissue in the aged transgenic mice may be due to the biology of this tissue in rodents. Accumulation of WAT in non-obese rodents is an age-dependent process and there is an increase in the proliferative capacity of these cells in older mice.[47]

TABLE 3. Clinical Correlates of aP2-DT-A Obesity-Resistant Mice

	Non-MSG-Treated Mice			MSG-Treated Mice		
	NT[a]	#1	#8	NT	#1	#8
Body weight (gm)	30.4 ± 2.1	35 ± 2.5	28.8 ± 3.9	41.5 ± 1.4	28.8 ± 1.1	28.2 ± 2.2
Body length (cm)	10.5 ± 0.2	ND[b]	ND[b]	9.8 ± 0.3	9.5 ± 0.17	9.5 ± 0.13
Fat pad weight (gm)[c]	0.44 ± 0.1	0.74 ± 0.14	0.35 ± 0.06	2.2 ± 0.5	0.21 ± 0.05	<0.1
Liver weight	1.5 ± 0.05	1.6 ± 0.08	1.5 ± 0.52	1.9 ± 0.25	4.0 ± 1.1	3.7 ± 0.69
Plasma triglycerides	86 ± 15	96 ± 25	57 ± 21	208 ± 56	461 ± 168	218 ± 26
Female fertility	+	+	+	−	+	+

[a] NT, nontransgenic.
[b] The overall body length of the non-MSG-treated transgenic animals did not differ significantly from nontransgenic mice. ND, not determined.
[c] Epididymal fat pad weight.

As discussed below, the DT-A toxin gene may have greater effects in proliferating adipocytes.

Although strains #1, #8, and #18 were not severely affected by the expression of the toxin transgene, they were completely resistant to the obesity-inducing effects of MSG. MSG treatment of neonatal mice results in the development of obesity at about the age of 6 to 10 weeks. No strain #18 transgenic mice injected neonatally with MSG survived past 1 week of age and most of these animals developed the chylous ascites seen in the high-expressing founders, while their non-transgenic littermates developed normally. In contrast, both the strain #1 and #8 mice treated with MSG were viable. At 6 weeks of age, the average total body weight of the MSG-treated nontransgenic mice began to surpass that of either MSG-treated transgenic strain of mice, and by 4 to 5 months of age the nontransgenic mice weighed on average greater than 10 grams more than their littermates (TABLE 3). This was primarily due to excess WAT weight in the MSG-treated nontransgenic mice (FIG. 5 and TABLE 3).

Thus, expression of the attenuated toxin gene had a dramatic effect on the ability of MSG to induce the expansion of adipocyte cell number and mass that occurs in obesity. Because MSG treatment results in increased adipocyte proliferation at an early age, the toxin gene may be more effective at causing fat cell dysfunction at a younger age. Alternatively, because there are a variety of metabolic alterations that occur in adipocytes during the progression to obesity, resulting in increased basal turnover in lipid synthesis and mobilization,[48] there could be an increase in the expression of a transgene under the transcriptional control of the aP2 lipid-binding protein gene. The observation that MSG treatment exacerbated the phenotype of the strain #18 animals, resulting in neonatal lethality, is consistent with either of these hypotheses. Moreover, the loss of adipose tissue that occurs with age in the non-MSG–treated transgenic mice may be the result of the increased proliferative capacity or altered metabolism of adipocytes seen in aging mice.

The strain #1 and #8 MSG-treated transgenic mice did not develop the massive adipose tissue mass associated with obesity, which therefore allowed us to begin to analyze the relationship between obesity and its clinical correlates. The MSG-treated mouse is mildly hypertriglyceridemic and hypercholesteremic and is stunted in body length; also, female, but not male mice, are infertile.[16] We therefore examined whether reduction of adipose mass in the context of the MSG-induced hypothalamic damage affected any of these parameters.

In the transgenic strains treated with MSG, there was no reversal of the hyperlipidemia (TABLE 3). Indeed, the levels were higher in the MSG-treated, non-obese transgenic mice than their nontransgenic obese siblings. Moreover, the hyperlipidemia became worse with age. This result indicates that the hyperlipidemia that accompanies MSG-induced obesity is not the consequence of

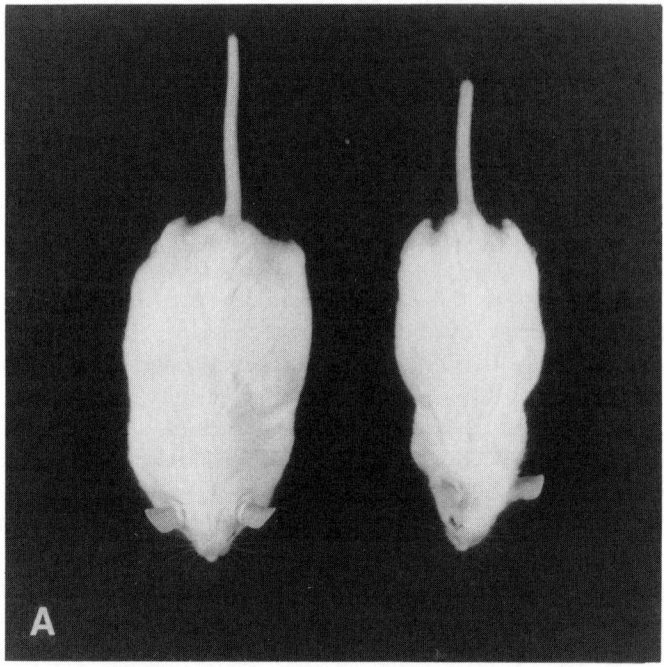

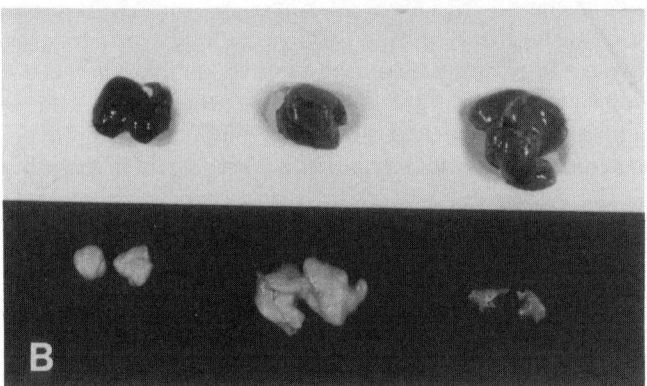

FIGURE 5. aP2-DTA transgenic mice are resistant to obesity. (**A**) MSG-treated transgenic mouse (*right*) and a nontransgenic littermate (*left*). (**B**) Liver (*top*) and epidydymal fat pad (*bottom*) from *left* to *right*: untreated aP2-DTA transgenic mouse, MSG-treated nontransgenic mouse, and MSG-treated aP2-DTA transgenic mouse.

the increased adipose mass. Because the MSG-treated obese animal is not hyperphagic[16] and even after denial of access to food this animal retains significant amounts of body fat,[49] these results suggest that the obesity seen in this model is the result rather than the cause of the elevated lipids in the plasma. That is, the elevated lipids may be the result of metabolic dysfunction caused by the MSG treatment and the expanded adipose mass may be a response to these elevated levels.

The major organ that metabolizes triglycerides in addition to fat is the liver, and hepatomegaly due to fatty liver is often seen in obese individuals. Not surprisingly, the MSG-treated non-obese transgenic mice developed severe hepatomegaly due to fatty liver, whereas the MSG-treated nontransgenic controls developed only moderate fatty liver and little if any hepatomegaly (TABLE 3 and FIG. 5). The more severe hepatomegaly seen in the transgenic mice was probably secondary to the greater hyperlipidemia that occurs because these animals were unable to expand their adipose mass.

Both the transgenic and nontransgenic mice treated with MSG sustained hypothalamic damage, because they had shortened body axis relative to that of untreated mice (TABLE 3). In spite of the hypothalamic dysfunction with regard to growth hormone regulation, the transgenic MSG-treated females were fertile and produced litters, in contrast to the MSG-treated obese nontransgenic females. Infertility in obese females is thought to result from aberrant endocrine control of ovulation and it was assumed that in the MSG-treated rodent, this was due to inappropriate gonadotropin-releasing hormone-signalling stemming from the hypothalamic damage.[16] Because reduction of adipose tissue mass in the transgenic mice restored fertility, it appears that the infertility in MSG-treated obese animals results from the excess adipose tissue itself. This could be due to exogenous steroid production by adipose tissue, resulting in high estrogen or androgen levels.[40]

CONCLUSIONS AND FUTURE DIRECTIONS

The identification of DNA sequences that direct tissue-specific expression in cultured cells and transgenic animals provides a way to alter tissue function in the context of the whole animal. We are applying this technique to the study of adipose tissue function. The study of the function of this tissue *in vivo* is particularly important to our understanding of the role of adipose tissue in systemic energy balance and in clinical disease. Such studies are important to a number of fields, including biology, pathophysiology and agriculture.

Pathological conditions that involve adipose tissue, such as obesity and lipodystrophy, are associated with a number of clinical conditions. The design of treatment for such disorders will be aided by understanding which of the clinical correlates of obesity are the direct result of the expanded adipose tissue mass and which are caused by the same genes or environmental factors that lead to increased adipose tissue storage. For example, obesity is the single most important risk factor associated with non-insulin-dependent diabetes mellitus (NIDDM); in humans, approximately 45% of males and 70% of females with this disorder are obese.[50] That obesity affects the onset of this disease is supported by the observation that diabetes is substantially improved by weight reduction. However, there is also evidence that the alterations in the activity of the autonomic nervous system that occur in obesity generated by some hypothalamic lesions result in increased pancreatic insulin release.[51] Individuals with NIDDM are usually hyperinsulinemic, especially at early stages of the disease, and this could in part be due to the hypothalamic lesion. Whether reduction of adiposity affects the hyperinsulinemia and subsequent diabetes development that occurs in genetically obese, diabetic rodents, such as the *db/db* mouse, is now an approachable research project.

In addition to creating mouse models of disease, we can now begin to alter the level of expression of genes thought to play key roles in the functioning of the adipocyte, such as adrenergic receptors. For example, activation of the β-class of adrenergic receptors stimulates the breakdown of lipid stores, whereas catecholamines that act through α_2 receptors stimulate lipogenesis.[52] It is thought that the ratio of β to α_2 receptors influences the regional variation of the ability of fat pads to respond to catecholamines and perhaps to store lipid.[53–56] Overexpression of the different classes of adrenergic receptors *in situ* will allow us to test whether this is the case and to create new experimental models of obesity and lipodystrophy.

Finally, the ability to alter adipose tissue storage of lipids and thus to directly manipulate fatness should be useful in agriculture. There has been considerable interest in developing genetic methods to alter the balance between lean and fat body mass in farm animals. Using transgenic technology, genetic alterations of fat cell metabolism that achieve a more economically useful balance are now possible.

REFERENCES

1. GORDON, J. W., G. A. SCANGOS, D. J. PLOTKIN, J. A. BARBOSA & F. H. RUDDLE. 1980. Genetic transformation of mouse embryos by microinjection of purified DNA. Proc. Natl. Acad. Sci. USA **77**: 7380–7384.
2. BRINSTER, R. L. & R. D. PALMITER. 1981. Somatic expression of herpes thymidine kinase in mice following injection of a fusion gene into eggs. Cell **27**: 223–231.
3. COSTANTINI, F. & E. LACY. 1981. Introduction of a rabbit beta-globin gene into the mouse germline. Nature **294**: 92–94.
4. WAGNER, E. F., T. A. STEWART & B. MINTZ. 1981. The human beta-globin gene and a functional viral thymidine kinase gene in developing mice. Proc. Natl. Acad. Sci. USA **78**: 5016–5020.
5. HARBERS, K., D. JAHNER & R. JAENISCH. 1981. Microinjection of cloned retroviral genomes into mouse zygotes: Integration and expression in the animal. Nature **293**: 540–542.
6. PALMITER, R. D. & R. L. BRINSTER. 1986. Germ-line transformation of mice. Ann. Rev. Genet. **20**: 465–499.

7. JAENISCH, R. 1988. Transgenic animals. Science **240:** 1468–1474.
8. HANAHAN, D. 1988. Dissecting multistep tumorigenesis in transgenic mice. Ann. Rev. Genet. **22:** 479–519.
9. MEISLER, M. H. 1992. Insertional mutation of 'classical' and novel genes in transgenic mice. TIG **8:** 341–344.
10. COLEMAN, D. L. 1979. Obesity genes: Beneficial effects in heterozygous mice. Science **203:** 663–665.
11. COLEMAN, D. L. 1982. Diabetes-obesity syndromes in mice. Diabetes **31:** 1–6.
12. MICHAUD, E. J., S. J. BULTMAN, L. J. STUBBS & R. P. WOYCHIK. 1993. The embryonic lethality of homozygous lethal yellow mice (Ay/Ay) is associated with the disruption of a novel RNA-binding protein. Genes & Dev. **7:** 1203–1213.
13. LEITER, E. H., W. G. BEAMER, L. D. SHULTZ, J. E. BARKER & P. W. LANE. 1987. Mouse models of genetic diseases. *In* Medical and Experimental Mammalian Genetics: A Perspective. V. A. McKusick *et al.*, Eds.: 221–257. Alan R. Liss. New York.
14. FRIEDMAN, J. M. & R. L. LEIBEL. 1992. Tackling a weighty problem. Cell **69:** 217–220.
15. BRAY, G. A. & D. A. YORK. 1979. Hypothalamic and genetic obesity in experimental animals: an autonomic and endocrine hypothesis. Physiol. Rev. **59:** 719–809.
16. OLNEY, J. W. 1969. Brain lesions, obesity and other disturbances in mice treated with monosodium glutamate. Science **164:** 719–721.
17. ZEZULAK, K. M. & H. GREEN. 1985. Specificity of gene expression in adipocytes. Mol. Cell. Biol. **5:** 419–421.
18. MATARES, V. & D. A. BERNLOHR. 1988. Purification of murine adipocyte lipid-binding protein. Characterization as a fatty acid- and retinoic acid-binding protein. J. Biol. Chem. **263:** 14544–14551.
19. FLIER, J. S., B. LOWELL, A. NAPOLITANO, P. USHER, B. ROSEN, K. S. COOK & B. SPIEGELMAN. 1989. Adipsin: Regulation and dysregulation in obesity and other metabolic states. Rec. Prog. Horm. Res. **45:** 567–580.
20. FLIER, J. S., K. S. COOK, P. USHER & B. M. SPIEGELMAN. 1987. Severely impaired adipsin expression in genetic and acquired obesity. Science **237:** 405–408.
21. RAUSCHER, F. J., III, L. C. SAMBUCETTI, T. CURRAN, R. J. DISTEL & B. M. SPIEGELMAN. 1988. A common DNA binding site for fos protein complexes and transcription factor AP-1. Cell **52:** 471–480.
22. DISTEL, R. J., H. S. RO, B. S. ROSEN, D. L. GROVES & B. M. SPIEGELMAN. 1988. Nucleoprotein complexes that regulate gene expression in adipocyte differentiation: direct participation of c-fos. Cell **49:** 835–844.
23. CHRISTY, R. J., V. W. YANG, J. M. NTABMI, D. E. GEIMAN, W. H. LANDSCHULZ, A. D. FRIEDMAN, Y. NAKABEPPU, T. J. KELLY & M. D. LANE. 1989. Differentiation-induced gene expression in 3T3-L1 preadipocytes: CCAAT/enhancer binding protein interacts with and activates the promoters of two adipocyte-specific genes. Genes & Devel. **3:** 1323–1335.
24. YANG, V. W., R. J. CHRISTY, J. S. COOK, T. J. KELLY & M. D. LANE. 1989. Mechanism of regulation of the 422(aP2) gene by cAMP during preadipocyte differentiation. Proc. Natl. Acad. Sci. USA **86:** 3629–3633.
25. HERRERA, R., H. S. RO, G. S. ROBINSON, K. G. XANTHOPOULOS & B. M. SPIEGELMAN. 1989. A direct role for C/EBP in gene expression linked to adipocyte differentiation. Mol. Cell. Biol. **9:** 5331–5339.
26. ROSS, S. R., R. A. GRAVES, A. GREENSTEIN, K. A. PLATT, H.-L. SHYU, B. MEL-

LOVITZ & B. M. SPIEGELMAN. 1990. A fat-specific enhancer is the primary determinant of gene expression for adipocyte P2 in vivo. Proc. Natl. Acad. Sci. USA **87:** 9590-9594.
27. GRAVES, R. A., P. TONTONOZ, S. R. ROSS & B. M. SPIEGELMAN. 1991. Identification of a potent adipocyte-specific enhancer: Involvement of an NF-1-like factor. Genes & Dev. **5:** 428-437.
28. GRAVES, R. A., P. TONTONOZ, S. R. ROSS & B. M. SPIEGELMAN. 1991. Identification and analysis of an adipose specific enhancer. *In* Obesity in Europe 91. G. Ailhaud *et al.*, Eds.: 155-161. J. Libbey & Co., Ltd. London.
29. ROTHWELL, N. J. & M. J. STOCK. 1979. A role for brown adipose tissue in diet-induced thermogenesis. Nature **281:** 31-34.
30. ROTHWELL, N. J., M. J. STOCK & D. STIBLING. 1982. Diet-induced thermogenesis. Pharmacol. Ther. **17:** 251-268.
31. HIMMS-HAGEN, J. 1990. Brown adipose tissue thermogenesis: interdisciplinary studies. FASEB J. **4:** 2890-2898.
32. KLAUS, S., A. CASSARD-DOULCIER & D. RICQUIER. 1991. Development of Phodopus sungorus brown preadipocytes in primary cell culture: Effect of an atypical beta-adrenergic agonist, insulin, and triiodothyronin on differentiation, mitochondrial development, and expression of the uncoupling protein UCP. J. Cell Biol. **115:** 1783-1790.
33. RICQUIER, D. & G. MORY. 1984. Factors affecting brown adipose tissue activity in animals and man. Clin. Endocrinol. Metab. **13:** 502-519.
34. REHNMARK, S., J. KOPECKY, A. JACOBSSON, M. NECHAD, D. HERRON, B. D. NELSON, M. OBREGON, J. NEDERGAARD & B. CANNON. 1989. Brown adipocytes differentiatioed in culture in vitro can express the gene for the uncoupling protein thermogenin: Effects of hypothyroidism and norepinephrine. Exp. Cell Res. **182:** 75-83.
35. ROSS, S. R., L. CHOY, R. A. GRAVES, N. FOX, V. SOLEVJEVA, S. KLAUS, D. RICQUIER & B. M. SPIEGELMAN. 1992. Hibernoma formation in transgenic mice and isolation of a brown adipocyte cell line expressing the uncoupling protein gene. Proc. Natl. Acad. USA **89:** 7561-7565.
36. ORNITZ, D. M., R. E. HAMMER, A. MESSING, R. D. PALMITER & R. L. BRINSTER. 1987. Pancreatic neoplasia induced by SV40 T-antigen expression in acinar cells of transgenic mice. Science **238:** 188-193.
37. CECI, J. D., R. M. KOVATCH, D. A. SWING, J. M. JONES, C. M. SNOW, M. P. ROSENBERG, N. A. JENKINS, N. G. COPELAND & M. H. MEISLER. 1991. Transgenic mice carrying a murine amylase 2.2/SV40 T antigen fusion gene develop pancreatic acinar cell and stomach carcinomas. Oncogene **6:** 323-332.
38. FOX, N., R. CROOKE, L.-H. S. HWANG, U. SCHIBLER, B. B. KNOWLES & D. SOLTER. 1989. Metastatic hibernomas in transgenic mice expressing an a-amylase-SV40 T antigen hybrid gene. Science **244:** 460-463.
39. KLAUS, S., L. CHOY, O. CHAMPIGNY, A.-M. CASSARD-DOULCIER, S. ROSS, B. SPIEGELMAN & D. RICQUIER. 1994. Characterization of the novel brown adipocyte cell line HIB 1B: Adrenergic pathways involved in regulation of uncoupling gene expression. J. Cell Sci. In press.
40. FOSTER, D. W. 1985. Eating disorders: Obesity and anorexia nervosa. *In* Textbook of Endocrinology :1081-1107. W. B. Saunders Co. Philadelphia.
41. ROSS, S. R., R. A. GRAVES & B. M. SPIEGELMAN. 1993. Transgenic mice resistant to obesity. Genes and Dev. **7:** 1318-1324.
42. MAXWELL, F., I. H. MAXWELL & L. M. GLODE. 1987. Cloning, sequence determination, and expression in transfected cells of the coding sequence for the tox 176 attenuated diptheria toxin A chain. Mol. Cell Biol. **7:** 1576-1579.

43. BREITMAN, M. L., H. ROMBOLA, I. H. MAXWELL, G. K. KLINTWORTH & A. BERSTEIN. 1990. Genetic ablation in transgenic mice with an attenuated diptheria toxin A gene. Mol. Cell. Biol. **10:** 474–479.
44. ROSS, S. R., R. A. GRAVES & B. M. SPIEGELMAN. 1993. Targeted expression of a toxin gene to adipose tissue: transgenic mice resistant to obesity. Genes & Dev. **7:** 1318–1324.
45. HERBERTSON, B. M. & M. E. WALLACE. 1964. Chylous ascites in newborn mice. J. Med. Genet. **1:** 10–23.
46. WALLACE, M. E. 1979. Analysis of genetic control of chylous ascites in ragged mice. Heredity **43:** 9–18.
47. AILHAUD, G., P. GRIMALDI & R. NEGREL. 1992. Cellular and molecular aspects of adipose tissue development. Ann. Rev. Nutr. **12:** 207–233.
48. CRANDALL, D. L. & M. DIGIROLAMO. 1990. Hemodynamic and metabolic correlates in adipose tissue: Pathyphysiologic considerations. FASEB J. **4:** 141–147.
49. CAMERON, D. P., L. CUTBUSH & F. OPAT. 1978. Effects of monosodium glutamate-induced obesity in mice on carbohydrate metabolism in insulin secretion. Clin. Exp. Pharmacol. Physiol. **5:** 41–51.
50. HARRIS, M. I. 1991. Diabetes Care **14:** 639–648.
51. BRAY, G. A., D. A. YORK & J. S. FISLER. 1989. Experimental obesity: A homeostatic failure due to defective nutrient stimulation of the sympathetic nervous system. Vitamins & Hormones **45:** 1–125.
52. ARNER, P. 1992. Adrenergic receptor function in fat cells. Am. J. Clin. Nutr. **55:** 228S–236S.
53. KATHER, H., K. ZOLLIG, B. SIMON & G. SCHLIERF. 1977. Human fat cell adenylate cyclase: Regional differences in adrenaline responsiveness. Eur. J. Clin. Invest. **7:** 595–597.
54. OSTMAN, J., P. ARNER, P. ENGFLEDT & L. KAGER. 1979. Regional differences in the control of lipolysis in human adipose tissue. Metabolism **28:** 1198–1205.
55. ARNER, P., L. HELLSTROM, H. WAHRENBERG & M. BRONNEGARD. 1990. Beta-adrenoceptor expression in human fat cells from different regions. J. Clin. Invest. **86:** 1595–1600.
56. ARNER, P. & F. LONNAVIST. 1987. Interactions betwen receptors for insulin and catecholamines in fat cells. *In* Recent advances in obesity research. J. Hirsch & E. van Hallei, Eds.: 204–211. J. Libbey & Co., Ltd. London.

Recombinant DNA Technology and Oral Medicine

HAROLD C. SLAVKIN

George and Mary Lou Boone Professor of Craniofacial Molecular Biology
Center for Craniofacial Molecular Biology
School of Dentistry
University of Southern California
2250 Alcazar Street, HSC, CSA-105
Los Angeles, California 90033

INTRODUCTION

We wish to suggest a structure for the salt of deoxyribonucleic acid (DNA). This structure has novel features which are of considerable biological interest.[1]

These modest interpretations made forty years ago have truly revolutionized our appreciation and growing understanding of a number of fascinating and complex processes within the life sciences. Indeed, in the original words of Francis Crick, we continue to live in a "biological revolution."[2] The discovery of the structure of DNA continues to provide seemingly unlimited promise towards advancing diagnosis, treatment, and the ultimate prevention of many human diseases. These advances are particularly important to oral medicine.

Oral medicine includes myriad conditions associated with the oral-dental-craniofacial complex including inherited or acquired congenital craniofacial anomalies (e.g., cleft lip, cleft palate; a number of first branchial arch syndromes such as DiGeorge, Treacher Collins, Rieger's and Pierre Robin syndromes; amelogenesis and dentinogenesis imperfecta), bacterial, viral, and fungal infections (e.g., dental caries, periodontal diseases, herpes simplex, HIV infection, *Candida albicans*), various forms of oral cancer and chronic diseases associated with an increasingly large proportion of Americans over the age of 65 years (e.g., osteoarthritis and rheumatism of the temporomandibular joint, chemosensory deficiencies, myopathies, osteoporosis, trigeminal neuralgia, and xerostomia).

This paper provides a selection from the numerous scientific advances in oral medicine related to the applications of recombinant DNA technology to the diagnosis, treatment, and prevention of craniofacial-oral-dental disorders. In no small measure, the new genetic paradigm for considering human oral diseases, and relative susceptibility or resistance to disease, is the direct consequence of the pioneering discoveries of the distinguished Francis Crick, James Watson and Maurice Wilkins who we are honoring in this landmark symposium: *DNA: The Double Helix Forty Years – Perspective and Prospective.*

PERSPECTIVE

A dominant theme in oral medicine has become the application of concepts and techniques developed in molecular genetics, clinical genetics, immunogenetics, developmental biology, and clinical diagnosis to the study of craniofacial-oral-dental disorders. It is anticipated that this approach will continue to provide DNA probes and antibodies that can be used as clinical diagnostic screening reagents for disorders, provide novel molecular approaches to gene-based therapeutics for soft- and hard-tissue wound-healing, create biomaterials for oral medicine applications, provide a molecular and developmental rationale for gene therapy studies, expand the molecular understanding of a number of first branchial arch syndromes, and increase public awareness and education for the prevention of human diseases.

This theme in oral medicine has evolved from several lines of evidence. First, the scientific advances resulting from the application of molecular biology approaches in such areas as immunological disorders, cancer, and aspects of hematology have provided a rationale and a number of strategies that have resulted in recombinant DNA technology's being applied to problems of craniofacial-oral-dental diseases.[3] Second, it is now evident that molecular biology has profoundly influenced basic and applied biomedical research by providing investigators with a variety of experimental strategies – ways of knowing and a large arsenal of techniques. As a consequence of the last few years, a number of advances have been made including: (i) the organization of the murine and human chromosomes, with the mapping of specific genetic sequences linked to selected human and mouse diseases; (ii) reverse genetic approaches to studies of gene function in organisms, organs, tissues, and cells; (iii) homologous recombination and the advent of "knock-out" transgenic animal models for human diseases; (iv) mutational analysis of congenital craniofacial-oral-dental dysmorphogenesis; (v) structural biology studies using computer-assisted molecular modeling of protein–protein and protein–nucleic acid interactions; (vi) computer modeling of genetic paradigms; (vii) computer-assisted microscopy, new microinjection technologies, myriad cell membrane and intracellular dyes, and a number of adenoviral and retroviral constructs suitable for labeling individual cells to enhanced resolution in cell fate and cell lineage studies; and (viii) increased international and national scientific activities in the field of craniofacial genetics and developmental biology.

Third, during the last decade, in no small measure the consequence of NIDR (National Institute of Dental Research) funding efforts in problems of oral medicine, there has been considerable "broadening the scope" and long-range research planning for the 1990s to advance a better understanding of craniofacial-oral-dental genetics, developmental biology, prokaryotic molecular genetics and the molecular biology of viruses, molecular genetic epidemiology, periodontal and pulp diseases, oral cancer, the molecular biology and inorganic chemistry of mineralized tissues (e.g., bone, cartilage, cementum, dentin, ectopic calcification, enamel), the molecular biology and physiology of neurosensory and chemosensory functions (i.e., taste and smell), as well as health promotion and disease prevention. Further, advances in the techniques of diagnosis, treatment and prevention of oral diseases have transferred to the delivery of oral health care.

These "ways of knowing" and a shared intellectual interest in the genetic regulatory processes which control the timing, position, and patterns of craniofacial-oral-dental development and maturation during the human lifespan, have led the biomedical research community to advance a molecular approach to a number of clinical problems in oral medicine related to structure and function in health and disease. Moreover, it is likely that the information gained through these investigations will lead to significant improvements in the diagnosis, treatment, and advancement of public awareness of prevention of craniofacial-oral-dental diseases.

Congenital Craniofacial-Oral-Dental Malformations

Congenital malformations represent a major public health issue in the United States. The existing treatment is extremely costly and often extends from birth through adolescence and includes the professional efforts of multidisciplinary teams consisting of parents and health professionals from the fields of dentistry, medicine, nursing, speech therapy, physical therapy, psychology and psychiatry, and social services. Mutations in regulatory and structural genes as well as chromosomal anomalies are associated with approximately 14 percent of the major craniofacial-oral-dental malformations. It is also now apparent that congenital malformations represent the consequence of mendelian as well as non-mendelian patterns of inheritance,[4,5] and also result from gene-environmental insults (e.g., Accutane [retinoic acid analogue], Dilantin, caffeine, alcohol)[6-8] during embryogenesis. Most prominent among the malformations are structural, functional and/or biochemical abnormalities of the skull, face, mouth, and teeth. Of all congenital malformations, approximately three-quarters affect the craniofacial-oral-dental and neck regions of the newborn infant.[4]

Approximately one in every 700 infants born alive are affected with cleft lip and/or palate, and approximately one in every 16,000 are affected with congenital craniofacial malformations other than cleft lip and/or cleft palate.[4]

These and other congenital craniofacial syndromes are caused by either maternal exposure to teratogens during the first three months of pregnancy or by the inheritance of one or more defective genes. The challenge is to understand the molecular controls for craniofacial-oral-dental morphogenesis in order to enhance diagnosis, therapeutics, and biomaterials and to promote disease prevention. Finally, some ten million Americans each year are afflicted with some type of acquired craniofacial malformation resulting from acute injuries to the craniofacial complex (e.g., those caused by automobile, sports, motorbike, and assault injuries) (data from Centers for Disease Control, 1992).

Recent landmarks from the last few years include the isolation of numerous genes responsible for the formation of craniofacial-oral-dental tissues. A genetic linkage map of the human genome was constructed that consists of 1416 loci, including 279 genes and expressed sequences.[9] At least 92% of the autosomal length of the genome and 95% of the X chromosome is estimated to be spanned by the linkage map. Since the maps have relatively high marker density and numerous highly informative loci, they can be used to map disease phenotypes, even for those with limited pedigree resources. In tandem, a genetic linkage map of the mouse with current applications and future prospects for models of human disease has been advanced.[10]

In light of congenital craniofacial malformations, a *Hox* code for first and second branchial arch morphogenesis has been formulated.[11] A large number of homeobox genes expressed during early craniofacial development have been isolated, sequenced, and mapped to chromosomes and, in some instances, functional analyses have been accomplished using over-expression and homologous recombination transgenic animals.[12-27] From the vertebrate paired box-containing *Pax* family, *Pax*-3 [chromosome 3(2q35)] and *Pax*-6 (chromosome 11p13) genes have been associated with Waardenburg's syndrome and aniridia, and additionally seven *Pax* genes have now been cloned and mapped to human chromosomes.[28] A candidate gene in autosomal dominant cases of cleft lip (with or without cleft palate) has been linked to chromosome 6, an association has been established between alleles of the TGF-alpha locus located at 2p13 and the occurrence of cleft lip,[29] and another example of cleft palate with ankyloglossia was linked to the q13-q21 region of the X chromosome.[30,31]

In addition, a genetic basis for several craniofacial syndromes has been established including: Treacher Collins syndrome(s) (mapped to chromosome 5 q31.3-q33.3,[32,33] associated with a deletion of chromosome 4p15.32-p14, caused by a translocation of chromosome 6p21.31/16p13.11); Rieger syndrome (mapped to the EGF gene on chromosome 4);[34] Greig syndrome (mapped to chromosome 7q13);[35] Waardenburg syndrome (a mutation in one or more Pax genes);[28] holoprosencephaly; and Van der Woude syndrome (linked to chromosome 1q). Furthermore, the amelogenin gene is located on the mouse X chromosome,[36] and the human Xp22.1-p22.3 and Y chromosomes provide a molecular basis for understanding X-linked amelogenesis imperfecta (AI);[37-40] the structural gene dentin phosphoprotein is mapped to human chromosome 1, and is not directly linked with dentinogenesis imperfecta (DI) types II and III on chromosome 4q11-4q21;[41] and loss of function mutations of the EGF gene in animal model studies may provide new opportunities to analyze Rieger's syndrome which maps to the EGF gene on chromosome 4.[34,42,43] Biomedical research scientists, advancing techniques and strategies of developmental biology and molecular genetics, have successfully mapped to human chromosomes approximately 70 genes related to craniofacial anomalies (e.g., various types of collagen, aggrecan, BMP1, BMP2A, BMP3 and BMP4, alkaline phosphatase, nidogen, retinoic acid receptors [RARs and RXRs], osteonectin, laminin, N-CAM, L-CAM, E-Cadherin), 30 genes related to dental tissue disorders (e.g., those involving bone morphogenetic proteins, enamelins, amelogenins, dentin sialoproteins, dentin phosphoprotein, osteopontin, decorin, biglycan), 20 genes related to clefting defects (e.g., those involving TGF-alpha), and 3 genes related to craniosynostosis. For the first time a mutation in the homeodomain of the human *Msx*-2 gene was found in a family affected with autosomal dominant craniosynostosis[15,44] (see TABLE 1).

From molecular embryology we learn that during initial neurulation and the establishment of the anteroposterior axis, a prepatterned, lineage-restricted cranial neural crest derived from rhombomeres segmentation pattern emerges from which the first branchial arches form and give rise to the midface and lower face (i.e., maxilla, mandible, oral cavity, salivary glands, tongue and dentition).[45-48]

Abrogation within these complex developmental processes results in a number of first branchial arch syndromes including mandibulofacial dysos-

TABLE 1. Recombinant DNA Technology and Craniofacial-Oral-Dental Disorders

Disorder	Chromosome Localization	Gene
Treacher Collins syndrome	5q32-33.2	Homeobox ?
Craniosynostosis, Boston type	5qter	*Msx*-2
Pfeiffer syndrome	8p	FGF receptor 1
Crouzon syndrome	10q	FGF receptor 2
Apert syndrome	10q	FGF receptor 2
Achondroplasia	4p	FGF receptor 3
Amelogenesis imperfecta	X	Amelogenin
Dentinogenesis imperfecta	4q21-25	?
Cleft lip susceptibility	2p13	TGF-alpha
	6	F13A Bl.Clot.Ftr
Rieger's syndrome	4	EGF
Waardenburg's syndrome	2q35	*Pax*-3
Aniridia	11p13	*Pax*-6

tosis (MFD). In terms of clinical genetics, it is becoming evident that these syndromes are highly heterogeneous (i.e., several different causes lead to the same dysmorphogenetic outcome). For example, a gene for MFD has been mapped at 5q31.3-5q33.3; however, MFD has also been mapped in patients with deletions (3) (p23p24.12) and deletions (4) (p15.32–p14), apparently balanced *de novo* translocation t(5;13) (q11;p11), and familial translocation t(6;16) (p21.31;p13.11).[49] Further, MFD is also part of some seven or more types of acrofacial dysostoses, some of which are causally defined as syndromes.[4,46-50] These complexities are becoming better understood through a paradigm in which early craniofacial morphogenesis is examined in the context of a "developmental field," in which developmental processes are temporally synchronized, spatially coordinated, and epimorphically hierarchical.[46-50] It is likely that advances in molecular studies of development will unravel the critical developmental problem of pattern formation. What is the molecular language for intercellular communication? How does the anticipated *Hox* code translate into metameric organization within the forming first and second branchial arch? How do homotypic and heterotypic intercellular communications provide instructions for temporally and positionally-restricted developmental events–instructions (e.g., transcription factors, oncogenes, growth factors and their cognate receptors, cell and substrate adhesion molecules [CAMs and SAMs], etc.) that provide the molecular basis for pattern formations during morphogenesis. Using the concepts and techniques from clinical genetics, molecular biology, and developmental biology (from *Drosophila* to human embryogenesis), biomedical research efforts will characterize the sequential, transient, or constitutive expression of a limited set of various gene products which are used in various combinations and/or permutations to form many parts of the developing organism (FIG. 1).

Animal Models for Human Oral Diseases

The recent advances in developing a dense genetic linkage map for the mouse genome as well as other technological advances in recombinant DNA technology now provide new opportunities for the genetic dissection of monogenic and polygenic traits.[10] Some of the available single-gene mouse models for human genetic disorders include those for beta-thalassemia (chromosome 11p15.5), X-linked cleft palate,[40] Duchenne muscular dystrophy (chromosome Xp21.3-p21.2), retinitis pigmentosa (chromosome 6p21.2-cen), osteoporosis (chromosome 3), microphthalmia (chromosome 6), and Waardenburg syndrome type 1 (chromosome 2q35-q37).[10] A transgenic mouse strain has recently been produced for X-linked cleft secondary palate syndrome,[40] which may represent the mouse counterpart to a human locus mutated in an X-linked cleft secondary palate syndrome.[30,31]

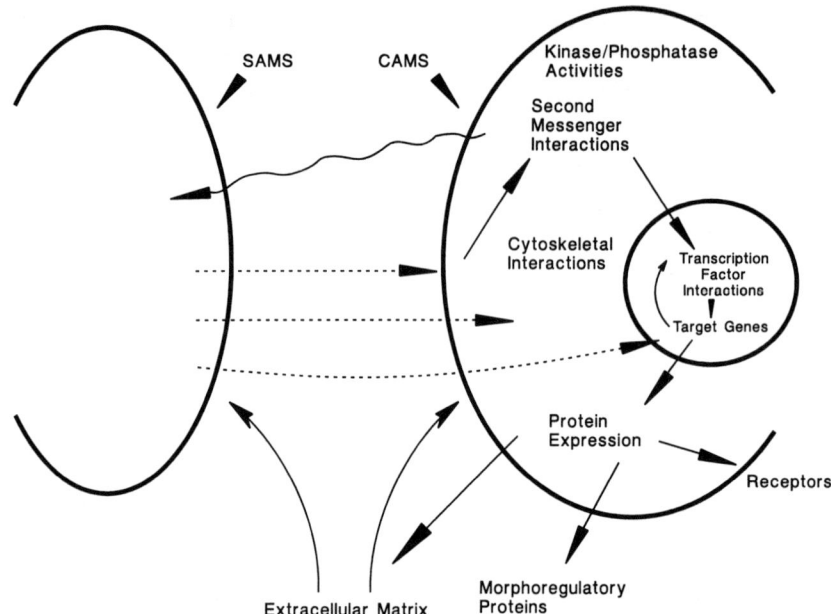

FIGURE 1. Scheme illustrating the hierarchy of molecular interactions related to craniofacial-oral-dental morphogenesis. It is becoming evident that vertebrate craniofacial morphogenesis is highly conserved, that there are redundant molecular controls for development, and that a number of time- and position-restricted cell–cell interactions mediate morphogenesis. This scheme depicts instructive cell–cell interactions in terms of epigenetic signalling between cells (e.g., growth hormone ligands binding to their cognate receptors, cell adhesion molecules [CAMs], and/or substrate adhesion molecules [SAMs] binding in a complementary manner), and a number of intracellular pathways involved in craniofacial-oral-dental development.

Animal models have also been produced that represent significant human viral diseases. For example, the development of disease and virus recovery in a transgenic mouse line containing HIV proviral DNA offers opportunities for the understanding of AIDS.[51] These transgenic mice carrying a subgenomic HIV-1 proviral construct lacking the *gag* and *pol* genes were found to develop proliferative epidermal lesions (epidermal hyperplasia) and benign papillomas.[52]

Another strategy is the use of genetically engineered cell lines for studies of human oral diseases. Recently, a genetically engineered bacterial toxin was used *in vitro* to lyse cells transfected with the HIV-1 virus without injury to adjacent normal cells. On the basis of the knowledge that HIV infection induces the production of the IL-2 receptor on the cell surface of transfected cells, a genetically altered diphtheria toxin was designed that specifically targeted the IL-2 receptor, ultimately leading to cell death.[53] Such strategies are being pursued using human cells *in vitro* for investigations of immune tolerance, immunochemistry, drug testing, and a number of other challenges in oral medicine.

Genetically Engineered Vaccines

There is growing scientific evidence of mucosal immune system in humans and that this system is shared by all mucosal sites including the oral cavity. This system develops in early infancy and appears to be extraordinarily significant in view of its role in prevention of infectious bacterial and/or viral disease.[54] Of particular interest are the recombinant vaccines designed to reduce dental caries (e.g., those directed against the oral streptococci *Streptococcus mutans*) and periodontal diseases in highly susceptible human populations. Promising research is in progress related to the development of vaccines against herpes simplex virus, human papillomavirus, and human immunodeficiency virus diseases with significant oral complications.

In humans, a primary infection of the lips, cornea, or genitalia with herpes simplex virus (HSV) type 1 or type 2 is often followed by the establishment of a latent ganglionic infection (as in the trigeminal ganglia with ophthalmic, maxillary, and mandibular divisions) that persists for the life of the individual. Approximately 80 percent of Americans have been exposed to HSV type 1, with a significantly smaller percentage expressing the disease recurrently. These latent infections often reactivate intermittently and give rise to recurrent herpetic lesions. Research studies determine that a vaccinia virus recombinant expressing herpes simplex virus type 1 glycoprotein D prevents latent herpes in a mouse animal model system.[55]

Advances in Biomaterials for Oral Medicine

A major challenge in oral medicine is the fabrication of biomaterials for replacement or augmentation therapy related to enamel, dentin, cementum, periodontal ligament, bone, cartilage, and oral mucosa. Recombinant DNA technology provides the strategy and methodology required to isolate, characterize, and express selected genes and/or gene products. This has recently led to the production of recombinant gene products such as bone morphogenetic proteins for cartilage, bone, and dentin induction.[56-59] Current estimates are that 24 million American women are affected with osteoporosis, and that a majority of individuals over the age of 50 are afflicted with some form of osteoarthritis or rheumatoid conditions. There are numerous challenges for somatic cell and gene therapeutic approaches in oral medicine.

Approximately 200 million dental fillings are placed annually in the United States. One major challenge for oral medicine is to develop bioceramics for dental restorative materials. For example, analogous to the reptilian and chick egg shell and barnacle bioceramics of calcium carbonate, the outer tooth-covering mineralized tissue (i.e., enamel) is also a bioceramic, but consists of highly ordered enamel crystals of calcium hydroxyapatite, the hardest tissue in the body. Recent investigations of biomineralization indicate that specific molecular interactions at inorganic–organic interfaces can result in the controlled nucleation and growth of inorganic crystals. Chemically characterized organic molecules in solution have been found to influence the morphology of inorganic crystals if there is molecular complementarity at the forming crystal–organic molecule interface. The recent progress in molecular studies of amelogenin and enamelin might provide opportunities for a biomimetic approach to the development of novel bioceramics for the clinical requirements of human enamel replacement therapy.[37,60]

Opportunities for Gene Therapy in Oral Medicine

Following the precedent for somatic cell gene therapy developed by French Anderson and his colleagues at the National Institute of Health in 1991 using the retroviral transfection of adenosine deaminase (ADA) gene for severe combined immunodeficiency caused by ADA deficiency, a number of additional clinical gene therapy trials, for cystic fibrosis, sickle cell anemia, and muscular dystrophy, for example, have begun.

Several studies have initiated the development of somatic cell gene therapy for the treatment of oral and other diseases. Studies have determined the utility of transfected keratinocytes derived from oral epithelium for the treatment of several oral diseases.[61] This strategy is potentially useful for the treatment of xeroderma pigmentosum or for the delivery of oral therapeutics. Recent studies by Bruce Baum and his colleagues at the NIDR have demonstrated the direct *in vivo* adenovirus-mediated gene transfer to salivary glands.[62] The adenovirus-mediated gene transfer strategy provides a number of opportunities to use gene therapeutics to treat intraoral diseases such as fungal infections (*Candida albicans* treated with Histatin) or xerostomia (dry mouth condition treated by methods to produce intraoral lubricants). In tandem, a number of molecular studies have contributed towards the development of artificial salivas.[63]

PROSPECTIVE

We are experiencing a literal scientific explosion in human and mouse gene mapping sparked by the advent of recombinant DNA technology and new types of genetic markers.[9,10] Recombinant DNA techniques have allowed the identification and mapping of DNA polymorphisms, which have provided a remarkable source of biologically interesting and clinically significant loci for the human and mouse genome maps. DNA markers were initially scored as restriction fragment length polymorphisms (RFLPs) on Southern (DNA) blots. More recently, molecular genetic problems in mouse and human studies are assayed by the polymerase chain reaction (PCR). The currently available and anticipated advances in the density of information within mouse and human gene maps provide myriad opportunities for developing animal models for human monogenic as well as polygenic craniofacial-oral-dental diseases, for advancing diagnosis and therapeutics in oral medicine,[64,65] and for realizing the ultimate goals for prevention of disease.

Moreover, the prospects for a number of different somatic cell gene therapy approaches and gene-mediated therapeutics in oral medicine are promising:

- Delivery of genetic therapeutics for oral fungal infections (e.g., *Candida albicans*);
- Production of oral mucosal lubricants for xerostomia;
- Production of system gene products from oral tissues;
- Genetic approaches to biomaterials such as enamel, dentin, cementum, and bone replacement therapy;
- Recombinant growth factors and cytokines for soft- and hard-tissue repair and regeneration;[55]
- Genetically engineered vaccines (e.g., to prevent dental caries, periodontal diseases, herpes simplex types 1 and 2; AIDS);[52,64]
- Genetically based therapeutics for oral cancer;[61,64-66] and
- Genetically based therapeutics for neurologic disorders (such as trigeminal neuralgia).

And finally, in my prospective view, some things should not change. In particular, the investigator-initiated creative explorations into the wonders of biological processes as exemplified by those of Watson, Crick and Wilkins must not be lost to the pragmatic urgencies of "relevance" and "cost effectiveness." The educational odyssey must continue to nurture wonder, curiosity, and originality. Each of us has a responsibility to keep that ideal alive and well. I'm reminded of a poignant description by François Jacob in his autobiography in which he describes Max Delbrück's symposium on viruses at Cold Spring Harbor in 1953 and heard Jim Watson's description of the structure of DNA, which he had worked out with Francis Crick.

> . . . For a moment, the room remained silent. There were a few questions. How, for example, during replication of the double helix, could the two chains entwined around one another separate without breaking? But no criticism. No objections. This structure was of such simplicity, such perfection, such harmony, such beauty even, and biological advantages flowed from it with such rigor and clarity, that one could not believe it untrue. By all indications, it was a turning point in the study of living things. It heralded an exciting period in biology.[67]

ACKNOWLEDGMENTS

I wish to acknowledge collaborative efforts with Brent Jacquet and Wayne Little of the Science Information Office of the National Institute of Dental Research (NIDR) in Bethesda, Maryland and thank them for their untiring efforts to provide current documentation for advances from recombinant DNA technology in oral medicine. Of course, I take full responsibility for the biased selections and interpretations that are discussed in this manuscript; these are presented to provide the reader with the general contours and texture of this emerging biomedical research adventure. Finally, I wish to acknowledge and thank my scientific colleagues from whom so much has been learned including Pablo Bringas, Yang Chai, Allan Fincham, Sharon Groff, Jan Hu, Larry Kedes, Ed Lau, Wen Luo, Mary MacDougall, Mark Mayo, Rob Maxson, Val Santos, Charles Shuler, Jim Simmer, Mal Snead, Maggie Zeichner-David, David Warburton, and Zena Werb.

REFERENCES

1. WATSON, J. D. & F. H. C. CRICK. 1953. Molecular structure of deoxypentose nucleic acids. Nature **171:** 738-739.
2. CRICK, F. 1982. DNA today. Perspect. Biol. Med. **25:** 512-517.
3. SLAVKIN, H. C. 1989. Splice of life: Toward understanding genetic determinants of oral diseases. Advan. Dent. Res. **3**(1): 42-57.
4. GORLIN, R. J., M. M. COHEN, JR. & L. S. LEVIN. 1990. Syndromes of the Head and Neck, 3rd Ed. Oxford University Press. New York.
5. GORLIN, R. J. & H. C. SLAVKIN. Embryology of the face. *In* Congenital Anomalies in Otolaryngology. T. L. Tewfik & V. Der Kaloustian, Eds. Oxford University Press. London. In press.
6. LAMMER, E. J., D. T. CHEN, R. M. HOAR, N. D. AGNISH, P. J. BENKE, J. T. BRAUN, C. J. CURRY, P. M. FERNHOFF, A. W. GRIX, I. T. LOTT, J. M. RICHARD & S. C. SUN. 1985. Retinoic acid embryopathy. N. Engl. J. Med. **313:** 837-841.
7. SULIK, K. K., M. C. JOHNSTON, P. A. DRAFT, W. E. RUSSELL & D. B. DEHART. 1986. Fetal alcohol syndrome and DiGeorge anomaly: Critical ethanol exposure period for craniofacial malformations as illustrated in an animal model. Am. J. Med. Genet. (Suppl) **2:** 97-112.
8. WEBSTER, W. S. 1989. Alcohol as a teratogen: A teratological perspective of the fetal alcohol syndrome. *In* Human Metabolism of Alcohol, Vol. 1. R. D. Bratt & K. Crow, Eds.: 133-155. CRC Press. Boca Raton, FL.
9. NIH/CEPH COLLABORATIVE MAPPING GROUP. 1992. A comprehensive genetic linkage map of the human genome. Science **258:** 67-86.
10. COPELAND, N. G., N. A. JENKINS, D. J. GILBERT, J. T. EPPIG, L. J. MALTAIS, J. G. MILLER, W. F. DIETRICH, A. WEAVER, S. E. LINCOLN, R. G. STEEN, L. D. STEIN, J. H. NADEAU & E. S. LANDER. 1993. A genetic linkage map of the mouse: current applications and future prospects. Science **262:** 57-66.
11. HUNT, P., J. WHITING, S. NONCHEV, M. SHAM, H. MARSHALL, A. GRAHAM, M. COOK, R. ALLEMANN, P. W. J. RIGBY, M. GULISANO, A. FAIELLA, E. BONCINELLI & R. KRUMLAUF. 1991. The branchial *Hox* code and its implications

for gene regulation, patterning of the nervous system and head evolution. Development **2** (Suppl): 63–77.
12. ASANO, M., Y. EMORI, K. SAIGO & K. SHIOKAWA. 1992. Isolation and characterization of a *Xenopus* cDNA which encodes a homeodomain highly homologous to *Drosophila Distal-less*. J. Biol. Chem. **267**: 5044–5047.
13. ALTABA, A. R. I. & D. A. MELTON. 1989. Interaction between peptide growth factors and homeobox genes in the establishment of antero-posterior polarity in frog embryos. Nature **341**: 33–38.
14. BASTIAN, H. & P. GRUSS. 1990. A murine *even-skipped* homologue, *Evx*-1, is expressed during early embryogenesis and neurogenesis in a biphasic manner. EMBO J. **9**: 1839–1852.
15. BELL, J. R., A. NOVEEN, Y. H. LIU, L. MA, S. DOBIAS, R. KUNDU, W. LUO, Y. XIA, A. LUSIS, M. L. SNEAD & R. MAXSON. 1993. Genomic structure, chromosomal location and evolution of the mouse *Hox*-8 gene. Genomics **16**: 123–131.
16. DOLLE, P., M. PRICE & D. DUBOULE. 1992. Expression of the murine *Dlx*-1 homeobox gene during facial, ocular, and limb development. Differentiation **49**: 93–99.
17. FERGUSON, M. W. J., A. MACKENZIE & P. T. SHARPE. 1992. *Hox*-8 expression and patterning of the developing mammalian tooth. *In* Chemistry and Biology of Mineralized Tissues. H. C. Slavkin & P. Price, Eds.: 377–384. Elsevier. Amsterdam.
18. JOWETT, A. K., S. VAINIO, M. W. FERGUSON, P. T. SHARPE & I. THESLEFF. 1993. Epithelial-mesenchymal interactions are required for *Msx*- and *Msx*-2 gene expression in the developing murine molar tooth. Development **117**: 461–470.
19. MACKENZIE, A., G. LEMMING, A. K. JOWETT, M. W. J. FERGUSON & P. T. SHARPE. 1991. The homeobox gene Hox 7.1 has specific regional and temporal expression patterns during early murine craniofacial embryogenesis, especially tooth development in vivo and in vitro. Development **111**: 269–285.
20. MACKENZIE, A., M. W. J. FERGUSON & P. T. SHARPE. 1992. Expression patterns of the homeobox gene, Hox-8, in the mouse embryo suggest a role in specifying tooth initiation and shape. Development **115**: 430–438.
21. MORRISON-GRAHAM, K., G. C. SCHATTEMAN, T. BORK, D. F. BOWEN-POPE & J. A. WESTON. 1992. A PDGF receptor mutation in the mouse (Patch) perturbs the development of a non-neuronal subset of neural crest-derived cells. Development **115**: 133–142.
22. NIETO, M. A., L. C. BRADLEY & D. G. WILKINSON. 1991. Conserved segmental expression of Krox-20 in the vertebrate hindbrain and its relationship to lineage restriction. Development **2** (Suppl): 59–62.
23. PAPALOPULU, N. & C. KINTER. 1993. *Xenopus Distal-less* related homeobox genes are expressed in the developing forebrain and are induced by planar signals. Development **117**: 961–975.
24. ROBINSON, G., S. WRAY & K. MAHON. 1991. Spatially restricted expression of a member of a new family of murine *Distal-less* homeobox genes in the developing forebrain. New Biologist **3**: 1183–1194.
25. STEPHENSON, D. A., M. MERCOLA, E. ANDERSON, C. WANG, C. D. STILES, D. F. BOWEN-POPE & V. M. CHAPMAN. 1991. Platelet-derived growth factor receptor alpha-subunit gene (PDGFRalpha) is deleted in the mouse *Patch* (*Ph*) mutation. Proc. Natl. Acad. Sci. USA **88**: 6–10.
26. WILKINSON, D. G., G. PETERS, C. DICKSON & A. P. MCMAHAN. 1988. Expression of the FGF-related proto-oncogene int-2 during gastrulation and neurulation in the mouse. EMBO J. **7**: 691–695.

27. WILKINSON, D. G., S. BHATT, M. COOK, E. BONCINELLI & R. KRUMLAUF. 1989. Segmental expression of Hox-2 homeobox-containing genes in the developing mouse hindbrain. Nature **341:** 405–409.
28. STAPLETON, P., A. WEITH, P. URBANEK, A. KOZMIK & M. BUSSLINGER. 1993. Chromosomal localization of seven PAX genes and cloning of a novel family member, PAX-9. Nature Genet. **3:** 292–298.
29. SASSANI, R., S. P. BARTLETT, H. FENG, A. GOLDNER-SAUVE, A. K. HAQ, K. H. BUETOW & D. L. GASSER. 1993. Association between alleles of the transforming growth factor-alpha locus and the occurrence of cleft lip. Am. J. Med. Genet. **45:** 565–569.
30. MOORE, G. E., A. IVENS, J. CHAMBERS, M. FARRALL, R. WILLIAMSON, D. C. PAGE, A. BJORSSON, A. ARNASON & O. JENNSON. 1987. Linkage of an X-chromosome cleft palate gene. Nature **326:** 96–99.
31. MOORE, G. E., R. WILLIAMSON, O. JENSSON, J. CHAMBERS, F. TAKAKUBO, R. NEWTON, M. A. BALACS & A. IVENS. 1991. Localization of a mutant gene for cleft palate and ankyloglossia in an X-linked Icelandic family. J. Craniofac. Genet. Devel. Biol. **11**(4): 372–376.
32. DIXON, M. J., J. DIXON, D. RASKOVA, M. M. LE BEAU, et al. 1992. Genetic and physical mapping of the Treacher Collins syndrome locus: Refinement of the localization to chromosome 5q32-33.2. Hum. Mol. Genet. **1**(4): 249–253.
33. DIXON, M. J., J. DIXON, T. HOUSEAL, M. BHATT, et al. 1993. Narrowing the position of the Treacher Collins syndrome locus to a small interval between three new microsatellite markers at 5q32-33.1. Am. J. Med. Genet. **52:** 907–914.
34. MURRAY, J. C., S. R. BENNETT, A. E. KWITEK, K. W. SMALL, W. L. M. ALWARD, J. L. WEBER & K. H. BUETOW. 1992. Linkage of Rieger syndrome to the region of the epidermal growth factor gene on chromosome 4. Nature Genet. **2:** 46–48.
35. VORTKAMP, A., M. GESSLER & K. H. GRZESCHIK. 1991. GL13 zinc-finger gene interrupted by translocations in Grieg syndrome families. Nature **352:** 539–540.
36. CHAPMAN, C., C. HEITZ, M. STEPHENSON, E. LAU & M. L. SNEAD. 1991. Linkage of amelogenin to the distal portion of the mouse X chromosome. Genomics **10:** 23–28.
37. LAU, E. C., T. K. MOHANDAS, L. J. SHAPIRO, H. C. SLAVKIN & M. L. SNEAD. 1989. Human and mouse amelogenin gene loci are on the sex chromosomes. Genomics **4:** 162–168.
38. LYLE, R., F. TAKAKUBA, R. WILLIAMSON, G. B. WINTER, A. H. BROOK, M. L. SNEAD, C. GIBSON, L. J. SHAPIRO & A. C. IVENS. Heterogeneity in the molecular basis of amelogenesis imperfecta. J. Med. Genetics. In press.
39. ALDRED, M. J., P. J. CRAWFORD, E. ROBERTS & N. S. THOMAS. 1992. Identification of a nonsense mutation in the amelogenin gene (AMELX) in a family with X-linked amelogenesis imperfecta (AIH1). Hum. Genet. **90:** 413–416.
40. WILSON, J. B., M. W. FERGUSON, N. A. JENKINS, L. F. LOCK, et al. 1993. Transgenic mouse model of X-linked cleft palate. Cell Growth Differ. **4**(2): 67–76.
41. MACDOUGALL, M., M. ZEICHNER-DAVID, J. MURRAY, M. CRALL, A. DAVIS & H. C. SLAVKIN. 1992. Dentin phosphoprotein gene locus is not associated with dentinogenesis imperfecta Type II and III. Am. J. Med. Genet. **50:** 190–194.
42. SHUM, L., Y. SAKAKURA, P. BRINGAS, W. LUO, M. L. SNEAD, M. MAYO, C. CROHIN, S. MILAR, Z. WERB, S. BUCKLEY, F. L. HALL, D. WARBURTON & H. C. SLAVKIN. 1993. EGF abrogation induced *fusilli*-form dysmorphogenesis of Meckel's cartilage during embryonic mouse mandibular morphogenesis in vitro. Development **118:** 903–917.

43. SLAVKIN, H. C. Rieger's syndrome revisited: Experimental approaches using pharmacologic and antisense strategies to abrogate EGF and TGF-alpha functions resulting in dysmorphogenesis during embryonic craniofacial morphogenesis. Am. J. Med. Genet. In press.
44. JABS, E. W., U. MULLER, X. LI, L. MA, W. LUO, I. HAWORTH, I. KLISAK, R. SPARKES, M. L. WARMAN, J. B. MULLIKEN, M. L. SNEAD & R. MAXSON. 1993. A mutation in the homeodomain of the human Msx 2 gene in a family affected with autosomal dominant craniosynostosis. Cell 75: 1–20.
45. GRAHAM, A. & A. LUMSDEN. 1993. The role of segmentation in the development of the branchial region of higher vertebrate embryos. In Blastogenesis, Normal and Abnormal. J. M. Opitz, Ed.: 99–108. Wiley-Liss. New York.
46. LE DOURARIN, N., C. ZILLER & G. COUL. 1993. Patterning of neural crest derivatives in the avian embryo: in vivo and in vitro studies. Devel. Biol. 159: 24–49.
47. OPITZ, J. M. Developmental field theory and the molecular analysis of morphogenesis. Am. J. Med. Genet. In press.
48. OPITZ, J. M. 1993. Blastogenesis and the "primary field" in human development. In Blastogenesis, Normal and Abnormal. J. M. Opitz, Ed.: 1–34, 99–108. Wiley-Liss. New York.
49. ARN, P. H., C. MANKINEN & E. W. JABS. 1993. Mild mandibulofacial dysostosis in a child with a deletion of 3p. Am. J. Med. Genet. 46: 534–536.
50. LAMMER, E. J. & J. M. OPITZ. 1986. The DiGeorge anomaly as a developmental field defect. Am. J. Med. Genet. (Suppl) 2: 113–127.
51. LEONARD, J. M., J. W. ABRAMCZUK, D. S. PEZEN, R. RUTLEDGE, J. H. BELCHER, F. HAKIM, G. SHEARER, L. LAMPERTH, W. TRAVIS, T. FREDRICKSON, A. L. NOTKINS & M. A. MARTIN. 1988. Development of disease and virus recovery in transgenic mice containing HIV proviral DNA. Science 242: 1665–1670.
52. KOPP, J. B., J. F. ROONEY, C. WOHLENBERG, N. DORFMAN, et al. 1993. Cutaneous disorders and viral gene expression in HIV-1 transgenic mice. AIDS Res. Hum. Retroviruses 9: 267–275.
53. FINBERG, R. W., S. M. WAHL, J. B. ALLEN, G. SOMAN, T. B. STREOM, J. R. MURPHY & J. C. NICHOLS. 1991. Selective elimination of HIV-1-infected cells with an interleukin-2 receptor-specific cytotoxin. Science 252: 1703–1705.
54. CIARDI, J. E., J. R. MCGHEE & J. M. KEITH. 1992. Genetically Engineered Vaccines: 301–314. Plenum Press. New York.
55. CREMER, K. J., M. MACKETT, C. WOHLENBERG, A. L. NOTKINS & B. MOSS. 1985. Vaccinia virus recombinant expressing herpes simplex virus type 1 glycoprotein D prevents latent herpes in mice. Science 228: 737–740.
56. CANALIS, E., J. PASH & S. VARGHESE. 1993. Skeletal growth factors. Crit. Rev. Eucaryotic Gene Express. 3: 155–166.
57. HOLLINGER, J. 1993. Strategies for regenerating bone of the craniofacial complex. Bone 14: 575–580.
58. RIPAMONTI, U., S. S. MA & A. H. REDDI. 1992. The critical role of geometry of porous hydroxyapatite delivery system in induction of bone by osteogenin, a bone morphogenetic protein. Matrix 12: 202–212.
59. WOZNEY, J. M., V. ROSEN, A. J. CELESTE, L. M. MITSOCK, M. J. WHITERS, R. W. KRIZ, R. M. HELWICK & E. A. WANG. 1988. Novel regulators of bone formation: Molecular clones and activities. Science 242: 1528–1534.
60. DEUTSCH, D., A. PALMON, L. W. FISHER, N. KOLODNY, J. D. TERMINE & M. F. YOUNG. 1991. Sequencing of bovine enamelin ("tuftelin"), a novel acidic enamel protein. J. Biol. Chem. 266: 16021–16028.
61. FENJVES, E. S., J. I. LEE, J. A. GARLICK, D. A. GORDON, D. L. WILLIAMS & L. B. TAICHMAN. 1990. Prospects for epithelial gene therapy. In DNA Damage

and Repair in Human Tissues. B. M. Sutherland & A. D. Woodhead, Eds: 215–230. Plenum Press. New York.
62. MASTRANGELI, A., B. O. O'CONNELL, W. ALADIB, P. C. FOX, B. J. BAUM & R. G. CRYSTAL. 1994. Direct in vivo adenovirus-mediated gene transfer to salivary glands. Am. J. Physiol. **266:** 1146–1155.
63. LEVINE, M. J. 1993. Development in artificial salivas. Crit. Rev. Oral Biol. Med. **4:** 279–286.
64. B. J. BAUM, C. J. BURSTONE, R. DUBNER, P. GOLDHABER, M. L. LEVINE, W. J. LOESCHE & V. TERRANOVA. 1989. Advances in diagnosis and detection of oral diseases. Adv. Dental Res. **3:** 7–15.
65. TAUBMAN, M. A., R. J. GENCO & J. D. HILLMAN. 1989. The specific pathogen-free human: A new frontier in oral infectious disease research. Adv. Dent. Res. **3:** 58–69.
66. STEELE, C., L. M. COWERT & E. J. SHILLITOE. 1993. Effects of human papillomavirus type 18-specific antisense oligonucleotides on the transformed phenotype of human carcinoma cell lines. Cancer Res. **53:** 2330–2338.
67. JACOB, F. 1988. The Statue Within: 271. Basic Books.

PART VII. DNA, ONCOGENES, AND CANCER

Introduction

SIR RICHARD DOLL
*Imperial Cancer Research Fund
Cancer Studies Unit
Harkness Building
Radcliffe Infirmary
Oxford OX2 6HE, England*

This session deals specifically with cancer, which should be of personal interest to many of us, even if we are not working in the field. For if present conditions of life were to persist unchanged–which is the one thing we can be certain will not happen–and if those attending this symposium were representative of the population of North America and Western Europe–which it equally certainly is not–more than a third of us would develop cancer and a quarter of us would die of it. There is, therefore, some pressure on us collectively to discover how the disease can be prevented or, if that proves too difficult, how it can be treated more effectively than it now is. In fact, we have already learnt how, in principle, the risk of cancer worldwide can be halved; or, rather, because the risk of the majority of forms of cancer increases with advancing age, and because our expectation of life is being extended by the rapid reduction in the risk and fatality of vascular disease, we are being faced with an increased lifetime risk of cancer. I should therefore have said that we have already learnt how the *age-specific* risk of cancer can be halved–which is something we have learnt, as it happens, without the aid of molecular biology.

Unfortunately, the fact that we have learnt how the age-specific risk could be so substantially reduced, does not necessarily mean that people will take the steps needed to bring it about–which include such unpopular behavior as smearing our bodies with grease when we go into the sun and avoiding what Dr. Meselson would doubtless describe as "mixing" with more than one member of the opposite sex. This, therefore, is one area where molecular biology can undoubtedly help; for it is beginning to enable us to classify people according to their susceptibility to particular types of cancer, knowledge of which may encourage the most susceptible to take appropriate steps to avoid the disease. Another, and I suspect more important area, is the understanding of the mechanisms by which the disease is produced, which may enable us to reduce risk by interfering with the mechanism, as, for example, by specific antiviral immunization. Cancer of the cervix provides a beautiful example, for we now know that by far the most cases are produced by the integration of certain types of the human papilloma virus (HPV) into the genome of the cervical mucosa cells, which then produce two protein products, E6 and E7, that form complexes, respectively, with normal cell p53 protein and the retinoblastoma gene RB products and interfere with their function. This explanation of the mechanism is clinched by the finding that the few cases of cervical cancer that do not contain the requisite HPV type have only mutated p53 mRNA. All that now has to be done is, therefore, to produce an effective method of immunization against the specific types of HPV to reduce the risk of cervical cancer by some 80–90%—which may, however, be easier said than done.

Then, of course, there is the half of the risk that we do not yet know reliably how to avoid; but it is, I believe, reasonable to hope that the combination of molecular biology with other, older sciences will provide clear indications how it can also be avoided, long before the 80th birthday of the happy event that we have been celebrating here.

The Molecular Basis of Oncogenes and Tumor Suppressor Genes

ROBERT A. WEINBERG

Whitehead Institute for Biomedical Research
9 Cambridge Center
Cambridge, Massachusetts 02142
Department of Biology
Massachusetts Institute of Technology
Cambridge, Massachusetts 02142

I will take up a theme that has already been introduced here by Sir Richard Doll, one that derives in part from his epidemiological work over the last 30 to 40 years. He advanced the notion that the process of creating cancer is one of multiple steps. When one quantifies the rates with which tumors occur at various ages, it seems that their appearance is due to a succession of four or five distinct stochastic events happening during a person's lifetime. The chance of sustaining a tumor becomes much larger the older one grows, being proportional to the fourth or fifth powers of elapsed lifetime.[1]

This realization had direct implications for those who were intent on rationalizing the formation of cancer at the level of cells and molecules. It meant that models that predicted the conversion of normal cells into tumor cells in one stroke were unrealistic. Rather, it became apparent that a whole series of steps would be required, indicating a long and complex evolution of cells from normalcy to malignancy. This paralleled the findings of pathologists, who found that within a specific organ, such as the colon, a whole series of distinct types of tissue architectures could be found having various degrees of abnormality. In the colon, this series begins with normal colonic epithelium and progresses to quasi-abnormal growths—hyperplasias, dysplasias—and ultimately to frankly neoplastic, invasive carcinomas. One presumed that the more normal cells serve as precursors for those having progressively greater degrees of malignant growth phenotypes.

Only in recent years did it become possible to begin to rationalize this multi-step process in terms of underlying molecular processes. We have begun to think that each of these steps, each of these conversions, could be demarcated by an underlying genetic change in the genome of an evolving tumor cell population. Each of these genetic changes would confer growth advantage on the cells that sustain them, such that each stage of tumor evolution would yield a cell population having a growth advantage over the cells that pre-existed it. As shown by Bert Vogelstein's work on colon carcinomas at Johns Hopkins, several of these conversions can be correlated with the presence of distinct genetic lesions present in the genomes of cell populations at specific stages of colon tumor progression.[2] This sharpens and refines our thinking, because it suggests that the rate-limiting steps that Doll hypothesized as long as 30 years ago represent rare genetic mutations in specific, definable genes, each of which confers on an evolving tumor cell population an incrementally greater growth-advantaging phenotype.

If we focus on the nature of the genes that are so affected during colon tumor pathogenesis, we note that Vogelstein has identified four such genes, mutation of which appears to play a central role in these interconversions. Among these is the *ras* oncogene, which a decade of work has shown is a growth-promoting gene in its normal incarnation in the cell.[3] Its product is normally involved in mediating the transmission of mitogenic signals. Mutations in this gene convert the normally benign, growth-promoting gene into one that is hyperactive and able to force unrelenting cell growth, resulting in the expansion of mutant cell clones within premalignant tissues.

However, when we examine the nature of the other three genes implicated in colon carcinogenesis, indicated here as FAP (sometimes known as APC), DCC, and p53, we see that these genes operate on an entirely different principle. This is made clear by the fact that the mutations that involve these genes in the process of tumorigenesis invariably wipe out gene function. Recall that the mutation that recruits the *ras* gene into the process of tumorigenesis is an activating mutation which potentiates its growth-promoting powers.

This leads to the idea that these other three genes—FAP, DCC and p53—normally act to constrain or suppress cell growth. When the cell loses their services through an inactivating gene mutation, the unconstrained proliferation of malignancy may then follow. This class of genes has been called tumor suppressor genes, a term indicating that their role is to constrain or limit normal cell proliferation and thereby prevent malignancy.[4]

If we were to generalize from the particular sequence of events seen in the colon, we might imagine that tumor suppressor genes like these are as important as oncogenes in the creation of human tumors. I will focus on these tumor suppressor genes now, for oncogenes will be discussed elsewhere in this volume.

By now we have evidence pointing to the existence of a large number of tumor suppressor genes, many of them mapping to distinct chromosomal loci. Each of these, when deleted or inactivated, appears to play an important part in triggering the pathogenesis of one or another tumor on a tissue-specific basis. Much of our thinking about tumor suppressor genes comes from study of a very rare cancer—the retinoblastoma tumor—which is seen in only 1 of 20,000 children. It appears in children up to the age of 5 or 6.

Until the end of the nineteenth century, this childhood tumor was invariably fatal. But with the development of techniques for diagnosing these tumors relatively early in their development and removing them surgically, there arose a whole cohort of young people who had been cured of retinoblastoma. As was first seen in Rio de Janeiro in the last decade of the nineteenth century, the offspring of those cured by surgery now sustained this otherwise very rare tumor with very high frequency. Indeed, the disease seemed to be passed through pedigrees in a fashion reminiscent of the behavior associated with a Mendelian dominant allele.[5]

Relatively little progress was made in the study of retinoblastoma until the early 1970s, when Alfred Knudson analyzed these familial retinoblastomas and a second, sporadic form of the disease seen in individuals who have never had any familial history of the disease. By integrating information on both kinds of retinoblastomas, he proposed a synthesis that described their common genetic origin.[6]

Knudson argued that in both familial and sporadic retinoblastomas, tumor cells must sustain two mutations in their genomes. At the time, the nature of the damaged genes was obscure to him. Moreover, he had no idea about the type of mutations that recruit these genes into the process of tumorigenesis, that is, whether they were inactivating mutations of the sort that wipe out gene function or mutations that potentiate gene function.

Knudson argued that in the case of sporadic retinoblastoma, the fertilized egg is genetically intact; however, two somatic mutations occurring during development then conspire to create the full-blown tumor.[6] Conversely, in the familial disease, he argued that one of these two essential mutations was already present in the fertilized egg and therefore implanted in all the cells of the developing retina. In such cells, only a single somatic mutation was required in order to reach this doubly mutated state, which he argued was sufficient to trigger tumor cell outgrowth.

This information *per se* gave little insight into the nature of the genes involved in this process. Relevant data began to accumulate in the late 1970s from microscopic studies showing that in tumor cells from a small fraction of retinoblastomas, the q14 band of chromosome 13 was missing. This showed clearly that the loss of genetic information contributed to the triggering of retinoblastomas. Perhaps this loss involved one of the genes that was postulated earlier by Knudson.

By the early '80s, it became apparent that the two genetic hits that Knudson had proposed were in fact two inactivating mutations, each of which wiped out a copy of a hypothesized retinoblastoma (RB) gene located on one of the paired copies of chromosome 13. This led in turn to the notion that when a cell has a single intact RB gene copy, it is phenotypically normal, but when both copies of the gene are lost, then the cell, now fully deprived of the growth-constraining powers of the RB gene, launches forth in a program of unrestrained growth.[7]

In 1986, a collaboration between researchers at the Massachusetts Eye and Ear Infirmary and my own group yielded a DNA probe for a region of chromosome 13 having many of the properties of the long-sought RB gene.[8] When this probe was used on Southern blotting, we could detect a series of six DNA fragments in normal human DNA. But in a number of retinoblastomas, we found that many of these DNA fragments were missing. Upon physical mapping of these various DNA fragments, we found that they encompassed a region of roughly 190 kilobases on chromosome 13. This DNA segment seemed to represent a single transcription unit that had many of the properties of the RB gene.

As soon became apparent, this gene operates via the activities of 105-kilodalton nuclear phosphoprotein.[9] This meant that the RB gene suppresses growth by interfering with some nuclear process. The RB protein has a multitude of phosphate sites on it, as visualized through phosphopeptide analysis. When we used antibodies against the RB protein in immunoprecipitation analysis of lysates from a variety of tumor samples, we found that it was absent in all retinoblastomas and, most unexpectedly, absent as well in a surprisingly high percentage of small cell lung carcinomas and bladder carcinomas.[10] This result echoed an oft-repeated experience of those studying the molecular biology of human cancer: a gene initially discovered in one tumor type often plays an important role in the pathogenesis of a fully unrelated kind of cancer.

The next important piece of this puzzle came from work going on at the time at Cold Spring Harbor in the laboratory of Ed Harlow. He and his collaborators were interested in studying the actions of human adenovirus, which in humans functions as an upper respiratory virus but acts as a potent tumor virus when injected into rodents. Adenovirus is tumorigenic by virtue of the E1A oncogene that it carries in its genome. Wishing to understand this oncogene, Harlow began a project in which he produced monoclonal antibodies against the E1A protein.

In doing so, he made the following provocative observations. Within lysates prepared from virus-transformed cells, the viral oncoprotein, as expected, could be immunoprecipitated by his monoclonal antibody. But in addition, he found a series of other proteins present in the immunoprecipitates. As he was able to show, the presence of these other proteins was attributable to the ability of the viral oncogene protein to form physical complexes with a group of pre-existing host cell proteins.[11]

Flowing from these observations was the notion that the viral oncoprotein is able to compromise the function of host cell proteins by forming complexes with them. Each of these host cell proteins, in turn, appeared to be an important regulator of cell proliferation. In effect, the viral oncoprotein seemed able to pull a whole series of important regulatory levers in the cell, each represented by one of these host cell proteins.[11] This led in turn to the conclusion that adenovirus succeeds in transforming cells through the ability of its oncoprotein to complex with pRB, compromising pRB function and mimicking the state seen in human tumor cells that have lost RB activity because of chromosomal gene mutation. This paradigm was soon extended in other laboratories, in that it was shown that the SV40 large T-oncoprotein, another DNA tumor virus, and the human papilloma virus type 16 both make oncoproteins able to form complexes with pRB.[12,13] This was a most satisfying outcome, in that it unified a whole series of ostensibly unrelated pathogenetic mechanisms of cell transformation.

Others soon showed that the phosphate groups of pRB were attached on a schedule that seemed to be closely tied with the progress of the cell through its growth cycle.[14] In the G1 phase of the cell's growth cycle, pRB is relatively underphosphorylated. Only later, toward the end of the G1 phase, as the cell prepares to replicate its DNA, are a number of additional phosphate groups attached to pRB. The protein remains in this hyperphosphorylated state until it emerges from cell division, on which occasion these phosphate groups are stripped off. This seemed most interesting, because it suggested that pRB was in some way involved in regulating the progression of the cell through its growth cycle.

Yet another provocative clue came from a second observation, which indicated that the viral oncoproteins specifically seek out and complex with the underphosphorylated, early G1 form of pRB.[15] Since the viral proteins focus their attention exclusively on this form of the pRB and ignore the phosphorylated form, this suggests that it is the underphosphorylated form that is active in suppressing growth. The hyperphosphorylated forms, conversely, may be inactive physiologically and therefore not worthy of the attentions of the viral oncoproteins. If the viral proteins can eliminate the underphosphorylated form of pRB from the cell, they might succeed in neutralizing the pool of pRB active in growth regulation.

We soon found that the phosphorylated state of the RB protein could be correlated with its ability to bind tightly to a series of other nuclear proteins. Thus, when cells were broken open by very gentle lysis methods, the phosphorylated forms of the RB protein would leak out of their nuclei; the underphosphorylated forms remained tightly tethered. Defective forms of the RB protein isolated from human tumors totally lost their ability to bind to isolated nuclei. In short, nuclear binding was tightly correlated with biological activity.[16]

The nuclear proteins to which pRB binds form a collective that we term operationally its "nuclear anchor." As a group, they appear to operate by promoting cell progression through G1. As first shown by Joseph Nevins in North Carolina, this cohort of nuclear proteins includes among others the E2F transcription factor, binding to which by pRB results in its functional inactivation.[17] The association between these growth-promoting proteins and pRB can be reversed by at least three processes that we know of: (1) The periodic phosphorylation of the RB protein, as occurs when the cell passes through the G1 phase of its growth cycle; as mentioned above, this causes pRB to lose its grip on these anchor proteins. (2) Alternatively, the cell may be invaded by a DNA tumor virus whose oncoprotein may then bind a cavity in the RB protein, termed the pRB "pocket"; this pre-emptive occupation of the pRB pocket precludes the RB protein from associating with its normal nuclear partners. (3) Finally, chromosomal gene mutation may create structurally defective forms of the pRB pocket. Since an intact pRB pocket is required for pRB association with many of these other proteins, this once again precludes complex formation and attendant growth suppression.

Such results fed our thinking about the mechanisms responsible for controlling pRB function by phosphorylation. As we now realize, this phosphorylation is driven by the machinery of the cell cycle clock, composed of cyclins and associated cyclin-dependent kinases (CDKs). In effect, the cyclins act as regulatory subunits of the CDKs, directing their catalytic activity toward specific substrates. These cyclins appear and disappear during defined windows of time within the cell cycle. The appearance of a specific cyclin and its association with an active CDK results in the phosphorylation of critical substrate proteins, modification of which then permits entrance by the cell into the next phase of its growth cycle.[18]

We now believe that cyclin E in association with a partner kinase drives the phosphorylation of pRB in mid/late G1. Cyclin A may be responsible for the entrance of the cell into its DNA synthetic phase, the S phase. The B cyclins appear to be responsible for triggering the entrance of the cell into mitosis.

Yet other cyclins, such as those of the D class, can form physical complexes with hypophosphorylated pRB. Provocatively, one of these—cyclin D1—is overexpressed in a number of tumors as a consequence of chromosomal translocation or gene amplification.[19] Cyclin D1, like the growth-promoting transcription factors, may be controlled by its association with hypophosphorylated pRB.[20] Recent evidence also suggests that it may contribute to pRB phosphorylation.[21] Together, these results place pRB in the middle of the cell cycle clock apparatus, which is already present in highly developed form in single-cell eukaryotes like yeast. This apparatus has been perpetuated almost unchanged over the billion and more years during which time single-cell organisms developed into higher metazoa like ourselves.

We now realize that the evolutionarily much more recently developed retinoblastoma protein has been interposed amidst this ancient clock machinery, on the one hand serving as the object of phosphorylation and functional inactivation by some of its cyclin:CDK complexes, and on the other hand acting as a regulator of yet other cyclins. When pRB is lost from this clock machinery, as happens during carcinogenesis, then the machinery races ahead, no longer held in check.

This is only the beginning of this tumor suppressor field, in that we now realize that the pRB has its own very special, indeed unique mechanism of action. But pRB is only one among many. There are more than a dozen other tumor suppressor proteins, each of which acts at a distinct site in the cell to constrain cell proliferation. As a consequence, this field of research can anticipate a decade and more of exciting work that will elucidate the mechanisms of action of these various suppressor proteins, illustrating how they fit into the complex circuitry of cell growth regulation and how they interdigitate with the well known oncogene proteins that will be further discussed in this book.

REFERENCES

1. DOLL, R. & R. PETO. 1981. The causes of cancer: Quantitative estimates of avoidable risks of cancer in the United States today. J. Natl. Cancer Inst. **66:** 1191–1308.
2. FEARON, E. R. & B. VOGELSTEIN. 1990. A genetic model for colorectal tumorigenesis. Cell **61:** 759–767.
3. BARBACID, M. 1987. Ras genes. Annu. Rev. Biochem. **56:** 779–827.
4. WEINBERG, T. A. 1991. Tumor suppressor genes. Science **254:** 1138–1146.
5. SPARKES, R. S. 1985. The genetics of retinoblastoma. Biochim. Biophys. Acta **780:** 95–118.
6. KNUDSON, A. G., JR. 1971. Mutation and cancer: Statistical study of retinoblastoma. Proc. Nat. Acad. Sci. USA **68:** 820–823.
7. DRYJA, T. P., W. CAVENEE, R. WHITE, J. M. RAPAPORT, R. PETERSEN, D. M. ALBERT & G. A. P. BRUNS. 1984. Homozygosity of chromosome 13 in retinoblastoma. N. Engl. J. Med. **310:** 550–553.
8. FRIEND, S. H., R. BERNARDS, S. ROGELJ, R. A. WEINGBERG, J. M. RAPAPORT, D. M. ALBERT & T. P. DRYJA. 1986. A human DNA segment with properties of the gene that predisposes to retinoblastoma and osteosarcoma. Nature **323:** 643–646.
9. LEE, W.-H., J.-Y. SHEW, F. D. HONG, T. W. SERY, L. A. DONOSO, L.-J. YOUNG, R. BOOKSTEIN & E. Y.-H. LEE. 1987. The retinoblastoma susceptibility gene encodes a nuclear phosphoprotein associated with DNA binding activity. Nature **329:** 642–645.
10. HOROWITZ, J. M., S.-H. PARK, E. BOGENMANN, J.-C. CHENG, D. W. YANDELL, F. J. KAYE, J. D. MINNA, T. P. DRYJA & R. A. WEINBERG. 1990. Frequent inactivation of the retinoblastoma antioncogene is restricted to a subset of human tumor cells. Proc. Natl. Acad. Sci. USA **87:** 2775–2779.
11. WHYTE, P., K. J. BUCKOVICH, J. M. HOROWITZ, S. H. FRIEND, M. RAYBUCK, R. A. WEINBERG & E. HARLOW. 1988. Association between an oncogene and an anti-oncogene: Retinoblastoma gene product. Nature **334:** 124–129.
12. LUDLOW, J. W., J. SHON, J. M. PIPAS, D. M. LIVINGSTON & J. A. DECAPRIO. 1990. The retinoblastoma susceptibility gene product undergoes cell cycle-dependent dephosphorylation and binding to and release from SV40 large T. Cell **60:** 387–396.
13. DYSON, N., P. M. HOWLEY, K. MUNGER & E. HARLOW. 1989. The human papilloma virus-16 E7 oncoprotein is able to bind to the retinoblastoma gene product. Science **243:** 934–937.
14. CHEN, P.-L., P. SCULLY, J.-Y. SHEW, J. Y. J. WANG & W.-H. LEE. 1989. Phos-

phorylation of the retinoblastoma gene product is modulated during the cell cycle and cellular differentiation. Cell **58**: 1193–1198.
15. LUDLOW, J. W. *et al.* 1989. SV40 large T antigen binds preferentially to an underphosphorylated member of the retinoblastoma susceptibility gene product family. Cell **56**: 57–65.
16. MITTNACHT, S. & R. A. WEINBERG. 1991. G1/S phosphorylation of the retinoblastoma protein is associated with an altered affinity for the nuclear compartment. Cell **65**: 381–393.
17. NEVINS, J. R. 1992. E2F: A link between the Rb tumor supressor protein and viral oncoproteins. Science **258**: 424–429.
18. XIONG, Y., H. ZHANG & D. BEACH. 1992. D type cyclins associate with multiple protein kinases and the DNA replication and repair factor PCNA. Cell **71**: 505–514.
19. SHERR, C. J. 1993. Mammalian G_1 cyclins. Cell **73**: 1059–1065.
20. DOWDY, S. F., P. W. HINDS, K. LOUIE, S. I. REED, A. ARNOLD & R. A. WEINBERG. 1993. Physical interaction of the retinoblastoma protein with human D cyclins. Cell **73**: 499–511.
21. EWEN, M. E., H. K. SLUSS, C. J. SHERR, H. MATSUSHIME, J.-Y. KATO & D. M. LIVINGSTON. 1993. Functional interactions of the retinoblastoma protein with mammalian D-type cyclins. Cell **73**: 487–497.

A Nuclear Tyrosine Kinase Becomes a Cytoplasmic Oncogene[a]

DAVID BALTIMORE,
RUIBAO REN, GENHONG CHENG,
KONSTANTINA ALEXANDROPOULOS,
AND PIERA CICCHETTI

The Rockefeller University
1230 York Avenue
New York, New York 10021-6399

It is a great pleasure to be here at this celebration. My whole scientific career has, of course, been shaped by the discovery in 1953 of the structure of DNA. At that time I was in high school, and hardly aware of it. But I do remember an incident in the late 1950s when, as a college student, I had the rare opportunity to drive Jim Watson from Cold Spring Harbor to a nearby airport on Long Island. At that time he said to me "There's a virus that's just been discovered that has only a small amount of DNA in it." It had to be SV40 or polyoma virus. "That such a virus is able to cause cancer means that a very small amount of genetic information is all that's required to cause cancer and we should be able to decipher that very quickly," he observed. Well, as usual, his insight was remarkable, because those viruses have played a central role in developing the notions of oncogenes; his time line, however, was a little short.

The development of oncogene research was enormously advanced by the discovery of the Src protein followed by the finding of its protein kinase activity in the late 1970s. At that point we were working on the Abelson murine leukemia virus, a virus we had chosen to work with because it was one of the few that would transform murine lymphocytes *in vitro*, thereby providing a particularly good model for investigating leukemogenesis. So we looked for a protein kinase in the Abelson protein (Abl) and found it. Having been a chemist, I first studied the linkage between the phosphate and the protein and discovered that there was something odd about it: it did not behave like any of the previously characterized protein kinase products, virtually all of which involved the phosphorylation of serine and threonine. Although the Src kinase had been identified as a threonine kinase, it was clear that the chemistry of the linkage made that conclusion unlikely.

---

[a] Prepared from a transcript of a paper delivered at the meeting by Dr. David Baltimore.

A review of the literature, especially the Russian literature, showed linkages of phosphates to a number of different amino acids. Phosphohistidine, phosphoarginine, phosphoaspartate, and phosphotyrosine were all known: we decided to go through them and see whether the chemistry of the linkage produced by the *Abl* kinase looked like those. The last one on our list was tyrosine, and it was last because we had earlier, working on poliovirus, discovered a phosphotyrosine linkage and figured that lightning never strikes twice in the same place. But, as you're all aware, lightning *did* strike twice, and when we got to the end of the list it turned out that the properties of phosphotyrosine were the same as the properties of the unknown product and that the Abl kinase was a protein-tyrosine kinase. At the same time Tony Hunter found that the Src kinase phosphorylated tyrosine and now a myriad of other protein-tyrosine kinases have been found.

A major focus of the research on oncogenes over the period since the late '70s has asked the question: How does phosphotyrosine allow a cell to become a tumor cell? We don't have a full answer to that question yet, although an enormous amount of progress has been made. What I want to discuss here are the ancillary motifs in oncogene proteins that work coordinately with the phosphotyrosine-producing kinase to allow for transformation of cells.

The cellular Abl gene is one that makes a protein of about 150,000 molecular weight, a big protein which is localized largely in the nucleus of cells. Some of it is always in the cytoplasm and some particularly is in the plasma membrane. The N-terminus protein of Abl looks a lot like Src. It is myristoylated, and thus binds to membranes, although the Abl in the nucleus also appears to be myristoylated. It has a little bit of unique sequence at its N-terminus and then two individual domains, SH2 and SH3, on the protein that have individual functions.

Following SH2 and SH3 are the kinase proper, a nuclear localization signal, a DNA-binding region, and an actin-binding region. These are elements that one does not find in Src or in most other tyrosine-specific protein kinases. In normal cells, as they are growing, c-Abl is not itself phosphorylated on tyrosine. Autophosphorylation, however, is a characteristic of most protein-tyrosine kinases and transforming variants of Abl as well as purified c-Abl do autophosphorylate, so the kinase activity of c-Abl must somehow be highly controlled and masked in the normal cell. The overexpression of c-Abl leads to inhibition of cell growth at the G1/S border of the cell cycle, presumably meaning that c-Abl plays some role in the process. If you knock out the Abl gene in mice, there is a variable lymphopenia and other effects, but the protein is not really necessary for cell life. It is not wholly necessary for the differentiation of any specific organ because you do get some almost normal mice, but it clearly plays an important role.

Finally, anything that turns c-Abl from a normal cellular protein to an oncogene, and there are lots of ways of doing that, changes it from a nuclear protein into a largely cytoplasmic protein in which there is a constitutively active kinase that phosphorylates itself as well as a lot of other proteins in the cell.

Now let me catch up for a minute on what's happened with protein-tyrosine kinase. There are now known two classes of tyrosine-specific protein kinases: those that pass through the membrane, having both an extracellular and intracellular portion, and those which have only an intracellular existence. Abl, Src, and their many relatives are proteins that are entirely intracellular, but there are many transmembrane kinases that have an extracellular binding function, binding to some hormone like epidermal growth factor (EGF), platelet-derived growth factor (PDGF), or insulin. Their intracellular domain is the kinase proper and also serves as an organizing site for sending signals into cells. When the EGF receptor or the PDGF receptor is activated by binding of ligand, inside the cell a large complex of signaling proteins becomes organized around the intracellular part of the receptor. These proteins signal to the cell through the ras pathway or phospholipase C pathway, or through novel lipid molecules, or in ways that have not entirely been characterized. These complexes are formed as a consequence of interactions that are at least partly mediated by the small SH2 and SH3 domains. They are 50–100 amino acids in size and bind, in the case of SH2, to the autophosphorylated sites on the protein. Actually what occurs is a transphosphorylation in a dimer. Each SH2 has its own binding specificity: some of the binding energy comes from interaction with a phosphorylated tyrosine and some from interactions with other amino acids. Basically, the first three amino acids C-terminal to the phosphotyrosine provide the specificity signal for binding of these complexes to proteins.

If that's what receptor tyrosine kinases do, what do non-receptor protein tyrosine kinases do? There are a lot of them and they have many different structural elements in them. Possibly these kinases organize similar signal transduction proteins around themselves: I will describe a bit of the evidence to show that that is true. The first indication is that the SH2 and SH3 regions, which are so important in organizing complexes around the receptor tyrosine kinases, are also present in the intracellular non-receptor protein tyrosine kinases: that suggests that the same principles will be at work. In fact SH2 and SH3 mean Src homology region 2 and 3 where Src homology region 1 is the kinase proper.

SH2 and SH3 are found in many different proteins, not only kinases but also other proteins that bind around the receptor tyrosine kinases like phospholipase C-gamma and GAP protein, which controls Ras function. There are cytoskeletal proteins like myosin and spectrin that also potentially link into these complexes as a consequence of SH3 interactions. And even transcription factors contain SH2 and SH3 domains, as in the case of ISGF3-alpha, which is a transcription factor that is interferon-responsive.

So SH2 and SH3 have very broad and important roles, and there's actually a third domain of this sort which I will introduce in a moment, although I will not say much about it; it has a similar distribution and a similar function. There are also proteins that consist solely of SH2 and SH3; they have no catalytic function, no other function known aside from having these little binding domains. These proteins, now called adapter proteins, act as a linker between one kind of protein and another, for instance between a receptor tyrosine kinase and a target protein.

I've introduced SH2 and SH3 as protein-interaction domains, SH2 being one that depends on phosphotyrosine, but I haven't said much about the specificity of SH3 because until quite recently nothing was known about SH3. A student of mine, Piera Cicchetti, now a couple of years ago, decided to try to understand the function of SH3 and, as a rationale for a relatively simple experiment, made the guess that SH3 would bind to some linear determinant, some linear string of amino acids. She hoped that they wouldn't be modified. If it was all true, then it should be possible to identify the binding site for SH3 by screening a library of expression clones in *E. coli* using a labeled SH3 as a probe. In that way she identified and cloned genes for mammalian proteins that would bind SH3.

That told us a number of things: that SH3 is a domain that can bind to sites on other proteins; that those sites are linear determinants because otherwise they would not have scored in this assay; and that the binding site cannot be a modified one. We also found that there is specificity, so that Abl SH3 is not equivalent to Src SH3 or Crk SH3 in its binding specificity. The proteins we found that bound to SH3 were not in the data base. The site of binding was proline-rich. One gene we got out of the screen was a complicated adapter protein, containing an SH2 domain of unique specificity. It also had a new domain, which we call the PH domain, because it was first found in the plextrin protein, and so it is called the plextrin homology domain. And it is again a widely distributed new domain found in adapter proteins and in signaling proteins, the same kinds of proteins that have SH2 and SH3 as well as binding sites for SH2 and SH3, and is presumably involved in the same sorts of interactions, although as yet there's no direct evidence for what PH binds to.

Having defined proline-rich binding sites for SH3, we decided to take a new tack, to look for proteins that interacted with Abl by a more physiological criterion. The one we chose was the so-called two hybrid assay, an assay in which one looks for interaction inside yeast between two mammalian proteins. One of these is the bait, the protein on which your attention is focused; the other is a library of proteins or fragments of proteins that are put into the yeast. From the library you choose genes whose products interact with the bait, Abl in this case. One of those turned out to be a clone of great excitement to us because it is a fragment of an adapter protein, an adapter protein that was known to be an oncogene and was characterized by Bruce Mayer and

Hidesaburo Hanafusa a number of years ago, the so-called Crk (pronounced crack) gene. You might expect that this comes from New York and it does! The Crk gene in mammalian cells comes in two forms: one has one SH2 and two SH3s, and in the other splicing removes one SH3. The latter is the one we found. And we knew just by looking at the structure of this clone that it must be interacting with Abl through an SH3 interaction with an SH3 binding site on Abl.

There are, in fact, three SH3 binding sites on Abl, again regions rich in proline. Crk SH3 binds quite tightly to site 1, weakly to site 2, and not to site 3. We then examined another adapter protein and found that Nck, a more complicated adapter protein, bound not to sites 1 or 2 but bound to site 3. Grb2, which has 2 SH3s in it, bound to all three sites.

We demonstrated these interactions in yeast and *in vitro* and mapped the sites of interactions. This provided a new landmark on the Abl protein, the sites mapping around the nuclear localization signal in the C-terminal half of Abl. We could also show that the interaction occurs inside cells. We coexpressed in cells Abl and Crk and asked whether they interact. To this end we used an epitope tag on Crk and picked up the interaction by immunoprecipitation followed by an *in vitro* kinase reaction. When we immunoprecipitated Abl, it came down with bound Crk so that in the immunoprecipitate Abl phosphorylated itself and phosphorylated Crk. If we mutated the binding sites on Abl, Crk no longer came down and was no longer phosphorylated. We therefore see these sites as having a new function, a function of bringing to the kinase its substrates. That function had been suggested previously for SH2 but had not been seen for SH3 and it's certainly not been seen in the non-receptor protein tyrosine kinase.

We've now looked at many SH3 binding proteins, binding to different SH3s. Our criteria for a plus in the matrix is that binding is 10-fold or more better than "minus." In this way we have defined six different specificity patterns and we have only looked at a small fraction of the SH3s, and presumably only a small fraction of SH3 binding sites. And so we suspect that this matrix will increase to give a complexity like that of epitopes interacting with antibodies, although with a less broad range of specificity.

The structures of SH2, SH3, and PH are now known. I think that between SH3 specificity, SH2 specificity and PH domain specificity, we can begin to see signaling proteins as interacting in highly specific interaction with each other. This is very important because up until now we've had great difficulty figuring out where the specificity of signaling comes from. When, for instance, a mitogen binds to the surface of the cell, we've seen too much evidence of non-specificity and I think we're now, by looking at the protein/protein interactions, finding the needed specificity.

To summarize: The N-terminal half of Abl is very like Src in having an SH2 domain and an SH3 domain abutting the kinase proper. This is what attracted our attention to the more general properties of SH2 and SH3. SH2 binds to phosphotyrosine-containing short peptide sequences with the non-phosphotyrosine residues providing specificity. SH3 binds to short linear sequences of amino acids, generally about 10 in length, with a generally excessive representation of proline, the only conserved motif being PXXP at the moment. Individual SH3 regions bind with high specificities to particular linear sequences and we know of at least six patterns of that sort now. Abl contains not only an SH3 domain, but also SH3 binding sites, greatly increasing its potential for interaction with signaling systems, and Crk and Nck are the proteins that seem to show the greatest specificity for binding to those sites.

One important thing about the SH3-mediated interactions is that Crk can become phosphorylated as a consequence of that interaction. I might say that Dr. Hanafusa's laboratory at the Rockefeller University, coming at this problem from his interest in Crk, has come to a similar conclusion. Abl may therefore function partly through the mediation of adapters in its normal function. Crk-mediated transformation, which has always been an obscure phenomenon, may involve an interaction with Abl.

Recombinant Toxins: New Therapeutic Agents for Cancer

IRA H. PASTAN, LEE H. PAI,
ULRICH BRINKMANN, AND
DAVID J. FITZGERALD

Laboratory of Molecular Biology
Division of Cancer Biology, Diagnosis and Centers
National Cancer Institute
National Institutes of Health
9000 Rockville Pike
Building 37, Room 4E16
Bethesda, Maryland 20892

INTRODUCTION

There is a desperate need for new therapies for cancer. Conventional chemotherapeutic agents have been successful in the treatment of leukemias, lymphomas, choriocarcinoma, and testicular cancer. They have also been useful in the adjuvant therapy of ovarian and breast cancer. However, no new agents have been discovered in the past ten years that have made a large impact on cancer death rates. A further difficulty with the current generation of chemotherapeutic agents is that we do not understand why these agents selectively kill cancer cells, so we do not know how to improve them. Some of these agents are toxic to rapidly dividing cells. Because leukemias and lymphomas grow very rapidly, they are particularly sensitive to such drugs. However, other mechanisms of action clearly exist, such as the induction of apoptosis.

Rational drug therapy will depend upon taking advantage of the properties of cancer cells. Because many cancers contain either mutant or overexpressed oncogenes, these molecules are attractive targets. Another target to exploit is the expression of certain differentiation antigens on the surface of cancer cells. It is now well established that many lymphomas and leukemias express differentiation antigens that are present on mature lymphocytes and monocytes, but are not present on bone marrow stem cells. Therefore, it should be possible to eliminate cancer cells and differentiated cells and have the normal cells but not the cancer cells replaced by bone marrow precursors.

In order to kill a cancer cell, the antibody or ligand which binds to the antigen or receptor expressed on the cancer cell surface must be armed with a cytotoxic moiety. Several choices are available; these are radioisotopes, cytotoxic drugs, and protein toxins. Unarmed antibodies have been shown to have some activity against human cancers, but this only occurs in rare cases. Apparently ADCC (antibody-dependent cellular cytotoxicity) and activation of complement are not sufficiently powerful cytotoxic mechanisms to affect most human cancers.

PSEUDOMONAS EXOTOXIN A

Our laboratory has chosen to direct a powerful bacterial toxin, *Pseudomonas* exotoxin A (PE), to cancer cells. There are several advantages in working with PE. One is that the gene has been cloned. A second is that the protein is readily made in *E. coli*. A third is that the three-dimensional structure of the toxin has been solved.[1] Therefore, mutant forms of the toxin with desirable properties can be designed based on the structure of PE and readily produced in *E. coli*.[2]

PE is composed of three major and one minor domain (FIG. 1). Domain Ia (amino acids 1–252) is the cell-binding domain. The toxin binds to the α_2-macroglobulin receptor, which is present on many different types of normal and cancer cells.[3] Domain II (amino acids 253–364) is the translocation domain. It enables the carboxyl terminus of the toxin (amino acids 280–613) to translocate into the cytosol, where it catalyzes the ADP-ribosylation and inactivation of elongation factor 2, permanently arrests protein synthesis, and kills the cells. Domain III (amino acids 400–613) is the ADP-ribosylating domain. The enzyme uses NAD^+ as a cofactor to ADP-ribosylate a modified histidine residue on elongation factor 2. Domain Ib (amino acids 365–399) has no known function and can be completely deleted without loss of toxin activity.

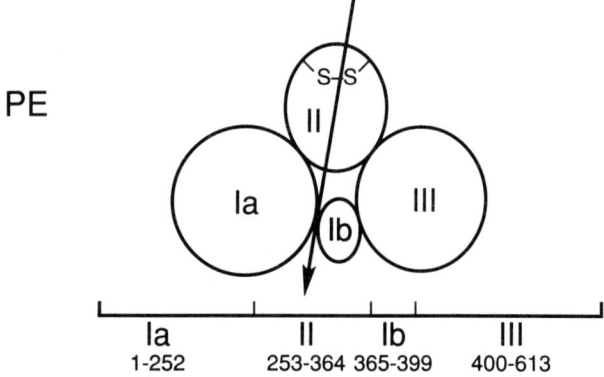

FIGURE 1. Schematic drawing of structure of *Pseudomonas* exotoxin A.

HOW PE KILLS CELLS

In order to kill a cell, PE must enter the cell by endocytosis and then it undergoes proteolytic cleavage generating a 37-kDa fragment (amino acids 280–613) composed of a portion of domain II and all of domain III. As well as ADP-ribosylating activity, this fragment contains an endoplasmic reticulum retention sequence at its carboxyl end (REDLK) and is, therefore, directed to the endoplasmic reticulum, where it can translocate to the cytosol and inactivate EF2. FIGURE 2 illustrates the steps in PE action.

IMMUNOTOXINS

Immunotoxins are cytotoxic moieties in which an antibody is linked to a toxin. Several different toxins and toxin derivatives have been used to prepare immunotoxins. These include ricin and other plant ribotoxins, as well as two bacterial toxins, diphtheria toxin and PE.[4] I will choose one immunotoxin that we have made using a recombinant form of PE to illustrate our approach to the construction of recombinant toxins. Our initial studies were carried out with chemical conjugates in which a truncated form of PE (LysPE38) was coupled to a monoclonal antibody (B3) that reacts with many forms of human cancers.[5] LysPE38 is a truncated form of PE in which domain Ia (amino acids 1–252) and amino acids 365–379 from domain Ib have been deleted and because of that this molecule does not bind to any cell. The molecular weight of this molecule is 38,000. In addition, a short peptide linker containing a readily accessible lysine residue has been inserted at the amino end of the protein to facilitate coupling to antibodies. Monoclonal antibody B3 reacts with a carbohydrate antigen in the Le^Y family that is present in many cancerous cells and only a limited number of normal cells.[5] It reacts with forms of colon cancer, breast cancer, ovarian cancer, lung cancer, stomach cancer, and esophageal cancer. An immunotoxin made from mAb B3 and NLysPE38 will selectively bind to all these carcinoma cells but not to antigen-negative normal cells. The carbohydrate with which B3 reacts is present on many different cell-surface glycoproteins as well as on glycolipids. Some of these glycoproteins are rapidly internalized and therefore bring the immunotoxin into the endocytic compartment where processing and translocation occur (FIG. 2).

TISSUE CULTURE EXPERIMENTS

The specificity of immunotoxins is initially examined using various cell lines which do or do not express the appropriate antigen. The antibody B3

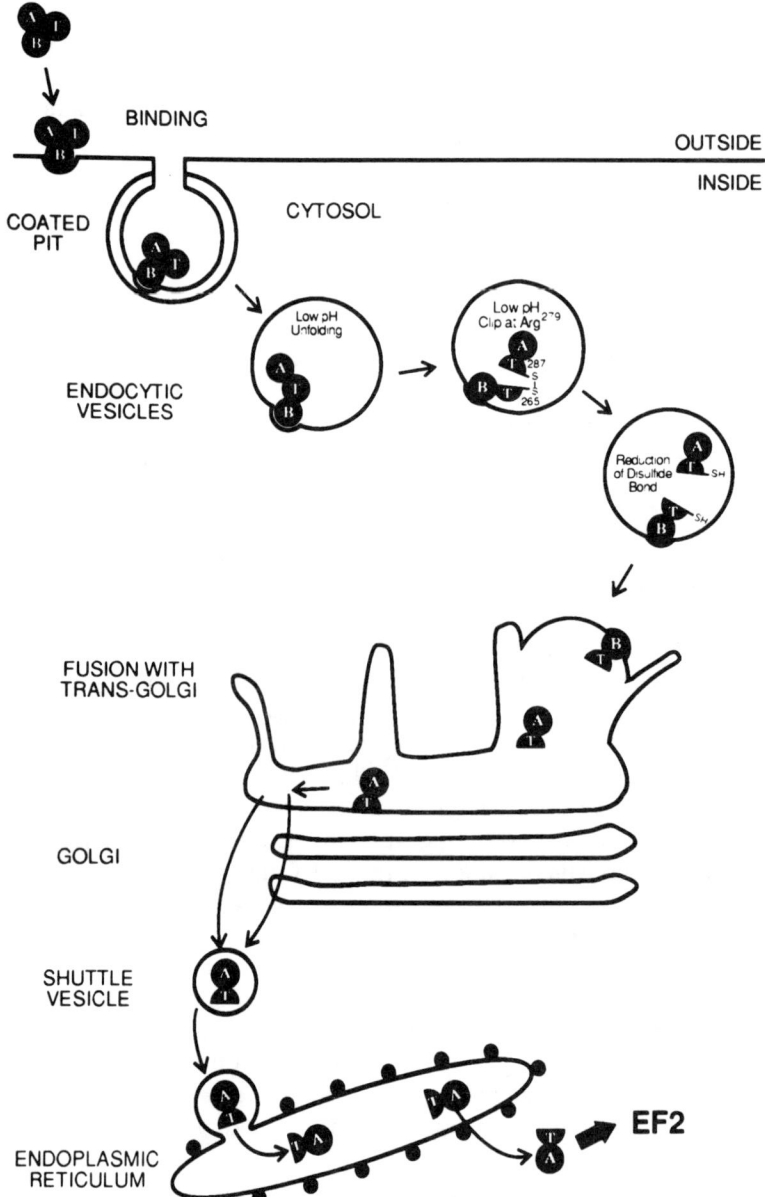

FIGURE 2. Model of endocytosis, processing, and translocation of PE or PE-derived endotoxins. B, T, and A indicate binding domain (Ia), translocation domain (II), and ADP-ribosylation domain (III), respectively. PE binds to a cell-surface receptor. The toxin–receptor complex is then internalized via coated pits into the endocytic vesicles, where the low-pH environment causes toxin unfolding and facilitates proteolysis. In addition, the 37-kDa fragment is released by reduction of the disulfide bond linking residues 265 and 287. Cleaved PE is trans-

TABLE 1. Cytotoxic Activity of B3-NLysPE38

Cell Line	Tumor	B3 Antigen	ID_{50} (ng/ml)
A431	Epidermoid carcinoma (vulva)	+++	3.0
CRL-1739	Gastric carcinoma	+++	2.0
MCF-7	Breast carcinoma	+++	5.0
LNCaP	Prostate carcinoma	+	10.0
KB	Epidermoid carcinoma (cervix)	−	>1000

NOTE: The activity of B3-NLysPE38 was examined in various human tumor cell lines. Inhibition of protein synthesis was used to measure the activity of each immunotoxin on various cell lines. Immunotoxin was diluted with 0.2% human serum albumin in PBS prior to addition to cells. Cells were seeded at 1.5×10^4 cells per milliliter in 96 well plates, incubated with immunotoxin at 37°C for 24 hr, and then assayed for incorporation of [^{3}H]leucine. The cytotoxic activity of B3-NLysPE38 was specific, because it was eliminated by the addition of excess B3 (250 μg/ml; data not shown). The cytotoxic activity of B3-NLysPE38 and the B3 antigen expression of various tumor cell lines is shown.

reacts with several different human cancer cell lines. These include cell line A431, an epidermoid carcinoma, cell line MCF7, an adenocarcinoma of the breast, and cell line CRL-1739, a carcinoma of the stomach. A convenient way to measure the activity of immunotoxins in cell culture is to incubate the cells with the immunotoxin overnight and then measure the decrease in protein synthesis that occurs as a consequence of immunotoxin action. The concentration of immunotoxin required to decrease protein synthesis to 50 percent in such assays is defined as the IC_{50}. The activity of LMB-1, the name given to the immunotoxin in which mAb B3 is conjugated to LysPE38, is shown in TABLE 1. It requires only a few nanograms per milliliter to give a significant cytotoxic effect. Cell lines without antigen are very resistant to the immunotoxin and even 1 μg/ml does not produce any cytotoxic effect. Further evidence of specificity is obtained by showing that an excess of unconjugated monoclonal antibody blocks the cytotoxic effect of the immunotoxin[6] because it competes with the immunotoxin for binding to the cell surface antigen.

ANIMAL EXPERIMENTS

Having shown that immunotoxins are active in tissue culture, we next wish to show that they are active in animals. To do this, human tumor cells are grown as xenografts subcutaneously in immunodeficient nude mice. When the xenograft has reached a readily measurable size, the therapy is ini-

---

ported to the Golgi apparatus and ultimately reaches the endoplasmic reticulum via a shuttle vesicle. Specific residues (REDLK) at the C-terminus mediate transport to and retention in the endoplasmic reticulum. Translocation of the 37-kDa fragment to the cytosol may occur in a reverse direction through translocation complexes or pores ordinarily used for translocating cellular proteins into the endoplasmic reticulum. (From Pastan et al.[24] Reproduced by permission.)

tiated. The immunotoxin is given intravenously every other day for three doses and the tumor size measured to assess response.[6] A typical experiment is shown in FIGURE 3. It is evident that LMB-1 produces a dose-related antitumor effect. Complete disappearance of the tumors occurs at 0.75 mg/kg. These animals were followed for several months and the tumor did not recur. In a separate series of experiments, animals bearing the breast cancer, MCF7, were also treated. This tumor will also completely regress when the animals are treated with LMB-1. In these experiments, specificity is demonstrated by showing that antibody alone does not affect the growth of the tumor and also that toxin coupled to an antibody that does not react with the tumor cells has no antitumor activity.

PRECLINICAL TOXICOLOGY

Monoclonal antibody B3 does not react with normal mouse tissues, but it does react with some normal human tissues. In humans, the B3 antigen is present on epithelial cells of the stomach, trachea, and bladder. Therefore, it is necessary to evaluate side effects that could occur in humans due to the immunotoxin's being directed towards normal tissues. One way to do this is to study an animal such as the cynomolgus monkey, whose cells also react with the antibody. Therefore, a series of studies were carried out in cynomolgus monkeys to determine how much immunotoxin could be safely given.[7] Two unexpected findings emerged from these studies. The first was

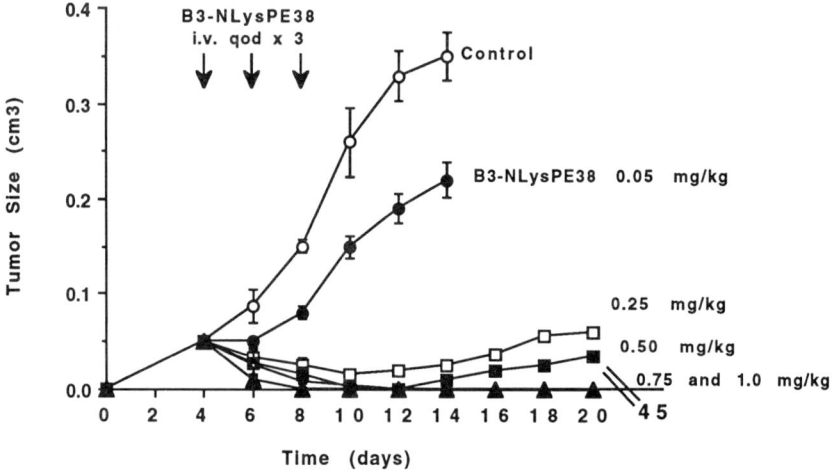

FIGURE 3. Antitumor effect of LMB-1 (B3-LysPE38). Nude mice bearing A431 tumors were injected three times intravenously with LMB-1, as indicated by the *arrows*. Tumor size was measured every other day.

that the maximum dose that could be given to monkeys (5 mg/kg) was greater than the maximum dose that could be given to mice (3 mg/kg). The second was that the dose-limiting toxicity in monkeys was due to a toxic action on the stomach (a predicted "target" for specific antibody-mediated toxicity of the immunotoxin); whereas, the dose-limiting toxicity in mice was liver damage (expected when the nonspecific activity of the PE moiety is limiting) (L. Pai, J. Tomashevsky, I. Pastan, unpublished data).

CLINICAL STUDIES

Because the preclinical toxicology indicated a sufficient therapeutic window between the MTD and immunotoxin concentrations required to kill carcinoma cells *in vitro* (TABLE 1) and in animals (FIG. 3), an investigational new drug application (IND) was filed with the FDA on March 22, 1993; and a clinical trial initiated in June 1993. This phase I trial is currently in progress.

COMPLETELY RECOMBINANT MOLECULES

Conventional immunotoxins like LMB-1 have several undesirable properties. LMB-1 is a large molecular weight protein (molecular weight ~200,000) and therefore penetrates tumors slowly. In addition, it is a heterogeneous protein, because the toxin is chemically linked to one of a variety of lysine residues present in different locations on the heavy and light chains of the antibody. Furthermore, there are several lysine residues on the toxin available for chemical coupling, although the lysine residue engineered into the amino terminus is the preferred location. Another problem is that it is difficult to make large amounts of immunotoxins, because one must make and purify the antibody and the toxin separately and then conjugate them together and finally purify the resulting product from side products of the conjugation.

We have used recombinant DNA techniques to produce recombinant toxins that circumvent all these problems. In these new molecules the antibody-combining site is fused directly to domains II and III of the toxin.[8,9] The structure of such a molecule is shown in FIGURE 4. To make such a protein, cDNAs encoding the heavy and light chain of the B3 antibody connected by a peptide linker are fused to a DNA element encoding domains II and III of the toxin. This molecule is termed B3(Fv)-PE38 or LMB-7. LMB-7 is produced in *E. coli* as insoluble, inactive, inclusion bodies. Active immunotoxin is obtained from inclusion bodies by dissolving and reducing them in guanidine/DTE and renaturation; the monomers can be purified to homogeneity by chromatographic techniques.[10] LMB-7 has been put through the same series of experiments as LMB-1. LMB-7 is selectively cytotoxic to cells

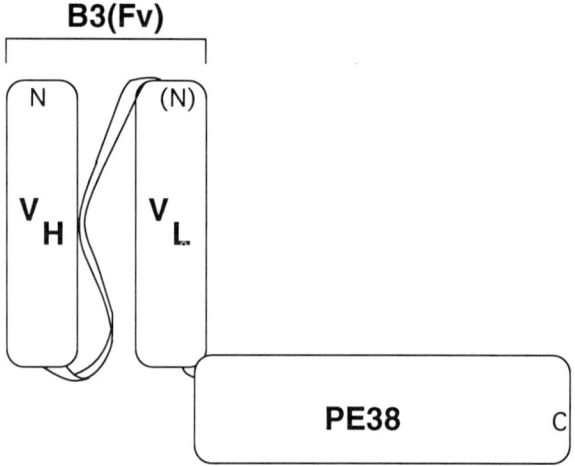

FIGURE 4. Structure of a recombinant single-chain toxin composed of the single-chain Fv fragment of mAb B3 fused to PE38 (LMB-7). V_H = variable region of the heavy chain of mAbB3; V_L = variable region of the light chain. V_H and V_L are connected to each other by a flexible peptide linker. V_L is fused at the C-terminus with a short "connector peptide" to the truncated form of PE. The V_H and V_L portions of the B3 antibody are held together by a p15 amino acid linker composed of three Gly4-Ser repeats.

bearing the B3 antigen and is several-fold more active than LMB-1. This increase in activity occurs despite the fact that the monovalent LMB-1 binds less well to target cells than the bivalent antibody used to make the immunotoxin. We assume the increased activity is due to better processing of the toxin within the target cells.

When evaluated in mice bearing human tumors, LMB-7 was also significantly more active than LMB-1. Complete tumor regressions could be obtained with as little as 0.063 mg/kg.

LMB-7 also has favorable pharmacokinetic properties. Because it is a small molecule it rapidly leaves the vascular system and penetrates into tumors and tissues. The $T_{1/2}$ of LMB-1 in the blood of mice and monkeys is about 10 hours, indicating its slow exit from the circulatory system. In contrast, LMB-7 leaves rapidly and has a $T_{1/2}$ of about 20 minutes. The high antitumor activity of LMB-7 is probably in part due to its rapid penetration into tumors.

FUTURE DIRECTIONS

One problem with immunotoxins is that they are composed of foreign proteins and neutralizing antibodies develop very rapidly—usually in about ten days. Therefore, repeated courses of such an agent are problematic. Initially, immunotoxins will be used as adjuvant therapy in patients who have undergone surgery or radiation. The fact that mice with human xenografts can be cured by therapy given over only four days gives us hope that we can achieve similar responses in patients with small tumors (metastases). Nevertheless, it would be desirable to be able to re-treat patients over several months. To do this, it may be possible to make mutant forms of LMB-7 with diminished immunogenicity. An alternative approach is to modify the surface of recombinant immunotoxins with polyethylene glycol, or a similar compound to prevent the immune system from recognizing the recombinant protein.

CONCLUSION

It is possible to make PE immunoconjugates and recombinant toxin molecules that target cancer cells and cause the complete regression of human cancers growing in mice. The clinical trials, which are just beginning, will determine whether or not these agents will be useful in the treatment of human cancer.

REFERENCES

1. ALLURED, V. S., R. J. COLLIER, S. F. CARROLL & D. B. McKAY. 1986. Structure of exotoxin A of *Pseudomonas aeruginosa* at 3.0 Angstrom resolution. Proc. Natl. Acad. Sci. USA **83**: 1320–1324.
2. HWANG, J., D. J. P. FITZGERALD, S. ADHYA & I. PASTAN. 1987. Functional domains of pseudomonas exotoxin identified by deletion analysis of the gene expressed in *E. coli*. Cell **48**: 129–136.
3. KOUNNAS, M. Z., R. E. MORRIS, M. R. THOMPSON, D. J. FITZGERALD, D. K. STRICKLAND & C. B. SAELINGER. 1992. The α_2-macroglobulin receptor/low density lipoprotein receptor-related protein binds and internalizes *Pseudomonas* exotoxin A. J. Biol. Chem. **267**: 12420–12423.
4. PASTAN, I., V. CHAUDHARY & D. J. FITZGERALD. 1992. Recombinant toxins as novel therapeutic agents. Annu. Rev. Biochem. **61**: 331–354.
5. PASTAN, I., E. T. LOVELACE, M. G. GALLO, A. V. RUTHERFORD, J. L. MAGNANI & M. C. WILLINGHAM. 1991. Characterization of monoclonal antibodies B1 and B3 that react with mucinous adenocarcinomas. Cancer Res. **51**: 3781–3787.
6. PAI, L. H., J. K. BATRA, D. J. FITZGERALD, M. C. WILLINGHAM & I. PASTAN. 1991. Anti-tumor activities of immunotoxins made of monoclonal antibody B3 and various forms of *Pseudomonas* exotoxin. Proc. Natl. Acad. Sci. USA **88**: 3358–3362.
7. PAI, L. H., J. K. BATRA, D. J. FITZGERALD, M. C. WILLINGHAM & I. PASTAN. 1992. Anti-tumor effects of B3-PE and B3-LysPE40 in a nude mouse model of human breast cancer and the evaluation of B3-PE toxicity in monkeys. Cancer Res. **52**: 3189–3193.
8. CHAUDHARY, V. K., C. QUEEN, R. P. JUNGHANS, T. A. WALDMANN, D. J. FITZGERALD & I. PASTAN. 1989. A recombinant immunotoxin consisting of two antibody variable domains fused to *Pseudomonas* exotoxin. Nature **339**: 394–397.
9. BRINKMANN, U., L. H. PAI, D. J. FITZGERALD, M. WILLINGHAM & I. PASTAN. 1991. B3(Fv)-PE38KDEL, a single chain immunotoxin that causes complete

regression of a human carcinoma in mice. Proc. Natl. Acad. Sci. USA **88:** 8616–8620.
10. BUCHNER, J., I. PASTAN & U. BRINKMANN. 1992. A method to increase the yield of properly folded recombinant fusion proteins, e.g., single-chain immunotoxins, from renaturations of bacterial inclusion bodies. Anal. Biochem. **205:** 263–270.

PART VIII. RECOMBINANT DNA AND BIOTECHNOLOGY

Introduction

RICHARD L. DAVIDSON

Department of Genetics
University of Illinois College of Medicine
Chicago, Illinois 60612

The subject matter of this section is recombinant DNA and biotechnology in the context of the application of the tools and concepts of molecular genetics to the world outside the basic research laboratory.

Biotechnology is developing into a major factor in many diverse sectors of the corporate world. This certainly includes the pharmaceutical and health care industries, but also includes areas such as agriculture and the environment. Prime examples of the impact of biotechnology in the field of human health include the production of insulin and growth hormone by means of genetically engineered microorganisms. The techniques of molecular genetics are also being applied increasingly for the diagnosis of human genetic diseases, and the implications of molecular genetics for therapeutic intervention are growing as rapidly as the numbers of genes associated with human disease that have been cloned.

Over the past 40 years since the elucidation of the structure of the DNA double helix, molecular genetics has come to dominate much of the biomedical research world. Enough has been accomplished already so that one can predict confidently that biotechnology will come to have as dominating a role in the corporate world.

In this section, three leaders of the biotechnology revolution, Dr. David Jackson (undergraduate student of James Watson at Harvard and pioneer of recombinant DNA technology with Nobel Laureate, Paul Berg, at Stanford), Professor John Baxter (cloner of the human growth hormone gene and founder of biotechnology companies), and Professor Leroy Hood (founding chair of the first department of biotechnology in the U.S. at the University of Washington) discuss the past, present, and future consequences of biotechnology.

DNA: Template for an Economic Revolution[a]

DAVID A. JACKSON

DuPont Merck Pharmaceuticals
Experimental Station
P.O. Box 80400
Wilmington, Delaware 19880-0400

I'd like to thank Dr. Donald Chambers and the other organizers of this meeting for the opportunity to speak with you on such an impressive occasion. As someone who has been concerned with both the implications and the applications of the structure of DNA for the last 30 years, and who was first introduced to molecular biology by Jim Watson himself, it's a special pleasure for me to participate in this celebration. I first met Jim Watson when I was a freshman at Harvard, where he gave a series of ten or twelve lectures on genetics in an introductory course in biology for biology majors that I was taking. He made genetics, both organismal and molecular, fascinating, describing work with chickens and mice as well as *E. coli* and T-4 in his talks. After taking Matt Meselson's and Paul Levine's course on genetics as a sophomore, I took Jim's course for graduate students and senior undergraduates on molecular genetics in 1962 and thus had the marvelous experience of being present in Jim's class on the day, almost exactly 31 years ago, that he won the Nobel Prize.

FIGURE 1 is a picture of that class taken on the morning that Jim won the prize. The caption on the board, which says, "Dr. Watson has just won the Nobel Prize," deserves some comment. This class was scheduled to start at 9:00 in the morning and Jim was always very punctual. That morning, however, he did not appear. Finally, at about 9:30, one of the graduate students got up and went down to the Biology Department office to find out what the story was; he came back into the room with a very strange expression on his face and, without saying a word, went up to the board and wrote the announcement of the prize's being awarded to Dr. Watson. About 10 minutes later Jim came into the room, and for about the next two hours proceeded to hold us spellbound recounting the story of the discovery of the structure of DNA. Several years later, at the time of the publication of *The Double Helix*, we all realized that what we had heard that morning had been the basis for

---

[a] This is an edited transcript of Dr. Jackson's presentation at the conference.

FIGURE 1. James Watson meeting his class on molecular genetics on the morning of the announcement of his winning, along with Francis Crick and Maurice Wilkins, the Nobel Prize in Physiology or Medicine.

the book, much of it word for word. I can't remember what Jim is saying in the photo that obviously we all thought was so funny, but given the subject matter, I suspect that it was a comment about Francis Crick. This picture appeared the next day on the front page of the *New York Times*, probably the only time my picture will appear there!

My topic here is the relationship of the discovery of the structure of DNA to the subsequent growth and development of what we now call biotechnology, but which I will also call – and I think more accurately – applied molecular genetics. I say this because biotechnology, properly speaking, is something much broader and much older than that which has flowed from the discovery of the structure of DNA. There is, however, a direct causal relationship between knowledge of the structure of DNA, along with the complementarity rules and the clear implication of the replication mechanism that was inherent in the structure proposed by Watson and Crick, and the subsequent development of recombinant DNA technology. Recombinant DNA technology is one of the three key technological developments relating to DNA that are the foundations of the widespread application of applied molecular genetics in science and industry. The other two key technologies are the development of DNA sequencing, including the recent high-throughput, machine-based sequencing technologies, and the development of facile methods for synthesizing DNA. Both of these are points I'll come back to.

I'd like first to discuss the relationship of knowing the structure of DNA to the subsequent application of that knowledge in terms of a metaphor–that of DNA as a *language*. Avery, MacLeod, and McCarty demonstrated that DNA embodied a language of supreme importance in biology. From their work one could fairly infer that DNA contained the instructions for all of the phenotypic characteristics of an organism. Watson, Crick, Wilkins, and their colleagues showed us the structure of the substance in which the language of DNA is written. They also discussed how that structure must be copied and transmitted to successive generations. Nirenberg, Khorana, Crick, Brenner, Yanofsky, and many other colleagues identified the individual *words* in the language of DNA and told us what they meant in terms of sequences of amino acid residues in proteins.

To be fluent in a language, one needs to be able to *read*, to *write*, to *copy*, and to *edit* in that language. The functional equivalents of each of those aspects of fluency have now been embodied in technologies to deal with the language of DNA. Gilbert and Sanger developed the tools for sequencing DNA, tools that allow us to "read" the language of DNA rapidly and accurately. Khorana, Letsinger, Crothers, and their colleagues developed synthetic chemical methods that have now been optimized to allow us to "write" quickly and accurately our own "text" in the language of DNA. The ability to create text is obviously an extremely important functional capability in any language. Arthur Kornberg and his many colleagues showed us how to "copy" existing texts in the language of DNA rapidly and accurately. Finally, Berg, Cohen, Boyer, and their many colleagues developed tools of recombinant DNA technology that are the basis of today's sophisticated and powerful ability to "edit" text in the language of DNA. To continue the analogy with language further, the ability to amplify recombinant DNA molecules by many orders of magnitude by the process of bacterial DNA replication of plasmid and phage vectors corresponds to the functions of printing and publishing in the language of DNA.

Knowledge of the structure of the DNA molecule has been key to much of today's fluency in the language of DNA. As I mentioned, the structure itself implies the copying mechanism and the complementarity rules are the basis for the formation of most recombinant DNA molecules.

I'd like now to turn briefly to the part of the story that I know best, because I was fortunate enough to be involved in it as a postdoctoral fellow in Paul Berg's laboratory, and examine it in a bit more detail. This is the story of how it became possible to develop recombinant DNA technology. Like most good stories, this one has a moral, and the moral concerns the importance of basic research to economic progress. Again, to recapitulate briefly some of what I've already said, the Avery and the Alfred Hershey laboratories showed us that DNA is the genetic material. Watson, Crick, and Wilkins showed us the structure of DNA and how it could be replicated by template copying. The labs of Arthur Kornberg, Bob Lehman, Dale Kaiser, and others taught us about the enzymology of DNA synthesis and replication, and the many different specific ways in which DNA can be degraded–methods that are very important parts of DNA enzymology.

The work of these investigators catalyzed a very important paradigm shift that occurred around the late '60s or early '70s, when it became recognized that DNA molecules could be treated as homogeneous biochemical reagents. The experiment that was done by Mickey Goulian and Arthur Kornberg in 1967, which was the complete synthesis of M13 DNA *in vitro*, was very important in this context, but not because of the way it was played in the press at the time, which was that it represented the synthesis of life in the test tube (which it did not). Rather, it was because these experiments taught a generation of biochemists and molecular biologists, as well as perhaps the public at large, that DNA molecules did not have to be considered as a sticky white amorphous substance that one drew out of an ethanol-water solution on the end of a glass stirring rod, but rather could be viewed as a homogeneous, precisely manipulatable biochemical reagent. This view transformed the way many people in the field looked at what could be done with DNA molecules. Consequently, by the end of the 1960s, there was a group of faculty members, postdoctoral fellows, and graduate students at Stanford and the University of California at San Francisco that both shared this transformed view of DNA as a precisely manipulatable biochemical reagent and had access to the highly purified enzymes and other state-of-the-art technologies that made precise manipulation of DNA feasible. Each of these people had an important contribution to make toward putting together all the parts of the technology that would eventually make it possible to construct recombinant DNA molecules by biochemical procedures.

It is not possible to list here everyone who contributed to these studies. Suffice it to say that in 1971, Peter Lobban, working as a graduate student in Dale Kaiser's lab and using bacteriophage P22 DNA molecules, and Bob Symons and I, working in Paul Berg's lab and using SV40 viral DNA and a λdvgal plasmid DNA molecule carrying all three genes from the galactose operon of *E. coli*, were able to create recombinant DNA molecules using a dA:dT homopolymeric tailing procedure. At about the same time, Vittorio Sgaramella, working in Joshua Lederberg's laboratory, accomplished the same objective using P22 DNA molecules and T4 ligase.

About a year after that, Janet Mertz and Ron Davis, both at Stanford, discovered that there was a much easier way to make recombinant DNA molecules. They discovered that the *EcoRI* restriction enzyme left overlapping cohesive ends, complementary segments of single-stranded DNA, when it made a staggered cut in the two strands of DNA. This meant that one could join DNA molecules very easily by cutting them with the *EcoRI* enzyme, allowing them to anneal and then ligating them. That fact was very important in the work that John Morrow then did with Herb Boyer, Stan Cohen, Annie Chang, and Bob Helling when they put the first gene from a higher organism, the ribosomal RNA gene from *Xenopus laevis*, into a plasmid DNA molecule and then put that recombinant DNA molecule back into a bacterial cell, where it was then amplified manyfold. This ability to isolate and amplify specific segments of any DNA molecule has revolutionized molecular biology.

The moral that I promised from this brief account of the development of recombinant DNA technology is that its development flowed almost entirely from a set of only loosely connected lines of basic research. Among these lines, the principal ones were:

- studies of DNA structure and physical chemistry;
- the enzymology of DNA synthesis and degradation;
- bacterial, phage, and plasmid genetics; and
- bacterial restriction/modification systems.

Some of these lines of investigation, such as bacterial restriction/modification systems, were really something of a backwater in the late 1960s. Yet the elegant work that was done in this area by Werner Arber, Hamilton Smith, Herb Boyer, Matthew Meselson, Rich Roberts, and their colleagues played a vital role enabling recombinant DNA technology to be done easily and quickly.

None of the people that I have talked with from this group had any idea at the time that they were doing work that would form the foundation of a new commercial technology, and that was certainly not their intention. Nevertheless that is what happened. I think that the lesson is clear: If one is free to do good science on important questions, knowledge that will be of great value will surely follow. This does not mean that there are or should be *unlimited* resources to pursue good science—we have to recognize that there are certainly other competing demands on resources in our society. But history teaches us that high-quality basic research will pay for itself many times over in the long term. So by the late 1970s there existed a technologically well supported and rapidly increasing fluency in the language of DNA. It was no coincidence that the growth of industrial biotechnology or applied molecular genetics began at about the same time—the late 1970s—about 5 years after the first development of recombinant DNA technology and at about the time that acceptable methods for doing DNA sequencing and synthesizing specific sequences of DNA chemically were coming into being. The development of this new kind of commercial biotechnology depended critically on the now-developed fluency in the language of DNA.

The pervasiveness of recombinant DNA technology in the biological sciences roughly 20 years after its invention is remarkable. TABLE 1 lists some of the major branches of biology in which recombinant DNA technology is playing a critical role. Virtually all aspects of genetics now utilize the technologies of applied molecular genetics, as does medicine increasingly for both di-

TABLE 1. A Selection of Scientific Disciplines Utilizing Applied Molecular Genetics

Genetics: virtually all aspects, including:	Cell biology
• microbial genetics	Biochemistry
• human, animal, plant genetics	Enzymology
• projects such as the Human Genome Project	Epidemiology
	Agricultural sciences: virtually all aspects dealing with organisms
• population genetics	
• pharmacogenetics	Systematics, taxonomy
Medicine	Evolutionary biology
• diagnosis	Ecology, environmental biology
• treatment	Anthropology
new agents	Paleontology
new modalities (gene therapy)	Archaeology
Microbiology	Polymer sciences: biopolymers
Immunology	

agnosis and treatment. Microbiology, immunology, cell biology, biochemistry, enzymology—all increasingly utilize these kinds of tools to do key experiments that could not have been done before. Epidemiology is being transformed by the ability to track organisms, particularly those of infectious diseases, at a level of resolution and sophistication not previously possible. The agricultural sciences—both animal and plant—are being affected in a huge way by DNA technology. Systematics and taxonomy are being transformed by the ability to look at the specific DNA sequences that determine the characteristics of an organism rather than having to look at those characteristics many steps removed from the actual sequence itself. Evolutionary biology, ecology, and environmental biology are being similarly affected, as are anthropology, paleontology, and archeology. Finally, and of interest to someone who worked for DuPont for a number of years, is the fact that DNA technology is having a significant impact on polymer sciences. There are now a number of labs in both academia and industry that are working on the production of polymers in microorganisms, both naturally occurring polymers from other species and synthetic polymers for which there is no known example in nature.

Equally remarkable is the depth and breadth with which applied molecular genetics has penetrated a vast array of commercial enterprises that utilize the biological sciences, as shown in TABLE 2. Of course, the impact of applied molecular genetics on the pharmaceutical and diagnostic industries is well known. The penetration of DNA technology into these businesses is both broad and deep at this point. The chemical industry is also being affected, both in terms of chiral syntheses and in terms of waste control and bioremediation. Agriculture and animal husbandry are being affected in many ways, not least by the increasingly widespread use of transgenic plants and animals. Engineered enzymes and microorganisms are used in food processing and in

TABLE 2. A Selection of Commercial Enterprises Utilizing Applied Molecular Genetics

Pharmaceuticals	Food processing
research	enzymes from recombinant
synthesis	microorganisms
pharmacogenetics	engineered enzymes
Diagnostics–for people, animals, and	Fermentation
plants	production of food ingredients and
DNA probes	flavoring agents by
PCR and other gene amplification	engineered microorganisms
technologies	production of industrial enzymes
enzymes	Household products
antibodies, engineered antibodies	enzymes from recombinant organisms
Chemicals	engineered enzymes
synthesis	Environmental cleanup
waste control	monitoring
bioremediation	bioremediation
Agriculture	Mining
stress-resistant plant varieties	concentration and recovery of metals
plant varieties with improved food value	bioremediation
plant varieties with improved commercial	Forensics
characteristics	individual identification: people, animals
Animal husbandry	paternity testing: people, animals
improved breeding stock	Entertainment–books, TV, movies (fact
production of human proteins	and fiction)

the fermentation industry. Engineered enzymes from recombinant organisms have made their way into household laundry detergents. Those concerned with environmental cleanup are increasingly contemplating utilizing DNA technology, both for monitoring and for bioremediation. The mining industry is using genetically engineered microorganisms and there are very important applications of this technology in a number of different areas of forensics. My list concludes with the entertainment industry despite the fact that we cannot yet produce recombinant books or movies. However, it is clear that DNA technology and its putative implications have stimulated much creative activity among authors and screenwriters over the past two decades.

In a week in which a movie (*Jurassic Park*) that was based on the polymerase chain reaction and cloning technology is mentioned in a Nobel Prize citation,[b] I think it is worthwhile taking seriously the impact of the entertainment industry on what people believe about DNA technology. After all it is not from conferences such as this that most of the people in this country learn about molecular genetics; rather it is from movies and books such as *Jurassic Park*, *The Boys from Brazil*, *The Andromeda Strain*, and *The Creature from 20,000*

---

[b] The announcement of Kary Mullis's being awarded the Nobel Prize for his part in the development of the polymerase chain reaction was made during the week of this conference.

Fathoms. It is worth paying attention to what is being said in popular literature and movies about genetic engineering, because implicit, if not explicit, in the premises of much of this material is a fundamentally incorrect view of what genetic manipulation can now or will in the foreseeable future be able to accomplish. But it is this world view that is influencing many beliefs in our society about the potential for genetic engineering.

The development of applied molecular genetics has also had a major financial impact in the capital markets of this country and in increasing funding for research and development in biotechnology. In 1991 plus 1992, nearly $6 billion in equity capital was raised by biotechnology companies. In the 12 years between 1981 and 1992, biotechnology companies also entered into approximately $8 billion of strategic alliances with larger, established companies. In 1992, the fifteen largest biotherapeutic companies spent $1.1 billion in R&D, about 40 percent of their $2.6 billion in sales. In comparison, in the same year the fifteen largest established pharmaceutical companies spent about $11 billion in R&D, which represented about 11 percent of their $102 billion in sales. Finally, the federal government's budget request for FY 1993 for what they define as biotechnology R&D totals another $4.3 billion, or about 6 percent of total governmental R&D spending.

I would like to close by turning to the third theme of my talk. In 1962 Thomas Kuhn published a book called *The Structure of Scientific Revolutions*, a book that has become a landmark in the study of the history and philosophy of science. In it Kuhn defines a scientific revolution as the intellectual developments in a given field of science that occur perhaps once every one or two centuries and that cause people to change important, often deeply held beliefs about the nature of the world as they perceive it. Scientific revolutions, as Kuhn uses the term, cause what he terms a "paradigm shift" in how the world is seen, first by members of the scientific discipline in question and eventually by people at large. An example of such a scientific revolution is the change brought about by the adoption of the Copernican theory of a heliocentric universe which replaced the Ptolemaic geocentric view of the universe that had obtained for hundreds of years previously. That change in world view based on this scientific revolution in astronomy had profound effects on both the social structure and the theological beliefs of the time.

Closer to our own time, the development of the quantum mechanics fundamentally changed our conception of the nature and relationship of matter and energy. The Heisenberg uncertainty principle put fundamental limits on how precisely the speed and position of objects could be measured. The theory of the wave particle duality of matter said that photons behaved as though they were somehow simultaneously both waves and particles. All of these ideas were very counterintuitive in terms of what had preceded them in physics.

I now wish to argue that fluency in the language of DNA has been the essential element in creating what is an ongoing Kuhnian scientific revolution in the biological sciences. I think that we are in the middle of that revolution right now. I would argue that the ability to read, write, and edit DNA is functionally unprecedented in human history. All we have ever been able to do before is to select among the various combinations of genes that the mechanisms of genetics have presented to us. And, while we have developed very powerful and very sophisticated selection procedures, selecting from among a set of alternatives over which one has almost no control is fundamentally different from being able to write and edit one's own text. My second point, then, is that the ability to write and edit DNA is the basis for a synthetic and a creative capability in biology that has not previously existed. Finally, I would argue that the capability of modifying the inherited characteristics of organisms in a directed way epitomizes Kuhn's notion of a scientific revolution, creating a paradigm shift in the way in which people view the world and their place in it. A statement that embodies this idea is a quote that I found in Arthur Kornberg's book on DNA synthesis: "Genetics has become a branch of chemistry." In fact we heard a good illustration of this statement in the elegant and deep genetic experiments that Matt Meselson described earlier at this conference.[c] These experiments, as others, rely absolutely on the ability to clone and sequence DNA.

What happens as Kuhnian scientific revolutions come to fruition? People see the world and their place in it in new ways. Economies change as fundamentally new technologies are introduced. In nineteenth century Germany and England, the development of synthetic organic chemistry broke down the previously assumed absolute barrier between the animate and the inanimate world. At the same time as it changed the perception of the world at the time, it also formed the basis for a new industry that had a major impact on the economies of both Germany and Great Britain for the next century. In our time, the development of solid-state physics has clearly transformed the world in which we live technologically, socially, and politically. For instance, the development and widespread dissemination of transistor radios 30 years ago opened up the entire Third World to a type of communication that had been impossible previously. The political consequences of this technological development have been striking.

What will the world be like in another 30 years or so when the scientific revolution in biology that has been initiated by the discovery of the structure of DNA is likely to be approaching maturity? It's impossible to know the answer in detail, but judging from the impact of past scientific revolutions the effects will be large, diverse, and positive. One thing is certain, however, and that is that 30 years from now we will have vastly more knowledge about ourselves and the biosphere in which we live than we do now. And that knowl-

---

[c] Regrettably, Dr. Meselson's contribution is not available for inclusion in this volume [*Ed.*].

edge will permit us to exercise a fundamentally different level of control over the biological world than we now have. That knowledge and that control can be utilized for enormous good or for significant harm, depending on how wise we are. Given the history of past scientific revolutions, I'm an optimist about the outcome of this one and align myself with Thomas Jefferson, who said, "Knowledge is power; knowledge is safety; knowledge is happiness."

The Molecular Biology of Thyroid Hormone Action

RALFF C. J. RIBEIRO,[a] JAMES W. APRILETTI,
BRIAN L. WEST, RICHARD L. WAGNER,[b]
ROBERT J. FLETTERICK,[b] FRED SCHAUFELE,
AND JOHN D. BAXTER

Metabolic Research Unit, HSW 1141
Department of Medicine, and the
[b]Department of Biochemistry and Biophysics
University of California, San Francisco
San Francisco, California 94143-0540

Most hormones act by binding to receptors that are either in the nucleus or on the surface of the cell. The receptors for thyroid hormones are nuclear receptors that belong to a large family of structurally related transcription factors that include the receptors for steroid hormones, vitamin D, and retinoids. Over the past 25 years, an enormous amount of data has accumulated on the mechanisms of action of thyroid hormones. This information has extended our knowledge of the mechanisms of hormone action and transcription in general, and has provided a better understanding of the role of hormones on physiology, differentiation, and development. In this report we summarize recent developments in the field of thyroid hormone action with special emphasis on contributions from work conducted in our laboratory.

HISTORICAL AND GENERAL ASPECTS OF THYROID HORMONE RECEPTORS

The importance of the thyroid gland was recognized more than a century ago, when the loss of thyroid function was associated with cretinism and myxedema.[1] Thyroid hormones were later identified as mediators of most systemic actions of the thyroid gland. The major form of thyroid hormone secreted by the gland is thyroxine (3,5,3′,5′-tetraiodo-L-thyronine, or T_4), but the gland releases a lesser quantity of 3,5,3′-L-triiodothyronine, or T_3; the latter is the major active form of thyroid hormone in the target tissues. Much of the T_3 is generated in the peripheral tissues by deiodination of T_4.[2,3]

---

[a] To whom correspondence should be addressed.

Thyroid hormones arrive at target tissues bound to plasma proteins and gain cellular access by mechanisms that may involve specific membrane transport proteins.[2]

Thyroid hormones have diverse actions in many vertebrate tissues, including effects on differentiation and development, thermogenesis, and metabolism (FIG. 1). Elevated or diminished thyroid hormone production seen in patients with hyperthyroidism or hypothyroidism, respectively, results in characteristic features that illustrate actions of these hormones.[2,3] For example, patients with hyperthyroidism have weight loss, low cholesterol levels, elevated body temperatures and tachycardia, whereas hypothyroidism provokes hair loss, hypertension, hypercholesterolemia, myxedema, and bradycardia. Although much was known about the derangements in the syndromes of thyroid dysfunction, very little was known about the cellular loci of thyroid hormone action.

Evidence for a nuclear site of thyroid hormone action began to accumulate in the 1960s and 1970s. It was noted first that effects of thyroid hormones on the liver were associated with changes in RNA levels.[4,5] Nuclear-localized thyroid hormone receptors (TRs) were identified in the early 1970s.[6] These TRs were tightly bound to transcriptionally active chromatin, even in the absence of hormone,[7-9] and were found also to be DNA-binding proteins.[10,11]

These data suggested that thyroid hormones regulate transcription of specific genes. Indeed, in 1977, with the availability of recombinant DNA techniques, we cloned the cDNA for a thyroid hormone–responsive gene product, rat growth hormone (rGH), and demonstrated that it was specifically regulated by T_3.[12,13] Shortly thereafter, we showed that thyroid hor-

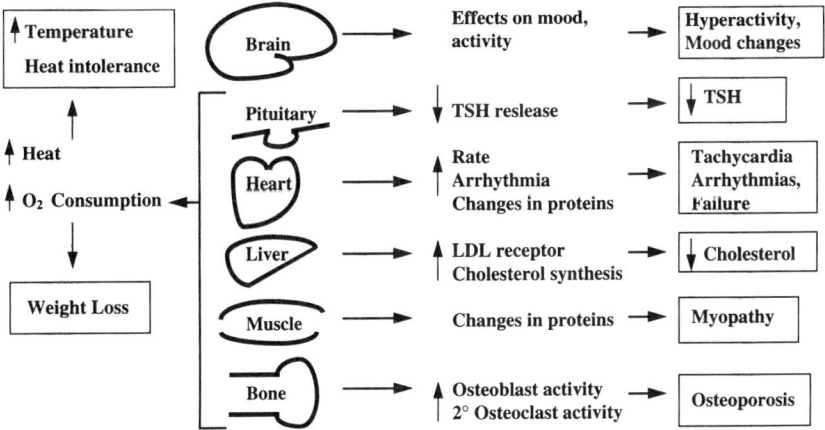

FIGURE 1. Actions of thyroid hormone. Effects of thyroid hormone excess are listed in *boxes*. LDL = low density lipoprotein; TSH = thyroid-stimulating hormone.

mones regulate approximately 1% of the expressed genes in rat pituitary tumor cells.[14]

In the early 1980s, with the cloning of TR-responsive chromosomal genes, such as that encoding rGH,[15] it became possible to study the promoters of TR-regulated genes. This allowed us[16] and others,[17,18] to use gene transfer experiments to define hormone response elements (HREs), which are sites on DNA that bind receptors and function as hormone-regulated transcriptional enhancer or repressor elements. Thyroid hormone response elements (TREs) were found by us[19,20] and others,[21-24] to bind specifically TRs and mediate T_3 action on the rGH gene.

In the early 1990s, the consensus is that TREs are composed of two or more half-sites related to the hexanucleotide AGGTCA.[25] The current model of T_3 action postulates that TRs can bind to TREs and regulate transcription either as monomers, homodimers, or heterodimers with auxiliary proteins, ordinarily retinoid X receptors (RXRs).[26] Binding of T_3 provokes conformational changes on these complexes altering their direct or indirect interactions with proteins of the basal transcription machinery, which then results in stimulatory or inhibitory effects on transcription.[27]

THE NUCLEAR HORMONE RECEPTOR SUPERFAMILY

cDNAs that encode nuclear receptors were cloned in the late 1980s.[28] This made it possible to express the cloned receptors in mammalian cells and bacteria, and to study their properties. The TR was found to share a number of properties with receptors for steroid hormones.[28] They bind both ligand and DNA, and affect transcription directly. The glucocorticoid receptor (GR) was the first steroid hormone receptor to be cloned.[29] Subsequent cloning of the estrogen receptor (ER) and progesterone receptor (PR) cDNAs revealed they are structurally related to the GR, with nucleotide and amino acid sequence homologies concentrated in the regions of the molecules that bind DNA and ligand [the DNA-binding (DBD) and ligand-binding (LBD) domains, respectively].[28] Cloning of related genes on the basis of nucleotide sequence homology identified a number of related proteins including that encoded by the viral oncogene v-erbA. Testing the proteins expressed from the new clones for hormone binding led to the discovery in 1986 that the cellular homologue of v-erbA, the protooncogene c-erbA, encoded a thyroid hormone receptor.[30,31]

Other homologous proteins subsequently found were the mineralocorticoid receptor (MR),[28] androgen receptor (AR),[32,33] vitamin D receptor (VDR),[28] all-*trans* retinoic acid receptors (RARs),[28] receptors for peroxisomal proliferators (PPARs),[34] and 9-*cis* retinoic acid receptors (RXRs).[35,36] A number of proteins for which no ligands have been found are now collectively called orphan receptors. Together these receptors form the nuclear hormone receptor gene superfamily, also termed for historical reasons, the steroid/thyroid hormone receptor gene superfamily, because the TR was the first nonsteroidal family member to be cloned.

NUCLEAR HORMONE RECEPTOR FUNCTION

Nuclear hormone receptors are single polypeptide chains composed of three major domains, the DBD, the LBD, and the amino-terminal domain (FIG. 2). These domains constitute modular regions that retain specific functions even when isolated from the rest of the protein.

The well-conserved DBD targets the receptor to its specific hormone response element. Crystallographic structures of isolated DBDs from several receptors have been determined.[37-39] The DBD contains two zinc finger motifs, in which each zinc ion is coordinated with four cysteines. Carboxy-terminal to each zinc finger is an alpha helix; these two alpha helices are perpendicularly oriented to each other. The alpha helix at the base of the first zinc finger contacts specific bases in the DNA's major groove. Three amino acids, referred to as the P-box, within this alpha helix are responsible for DNA-binding site discrimination. Homology of the P-box among DBDs of superfamily members has allowed them to be grouped into subfamilies that recognize similar DNA sequences. Thus, the P-boxes are similar among VDRs, RARs, RXRs, ERs, and orphan receptors, and among the GR, MR, PR, and AR.[25] Exchanging the P-box amino acids can alter binding specificity from a GR- to a TR-type recognition.[40] The DBD also has dimerization functions that enable the receptors to discriminate the orientation and spacing of their DNA half-sites and that influence whether they bind DNA as monomers, homodimers, or heterodimers.[25]

The LBD is less conserved than the DBD. This may reflect the varied nature of the ligands that bind to these receptors. In addition to hormone

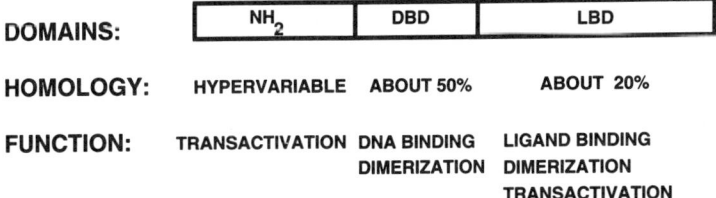

FIGURE 2. Schematic representation of the three major domains of the nuclear receptors. NH_2 = amino terminal domain; DBD = DNA-binding domain; LBD = ligand-binding domain. Homology refers to the amino acid sequence conservation among the nuclear hormone receptor superfamily. Function refers to the diverse activities of the various domains.

binding, the LBD has functions that include dissociation from heat-shock proteins, nuclear translocation, dimerization, DNA binding, and transcriptional activation.[41] The dimerization interface present in the LBD of TRs and other nuclear hormone receptors is composed of nine discontinuous heptad repeats, reminiscent of a leucine zipper motif; this structure may provide a hydrophobic interface for protein–protein interactions required for homo- and heterodimerization.[42] The transcriptional activation function in the LBD is imparted by conserved sequences that allow the DNA-bound receptor to activate transcription.

The amino-terminal domain varies markedly in size and is not conserved among nuclear receptor family members. This domain in the rat TRs is 52, 106, or 159 amino acids, in the α, $\beta1$ and $\beta2$ isoforms, respectively,[43] and is much smaller than that for some other receptors, such as the human MR, which contains 603 amino acids.[28] The amino-terminal domain can exhibit transcriptional activation functions for several family members. Deletion of this domain may[44] or may not[45] affect TR transactivation. Replacement of the smaller TR amino-terminal domain with that of the larger GR domain can result in a receptor that is more active than the wild-type TR.[45] The amino terminal domain can also impart differential DNA-binding properties to three isoforms of an orphan receptor, RORα,[46] that has distinct amino terminal domains; these data suggest that DNA-binding properties of nuclear receptors may be in some cases influenced by their amino terminal domains.

CHARACTERISTICS OF THYROID HORMONE RECEPTORS

There are two different TR genes, α and β, that share strong homology in the DBD and LDB and are found in humans on chromosomes, 17 and 3, respectively.[43] Alternative splicing at the 3' end of the TRα RNA results in two TRα isoforms, $\alpha1$ and $\alpha2$, that differ in the ligand binding domain.[43] Alternative 5'-exon splicing of the TRβ RNA generates two TRβ isoforms, $\beta1$ and $\beta2$, that have unique amino terminal regions.[43]

The functional consequences of the various TR isoforms are currently being unraveled. The $\alpha1$, $\beta1$ and $\alpha2$ isoforms are expressed in nearly all tissues, albeit at different levels,[26] whereas the $\beta2$ isoform has been detected only in pituitary and certain areas of the brain.[47,49] In the rat, the heart contains approximately equal quantities of the $\alpha1$ and $\beta1$ isoforms, the liver contains more of the $\beta1$ isoform, and the $\alpha1$ and $\alpha2$ are the dominant isoforms in the rat brain.[48,49] TR isoform expression also shows a specific developmental pattern. TR$\alpha1$ is expressed earlier than TR$\beta1$.[50] The relative roles of the various TR receptor isoforms are not defined, but different TR isoforms are likely to have specific roles both in development and in the adult.

In most cases studied, the actions of the TR$\alpha1$, TR$\beta1$ and TR$\beta2$ isoforms are similar, but there are exceptions. TRβ1 is reported to regulate specifically T3-responsive genes during cerebellar development[51] and auditory neurogenesis,[52] and the thyrotropin-releasing hormone gene in hypothalamus.[53] TRα1, but not TRβ1, has been reported to activate transcription on a novel TRE found in the long terminal repeat of the Rous sarcoma virus genome, an effect mediated through its unique amino terminal domain.[44]

The TRα2 isoform does not bind thyroid hormone, but can inhibit T3-stimulated transcription by other TR isoforms. This inhibition is reported to be due to TRα2 competition for occupancy of TREs[54] and/or for TRα2 competition for limiting heterodimerizing proteins important for TR function.[55]

PROPERTIES OF TRs PURIFIED FROM NATURAL SOURCES

Much information about TRs has been derived from studies of recombinant proteins. Data from endogenous TRs has been limited because of their low abundancy. Nevertheless, such studies are important because they characterize the receptor found naturally and may reveal properties/modifications acquired by TRs *in vivo* that may not be present in recombinant TRs. Prior to cloning of the TR cDNA, we partially purified TRs from rat liver.[56] Despite the very low TR abundance (circa 5,000 molecules per hepatocyte), we obtained TRs more than 500-fold purified (approximately 1% pure) and investigated their properties. We found that the TR binds specifically to several sites on the rGH promoter.[19,20] The rat liver TRs also co-purified with a non-TR protein that formed heterodimers with the TRs on DNA.[57] TR binding to DNA is enhanced by this factor, which was termed RXRα-related factor (RXRα-RF) because of its similarities with recombinant human RXRα (hRXRα).[57]

With the use of TR isoform-specific antibodies, we demonstrated that the endogenous rat liver DNA-binding TRs are mostly of the TRβ1 isoform,[57] whereas the dominant DNA-binding TR isoform in pituitary GC cells, another major site of T3 action, was TRβ2 (unpublished). We also found that the predominant TR heterodimerizing partner in GC cells was related to RXRβ (unpublished). These observations support the notion that distinct isoforms of both the TR and its heterodimerizing partner may arise in different tissues and play a role in tissue-specific T3 responses.

RECOMBINANT OVEREXPRESSION OF TRs

We have produced large quantities of recombinant TRs for functional and structural characterization. Expression systems using mammalian cells, yeast, baculovirus, and bacteria were compared. Baculovirus gave excellent TR expression, but it was difficult to purify TRs from this source. We also had difficulty expressing TRs in mammalian cells. By contrast, we could overexpress the ER in mammalian cells[58] provided that care was taken to eliminate traces of estrogen activity in the medium. This may be because unliganded ERs are associated with heat-shock proteins and thus cannot bind to DNA or regulate transcription. In contrast, unliganded TRs are bound to chromatin and may interact/interfere with other transcription factors, and, when in excess, exert detrimental effects on expression of genes important for cell function. In support of this notion, we found that administration of estradiol to ER-overexpressing cells, which allows ER to dissociate from heat-shock proteins and bind to DNA (and/or to transcription factors), results in cell death.[58]

Bacterial expression was found to be the most convenient method to overexpress TR and to yield the best results. We expressed full-length TRβ1 and the LBD of TRα1 in *Escherichia coli* and purified them to homogeneity (>99% pure) using a modification of the techniques we developed for purifying TRs from rat liver.[59,60] These proteins bound T_3 and analogues with the same relative affinities as native TRs isolated from rat liver. The findings with the isolated LBD underscore the modular nature of the various TR domains and indicate that ligand binding in solution is independent of other receptor domains.

TARGET GENE RECOGNITION BY TRs

TRs regulate transcription when bound to TREs in the promoter region of target genes. The consensus sequence in the TRE is the hexanucleotide "half-site" AGGTCA, which binds a TR monomer.[25] In naturally occurring genes, most TREs consist of two or more half-sites oriented with respect to each other either as direct repeats (AGGTCA"n"AGGTCA), palindromes (AGGTCA"n"TGACCT), or inverted palindromes (TGACCT"n"AGGTCA) (FIG. 3). These half-sites are spaced by a variable number "n" of nucleotides and the spacing required for TR binding varies, depending on the orientation of the half-sites. For example, there is optimal TR binding to direct repeats when four nucleotides separate the half-sites (DR-4), whereas optimal TR dimer binding to palindromes occurs when the half-sites are unspaced.[61,62] TRs can bind to these various TREs as monomers and/or homodimers, or as heterodimers with RXRs.[25]

We have examined binding of bacterially expressed and *in vitro*-translated TRβ1 to a variety of TREs (see FIGURE 3 for sequence of TREs). The binding characteristics were similar irrespective of the source of TRs.[57,59] TRs bind to an isolated half-site (F2M) or an imperfect inverted palindrome (EM1) as

Direct Repeats

DR-4　AGCTTC**AGGTCA**CAGG**AGGTCA**GAG

Palindromes

TREpal　ATATTC**AGGTCATGACCT**GAATAT

vitA2-ERE　TC**AGGTCA**CAG**TGACCT**GATCAAAGTTAATGTAACCTCAACCTGGA

Inverted Palindromes

F2　AGCTTAT**TGACCC**CAGCTG**AGGTCA**AGTTACG

EM1pal　AAAA**TGACCC**TATA**GGGTCA**GTG

Imperfect Inverted Palindrome

EM1　AGACACAC**TGACCC**TAAGGTCCTTTTATAGAGCT

Single Half-site

F2M　AGCTTAGTTACTTATGCTG**AGGTCA**AGTTACG

FIGURE 3. Sequence of thyroid response elements. Oligonucleotide sequences represent the diverse number, orientation, and spacing of the half-sites. Half-sites are represented by the hexanucleotides AGGTCA is in *larger letters* and marked by the *arrows*. The *arrows* represent the orientation of the half-sites. The *dashed arrow* marks the imperfect half-site of EM1.

monomers, to palindromic half-sites as either monomers or homodimers regardless of whether $n = 0$ (TREpal) or $n = 3$ (vitA2-ERE), and exclusively as homodimers to DR-4 or to inverted palindromes spaced by four (EM1pal) or six (F2) nucleotides.[57,59]

Scatchard analyses of the various reactions containing progressively increasing concentrations of DNA-binding sites with a fixed content of TRs indicate that more TRs in the reaction mixture are capable of binding to inverted palindromes (as homodimers) than to palindromic TREs (as monomers and homodimers);[59] similar findings have subsequently been reported by others.[63] These observations indicate that in the reaction most of the TRs that form homodimers on inverted palindromes cannot form homodimers on palindromes. Therefore, there is heterogeneity in the TR population with respect to this property, and the efficiency of binding of recombinant TRβ1 to the diverse TREs and the conformation adopted by TRs on them (monomers and/or homodimers) depends on the architecture of TRE half-sites.

The pattern of binding of endogenous partially purified rat liver TRs (containing the endogenous RXRα-RF) to TREs differed from that with recombinant TRs.[57,59] TR-RXRα-RF heterodimers bound each TRE with roughly equal efficiency, irrespective of the number, orientation, and spacing of the half sites. These data suggest that heterodimer formation obliterates some of the DNA-structure selectivity exhibited by TRs in the absence of a heterodimerizing partner. The role of TR heterodimerization with RXRs and other proteins in target gene recognition is also addressed below.

Target gene recognition by TRs can also be influenced by ligand binding (see below) and post-translational modifications of TRs. Indeed, we recently demonstrated that phosphorylation of TRs can selectively increase their homodimeric binding to TREs irrespective of the arrangement of the half-sites.[64]

HETERODIMERIZATION OF TRs WITH NUCLEAR HORMONE RECEPTORS

Heterodimerization with RXRs

RXRs form heterodimers with TRs, VDRs, or RARs; these heterodimers bind DNA more effectively than the individual homodimers.[65] Most tissues investigated thus far have been shown to express at least one of the three different RXR isoforms (α, β and γ) that are produced by RNAs transcribed from three different genes.[66-68]

We found that hRXRα and RXRα-RF show both similarities and differences. They both cross-react with antisera to hRXRα, and form heterodimers on DNA with TRs, VDRs, and RARs but not ERs.[57] RXR-RF differs from hRXRα in three respects. First, the enhancement of recombinant TRβ1 binding to DR-4 and palindromes is much greater (6–8-fold) with RXRα-RF than with hRXRα. Second, hRXRα, but not RXRα-RF, greatly reduces recombinant TRβ1 binding (70–80%) to inverted palindromes. These two differences could reflect either qualitative distinctions between the two molecules or an overall quantitatively greater activity of RXRα-RF. The third difference is qualitative in nature. Recombinant hRXRα greatly enhances recombinant TRα2 binding to DR-4, whereas RXRα-RF (in quantities that greatly enhanced binding of TRβ1 to DR-4) shows no enhancement of this binding. This finding may be relevant to understanding mechanisms of TRα2 inhibition of T$_3$-stimulated transcription because it suggests that formation of heterodimers between TRα2 and RXRs may not occur *in vivo*. Given all the data, it seems most likely that RXRα-RF is a post-translationally modified form of rat RXRα. A less likely possibility is that RXRα-RF is a novel protein. The finding that properties of endogenous RXRs may differ from recombinant RXRs reinforces the need to characterize the properties of proteins in their native state in order to fully assess their role within the cell. The data also indicate a pattern with RXRα-RF that is not evident with recombinant RXR, namely, that heterodimerization with the endogenous RXR species provokes an overall enhancement of TR-DNA recognition; this could promote wider

DNA recognition by TRs and expand the thyroid hormone transcriptional influence in the cell.

Heterodimerization with Other Nuclear Hormone Receptors

Formation of heterodimers provides a means to diversify gene regulation by TRs. TRs heterodimerize in solution and on DNA not only with RXRs but also with certain other nuclear receptor superfamily members, including PPAR, VDR, and RAR.[69–72] Thus, these interactions can expand not only the DNA sites that attract the TRs to the various promoters, but also the number of different protein–protein interactions available for regulating promoter activity. Although the overall pattern of DNA binding of these heterodimers has not been defined to the extent as those established with RXRs, there are indications that these heterodimerizations affect TR action through differential preferences for specific TR isoforms and for novel nonconsensus TREs. Thus, it has been reported that TRβ1-PPAR heterodimers, but not TRα1-PPAR heterodimers, can activate transcription through a DR-2 [ordinarily a retinoic acid response element (RARE), not a TRE], but not through the consensus DR-4.[69]

DNA structure, as shown for direct repeats, can also specify the receptor polarity in heterodimer formation, and this receptor polarity can alter the sensitivity of a promoter to hormone concentration. For example, VDR-TR heterodimers assemble with opposite polarities on DR-3 (a VDR response element) and DR-4 (a TRE).[71] VDR occupies the 3' half-site on DR-3, whereas TR occupies the 3' half-site on DR-4. Although both ligands stimulate transcription, activation requires ten-fold-higher ligand concentrations for the receptor that occupies the 5' site. Similarly, RXR-RAR heterodimers activate transcription on DR-5, but not on DR-1 elements.[73] The 3' half-site is occupied by the RAR on the DR-5 element, and by the RXR on the DR-1 element. It was also shown that formation of RAR-RXR heterodimers on the DR-1 element prevents RXR from binding to its ligand, 9-*cis*-retinoic acid. These observations suggest that heterodimer interfaces vary according to receptor polarity, which may affect not only access of ligand to receptor, but also their ability to contact other proteins and activate transcription.

TR REGULATION OF RAT GROWTH HORMONE GENE EXPRESSION

The TRE functions in the context of promoters that are specific to each gene. TR-regulated promoters contain binding sites for other transcription factors in an arrangement that is unique for each gene. Different cell types vary in the presence and/or levels of these various transcription factors, including those that form heterodimers or have other interactions with the TRs. Therefore, the function of a given TRE can only be known when considered in the context of the specific promoter in which it is located and the cell type in which the promoter is active.

As stated above, expression of the rGH gene is under thyroid hormone control. Studies of the TREs found in the rGH gene promoter by our group[19,20] and others[21-24] have placed the rGH gene promoter as one of the most extensively studied T_3-responsive promoters. We and others have utilized cultured rat pituitary (GC, GH3, GH1 lines) cells that express rGH to study expression of the rGH gene. An outgrowth of these studies were other investigations of the mechanisms by which the tissue-specific expression of the rGH promoter are regulated. We have more recently reconstituted rGH promoter activity in monocytic U937 cells that do not naturally express the gene.[74] Unlike GH and GC cells, these cells have the advantage that they do not express pituitary-specific factors, such as Pit-1, that regulate rGH promoter activity, and thus they provide a better background into which necessary factors can be incorporated.

The transcription factors known to bind to and influence rGH promoter activity in the pituitary cells are shown in FIGURE 4. Several of these factors, Pit-1 (GHF-1), SP1, and TRs have been well characterized. We characterized one other factor, GHF3, and discovered that it bound DNA complexed with other proteins,[75] which may include cEBP proteins (unpublished). Pit-1, which is expressed specifically in pituitary cells of the GH, prolactin, and thyroid-stimulating hormone lineages, has been heralded as the most important factor in regulating rGH promoter activity, in contrast to the other factors which are also found in non-pituitary cells.[76,77] However, the influence of cEBP proteins in some cases is roughly equivalent to that of Pit-1 (unpublished).

The rGH promoter is also regulated by inducers of protein kinase A (forskolin) and protein kinase C [(phorbol 12-myristate 13-acetate (PMA)].[74] The factors responsible for the forskolin and PMA activation of the rGH promoter have not been well defined. Utilizing nonpituitary U937 cells described above, substantial expression of the rGH promoter has been obtained with a minimum of three additions: forskolin, PMA, and vectors that express either the TR or Pit-1. In fact, under these conditions, promoter activity is greater with TR than with Pit-1 expression.[74] These observations demonstrate that (i) Pit-1 is not as necessary for GH gene expression as is commonly believed, although it is an important factor; (ii) the TRs can have a major role in tissue-specific expression of a gene; and (iii) both the TR and Pit-1 can act independently of each other.

Coexpression of Pit-1 and TR in the presence of forskolin and PMA results

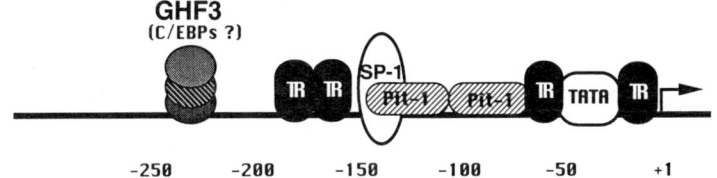

FIGURE 4. Transcription factors known to interact with the rGH promoter. TATA = TATA box factors. Sp1, TR, and Pit-1 refer to transcription factors described in the text. GHF-3 now appears to contain cEBP-related proteins (see text). The *numbers* represent the position of the nucleotides upstream of the transcription start site.

in a markedly synergistic activation of the promoter.[74] Several mechanisms could account for the synergistic effects of Pit-1 and TR on the rGH promoter. For example, they could bind independently to the rGH promoter and independently recruit different, complementary components of the basal transcription complex to the transcription initiation site. On the other hand, they could bind to the DNA cooperatively and act synergistically in the formation of a single, transcriptionally active complex. We have not observed cooperative Pit-1 and TR binding to the rGH promoter *in vitro*, but this does not exclude cooperative binding *in vivo*. We have, however, observed that Pit-1 and TR interact with each other *in vitro* through domains that are required for their transcriptional activity (unpublished). Although the interpretations of these data are unclear, they could imply that the synergy between the TR and Pit-1 involves direct contacts between the factors that facilitate the formation of properties essential to the transcription activation function.

Other synergies have also been noted with activation of the rGH promoter in U937 cells (unpublished). Co-expression of RAR and TR in cells that also express Pit-1 and treatment with forskolin and PMA results in a further synergistic activation (unpublished). In contrast, RXR expression decreases the synergistic activation observed with TR and Pit-1 in cells treated with forskolin and PMA (unpublished). These studies reinforce the notion that transcription factors that regulate rGH promoter activity operate cooperatively rather than in isolation, and that TR function is influenced heavily by other transcription factors (e.g., Pit-1) that either bind independently to other sites within the promoter or form heterodimers with TR.

TR EFFECTS ON THE HUMAN GROWTH HORMONE AND CHORIONIC SOMATOMAMMOTROPIN GENES

Human growth hormone (hGH) and chorionic somatomammotropin (hCS, also termed placental lactogen) genes are highly homologous, having approximately 95% identical nucleotide sequences. In spite of this high degree of homology, hCS is expressed only in the placenta, whereas hGH is exclusively expressed in the pituitary.

These two genes also differ in their responsiveness to T_3 in rat pituitary tumor (GC cells).[78,79] The intact hCS gene is positively and the intact hGH gene negatively regulated by T_3. This differential T_3 responsiveness is due to subtle structural differences in a TRE located between nucleotides −64 and −44 of the promoter of the hGH and hCS promoters.[80] Binding analyses showed that TRs bind the hCS-TRE with higher affinity than the corresponding hGH-TRE, but more importantly, we showed that the TR binding induced a more pronounced DNA bending with the hCS-TRE.[80] We address the implications of TR-induced DNA bending later, in the section on the effects of TR on DNA conformation.

In contrast to the entire hGH gene, a hybrid gene containing the hGH 5'-flanking region fused to the chloramphenicol acetyl transferase gene was not T_3-responsive in GC cells.[81] Further studies showed that an element in the gene's 3'-untranslated/3'-flanking DNA was necessary for T_3 to exert its negative effects on the expression of the hGH gene in transfected GC cells.[82] Although TRs bind to this region of the hGH gene, the mechanisms of action of this novel element in the 3'-untranslated/3'-flanking DNA of the gene have not been defined.

MECHANISMS OF TRANSCRIPTION REGULATION BY TRs

The mechanisms of transcription regulation by TRs and other members of the nuclear receptor superfamily are just being unravelled, and it is likely that there are direct and indirect mechanisms (FIG. 5).

Nuclear receptors can indirectly regulate transcription by altering chromatin structure such that other transcription factors can access regulatory elements and influence transcription. This indirect mechanism was shown to occur with the GR. Binding of liganded GR to the glucocorticoid response element of the mouse mammary tumor virus promoter induces nucleosome repositioning that allows the binding site for another transcription factor, NF1, to become exposed and available for NF1 binding and transcription activation.[83] DNA bending, as first described in our laboratory for the TR, could be another way whereby nuclear receptors alter chromatin structure and indirectly promote binding of other transcription regulatory factors.

Nuclear hormone receptors can also influence transcription through direct interactions with proteins of the transcription machinery. These interactions could function to position RNA polymerase at the transcription start site and release it to transcribe through the gene.

TR binding to these proteins has been examined using columns to which various proteins are attached, and to which radiolabeled proteins bind. Uti-

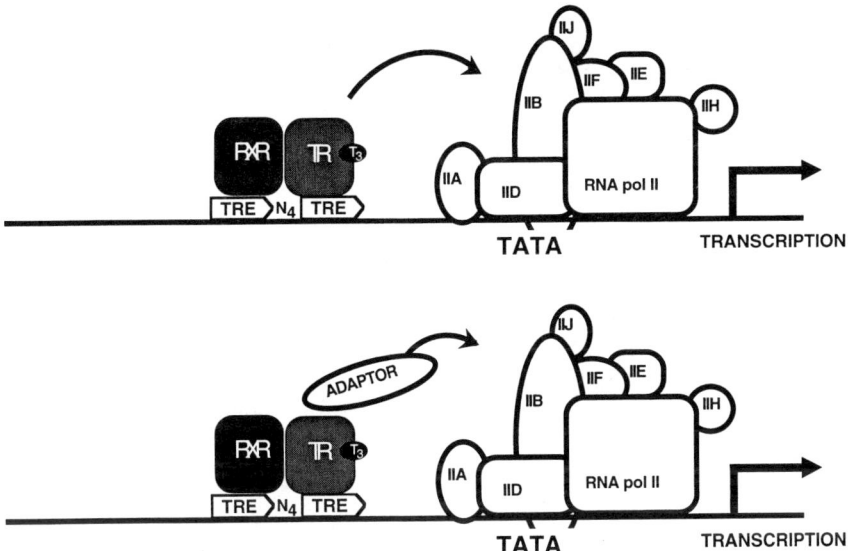

FIGURE 5. Diagram representing models of transcription activation by TR hypothetical interactions with proteins of the basal transcriptional machinery (BTM). The diagram depicts the possible interactions (direct, *top panel*; or indirect, through an adaptor, *bottom panel*) of the TR with the components of the BTM, such as TF IIB, IID and IIE (see text). These are oversimplified models. TR is represented heterodimerized with RXR on DR-4 (TRE direct repeats spaced by 4 base pairs). Both TR and RXR can contact simultaneously or independently components of the BTM or the adaptor. One alternative model, not represented in the diagram, is that TR may alter nucleosome structure (see text for description of DNA bending), and facilitate access of activators to components of the BTM. In this model TR may or may not participate in the interactions between activators and the components of the BTM.

lizing these assays, we found that TR interacts *in vitro* with several different protein components of the basal transcriptional apparatus. This apparatus comprises some 25–30 known proteins (not including the RNA polymerase II complex), with seven activities required for basal transcription: TF IIA, IIB, IID, IIE, IIF, IIH, and IIJ.[84] The TR interacts with TFIIB, TFIID and the small subunit of TFIIE, but not to a protein termed DR (which binds to TFIID) or the large subunit of TFIIE (unpublished). The TR/TFIIB interaction has been most intensively investigated.[85-86] Studies of mutated TFIIB molecules suggest that two different regions within TFIIB bind specifically to different regions of TRβ1 (R. Forde and F. Schaufele,[86] unpublished). The carboxy terminal site of TFIIB interacts with the amino terminal of TR, whereas the amino terminal site of TFIIB interacts with the TR-LBD in a hormone-dependent fashion. A model has been proposed in which transcription repression by unliganded TRs was caused by interference with TFIIB function, thereby inhibiting the assembly and initiation of transcription.[85] In this model, repression is superseded by the second, ligand-specific TR/TFIIB interaction which activates TFIIB function.

On the basis of these studies, we see that it is possible that TRs may also contact other general transcription factors and/or coactivator proteins. Coactivators link upstream transcription factors, such as TRs, to general transcription factors such as TFIIB.

EFFECTS OF TRs ON DNA CONFORMATION

We initially noted that certain rGH promoter mutations that abolished the T_3 response resulted in only minor effects on the receptor's binding to DNA; however, the TR-DNA complexes with the mutated DNA migrated faster than the TR wild-type DNA complexes in gel retardation assays.[20] This was the first indication that the TR or any member of the nuclear receptor superfamily could bend DNA. The data also suggested that the bending might be relevant to thyroid hormone action. We subsequently found that the TR bound to both the hCS and hGH promoters in the region corresponding to the TRE of the hCS promoter (discussed above), and that the binding was nearly identical to both promoters.[80] However, again, the migration of the complexes with the two promoters differed slightly in gel retardation assays.

Circular permutation analyses subsequently confirmed that the TRs bend both the hCS and hGH promoter DNAs, but that the bending of the T_3-responsive hCS promoter was greater than that with the hGH promoter.[80] We also found that rat liver TRs bend DNA.[80] These earlier studies were performed with TRs purified from rat liver, such that the binding and bending was due to TR-RXR-RF heterodimers. It now appears that TRs present either as monomers or heterodimers can bend DNA.[87]

These initial findings have now been extended to other nuclear hormone receptor superfamily members,[88,89] and it appears that a general property of the family is the ability to bend DNA. This bending could participate in receptor function through inducing changes in chromatin conformation that could influence the interactions of transcription factors and/or overall chromatin conformation.

THYROID HORMONE-INDUCED CHANGES IN THE PROPERTIES OF TRs

As discussed above, a critical aspect of transcriptional control by TRs and superfamily members in general is ligand-induced conformational changes in receptors. Therefore, it is critical to define these changes with isolated receptors. The first physical data suggesting that ligand binding influences the conformation of the TR was obtained by Silva and co-workers,[90] who noted that liganded TR elutes at a lower ionic strength from an ion exchange (DEAE-Sephadex) column than does unliganded TR. We subsequently observed that TR–T$_3$ complexes elute at a higher ammonium sulfate concentration from a mild hydrophobic interaction HPLC column than do unliganded TRs.[56] These data indicate that T$_3$ decreases the net hydrophobicity of the receptor. Aqueous two-phase partitioning studies also showed that the TR becomes less hydrophobic upon T$_3$ binding.[91] Binding hormone to the TR increases the protein's resistance to digestion by trypsin and chymotrypsin[92,93] and also causes a change in the circular dichroism spectrum.[94] We[59] and others[62] also found that T$_3$ alters the gel electrophoresis mobility of DNA-bound TR in a manner consistent with a conformational change.

We recently found that the T$_3$-induced conformational changes observed in the intact receptor could also be observed with a 33-kDa protein fragment (Met^{122}-Val^{410}) of rat TRα1 that was expressed from a vector in *E. coli*.[60] This truncated TR contains the entire LBD with the 289 carboxyl-terminal amino acids (Met^{122}-Val^{410}) of rat TRα1, but is deleted in the DBD and the amino-terminal domains.[60] This segment binds T$_3$ with a Kd of 0.06 nM, identical to that for the full-length rat TRα1.

The 33-kDa fragment prebound to [^{125}I]T$_3$ eluted earlier than did unliganded TR-LBD during hydrophobic interaction, DEAE anionic exchange, and heparin affinity chromatographic steps, like the full-length receptor. Thus, the ligand-induced conformational changes can occur in the isolated LBD and independent of other TR domains; these changes may be relevant for transmitting the T$_3$ transcriptional response. Other TR-binding ligands were also examined. When tested on the hydrophobic interaction column, the different T$_3$ analogues varied in the extent of this column shift, with T$_3$ = Triac > isopropyl-T$_2$ > DIMIT > T$_4$. The differences in ligand-induced chromatographic behavior imply that the LBD can exist in several different conformational states, and that the induction of conformational changes and the extent of these is affected by the structure of the ligand. It has been reported that the function of a protein can be differentially affected by multiple effector-induced conformational states.[95] Thus, these observations with the TR may be relevant to antagonist and partial agonist interactions with the TR, and may imply qualitative differences between various ligands that bind to the TR. It is noteworthy that *in vivo* it is thought that T$_3$ is the major ligand that mediates thyroid hormone action, even though T$_4$ is the major hormone secreted by the gland. Perhaps the influence of T$_4$ on the TR is not as great as that of T$_3$. In this regard, we found that T$_3$ and T$_3$ analogues differentially enhanced bacterially expressed TR binding to the imperfect inverted palindrome EM1. At saturating levels Triac had a greater effect than T$_3$, T$_4$, and reverse T$_3$ on inducing TR-DNA binding. Thus, under these conditions, all of the T_3 agonists enhance bacterially expressed TR binding to DNA, but the magnitude of the effect varied according to the ligand, implying that there is a correlation between the affinity of the ligand for binding TR and the extent to which the ligand can stimulate the TR binding to DNA.

We also found that T_3 alters the way that TR monomers and dimers bind to DNA and that these effects are TRE half-site–dependent.[59] T_3 stimulated and increased the gel mobility of monomers irrespective of the half-site arrangement, but had diverse effects on dimers. For example, whereas T_3 greatly inhibited formation of TR homodimers on DR-4, it greatly enhanced them on TREpal. Dissociation of homodimers, also reported by other investigators,[96-98] and augmentation of monomers may indirectly concur for heterodimer formation. T_3 can also enhance heterodimer formation, although the effect as studied in gel mobility shifts is modest.[59] These findings argue that ligand binding does change the conformation of TR–DNA complexes and also affects the configuration they adopt on DNA, with potential implications for transcription regulation.

PROGRESS TOWARDS DETERMINATION OF THE THREE-DIMENSIONAL STRUCTURE OF THE TR-LBD

Knowledge of the three-dimensional structure of the TR, liganded and unliganded, would advance the understanding of TR biology in several important areas. The three-dimensional structure of the DBD for several of the receptors has been determined, and has greatly expanded our knowledge of the mechanisms of nuclear receptor binding to DNA. However, the structure of the LBD has not been determined for any of the superfamily members. Information about the LBD will provide insights into mechanisms of hormone binding, but also of the mechanisms that involve receptor homo- and heterodimerization. Regarding the LBD of TR, such structural information should identify the important residues involved in ligand binding as well as the key contacts made by the various T_3 analogues. This information will permit the rational design of thyroid hormone antagonists and agonists with receptor isoform specificity. Furthermore, a comparison of the structures of the receptor in the liganded and unliganded states would allow examination of the conformational changes that accompany hormone binding, and, most importantly, how these ligand-induced structural changes can influence transcription. Finally, a structure of the DNA-bound receptor as a homodimer or as a heterodimer with RXR would identify the interfaces involved in these interactions and further our understanding of the role played by dimerization on the recognition of TREs with various orientation and spacing of TRE half-sites.

We have recently determined the three-dimensional structure of the TR-LBD by X-ray crystallography. We utilized the 33-kD rat $TR\alpha 1$-LBD mentioned above, which is deleted in the amino-terminal domain and the DBD. The TR-

LBD was purified to >99% homogeneity and crystals from it were prepared for X-ray diffraction studies.[99] X-ray diffraction using a high-intensity X-ray beam provided by synchrotron radiation–generated data measured to 2.2Å resolution for the TR-LBD with T_3 and two analogs of T_3, 3,5-dibromo-3'-isopropyl-L-thyronine and 3,5-dimethyl-3'-isopropyl-L-thyronine. The TR-LBD structure has been obtained by suitable manipulations of these diffraction data and combining X-ray data from heavy atom derivatives to provide the phases needed for the reconstruction. The TRα1 LBD is comprised mostly of alpha helices. The ligand is fitted into this protein into a deep pocket in which the amino group of the hormone is facing towards the outside solvent. Analysis of the structure also suggests that the TR may assume different conformations to fit various ligands. This is consistent with the diverse ligand-induced conformational states observed on the TR-LBD during the ion-exchange or hydrophobic interaction chromatography purification steps as discussed above.

REFERENCES

1. WERNER, S. C. 1991. History of the thyroid. In The thyroid: A Fundamental and Clinical Text. L. E. Braverman & R. D. Utiger, Eds. Vol. **1**: 3–6. J. B. Lippincott. Philadelphia, PA.
2. GREENSPAN, F. S. 1994. The thyroid gland. In Basic & Clinical Endocrinology, 4th ed. F. S. Greenspan & J. D. Baxter, Eds. Vol. **1**: 160–226. Appleton & Lange. Norwalk, CT.
3. UTIGER, R. D. 1995. The thyroid: Physiology, thyrotoxicosis, hypothyroidism, and the painful thyroid. In Endocrinology and Metabolism, 3rd ed. P. F. Felig, J. D. Baxter & C. A. Frohman, Eds. Vol. **1**: 435–519. McGraw-Hill. New York, N.Y.
4. TATA, J. R. 1963. Inhibition of the biological action of thyroid hormones by actinomycin D and puromycin. Nature **197**: 1167–1168.
5. TATA, J. R. 1966. Ribonucleic acid synthesis during the early action of thyroid hormones. Biochem. J. **98**: 604–620.
6. OPPENHEIMER, J. H., D. KOERNER, H. L. SCHWARTZ & M. I. SURKS. 1972. Specific nuclear triiodothyronine binding sites in rat liver and kidney. J. Clin. Endocrinol. Metab. **35**(2): 330–333.
7. SPINDLER, B. J., K. M. MACLEOD, J. RING & J. D. BAXTER. 1975. Thyroid hormone receptors: Binding characteristics and lack of hormonal dependency for nuclear localization. J. Biol. Chem. **250**: 4113–4119.
8. MACLEOD, K. N. & J. D. BAXTER. 1975. DNA binding of thyroid hormone receptors. Biochem. Biophys. Res. Commun. **62**(3): 577–583.
9. MACLEOD, K. M. & J. D. BAXTER. 1976. Chromatin receptors for thyroid hormones. Interactions of the solubilized proteins with DNA. J. Biol. Chem. **251**(23): 7380–7387.
10. CHARLES, M. A., G. U. RYFFEL, M. OBINATA, B. J. MCCARTHY & J. D. BAXTER. 1975. Nuclear receptors for thyroid hormone: evidence for nonrandom distribution within chromatin. Proc. Natl. Acad. Sci. USA **72**(5): 1787–1791.
11. LEVY, B. & J. D. BAXTER. 1976. Distribution of thyroid and glucocorticoid hormone receptors in transcriptionally active and inactive chromatin. Biochem. Biophys. Res. Commun. **68**(4): 1045–1051.
12. MARTIAL, J. A., P. H. SEEBURG, D. GUENZI, H. M. GOODMAN & J. D. BAXTER.

1977. Regulation of growth hormone gene expression: synergistic effects of thyroid and glucocorticoid hormones. Proc. Natl. Acad. Sci. USA **74**(10): 4293-4295.
13. MARTIAL, J. A., J. D. BAXTER, H. M. GOODMAN & P. H. SEEBURG. 1977. Regulation of growth hormone messenger RNA by thyroid and glucocorticoid hormones. Proc. Natl. Acad. Sci. USA **74**(5): 1816-1820.
14. IVARIE, R. D., J. D. BAXTER & J. A. MORRIS. 1981. Interaction of thyroid and glucocorticoid hormones in rat pituitary tumor cells. Specificity and diversity of the responses analyzed by two-dimensional gel electrophoresis. J. Biol. Chem. **256**(9): 4520-4528.
15. BARTA, A., R. I. RICHARDS, J. D. BAXTER & J. SHINE. 1981. Primary structure and evolution of rat growth hormone gene. Proc. Natl. Acad. Sci. USA **78**(8): 4867-4871.
16. KARIN, M., A. HASLINGER, H. HOLTGREVE, G. CATHALA, E. SLATER & J. D. BAXTER. 1984. Activation of a heterologous promoter in response to dexamethasone and cadmium by metallothionein gene 5'-flanking DNA. Cell **36**(2): 371-379.
17. CHANDLER, V. L., B. A. MALER & K. R. YAMAMOTO. 1983. DNA sequences bound specifically by glucocorticoid receptor in vitro render a heterologous promoter hormone responsive in vivo. Cell **33**(2): 489-499.
18. HYNES, N., A. J. VAN OOYEN, N. KENNEDY, P. HERRLICH, H. PONTA & B. GRONER. 1983. Subfragments of the large terminal repeat cause glucocorticoid-responsive expression of mouse mammary tumor virus and of an adjacent gene. Proc. Natl. Acad. Sci. USA **80**(12): 3637-3641.
19. LAVIN, T. N., J. D. BAXTER & S. HORITA. 1988. The thyroid hormone receptor binds to multiple domains of the rat growth hormone 5'-flanking sequence. J. Biol. Chem. **263**: 9418-9426.
20. NORMAN, M. F., T. N. LAVIN, J. D. BAXTER & B. L. WEST. 1989. The rat growth hormone gene contains multiple thyroid response elements. J. Biol. Chem. **264**: 12063-12073.
21. FLUG, F., R. P. COPP, J. CASANOVA, Z. D. HOROWITZ, L. JANOCKO, M. PLOTNICK & H. H. SAMUELS. 1987. cis-acting elements of the rat growth hormone gene which mediate basal and regulated expression by thyroid hormone. J. Biol. Chem. **262**: 6373-6382.
22. GLASS, C. K., R. FRANCO, C. WEINBERGER, V. R. ALBERT, R. M. EVANS & M. G. ROSENFELD. 1987. A c-erb-A binding site in rat growth hormone gene mediates trans-activation by thyroid hormone. Nature **329**: 738-741.
23. KOENIG, R. J., G. A. BRENT, R. L. WARNE, P. R. LARSEN & D. D. MOORE. 1987. Thyroid hormone receptor binds to a site in the rat growth hormone promoter required for induction by thyroid hormone. Proc. Natl. Acad. Sci. USA **84**: 5670-5674.
24. BRENT, G. A., J. W. HARNEY, Y. CHEN, R. L. WARNE, D. D. MOORE & P. R. LARSEN. 1989. Mutations of the rat growth hormone promoter which increase and decrease response to thyroid hormone define a consensus thyroid hormone response element. Mol. Endocrinol. **3**: 1996-2004.
25. GLASS, C. K. 1994. Differential recognition of target genes by nuclear receptor monomers, dimers, and heterodimers. Endocr. Rev. **15**(3): 391-407.
26. BRENT, G. A. 1994. The molecular basis of thyroid hormone action. N. Engl. J. Med. **331**(13): 847-853.
27. SHEPARD, A. R. & N. L. EBERHARDT. 1993. Molecular mechanisms of thyroid hormone action. Clin. Lab. Med. **13**(3): 531-541.

28. EVANS, R. M. 1988. The steroid and thyroid hormone receptor superfamily. Science **240:** 889–895.
29. HOLLENBERG, S., C. WEINBERGER, E. ONG, G. CERELLI, A. ORO, R. LEBO, E. THOMPSON, M. ROSENFELD & R. EVANS. 1985. Primary structure and expression of a functional human glucocorticoid receptor cDNA. Nature **318**(6047): 635–641.
30. WEINBERGER, C., C. C. THOMPSON, E. S. ONG, R. LEBO, D. J. GRUOL & R. J. EVANS. 1986. The c-erb-A gene encodes a thyroid hormone receptor. Nature **324:** 641–646.
31. SAP, J., A. MUNOZ, K. DAMM, Y. GOLDBERG, J. GHYSDAEL, A. LEUTZ, H. BEUG & B. VENNSTROM. 1986. The c-erb-A protein is a high-affinity receptor for thyroid hormone. Nature **324:** 635–640.
32. CHANG, C. S., J. KOKONTIS & S. T. LIAO. 1988. Molecular cloning of human and rat complementary DNA encoding androgen receptors. Science **240**(4850): 324–326.
33. LUBAHN, D. B., D. R. JOSEPH, P. M. SULLIVAN, H. F. WILLARD, F. S. FRENCH & E. M. WILSON. 1988. Cloning of human androgen receptor complementary DNA and localization to the X chromosome. Science **240**(4850): 327–330.
34. GREEN, S. & W. WAHLI. 1994. Peroxisome proliferator-activated receptors: finding the orphan a home. Mol. Cell. Endocrinol. **100**(1–2): 149–153.
35. HEYMAN, R. A., D. J. MANGELSDORF, J. A. DYCK, R. B. STEIN, G. EICHELE, R. M. EVANS & C. THALLER. 1992. 9-*cis* retinoic acid is a high affinity ligand for the retinoid X receptor. Cell **68:** 397–406.
36. LEVIN, A. A., L. J. STURZENBECKER, S. KAZMER, T. BOSAKOWSKI, C. HUSELTON, G. ALLENBY, J. SPECK, C. KRATZEISEN, M. ROSENBERGER & A. LOVEY. 1992. 9-*cis* retinoic acid stereoisomer binds and activates the nuclear receptor RXR alpha. Nature **355:** 359–361.
37. HARD, T., E. KELLENBACH, R. BOELENS, B. A. MALER, K. DAHLMAN, L. P. FREEDMAN, J. CARLSTEDT-DUKE, K. R. YAMAMOTO, J. A. GUSTAFSSON & R. KAPTEIN. 1990. Solution structure of the glucocorticoid receptor DNA-binding domain. Science **249:** 157–160.
38. SCHWABE, J. W., D. NEUHAUS & D. RHODES. 1990. Solution structure of the DNA-binding domain of the oestrogen receptor. Nature **348:** 458–461.
39. LUISI, B. F., W. X. XU, Z. OTWINOWSKI, L. P. FREEDMAN, K. R. YAMAMOTO & P. B. SIGLER. 1991. Crystallographic analysis of the interaction of the glucocorticoid receptor with DNA [see comments]. Nature **352:** 497–505.
40. YEN, P. M., M. IKEDA, E. C. WILCOX, J. H. BRUBAKER, R. A. SPANJAARD, A. SUGAWARA & W. W. CHIN. 1994. Half-site arrangement of hybrid glucocorticoid and thyroid hormone response elements specifies thyroid hormone receptor complex binding to DNA and transcriptional activity. J. Biol. Chem. **269**(17): 12704–12709.
41. CARLSTEDT-DUKE, A. W. J., M. GOTTLICHER, S. OKRET & J.-A. GUSTAFSSON. 1995. Molecular mechanisms of hormone action: Regulation of target cell function by the steroid hormone receptor supergene family. *In* Endocrinology and Metabolism, 3rd ed. P. F. Felig, J. D. Baxter & C. A. Frohman, Eds. Vol. **1:** 169–199. McGraw-Hill. New York.
42. FORMAN, B. M. & H. H. SAMUELS. 1990. Interactions among a subfamily of nuclear hormone receptors: The regulatory zipper model. Mol. Endocrinol. **4:** 1293–1301.
43. LAZAR, M. A. 1993. Thyroid hormone receptors: Multiple forms, multiple possibilities. Endocr. Rev. **14:** 184–193.

44. SAATCIOGLU, F., T. DENG & M. KARIN. 1993. A novel cis element mediating ligand-independent activation by c-ErbA: Implications for hormonal regulation. Cell **75**(6): 1095–1105.
45. THOMPSON, C. C. & R. M. EVANS. 1989. Trans-activation by thyroid hormone receptors: functional parallels with steroid hormone receptors. Proc. Natl. Acad. Sci. USA **86**: 3494–3498.
46. GIGUERE, V., M. TINI, G. FLOCK, E. ONG, R. M. EVANS & G. OTULAKOWSKI. 1994. Isoform-specific amino-terminal domains dictate DNA-binding properties of ROR-alpha, a novel family of orphan hormone nuclear receptors. Genes & Dev. **8**(5): 538–553.
47. HODIN, R. A., M. A. LAZAR, B. I. WINTMAN, D. S. DARLING, R. J. KOENIG, P. R. LARSEN, D. D. MOORE & W. W. CHIN. 1989. Identification of a thyroid hormone receptor that is pituitary-specific. Science **244**: 76–79.
48. BRADLEY, D. J., H. C. TOWLE & W. S. YOUNG. 1992. Spatial and temporal expression of alpha- and beta-thyroid hormone receptor mRNAs, including the beta 2-subtype, in the developing mammalian nervous system. J. Neurosci. **12**: 2288–2302.
49. SCHWARTZ, H. L., K. A. STRAIT, N. C. LING & J. H. OPPENHEIMER. 1992. Quantitation of rat tissue thyroid hormone binding receptor isoforms by immunoprecipitation of nuclear triiodothyronine binding capacity. J. Biol. Chem. **267**: 11794–11799.
50. RODD, C., H. L. SCHWARTZ, K. A. STRAIT & J. H. OPPENHEIMER. 1992. Ontogeny of hepatic nuclear triiodothyronine receptor isoforms in the rat. Endocrinology **131**: 2559–2564.
51. STRAIT, K. A., L. ZOU & J. H. OPPENHEIMER. 1992. Beta 1 isoform-specific regulation of a triiodothyronine-induced gene during cerebellar development. Mol. Endocrinol. **6**: 1874–1880.
52. BRADLEY, D. J., H. C. TOWLE & W. S. YOUNG, 3RD. 1994. Alpha and beta thyroid hormone receptor (TR) gene expression during auditory neurogenesis: Evidence for TR isoform-specific transcriptional regulation in vivo. Proc. Natl. Acad. Sci. USA **91**(2): 439–443.
53. LEZOUALC'H, F., A. H. HASSAN, P. GIRAUD, J. P. LOEFFLER, S. L. LEE & B. A. DEMENEIX. 1992. Assignment of the beta-thyroid hormone receptor to 3,5,3'-triiodothyronine-dependent inhibition of transcription from the thyrotropin-releasing hormone promoter in chick hypothalamic neurons. Mol. Endocrinol. **6**(11): 1797–1804.
54. KATZ, D. & M. A. LAZAR. 1993. Dominant negative activity of an endogenous thyroid hormone receptor variant (alpha 2) is due to competition for binding sites on target genes. J. Biol. Chem. **268**(28): 20904–20910.
55. NAGAYA, T. & J. L. JAMESON. 1993. Distinct dimerization domains provide antagonist pathways for thyroid hormone receptor action. J. Biol. Chem. **268**(32): 24278–24282.
56. APRILETTI, J. W., J. D. BAXTER & T. N. LAVIN. 1988. Large scale purification of the nuclear thyroid hormone receptor from rat liver and sequence-specific binding of the receptor to DNA. J. Biol. Chem. **263**: 9409–9417.
57. RIBEIRO, R. C. J., J. W. APRILETTI, P. M. YEN, W. W. CHIN & J. D. BAXTER. 1994. Heterodimerization and deoxyribonucleic acid-binding properties of a retinoid X receptor-related factor. Endocrinology **135**(11): 2076–2085.
58. KUSHNER, P. J., E. HORT, J. SHINE, J. D. BAXTER & G. L. GREENE. 1990. Construction of cell lines that express high levels of the human estrogen receptor and are killed by estrogens. Mol. Endocrinol. **4**: 1465–1473.

59. RIBEIRO, R. C. J., P. J. KUSHNER, J. W. APRILETTI, B. L. WEST & J. D. BAXTER. 1992. Thyroid hormone alters in vitro DNA binding of monomers and dimers of thyroid hormone receptors. Mol. Endocrinol. **6:** 1142–1152.
60. APRILETTI, J. W., J. D. BAXTER, K. H. LAU & B. L. WEST. 1995. Expression of the rat α1 thyroid hormone receptor ligand binding domain in *Escherichia coli* and the use of a ligand-induced conformation change as a method for its purification to homogeneity. Protein Express. Purif. **6:** in press.
61. UMESONO, K., K. K. MURAKAMI, C. C. THOMPSON & R. M. EVANS. 1991. Direct repeats as selective response elements for the thyroid hormone, retinoic acid, and vitamin D3 receptors. Cell **65:** 1255–1266.
62. FORMAN, B. M., J. CASANOVA, B. M. RAAKA, J. GHYSDAEL & H. H. SAMUELS. 1992. Half-site spacing and orientation determines whether thyroid hormone and retinoic acid receptors and related factors bind to DNA response elements as monomers, homodimers, or heterodimers. Mol. Endocrinol. **6:** 429–442.
63. WILLIAMS, G. R., A. M. ZAVACKI, J. W. HARNEY & G. A. BRENT. 1994. Thyroid hormone receptor binds with unique properties to response elements that contain hexamer domains in an inverted palindrome arrangement. Endocrinology **134**(4): 1888–1896.
64. SUGAWARA, A., P. M. YEN, J. W. APRILETTI, R. C. RIBEIRO, D. B. SACKS, J. D. BAXTER & W. W. CHIN. 1994. Phosphorylation selectively increases triiodothyronine receptor homodimer binding to DNA. J. Biol. Chem. **269**(1): 433–437.
65. YU, V. C., C. DELSERT, B. ANDERSEN, J. M. HOLLOWAY, O. V. DEVARY, A. M. NAAR, S. Y. KIM, J. M. BOUTIN, C. K. GLASS & M. G. ROSENFELD. 1991. RXR beta: A coregulator that enhances binding of retinoic acid, thyroid hormone, and vitamin D receptors to their cognate response elements. Cell **67:** 1251–1266.
66. MANGELSDORF, D .J., U. BORGMEYER, R. A. HEYMAN, J. Y. ZHOU, E. S. ONG, A. E. ORO, A. KAKIZUKA & R. M. EVANS. 1992. Characterization of three RXR genes that mediate the action of 9- cis retinoic acid. Genes & Dev. **6:** 329–344.
67. SUGAWARA, A., P. M. YEN, D. S. DARLING & W. W. CHIN. 1993. Characterization and tissue expression of multiple triiodothyronine receptor-auxiliary proteins and their relationship to the retinoid X-receptors. Endocrinology **133**(3): 965–971.
68. GIGUERE, V. 1994. Retinoic acid receptors and cellular retinoid binding proteins: complex interplay in retinoid signaling. Endocr. Rev. **15**(1): 61–79.
69. BOGAZZI, F., L. D. HUDSON & V. M. NIKODEM. 1994. A novel heterodimerization partner for thyroid hormone receptor. Peroxisome proliferator-activated receptor. J. Biol. Chem. **269**(16): 11683–11686.
70. GLASS, C. K., S. M. LIPKIN, O. V. DEVARY & M. G. ROSENFELD. 1989. Positive and negative regulation of gene transcription by a retinoic acid-thyroid hormone receptor heterodimer. Cell **59:** 697–708.
71. SCHRADER, M., K. M. MULLER, S. NAYERI, J. P. KAHLEN & C. CARLBERG. 1994. Vitamin D3-thyroid hormone receptor heterodimer polarity directs ligand sensitivity of transactivation [see comments]. Nature **370**(6488): 382–386.
72. SCHRADER, M. & C. CARLBERG. 1994. Thyroid hormone and retinoic acid receptors form heterodimers with retinoid X receptors on direct repeats, palindromes, and inverted palindromes. DNA Cell Biol. **13**(4): 333–341.
73. KUROKAWA, R., J. DIRENZO, M. BOEHM, J. SUGARMAN, B. GLOSS, M. G. RO-

SENFELD, R. A. HEYMAN & C. K. GLASS. 1994. Regulation of retinoid signalling by receptor polarity and allosteric control of ligand binding. Nature **371**(6497): 528–531.
74. SCHAUFELE, F., B. L. WEST & J. D. BAXTER. 1992. Synergistic activation of the rat growth hormone promoter by Pit-1 and the thyroid hormone receptor. Mol. Endocrinol. **6**: 656–665.
75. SCHAUFELE, F., J. A. CASSILL, B. L. WEST & T. REUDELHUBER. 1990. Resolution by diagonal gel mobility shift assays of multisubunit complexes binding to a functionally important element of the rat growth hormone gene promoter. J. Biol. Chem. **265**(24): 14592–14598.
76. BRIGGS, M. R., J. T. KADONAGA, S. P. BELL & R. TJIAN. 1986. Purification and biochemical characterization of the promoter-specific transcription factor, Sp1. Science **234**(4772): 47–52.
77. LANDSCHULZ, W. H., P. F. JOHNSON, E. Y. ADASHI, B. J. GRAVES & S. L. MCKNIGHT. 1988. Isolation of a recombinant copy of the gene encoding C/EBP. Genes & Dev. **2**(7): 786–800.
78. CATTINI, P. A., T. R. ANDERSON, J. D. BAXTER, P. MELLON & N. L. EBERHARDT. 1986. The human growth hormone gene is negatively regulated by triiodothyronine when transfected into rat pituitary tumor cells. J. Biol. Chem. **261**: 13367–13372.
79. CATTINI, P. A., M. KLASSEN & M. NACHTIGAL. 1988. Regulation of human chorionic somatomammotropin gene expression in rat pituitary tumour cells. Mol. Cell. Endocrinol. **60**: 217–224.
80. LEIDIG, F., A. R. SHEPARD, W. G. ZHANG, A. STELTER, P. A. CATTINI, J. D. BAXTER & N. L. EBERHARDT. 1992. Thyroid hormone responsiveness in human growth hormone-related genes. Possible correlation with receptor-induced DNA conformational changes. J. Biol. Chem. **267**: 913–921.
81. CATTINI, P. A. & N. L. EBERHARDT. 1987. Regulated expression of chimaeric genes containing the 5′-flanking regions of human growth hormone-related genes in transiently transfected rat anterior pituitary tumor cells. Nucleic Acids Res. **15**(3): 1297–1309.
82. ZHANG, W., R. L. BROOKS, D. W. SILVERSIDES, B. L. WEST, F. LEIDIG, J. D. BAXTER & N. L. EBERHARDT. 1992. Negative thyroid hormone control of human growth hormone gene expression is mediated by 3′-untranslated/3′-flanking DNA. J. Biol. Chem. **267**: 15056–15063.
83. BEATO, M. 1991. Transcriptional control by nuclear receptors. FASEB J. **5**(7): 2044–2051.
84. ZAWEL, L. & D. REINBERG. 1992. Advances in RNA polymerase II transcription. Curr. Opin. Cell Biol. **4**(3): 488–495.
85. FONDELL, J. D., A. L. ROY & R. G. ROEDER. 1993. Unliganded thyroid hormone receptor inhibits formation of a functional preinitiation complex: Implications for active repression. Genes & Dev. **7**: 1400–1410.
86. BANIAHMAD, A., I. HA, D. REINBERG, S. TSAI, M. J. TSAI & B. W. O'MALLEY. 1993. Interaction of human thyroid hormone receptor beta with transcription factor TFIIB may mediate target gene derepression and activation by thyroid hormone. Proc. Natl. Acad. Sci. USA **90**(19): 8832–8836.
87. KING, I. N., T. DE SOYZA, D. F. CATANZARO & T. N. LAVIN. 1993. Thyroid hormone receptor-induced bending of specific DNA sequences is modified by an accessory factor. J. Biol. Chem. **268**: 495–501.
88. LU, X. P., N. L. EBERHARDT & M. PFAHL. 1993. DNA bending by retinoid X receptor-containing retinoid and thyroid hormone receptor complexes. Mol. Cell. Biol. **13**(10): 6509–6519.

89. NARDULLI, A. M. & D. J. SHAPIRO. 1993. DNA bending by nuclear receptors. Receptor **3**(4): 247–255.
90. SILVA, E. S., H. ASTIER, U. THAKARE, H. L. SCHWARTZ & J. H. OPPENHEIMER. 1977. Partial purification of the triiodothyronine receptor from rat liver nuclei. Differences in the chromatographic mobility of occupied and unoccupied sites. J. Biol. Chem. **252**: 6799–6805.
91. ICHIKAWA, K., K. HASHIZUME, S. FURUTA, T. OSUMI, T. MIYAMOTO, K. YAMAUCHI, T. TAKEDA & T. YAMADA. 1990. Human c-erb A protein expressed in *Escherichia coli*: changes in hydrophobicity upon thyroid hormone binding. Mol. Cell. Endocrinol. **70**: 175–184.
92. ICHIKAWA, K. & L. J. DEGROOT. 1986. Separation of DNA binding domain from hormone and core histone binding domains by trypsin digestion of rat liver nuclear thyroid hormone receptor. J. Biol. Chem. **261**: 16540–16545.
93. BHAT, M. K., C. PARKISON, P. MCPHIE, C. M. LIANG & S. Y. CHENG. 1993. Conformational changes of human beta 1 thyroid hormone receptor induced by binding of 3,3′,5-triiodo-L-thyronine. Biochem. Biophys. Res. Commun. **195**(1): 385–392.
94. TONEY, J. H., L. WU, A. E. SUMMERFIELD, G. SANYAL, B. M. FORMAN, J. ZHU & H. H. SAMUELS. 1993. Conformational changes in chicken thyroid hormone receptor alpha 1 induced by binding to ligand or to DNA. Biochemistry **32**: 2–6.
95. BROWNER, M. F., E. B. FAUMAN & R. J. FLETTERICK. 1992. Tracking conformational states in allosteric transitions of phosphorylase. Biochemistry **31**(46): 11297–11304.
96. YEN, P. M., D. S. DARLING, R. L. CARTER, M. FORGIONE, P. K. UMEDA & W. W. CHIN. 1992. Triiodothyronine (T3) decreases binding to DNA by T3-receptor homodimers but not receptor-auxiliary protein heterodimers. J. Biol. Chem. **267**: 3565–3568.
97. ANDERSSON, M. L., K. NORDSTROM, S. DEMCZUK, M. HARBERS & B. VENNSTROM. 1992. Thyroid hormone alters the DNA binding properties of chicken thyroid hormone receptors alpha and beta. Nucleic Acids Res. **20**: 4803–4810.
98. MIYAMOTO, T., S. SUZUKI & L. J. DEGROOT. 1993. High affinity and specificity of dimeric binding of thyroid hormone receptors to DNA and their ligand-dependent dissociation. Mol. Endocrinol. **7**: 224–231.
99. MCGRATH, M. E., R. L. WAGNER, J. W. APRILETTI, B. L. WEST, V. RAMALINGAM, J. D. BAXTER & R. J. FLETTERICK. 1994. Preliminary crystallographic studies of the ligand-binding domain of the thyroid hormone receptor complexed with triiodothyronine. J. Mol. Biol. **237**(2): 236–239.

Human and Mouse T-Cell Receptor Loci: Genomics, Evolution, Diversity, and Serendipity[a]

LEROY HOOD,[b] LEE ROWEN,[b]
AND BEN F. KOOP[c]

[b] Department of Molecular Biotechnology
University of Washington, FJ-20
Seattle, Washington 98195

[c] University of Victoria
Department of Biology
Center for Environmental Health
9865 West Saanich Road
Sidney, British Columbia V8L 3S1, Canada

Jim Watson has made singular contributions to the Human Genome Project. As the program's first NIH Director, Jim lent it credibility, won over critics, and, most importantly, persuaded outstanding young people to gamble their careers on genomics, and then listened to how they felt the project should be done. With typical intuitive brilliance, Jim pushed the genome program on the ethical, legal, and social issues, attacking head-on one of its most potentially divisive issues. Moreover, with great success, Jim almost single-handedly directed a program, during its initial years, which will fundamentally change science and medicine as we move into the twenty-first century. Launching the Genome Project may be one of Jim's most enduring contributions.

The major challenge in contemporary biology and medicine lies in deciphering and manipulating biological information. Biological information falls into three distinct categories. The biological information of our chromosomes is displayed in linear arrays comprised of a four-letter DNA alphabet. Embedded in these digital or one-dimensional chromosomal strings is the information for perhaps the most fascinating of all biological processes—development. In humans, most of the 100,000 or so genes are expressed as proteins, linear strings composed of a 20-letter alphabet which fold to generate the

---

[a] This work was supported by the National Institutes of Health, the Department of Energy, and the National Science Foundation.

three-dimensional molecular machines of life. Variations among humans in this one-dimensional information lead to polymorphic traits that provide unique individuality and predispositions to human disease. One of the major problems of modern biology is the "protein folding" problem—how the linear order of amino acid subunits directs the folding of proteins into precise three-dimensional shapes. Thus, the second type of biological information, the three-dimensional shapes of proteins, provides insight into how proteins operate as biological machines. The third type of biological information arises from the fact that most of the fascinating properties of higher organisms are composed of complex systems or networks of molecules and/or cells. From biological complexity come the fascinating emergent properties which lie at the core of human growth, development, and behavior. For example, the properties of intelligence, memory, learning, and emotional stability arise from the complex network of 10^{11} neurons with 10^{18} connections. Understanding all of the properties of a single neuron will not provide any deep insights into these emergent properties, as they are a consequence of the network as a whole. A thesis that will emerge from this paper is that defining the sequences of one-dimensional information in our chromosomes will provide fundamental tools for deciphering the three-dimensional information of proteins and the four-dimensional information of complex biological systems and networks.

DECIPHERING CHROMOSOMAL LANGUAGES

Chromosomes are organelles that deal with biological information. Chromosomes have the capacity to store, duplicate, express, and evolve information. As a consequence, they encode a variety of different types of information, the so-called "languages" of the chromosomes. In a sense, the major objective of the Human Genome Project is to generate the sequences of all human (and model organism) chromosomes so that these languages may be deduced and their relevances to biological systems deciphered.

Two powerful tools have emerged for translating the languages of the chromosomes. First, related sets of biological information are encoded in tandemly arrayed multigene families (e.g., immune receptors, cell adhesion receptors, growth receptors, growth hormones, and developmental proteins).[1] Not only do multigene families represent an efficient way to extract information from our chromosomes, but their very organization reflects the evolutionary mechanisms that led to their creation. Second, biological information is conserved across species. Accordingly, an important aspect of deciphering the languages of the chromosomes will be comparative species analyses of interesting chromosomal regions to identify the conserved sequences that presumably encode biological information.[2] These conserved regions can represent genes or regulatory sequences, or they can be associated with the variety of functions that chromosomes must carry out as information organelles. Comparative analyses of chromosomes from different species will, in time, allow us to identify some of the long-range patterns that must be intrinsic to certain chromosomal languages. For example, chromosomes have a striking ability to undergo compression. In each human cell, more than two meters of DNA can be compactly folded into 24 small chromosomal packages that are smaller than the dot of a pencil. This compaction process arises as a consequence of coiling and supercoiling, which is directed by the interaction of proteins with the DNA. These interactions will probably be reflected as interesting long-range constraints in DNA sequence that will, in time, be deciphered. The comparative analyses of large chromosomal regions will also lead us into fundamental insights regarding the evolutionary histories of organisms and the divergences of their multigene families.

SEQUENCE ANALYSIS OF THE T-CELL RECEPTOR FAMILIES

We have chosen to compare the DNA sequences of the T-cell receptor families of human and mouse.[3-6] These organisms diverged during the mammalian radiation 80 million years ago and, in general, have relatively highly conserved homologous gene regions (approximately 70%) and quite divergent intergenic regions.[7] Thus, mouse/human sequence comparisons should allow us to identify conserved chromosomal regions that reflect biological information.

The T-cell receptors of mouse and humans are encoded in three distinct multigene families—α/δ, β, and γ (FIG. 1).[8,9] There are two major types of heterodimeric T-cell receptors—α/β and γ/δ. The T-cell receptors contain a variable domain that recognizes foreign antigens and a constant domain that fixes the T-cell receptor to the immune cell. The variable domains are encoded by a multiplicity of gene segments (variable [V], joining [J], and for the β and δ chains, diversity [D]). These gene segments rearrange and join together in contiguous apposition during T-cell development to encode a variable gene. RNA splicing joins the V and C genes at the mRNA level before translation into polypeptides. T-cell receptor loci range in size from 0.2 to approximately 1 megabase.[8]

The basic strategies we have employed to obtain and analyze DNA are diagrammed in FIGURE 2. Large-insert DNA clones (BAC [bacterial artificial chromosomes] or YAC [yeast artificial chromosomes]) are obtained to cover the locus. The YAC or BAC clones are subcloned into smaller cosmid clones, which are then mapped in a dense array across the large-insert clone. Then a minimum tiling path (e.g., clones overlapping minimally) is selected across the cosmid clones and each of the cosmid clones on this path is converted

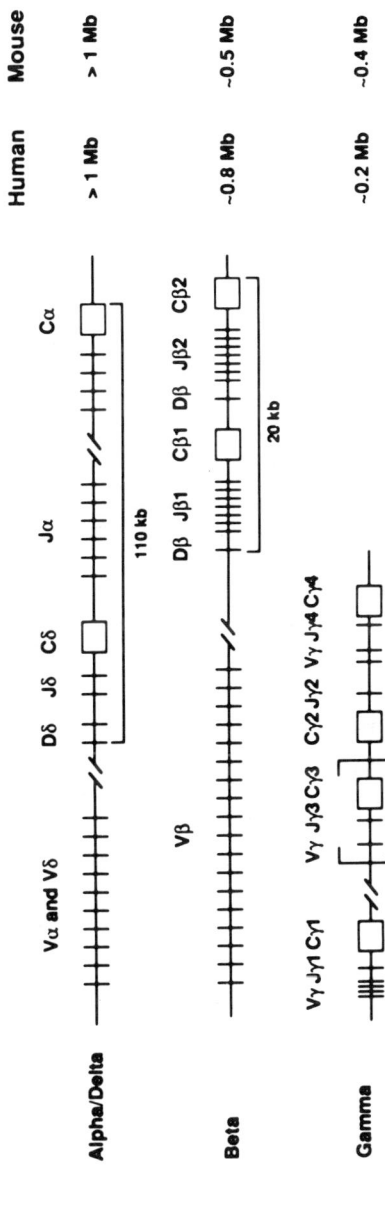

FIGURE 1. Schematic diagram of the three TCR loci of humans and mice. The δ gene family is entirely contained within the α locus. The approximate lengths of these loci are indicated. *Vertical lines* and *open boxes* represent gene segment and gene coding regions, respectively. (From Hood *et al.*[17] Reproduced by permission.)

Gene Family Sequencing Strategy - Random or Shotgun

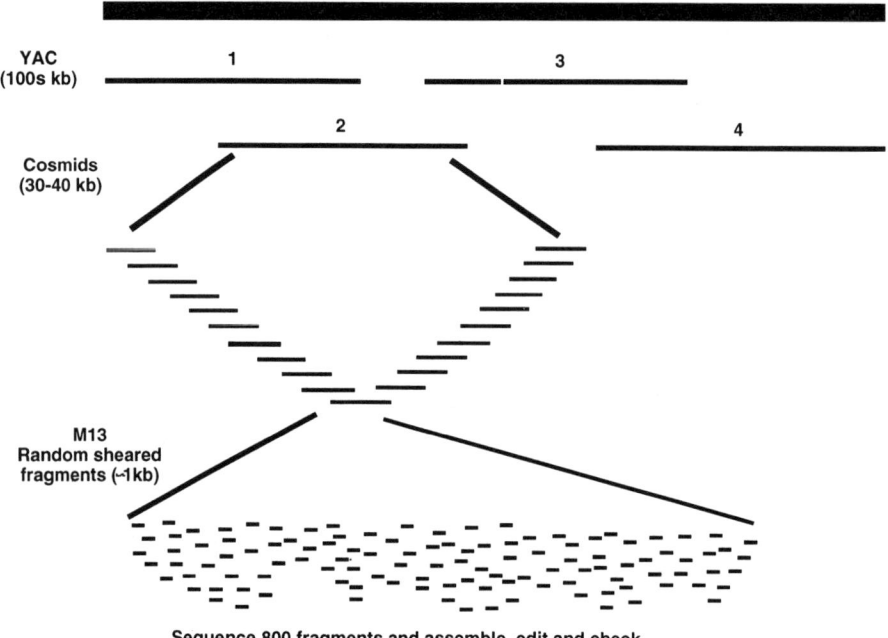

FIGURE 2. A schematic diagram of the mapping and sequencing strategies employed for the T-cell receptor loci. YAC indicates yeast artificial chromosome. The approximate sizes of the inserts are indicated.

by random shearing to small 1kb fragments which are subclone into the phage M13.[10] An automated fluorescent DNA sequencer is then employed to sequence approximately 800 of these clones for 450 base pairs, covering the entire cosmid sequence on average eight times (8-fold coverage). A computer is used to assemble the smaller M13 DNA strings into the larger cosmid string. With an 8-fold coverage of sequence across the cosmid insert, about 90% of the time the cosmid clones can be assembled without gaps (L. Rowen, unpublished observation). This approach provides overlapping sequences in adjacent cosmid clones. If the cosmid clones come from the same haplotype (maternal or paternal), the differences in the overlapping regions reflect the error rate, which is roughly 1 in 5,000 base pairs. If the overlapping cosmid clones come from different haplotypes, the differences reflect the rate of polymorphism, which averages 1 in 700 base pairs.

We have analyzed in mouse and human two regions from the three T-cell receptor loci (FIG. 1). About 100 kb has been compared in the $C\delta C\alpha$ region.[3-5] The human β locus has been completely sequenced (685 Kb) and compared against the partially analyzed (400 Kb) mouse β locus (L. Rowen, unpublished).

CδCα REGION

The CδCα region in human and mouse was of considerable interest because the number of Jα gene segments encoded within this region was unknown at the initiation of this project. This project illustrates the power of interspecies comparative analyses.

The CδCα is Highly Conserved

A dot matrix analysis of the mouse and human sequences in this region led to a surprising result (FIG. 3). The entire 100-kb region shows a striking degree of similarity. Indeed, when one analyzes other large stretches of genomic sequence, no such striking conservation is seen.[5] After appropriate alignment of the sequences to take into account the insertion of species-specific repetitive elements, the overall sequence homology is 71%.[4] The level of sequence homology for the coding regions is not strikingly different from that of the non-coding regions. Moreover, the coding regions constitute just 5.6% of this 100-kb stretch. Accordingly, this conservation of DNA sequence across the large region can have two possible explanations. First, it could be stochastic or inadvertent, consequence of the fact that by chance certain large regions of our chromosomes may be highly conserved. We think this explanation is extremely unlikely because the probability of such large-scale conservation would appear to be very low. Second, this conservation may reflect some interesting new type of biological information. We have no idea what this biological information might be and, indeed, whether it is specific to the T-cell receptor α/δ locus or, alternatively, whether it reflects some general chromosomal feature that requires conservation. Two important points emerge from these comparisons. A species comparison did identify a highly conserved and extended region, presumably reflecting some type of biological information. In addition, one must be cautious with the glib generalization that all DNA, apart from the genes, is junk. Indeed, in this particular case, it would be hard to argue that the 95% of the non-coding sequence is junk.

Identifying the T-Cell Receptor Elements

The major challenge for gene identification in this region was the delineation of the Jα gene segments. The Vα gene segment, as well as the Cα and Cδ genes, had been previously sequenced. To identify the Jα gene segments, we constructed an artificial linear sequence of all the known Jα gene segments defined by cDNA analyses for human and mouse. This sequence was analyzed against the mouse CδCα region by dot matrix analysis and 50 Jα gene segments were identified.[3] The new Jα gene segments from this analysis were

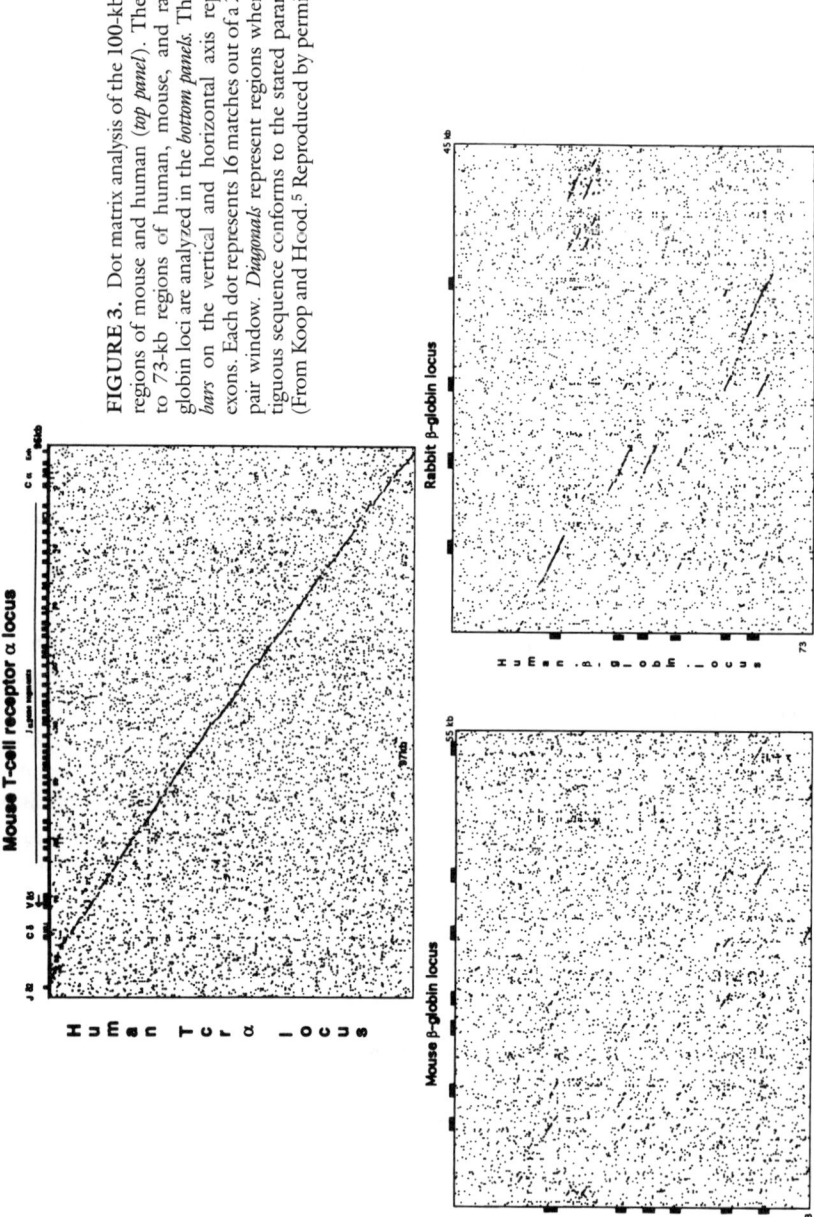

FIGURE 3. Dot matrix analysis of the 100-kb Cδ&Cα regions of mouse and human (*top panel*). The 45-kb to 73-kb regions of human, mouse, and rabbit β globin loci are analyzed in the *bottom panels*. The *raised bars* on the vertical and horizontal axis represent exons. Each dot represents 16 matches out of a 20 base pair window. *Diagonals* represent regions where contiguous sequence conforms to the stated parameters. (From Koop and Hood.[5] Reproduced by permission.)

added to the linear cDNA Jα gene sequence and these were compared by dot matrix analysis against the human Jα region. Sixty-one human Jα gene segments were identified.[4] We then searched for the 11 additional human Jα gene segments in homologous regions of the mouse CδCα sequence and found in each case orthologous counterparts.[5] All 11 additional mouse Jα elements were, in fact, pseudogenes. Thus, this cross-species comparison of the Jα elements gave us the capacity to define the 61 Jα gene segments present in both the mouse and the human α T-cell receptor loci.

An analysis of the 61 elements that mediate DNA rearrangements or RNA splicing for each Jα gene segment was carried out. Characteristic motifs were identified and, curiously enough, the T-cell receptor DNA rearranging elements had a single base that distinguished them from their antibody gene segment counterparts.[5,11,12] There is a possibility that this variation could encode a subtle difference in the way the gene segment rearrangement machinery operates in B cells and T cells. This possibility is now being explored.

Organizational Conservation of Orthologs

Each of the 61 mouse Jα gene segments has an orthologous counterpart in the 61 human Jα sequences.[5] On average, the orthologs are 71% similar to one another, whereas they exhibit a similarity to other Jα gene segments within the same species of less than 50%. The order of the orthologs is perfectly conserved in the two species (FIG. 4). This remarkable conservation of organization is a consequence of the overall 71% homology that has been observed across the comparative analysis of the CδCα region. Possible explanations for this striking conservation have been discussed earlier.

cDNA Analyses Fail to Identify Many Jα Elements

Approximately 200 α cDNAs have been sequenced.[5] From these cDNA clones, only 34 and 35 Jα gene segments, respectively, were defined initially for mouse and human α/δ loci. Indeed, about ten of those Jα gene segments emerged from rare T-cell clones that had been selected because of interesting biological functions. Accordingly, if one looks at the normal pool of T-cell messenger RNA by cDNA analysis, less than half of the Jα gene segments could be easily delineated. The important point to stress is that germline or chromosomal analyses of multigene families is necessary if one is to delineate in an efficient and unambiguous manner all of the members of the family.

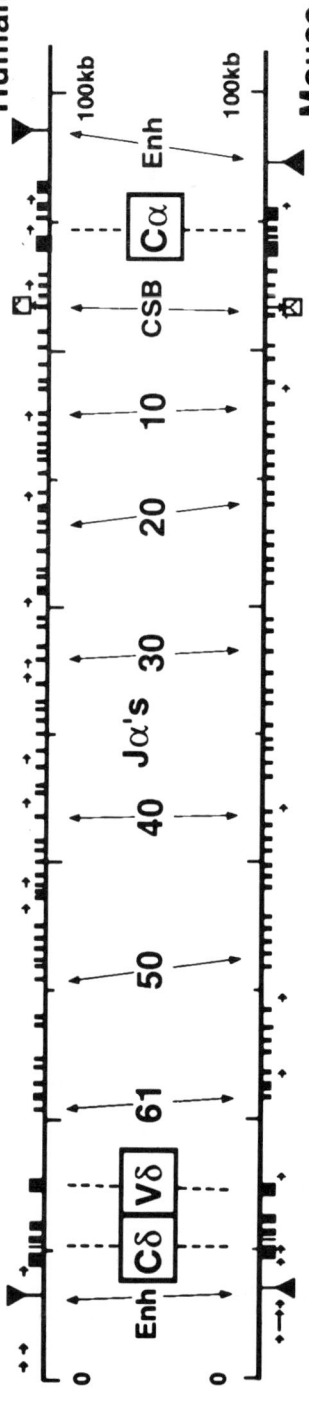

FIGURE 4. Schematic diagram of the Cδ&Cα regions of human and mouse. → = repeats; ▲ = enhancer sequences; CSB = conserved sequence block; ■ = gene segments. The conserved sequence block represents a highly conserved non-coding region in both species, possibly a new regulatory sequence.[3] (From Koop and Hood.[5] Reproduced by permission.)

Pseudo Jα Gene Segments

A question arises as to how many of these Jα gene segments from human and mouse are pseudogenes. Pseudo Jα gene segments might arise as a consequence of five distinct lesions: (1) the DNA rearrangement signal is nonfunctional; (2) the RNA splicing signal is defective; (3) a chain termination codon may occur in the Jα coding region; (4) a reading frame shift may occur in the Jα coding region; and (5) a non-functional amino acid substitution may occur in the Jα region. Obviously, any Jα gene segment that is identified as a cDNA is functional with regard to DNA rearrangement and RNA splicing. Analysis of the 61 human and 61 mouse Jα gene segments by these criteria leads to the conclusion that approximately one-third of them appear to be pseudogenes.[3,4]

Identification of Non-T-Cell Receptor Elements

We used three approaches to attempt to identify non-T-cell receptor gene elements.[5] First, we carried out a similarity analysis of the CδCα region against the database of DNA sequences. This analysis identified various T-cell receptor elements, but failed to identify any non-T-cell receptor genes. Second, we analyzed the region with a gene-finding program, GRAIL, that identifies exons by virtue of codon expression asymmetries and several special boundary conditions which delineate coding and non-coding sequences.[13] The GRAIL program identified the major T-cell receptor elements, apart from the Jα gene segments which were too small for detection, and in each species identified six to seven additional potential candidate exons. Strikingly, none of these candidate exons were conserved across the two species, rendering them unlikely as gene candidates. Third, we examined all open reading frames larger than 300 base pairs in the CδCα region of those species. In the mouse, there are 69 such open reading frames, and in the human, 58. Apart from the previously defined T-cell receptor elements, none of these open reading frames were in orthologous alignment and, accordingly, none were viewed as likely exon candidates. Thus, the species comparison has provided a powerful tool for delineating the nature of coding regions in sequences that can be analyzed comparatively.

Regulatory Elements

Five regulatory elements have been defined across the CδCα region (TABLE 1).[5] Two of these elements, the 5′ Cδ and the 3′ α enhancers, are highly conserved. The two silencer elements (3′ to the Cα gene)[14] are not conserved across species. The 5′ Cα enhancer has been deleted in the mouse α/δ locus.

TABLE 1. Putative Regulatory Sequences in the CδCα Regions of Human and Mouse

Regulatory Sequences	Degree of Conservation
3' α enhancer[a]	84% (250 bp); 95% (50 bp)
5' δ enhancer[a]	70% (340 bp); 80% (30 bp)
Silencer I	Not conserved
Silencer II	Not conserved
5' α enhancer	Deleted in mouse

[a] Identifiable in both species.

Thus, only two of the five regulatory elements have been highly conserved. Accordingly, the 5' Cα enhancer is either mouse-specific or not a real control element. There are several explanations for the failure of the silencer elements to be conserved. First, the DNA motifs' binding protein silencing factors may be highly degenerate and not recognized by our similarity analyses. Second, the silencers may have evolved in a species-specific manner. Finally, the silencers may not be true regulatory elements, but rather reflect complex *in vitro* artifacts. Two important points emerge. Comparative species analyses of control regions will be useful in delineating those which require further analyses (e.g., species-specific control elements appear unlikely). Alternatively, certain candidate regulatory sequences can be identified through cross-species conservation of orthologous sequences as potential control elements. One such example is the highly conserved non-coding sequence block lying next to the 5' side of the Cα gene.[5]

Tools That Emerge From a Complete Sequence Analysis of the α Elements

A complete knowledge of all the Jα gene segments and their flanking sequences has permitted us to design unique PCR primers for each of the Jα gene segments in human and mouse.[3,4] Thus, each Jα exon can be interrogated by PCR in an individual manner (FIG. 5). This raises striking possibilities for exploring the developmental expression of these gene elements, as well as studying the influence of immunization and tolerization on the T-cell receptor repertoire. Thus, the ability to interrogate individual members of a multigene family as a consequence of knowing the complete sequence of the multigene family generates powerful new possibilities for molecular exploration of these important metazoan chromosomal features.

A second powerful tool emerges from the distribution of simple repeat sequences, dinucleotides, trinucleotides, or tetranucleotides, across the CδCα region. Simple repeat sequences are generally highly polymorphic within a spe-

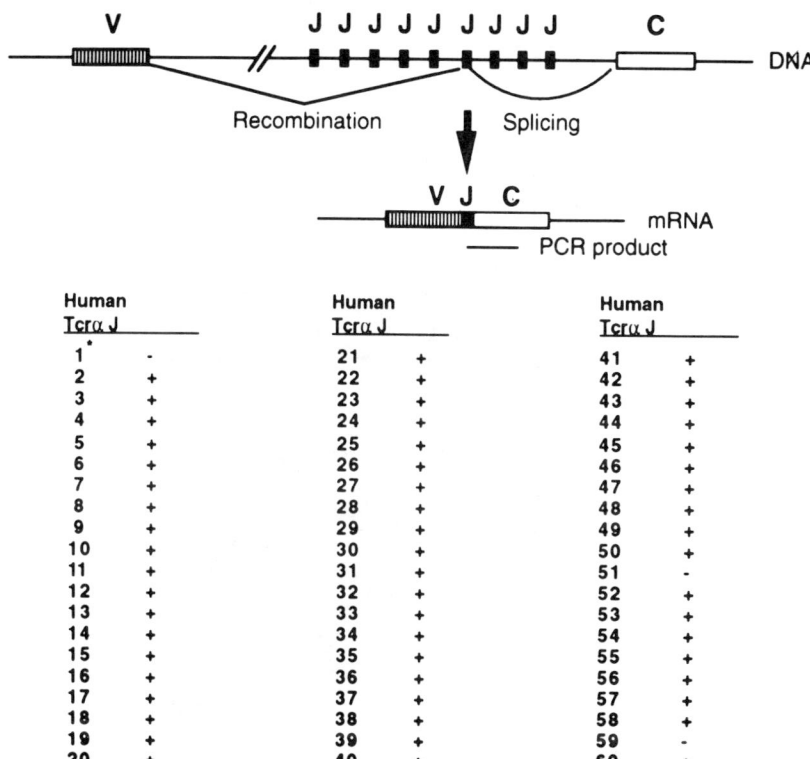

FIGURE 5. Schematic figure indicating the PCR assay for human Jα/Cδ mRNA. The success of the assay is indicated for each Jα gene segment.[17]

cies because of the propensity of the DNA polymerase to slip during the copying of these sequences. Indeed, we have selected three simple repeats from the mouse CδCα region, made PCR primers on either side of these repeats, and interrogated their polymorphisms in a series of inbred and wild strains of mice (FIG. 6). These simple repeat or microsatellite markers demonstrate a striking degree of polymorphism. Thus, genetic markers can be placed virtually at will across the entire span of a complex multigenic family. In some multigene families, it is important to identify dense arrays of genetic markers across the gene family because there may be a multiplicity of apparent hot spots of recombination within the family.[15] Thus, a multiplicity of genetic markers are necessary to ensure that each adjacent marker is at least in partial linkage disequilibrium with its neighboring counterparts. This coverage is important in attempts to assess whether individual members of a multigene family exhibit polymorphisms that predispose to disease.

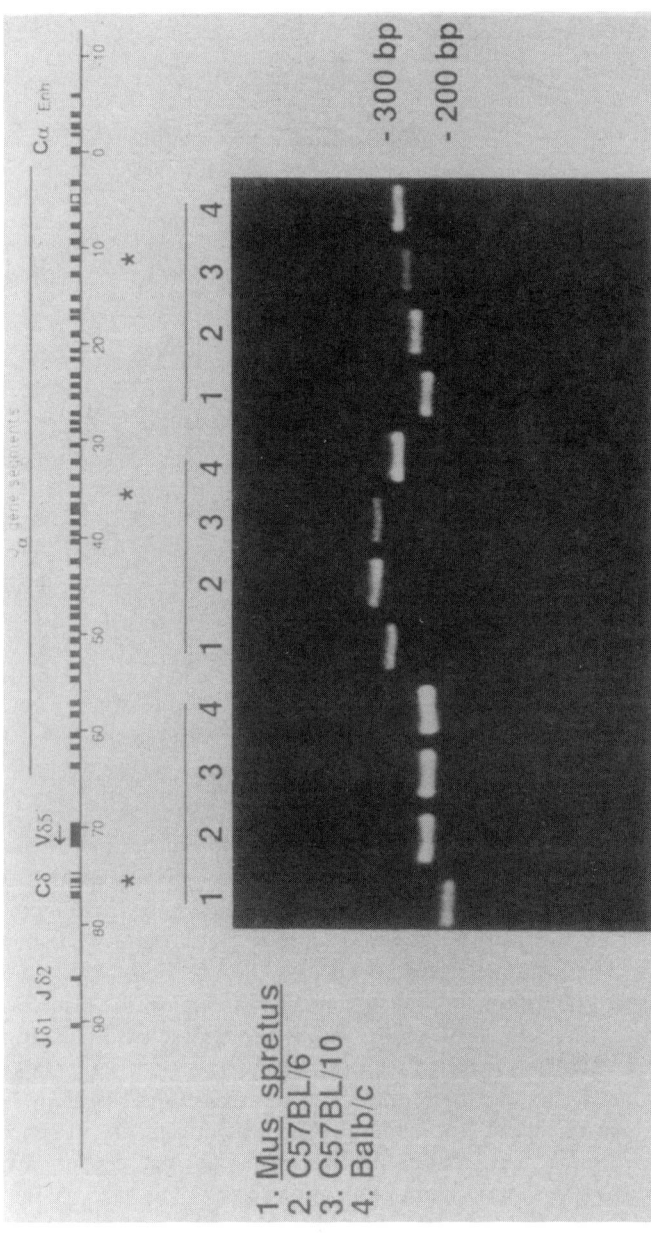

FIGURE 6. Location and analysis of simple repeat polymorphisms in the mouse CδCα region. The *top line* indicates the position of coding regions within the 95-kb of mouse sequence. The *asterisk* marks the location of the repeats tested for polymorphism. The level of polymorphism in each of the three repeats is shown below the physical map. The specific mouse strains used are indicated by the numbers above the gel, and the size of each of the bands can be determined from size markers positioned alongside the gel. (From Koop *et al.*[3] Reproduced by permission.)

THE HUMAN β T-CELL RECEPTOR FAMILY

The human β locus extends over about 685 kilobases of DNA (Lee Rowen, unpublished observations). It contains 67 Vβ, 2 Dβ, 14 Jβ, and 2 Cβ elements. About one-third of these T-cell receptor β elements are pseudogenes.

Homology Units

A careful analysis of this locus demonstrates that it comprises five major types of homology units (FIG. 7). A homology unit is defined as a section of a multigene family that has been duplicated two or more times. These homology units may range in length from 1 to 20 or more kilobases (kb = 1,000 base pairs). These duplications may occur in tandem arrays, or they may be distributed throughout the multigene family. Three different sets of V gene segments, including two or three elements, have been duplicated in both a tandem and a distributed manner. The D, J, and C elements constitute a fourth homology unit that has been duplicated in a tandem manner. The fifth homology unit is 10 kb in length and has been duplicated 5-fold in a tandem array that lies between the 3'-most variable gene segment and the 5'-most D gene segment. By analyzing the similarities of the individual homology units, one can begin to reconstruct the molecular archeology of this locus.

Trypsinogen Genes

The five 10 kb homology units in the Vβ Dβ region occupy most of this region. Embedded within each of these tandem repeats is a 4 kb gene that encodes a trypsinogen. These homology units are 90–94% homologous to one another and constituted an enormous technical challenge for large-scale DNA sequencing (L. Rowen, unpublished observation). Three of these genes appear to be functional, and two are pseudogenes. Two of the three functional trypsinogen genes are expressed in the pancreas. Curiously enough, there is a third trypsinogen messenger RNA expressed in the pancreas that is not represented by any of the three functional genes.

A Translocation of a Portion of the β Family from Chromosome 7 to Chromosome 9

We carried out an *in situ* hybridization analysis using a cosmid clone with the trypsinogen gene to examine sites in the human genome where the third functional trypsinogen gene might map (B. Trask, unpublished observation).

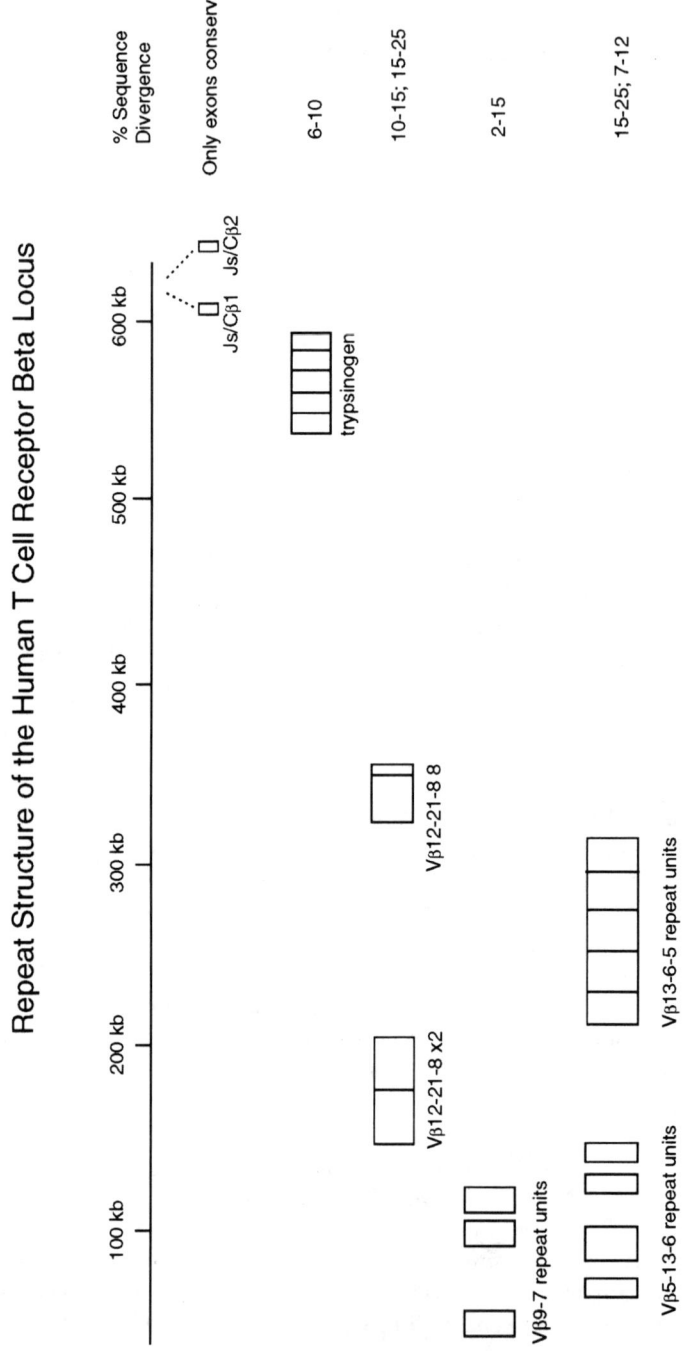

FIGURE 7. A schematic illustration of the five types of homology units present in the human β locus (see text).

Surprisingly, one or more new trypsinogen genes are located on chromosome 9–a locus that is distinct from the β T-cell receptor family located on chromosome 7. We have cloned and sequenced the trypsinogen gene from chromosome 9 and it turns out to encode the third messenger RNA expressed in the pancreas. Our results have enlarged the scope of the translocation of the β locus reported while these studies were in progress.[16]

Thus, three striking surprises emerged from the study of homology units in the human β locus. First, there appears to be an apparently inadvertent association of trypsinogen genes in the midst of the T-cell receptor β locus. One wonders whether this association is trivial or significant. Second, the translocation of a portion of the β locus from chromosome 7 to chromosome 9 gives an indication of how multigene families may be born. Six or seven Vβ gene segments were also duplicated and then translocated in this chromosomal rearrangement event (L. Rowen, unpublished observations).[16] Finally, the homology units enumerated here comprise almost 50% of the β T-cell receptor locus. This presents a striking technical challenge for large-scale DNA sequencing.

EVOLUTIONARY COMPARISONS OF THE T-CELL RECEPTOR β LOCI

The evolutionary analysis of animals which diverged at differing times from the human evolutionary line can provide useful insights into the nature of the evolutionary events that have shaped the T-cell receptor β family.

Comparison with the Mouse β Locus

As mentioned earlier, the mouse diverged from the human evolutionary line about 80 billion years ago. We have a partial analysis of the mouse β locus (L. Rowen, unpublished data). There are approximately 35 mouse Vβ gene segments in contrast to 67 human Vβ gene segments (FIG. 8). The human Vβ gene segments have all expanded as a consequence of the duplication of the homology units for variable gene segments described above. Thus, the Vβ information content in this family has nearly doubled in the human evolutionary line.

Comparison with Primate β Loci

To examine in more detail the specific events that occur within this gene family, we examined in four primates the Vβ 8.1 and Vβ 8.2 gene segments

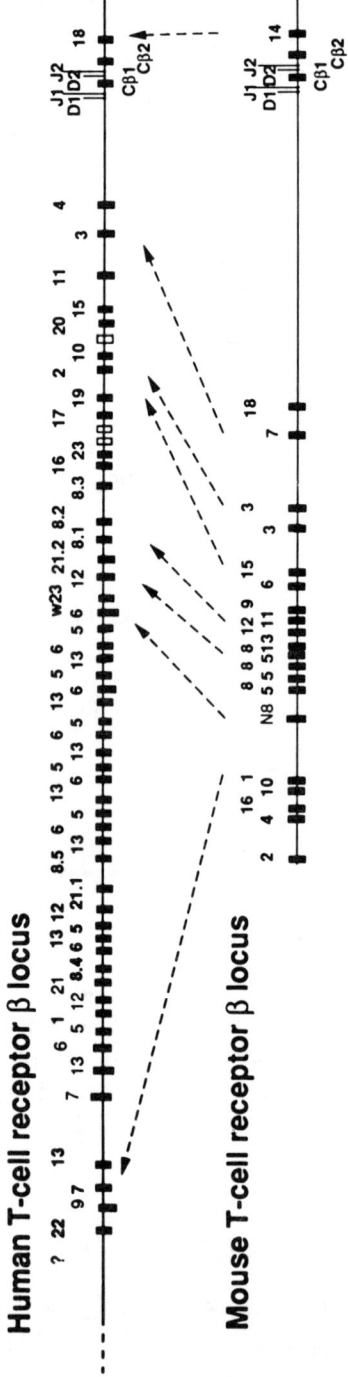

FIGURE 8. Schematic diagram illustrating homology between the human and mouse β gene families. The lines connecting these gene families indicate orthologous Vβ sequences. The human β family has undergone a 2-fold gene segment expansion as compared to that of its mouse counterpart. (From Hood *et al.*[17] Reproduced by permission.)

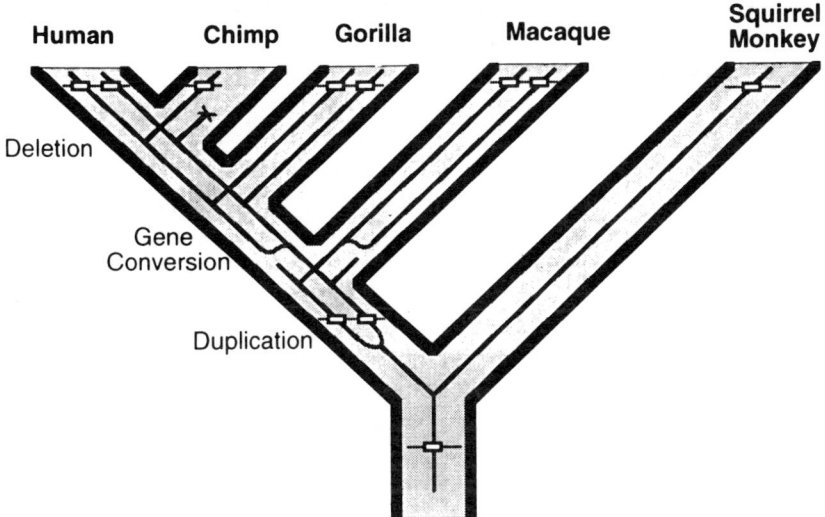

FIGURE 9. Genealogical tree for the 3' end of the Vβ 8.1–8.2 segment in five primates. The gene phylogenetic tree is presented within the constraints of a well-documented species phylogeny. Deletion, duplication, and gene conversion events are required to make both the species and gene phylogenies consistent with each other. (From W. Funkhauser *et al.* Manuscript, in preparation.)

(W. Funkhauser and B. F. Koop, unpublished data). The two are linked together in a 2.3-kb fragment. PCR primers were prepared and used to amplify, isolate, and sequence the corresponding regions from the chimp, gorilla, rhesus monkey, and squirrel monkey. The time of divergence of these animals from the human evolutionary line ranges between 4 million and 25 million years. An analysis of the nucleotide substitutions in these genes leads to a relatively conventional genealogic tree for primates. However, if one places on the genealogic tree the major genetic events that have shaped the family in the last 25 million years, that is, gene duplication, gene deletion, and gene conversion, quite a dynamic picture of gene expansion, contraction, and conversion emerges (FIG. 9). After the divergence of the evolutionary lines that led to squirrel monkey, on the one hand, and the other primates, on the other hand, a gene duplication occurred. After this time, two gene conversion events occurred—one in the evolutionary branch to the rhesus monkey, and a second in the branch leading to humans, chimps, and the gorilla. Subsequently, a gene-deletion event occurred in the chimp branch. Thus, the picture that emerges is one of enormous dynamism with gene expansion and contraction and gene conversion playing major events in altering the information in human β T-cell receptor gene families. Remember that we have examined only two of the 67 Vβ gene segments in this analysis.

Comparison of the Chicken β Locus

The presence of trypsinogen genes in the midst of the T-cell receptor β family has raised a question of whether this curious association is physiologically important. One approach to examining whether the association of apparently unrelated genes is important is to determine whether they are associated in species that diverged in ancient times. In this regard, we have isolated chicken trypsinogen genes by PCR cloning through highly conserved gene regions. The β T-cell receptor genes and trypsinogen genes are closely linked in the chicken evolutionary line, which diverged from its human counterpart 300 million to 350 million years ago (K. Wang, unpublished data). Accordingly, this linkage relationship has been conserved across much of metazoan evolution. It raises a fascinating question as to whether this relationship is highly conserved as a consequence of functional and/or regulatory constraints.

GENOMICS AND BIOLOGY

The tools of genomics are rapidly being used to enumerate the 100,000 or so human genes. This presents a conundrum for biology, namely, how does one go from gene structure to function? In the past, biologists generally started with a functional assay and went through the protein to the gene. Accordingly, function was generally linked to gene structure. In the future, biologists will increasingly be forced to go in the opposite direction because of the powerful tools of genomics—large-scale genetic mapping and large-scale DNA sequencing. Gene knock-outs, transgenic animals, and homologous recombination all present opportunities to study the functions of unknown genes. However, the insights of genomics will increasingly provide important tools for discerning probable functions from gene structure.

Motifs

One of the most important advances that will occur over the next 15 years, as the Human Genome Project is completed and the 100,000 or so human genes are delineated, is to fill in the lexicon of motifs that are the building block components of genes. Motifs are conserved, but degenerate patterns of sequence ranging in size from 5–10 to several hundred nucleotides (e.g., zinc fingers [~90 nucleotides]) that encode recurring protein structural motifs and often have functional implications. The computational challenge in identifying motifs is to determine the degenerate nature of the sequences they employ (TABLE 2). However, with more powerful computational tools, we should be able to delineate the hundreds of distinct motifs. Once delineated,

TABLE 2. A Hypothetical Biological Motif

Hy-Ac-F/Y-X-X-X-S/T-X-X-X-Hy
X = any amino acid
Hy = hydrophobic residue
Ac = acidic residue
F, Y, S, and T = phenylalanine, tyrosine, serine and threonine, respectively
F/Y = either serine or threonine

we can check these motifs against the library of known three-dimensional structures to identify their structural counterparts, and even their functional implications. In time we should be able to take virtually any gene and break it down into its component motifs. These should place significant constraints on both the structure and function of the corresponding protein. One of the unsolved problems in contemporary biology is the protein-folding problem. A variety of approaches have been employed to attack this problem, including theoretical calculation of energy or potential minimization, *in vitro* mutagenesis, and the computational analysis of known three-dimensional structures. In the future, the motifs derived from genes identified by genome programs will provide fundamental folding constraints and these, in conjunction with the other approaches, should help solve the protein folding problem, perhaps along the same timescale as the completion of the genome project – about 15 years.

The Regulatory Code

A second challenging problem relates to the complex systems and networks of genes that are interconnected to complex biological functions. Curiously enough, the genome project may contribute fundamental insights into this problem as well. As the structure of 100,000 or so genes is determined, the regions that regulate these genes (such as promoters, silencers, or enhancers) will be delineated and their interactions with protein factors will be studied. In time, the "regulatory" code will be deciphered through an analysis of the regulatory sequences and the proteins that interact with them. The regulatory code is concerned with three major parameters – the amplitude of gene expression, as well as *where* in space (tissue) and *when* in time (development). Once these parameters can be deciphered with regard to each gene, then the interconnections of gene products in cellular compartments and complex cell systems can begin to be deciphered. There will be a time when one merely types the word *liver* into the computer to obtain a display of all the genes that are expressed in the liver. By virtue of having deciphered the regulatory language for each of the human genes, we will be able to determine which genes operate in related biological tissues or networks. The difficult challenge will be to delineate their interrelated functions—this is the realm of biology, not genomics.

Evolution

Large-scale sequence analyses will provide fundamental insights into the nature of molecular evolution, both in mechanistic terms, and by delineating the molecular archeology of gene families and chromosomal regions. The power of a multiplicity of evolutionary analyses is evident in the studies described above for the T-cell receptor gene families.

Languages of the Chromosomes and Theoretical Biology

Chromosomes, as noted earlier, are organelles for information handling. They carry out a variety of discrete functions: They (1) encode genes; (2) contain regulatory sequences; (3) replicate DNA; (4) segregate DNA during mitosis or miosis (centromeres); (5) compact and decompact DNA; (6) mutate, duplicate, delete, and convert DNA (the mechanisms for evolution); and (7) maintain chromosome lengths (telomeres). Each of these functions places informational constraints upon our chromosomes. In a sense, these constraints represent distinct and sometimes overlapping languages. Deciphering some of these languages, and others that we do not have the imagination to envision, will require increasingly powerful computational and mathematical tools. Indeed, it is our belief that a powerful new field of computational biology will emerge over the next ten years populated by computer scientists and applied mathematicians: They will use their powerful tools to decipher the one-dimensional languages of our chromosomes, the three-dimensional information of our proteins, and the biological complexity arising from complex biological systems and networks. Indeed, computational biology will create a new field of theoretical biology, which will formulate models and predictions to be tested by experimental biologists. The roles of theoretical and experimental biology will parallel those of their physics counterparts. The challenge is to determine how the language barriers that separate different disciplines, in this case biology, computer science, and mathematics, can be breached so that scientists can collaborate in a truly effective manner. A similar challenge lies in training students to speak the languages of the different disciplines so as to bring their powerful tools to focus on the hard problems of biology. Thus, the technical, informational, and biological challenges of genomics are providing a powerful impetus toward the practice of interdisciplinary science.

SERENDIPITY

The analysis of the human and mouse T-cell receptor loci has revealed several serendipitous surprises. The highly conserved CδCα region poses fascinating questions as its implied chromosomal function. The association of trypsinogen genes with the β T-cell receptor elements over at least 350 million years raises the possibility that selection has maintained this linkage because of regulatory or functional constraints. The rich archeological history that can be unraveled with a characterization of the β homology units, as well as their extensive nature (50% of the locus), was unexpected. The strikingly dynamic nature of just two Vβ gene segments in primate evolution with regard to gene duplication, deletion, and conversion was also surprising. Finally, although the comparative species analyses were carried out to help identify genes, the power of these analyses in revealing biological information was unexpected and gratifying. It is clear that serendipity will continue to play a striking role in genomic analyses of the future.

ACKNOWLEDGMENTS

We thank Tawny Biddulph for assistance with the manuscript.

REFERENCES

1. HOOD, L., J. CAMPBELL & S. ELGIN. 1975. The organization, expression and evolution of antibody genes and other multigene families. Annu. Rev. Genet. **9**: 305–353.
2. HOOD, L., B. F. KOOP, J. GOVERMAN & T. HUNKAPILLER. 1992. Model genomes: The benefits of analyzing homologous human and mouse sequences. Trends Biotech. **10**: 19–22.
3. KOOP, B. F., R. K. WILSON, K. WANG, B. VERNOOIJ, D. ZALLER, C. L. KUO, D. SETO, M. TODA & L. HOOD. 1992. Organization, structure, and function of 95 kb of DNA spanning the murine T-cell receptor Cα/Cδ region. Genomics **13**: 1209–1230.
4. KOOP, B. F., L. ROWEN, K. WANG, C. LAM KUO, D. SETO, J. A. LENSTRA, S. HOWARD, W. SHAN, E. WILKE & L. HOOD. 1994. The human T cell receptor Cα/Cδ region: Organization, sequence and evolution of 97.6 kb of DNA. Genomics **19**: 478–493.
5. KOOP, B. F. & L. HOOD. 1994. Striking sequence similarity over almost 100 kilobases of human and mouse T-cell receptor DNA. Nature Genetics **7**: 48–53.
6. SLIGHTOM, J. L., D. R. SIEMIENIAK, L. C. SIEU, B. F. KOOP & L. HOOD. 1994. Nucleotide sequence analysis of 77.7 kb of the human Vβ T-cell receptor gene locus: Direct primer-walking using cosmid template DNAs. Genomics **20**: 149–168.
7. LI, W.-H. & D. GRAUR. 1991. Fundamentals of Molecular Evolution **70**: 71. Sinauer Associates. Sunderland, MA.

8. LAI, E., R. K. WILSON & L. E. HOOD. 1990. Physical maps of the mouse and human immunoglobulin-like loci. Advan. Immunol. **46:** 1–59.
9. DAVIS, M. M. 1990. T cell receptor gene diversity and selection. Annu. Rev. Biochem. **59:** 475–496.
10. ROWEN, L. & B. F. KOOP. 1994. Zen and the art of large scale genomic sequencing. *In* Automated DNA Sequencing and Analysis Techniques. **25:** 167–174. J. C. Venter, Ed. Academic Press. San Diego, CA.
11. ALT, F. W., *et al.* 1992. V(D)J recombination. Immunol. Today **13:** 306–314.
12. LEWIS, S. & M. GELLERT. 1989. The mechanism of antigen receptor gene assembly. Cell **59:** 585–588.
13. UBERBACKER, E. C. & R. J. MURAL. 1991. Locating protein coding region in human DNA sequences using a neural network–multiple sensor approach. Proc. Natl. Acad. Sci. USA **88:** 11262–11264.
14. WINOTO, A. & D. BALTIMORE. 1989. α/β lineage specific expression of the α T cell receptor gene by nearby silencers. Cell **59:** 649–655.
15. NICKERSON, D. A., C. WHITEHURST, C. BOYSEN, P. CHARMLEY, R. KAISER & L. HOOD. 1992. Identification of clusters of biallelic polymorphic sequence-tagged sites (pSTSs) that generate highly informative and automatable markers for genetic linkage mapping. Genomics **12:** 377–387.
16. ROBINSON, M. A., M. P. MITCHELL, S. WEI, C. E. DAY, T. M. ZHAO & P. CONCANNON. 1993. Organization of human T cell receptor β chain genes: Clusters of Vb genes are present on chromosomes 7 and 9. Proc. Natl. Acad. Sci. USA **90:** 2433–2437.
17. HOOD, L., B. F. KOOP, L. ROWEN & K. WANG. 1993. Human and mouse T cell receptor loci: The importance of comparative large scale DNA sequence analyses. Cold Spring Harbor Symp. Quant. Biol. **LVIII:** 339–348.

PART IX. THE DOUBLE HELIX: PROSPECTIVE

Introduction

DONALD A. CHAMBERS

*Department of Biochemistry, and
Center for Molecular Biology of Oral Diseases
University of Illinois at Chicago,
1819 W. Polk Street
Chicago, Illinois 60612*

As this volume attests, the growth of the "new biology" born of the elucidation of the nature of DNA, has given rise to dramatic advances in the biomedical sciences. Molecular medicine is not a vision for the future, but is at hand as our intrepid gene hunters identify genetic lesions of disease, develop new diagnostics, and achieve mechanistic understandings that will yield new, rational "molecular therapies." What will be the future of the biomedical revolution and what are the philosophical and ethical considerations and constraints of the continuing advance of this science, both for the scientist and the community? These are the major questions to be addressed in this section by Sir Walter Bodmer, geneticist and cancer researcher, and Horace Freeland Judson, historian and critic of molecular biology.

Finally, at the end of this section, we include a review of the entire conference, providing summaries of contributions for which manuscripts were not available, as well as an overview of the contents of this book.

Where Will Genome Analysis Lead Us Forty Years On?

SIR WALTER BODMER

Imperial Cancer Research Fund
P. O. Box No. 123
Lincoln's Inn Fields
London WC2A 3PX, England

INTRODUCTION

Forty years ago, almost to the day, I started as an undergraduate at Clare College in Cambridge, Jim Watson's College, six months after the famous paper was published. I had never heard about DNA nor about genetics. Two years later I had my first taste of genetics from Srb and Owen's famous textbook,[1] but that hardly mentioned DNA. I first heard about the structure of DNA from R. A. Fisher in an introduction to an abstruse series of lectures on mathematical genetics. Nowadays, "genetics" and "DNA" are household words.

To be "prospective" about the double helix is to open the door to blatant speculation. Those of you young enough to be around and mentally alert in forty years' time, about one-third of the way through the twenty first century, may laugh or cry at how right or wrong we are in our predictions. By its very nature the future of science is largely unpredictable, so that many of my predictions are simply items on a wish list. I start, inevitably, on safe and solid ground, only gradually preparing for my later flights of fancy.

DISEASE GENES AND THE HANDBOOK OF MAN

Well within the next forty years, perhaps even within the next ten, essentially all the human genes (and so effectively all vertebrate genes) will have been found, sequenced, and mapped. No doubt soon, too, the complete human genome will have been sequenced, I believe probably confirming that there is not a lot of interest in the stretches between genes, except perhaps some intriguing evolutionary relics. This overall information will enable genetic analysis of essentially any human difference. First to be elucidated will, of course, be the simply inherited diseases, however rare. Slowly, but also surely will come the identification of relatively common disease susceptibility genes with low penetrance. Linkage disequilibrium, namely population association between a genetic marker and the disease susceptibility, then becomes the most powerful tool for fine-mapping of such susceptibility genes. This should place markers within 0.5% recombination fraction, or about 500,000 base pairs, of a disease gene. Detecting disease susceptibility by linkage disequilibrium makes the strongest case for a high-density DNA polymorphism map. Within a 1M bp region there may be nor more than 20 or so candidate genes, some of which might well be excluded as extremely unlikely to be relevant to the disease susceptibility being studied.[2] The highly automated DNA sequencing of the future should make light work of sequencing, say, twenty genes in twenty individuals in order to identify the DNA sequence variation that matters.

The aim of identifying disease susceptibility genes is to target effective preventative measures to those individuals with a given inherited susceptibility, aiming to bring their disease profile back to the norm. Perhaps within forty years we can hope, for example, for corrective measures for individuals carrying the Huntington's disease gene, Fragile X, or the Alzheimer's-associated apolipoprotein E4 variant,[3] and no doubt other similar variants. Drugs or diet will help those at high risk of heart disease from angiotensin-converting enzyme and other variants, while screening and other preventive measures will help those carrying common cancer susceptibility variants, and no doubt some form of specific immunosuppression will help those with an inherited susceptibility to allergies, including hay fever and asthma. Already there are hints of a localization for a gene for dyslexia,[4] and genes for childhood partial deafness will be found. At the very least, genetic screening could then identify those children needing special help at school at an early stage, avoiding the time it now seems to take to distinguish between poor performance and ability, and such specific disabilities.

The biggest disease challenge will be to identify the genes for mental diseases such as schizophrenia and manic depression. Gene mapping may turn out to be the only, or at least the most penetrating, route to a better understanding of mental disease. The Fragile X syndrome is a start. Will the recently described X-linked gene for male sexual orientation be another example?[5] If there really is a correlation between artistic talent and sexual orientation, will studies, such as those already carried out on homosexual sib pairs, identify one or more gene variations influencing those for artistic ability? How will it then be possible to determine whether, say, a hormonal treatment which brings back heterosexual behavior has a dampening effect on artistic talent?

PREVENTION AND TREATMENT

The ease with which DNA from many individuals can be sampled (even now by just a simple mouthwash) and sequenced will usher in a new era of molecular epidemiology. More and more studies of disease etiology will include the study of association with specific genetic variations. Already in the cancer field there is an interest in, for example, the distribution of oncogene mutations such as p53, ras, Rb, and APC in the same type of tumor in different parts of the world subject to different environments, for example, diets. There could also be differences between similar tumors in different locations, such as rectal versus left- and right-sided colon carcinomas.

All these possibilities for the applications of genetic diagnosis at the DNA level will create an extraordinary demand for DNA sequencing, in humans at least. One hundred base pairs per year for 0.5% of the world's population is already about one genome's worth of sequencing. I have no doubt that in the years to come, DNA diagnosis will require several human genomes' worth of DNA sequences every year.

Sooner or later there will be an effective HIV vaccine. Molecular biology will, no doubt, also produce many new vaccines for other infectious diseases, including especially bacterial diseases, such as tuberculosis, which may return in increasing numbers as they become resistant to all available antibiotics. Immunotherapy of cancer is a burgeoning field, finally coming into its own following our much better understanding of both the nature of the immune system, especially T-cell specificity, and the nature of the genetic changes in cancer cells, which are the potential targets for immune cells. DNA-based cancer immunotherapies, and indeed quite possibly vaccines (for example, against p53, ras and other oncogene mutations) may become commonplace.[6]

Many vaccines may simply be directly DNA or perhaps DNA wrapped in a liposome coat, rather than carried by a live or attenuated virus or bacterium. Proper regulation of DNA for therapeutic uses is already an issue. Will every aliquot used need to have its DNA sequence confirmed, or is it enough to do that on a batch basis?

There will be new drugs galore, not so much from gene products themselves but overwhelmingly from the discovery of new chemical entities by an extension of traditional pharmacology through assays that inhibit or stimulate the functions of newly discovered genes and their products. The range of receptors, extracellular and cytoplasmic enzymes, signaling proteins, transcription factors, and the like that are being discovered through the identification of new genes is extraordinary. And almost everyone is a potential target for the development of new drugs. Many, perhaps, may be used effectively without a full understanding of their function, such as was the case for aspirin for nearly a century.

As protein structures are determined at an increasing rate, will it become possible to automate the process of drug design by using sophisticated expert-system–based, neural network, or other new software. Indeed, will even the organic synthesis of molecules with a required shape itself become automatable? The limiting factor in the introduction of new drugs may not so much be the discovery of effective ones, but the proper validation and testing of each new drug. This is already the most expensive phase of getting a drug to the market.

How widespread will somatic gene therapy be? How far will it extend beyond treatment of rare inherited deficiency diseases and of cancers, by inserting genes that enable selective drug killing or that stimulate the immune system to recognize the differences between cancer and normal cells? Perhaps viral vectors or other DNA delivery systems will improve to be both close to 100% effective and 100% selective with respect to cell type, or at least expression within a given cell type. In that case, somatic gene therapy may have much wider applications, even contributing to prevention of relatively common diseases by, for example, controlling blood cholesterol levels.

AGRICULTURE

The potential for applying genome analysis and its associated technology is as great for nonhuman species as for humans. But the impetus for doing so is different. There is clearly a huge opportunity for improvement of agricultural crops and economically important animals. The range of species farmed at present is remarkably narrow. Will many novel species be tamed and modified for farming, with a view to optimizing dietary requirements and satisfying the wide variety of human tastes?

Plant genomes from different species are turning out to be surprisingly homologous. That, no doubt, increases the potential for transgenesis.

How much plant breeding in future will be by transgenesis? There is no doubt that as the human genome is solved, so will be that of all domesticated mammals, many other useful species such as yeasts, and troublesome species such as the mosquito, as well as a wide variety of plant species.

TECHNOLOGY

When I started in the field of molecular biology in 1961 there were two DNAses known. One you could buy off the shelf, but the other you had to make for yourself. Now, there are, of course, not only a range of hundreds of enzymes and other reagents, but also simple kits to help you make DNA or messenger RNA, translate proteins, and clone and PCR any sequence at will. Laboratory experimental work almost becomes a matter of stringing together a series of kits in the right order.

The machines used in the laboratory have increased enormously in sophistication: no longer is just a complex ultracentrifuge available, but automated DNA and protein synthesizers and sequencers, and a wide range of robots now exist for standard laboratory techniques. There seems little doubt that much of the work of molecular biology in the laboratory could be automated. At the same time, I am sure that novel, and perhaps simplifying, techniques will be developed. One need only think of the extraordinary polymerase chain reaction and its ramifications, and the simplification that came from using gel electrophoresis to determine the size of DNA fragments.

Apart from the obvious benefits of automation and robotics, there is a clear and special need for rapid DNA sequencing and characterization. There is also a need for detecting differences between DNA stretches from different sources, for example, different individuals, or cancer versus normal cells, as well as for improving the sensitivity of detection of DNA sequence differences. This is especially important, for example, in the potential use of identification of mutants, such as in APC genes, as signals of the presence of an early cancer. Some progress has already been made in the detection of various mutants in DNA isolated from the stool of patients with colorectal carcinoma. If the sensitivity and specificity of DNA variant detection are high enough, then this approach to screening for early cancer could become widespread and, therefore, a most important application of DNA-based analysis to the population at large.

FUNCTION

The biggest challenge is not to find the genes and identify significant variations in DNA sequence, but to analyze their function. When in 1980 I spoke to the American Society of Human Genetics about genome analysis and what it could achieve, my subtitle was "When the sequence is known what will it mean?"[7] This will remain the challenge long after all the human genes have been mapped and sequenced, and even when the protein structures have all been determined.

The determination of protein structure is undoubtedly at present a limiting factor in going from the linear DNA sequence to function. The more structures are determined and the relationship between sequence and structure becomes understood, the easier it will become to obtain reliable predictions of protein structure simply from a DNA sequence. The relative rate of determination of protein structures lags behind the rate of identification of new genes, and so new proteins, probably by at least two orders of magnitude. Will it ever be possible to automate fully the production of crystals and their analysis by X-ray crystallography? Or will NMR technology improve to make this the technique of choice even for larger proteins? After all, with modern molecular biological techniques and the developments that must be expected in the future, producing a significant amount of any given protein is no longer a major problem.

Beyond the challenge of determining the structure of individual proteins lies that of establishing the precise molecular structure of large aggregates of proteins in their functional configuration, often at the surface of, or in association with, a cellular membrane. Even such comparatively straightforward problems as the structure of the replicating machinery, of transcription complexes, or of the complex membrane assemblies surrounding immune function molecules such as the T receptor and the HLC Class I and Class II products, are far from being solved. How much further is there still to go before we have a complete, understandable molecular model of the cell?

Production of transgenic animals, especially, for example, mice with knockout mutations in a particular gene, has become a relatively common and powerful way to examine the function of a gene. But here again the scale on which this technology can be applied is very limited relative to the rate of discovery of new genes. A full investigation of each "knockout" mouse is a major study in itself. Furthermore, it may be, indeed is likely to be, necessary to study combinations of such mutants rather than each, one at a time. Undoubtedly, the easier it becomes to make suitable transgenic strains of animals, the more widespread will become the use of the technique. It would be convenient to be able to hand someone half a dozen new genes and ask for a suitable number of knockouts for each to be produced within a month or two! Will even these procedures become automatable? There can be no doubt that in nearly all aspects of genome analysis, scale of activity is important, and so increasing the rate of analysis, and broadening the database, may in themselves lead to significant new insights.

Will it become possible, by analogy with the study of complexes of proteins, to study complex cellular systems *in vitro*? How long will it take, for example, to reproduce the microenvironment in the bone marrow, in a lymph node, in the spleen, or in the thymus *in vitro*, and so to generate *in vitro* all the aspects of a properly developing immune system? Will it be possible to produce true cellular neural networks at will in the laboratory that carry out defined functions?

DEVELOPMENT AND THE BRAIN

The two major general challenges to biological understanding are (1) differentiation, development, and morphogenesis, and (2) the function of the brain, including memory, perception, and consciousness. There can be no doubt that genome analysis is making, and will continue to make, major contributions to the understanding of development and the brain.

The paradigm for understanding differentiation and development is to work out the instructional system, and its controls, for switching genes on and off. There is already an incredible zoo of transcription factors that interact with each other and that no doubt form the basis for this instructional network. It always seems possible to find a new transcription factor that takes the control one step further back. But how does it all start? One is reminded of Jonathan Swift's dictum more than 250 years ago;

> So naturalists observe a flea
> Have smaller fleas that on him prey;
> And these have smaller fleas to bite him
> And so proceed *ad infinitum*.

Perhaps the answer is that transcription factors operate in cycles, or epicycles which interact with each other. Here is another case for expert systems software to explore the logic of transcription factors and the control of gene expression.

Tissue architecture and cell movements are beginning to be explained by adhesion molecules, of which there are now at least four different sorts. Couple this with the morphogens and Lewis Wolpert's ideas on positional information, and many of the ingredients needed to explain differentiation, development, and morphogenesis are already at hand.[8]

By comparison, as far as I am aware, our understanding of the fundamental functioning of the brain is still at a relatively primitive level. A major contribution will be made by identifying all those genes which specifically function in the brain. There are already good examples of this in the various receptors for neurotransmitters. Furthermore, genetic analysis may make a major contribution to the understanding of subtle differences in brain functions associated with mental disease and, ultimately, aspects of normal behavior. The genetic contribution to understanding schizophrenia and depressive states may, for example, have a significant impact on the understanding of normal brain functions.

NORMAL VARIATION

The human applications of genome analysis and positional cloning have so far been largely directed towards the understanding of disease. But the same approaches can elucidate the genetics of normal individual variation. The striking similarity between identical twins at all ages is evidence of the overwhelming genetic component to the determination of facial features. The molecular genetics of the face is a fascinating challenge that will surely be largely solved within the next forty years. Complex computer-pattern recognition programs can analyze facial features and help to classify them into categories suitable for genetic investigation. Chromosomal abnormalities, such as Down's syndrome, can provide clues as to where to look for genes for particular sorts of facial features. There may be one or more genes on chromosome 21 which, when overexpressed, influence facial features (but with no necessary correlation with behavior) and this effect could be mimicked by sequence variation in these genes. How long will it be before we can effectively reconstruct the face from a sample of DNA? Will that be the future source for identity parades [lineups] and for televising the features of wanted criminals or missing persons?

The genetic determination of skin color, in particular a fair complexion, is a major susceptibility factor for skin cancer, including melanoma, as shown by the extraordinarily high incidence of skin cancer and melanoma in white-skinned people who live in the tropics, as compared to, say, Indians and Africans. Fair skin and complexion probably evolved among the pre-Neolithic people of Northern Europe as a response to lack of sunlight, and so lack of vitamin D. It seems reasonable to assume that the genes for fair complexion were then spread southwards by sexual selection favoring the blond phenotype. There is a good practical case for identifying the genetic variation that controls cell pigmentation, eye pigmentation, and hair color and texture, because of its implications for cancer. Once found, it will be interesting to trace the frequencies of these genes in different populations in Europe, and elsewhere.

A difference in the ability to taste the chemical phenylthiocarbamide (PTC) was an early, classical normal polymorphism. This must be just the tip of the iceberg of a wide range of individual differences in taste and smell. Recently, a large cluster of olfactory genes has been identified[9] and so, soon, there will be a molecular basis for the old saying *de gustibus non est disputandum*, that is, there can be no dispute about tastes. There is already some evidence to suggest that minor variations in the pigment genes that are defective in color-blind individuals may contribute in an analogous way to individual differences in color perception.[10]

Studies in twins, and even common observation, clearly indicate genetic components to behavior. This does *not*, of course, mean that there are specific genes for particular human behaviors, such as for high moral quality, for musical and mathematical ability, or for achievement in sport. What *is* meant is that there will be genetic differences that affect the probability of an individual's having a certain behavior, depending on the environmental stimuli. It is impossible to believe that Mozart's genius was not to a large extent genetically determined. And, indeed, there are examples of notable musical families, such as the Bachs. But I have little doubt that had Mozart been adopted by a Bedouin tribe in the middle of the Sahara he would not have composed the way he did—nor do I believe that I would have composed like Mozart, had I been adopted by Mozart's parents. The same could well be said of mathematical ability.

The challenge to understanding the genetics of such complex behaviors is first to analyze the components of these behaviors and so to identify those components that could be studied objectively and provide the basis for the search for specific associated genetic variants. Memory, for example, perhaps of different types, must often play a role in high musical ability, as well as in being a chess champion. Perfect pitch, which has been shown to occur in families, is another potentially important component of musical ability.[11] If we could identify the genes that gave rise to a perfect pitch or an enhanced analytical ability, would this not be a contribution, and possibly an important contribution, to understanding the function of the brain?

POPULATION STUDIES AND EVOLUTION

All humans, apart from identical twins, differ genetically in all sorts of ways. These differences can now be identified at the DNA level. Studies of the distribution of classical marker polymorphisms in different populations have already told us that most genetic variation can be found within any population and, in comparison, the differences between populations are relatively minor. There are, however, often genetic differences between populations which are very obvious, such as in facial features, body build, skin color, and hair texture.

The study of the distribution of genetic markers in different populations has already told us a great deal about their interrelationships. It has supported the view, which also comes from archaeologists, that modern *Homo sapiens* migrated out of Africa and spread throughout the world only between 100,000 and 200,000 years ago. Detailed studies in Europe, following an original observation on the pattern of frequency distribution of the Rh-negative blood group variant, led Luca Cavalli-Sforza and his colleagues to propose the hypothesis that the gene-frequency gradients observed across Europe could be explained by the outward migration of the Neolithic people from the Middle East, starting about 10,000 years ago.[12] Further studies have suggested that, for example, the Basques and Sardinians may still retain a strong element of the genetic composition of Paleolithic populations. These groups might have been exceptions to the rule that agriculture advanced with the migration of people, rather than the spread of ideas. Similarly, such data suggest that the Celts of the periphery of the United Kingdom, namely the people of Cornwall, Wales, Ireland and Scotland, may, in fact, be a mixture between the pre-Neolithic population of Britain and the last wave of advance of people from Central Europe, who brought farming to the British Isles.[13]

So far, however, the data are relatively imprecise because of limitations to the number of individuals sampled and the number of genetic markers studied. Now, with an enormous range of polymorphisms to be studied at the DNA level and as the techniques for analysis become automated, it should be possible to establish the origins of different human populations and their patterns of migration and movement with enormously increased precision. In this way, for example, we should eventually be able to answer the question as to whether the genes for fair complexion migrated south in Europe more quickly because of sexual selection than did neutral unselected genes. We should, also, in the course of such studies, be able to obtain the pattern of distribution of disease gene variants, both for rare inherited diseases such as cystic fibrosis, and for common disease susceptibilities of which the HLA types such as B27 and its striking association with ankylosing spondylitis are already examples.

When we have established the book of man we shall also, effectively, have the book of the chimpanzee. Taken to its logical conclusion the comparison of these two books should tell us which are the differences in gene sequence that distinguish *Homo sapiens* from a chimpanzee. Many of the differences between chimpanzee and human DNA sequences may be of no particular functional significance. But it would surely be fascinating to known which differences were critical for the evolution of *Homo sapiens* from their common ancestor among the higher primates.

Establishing the gene content and genetic maps of a wide range of different species should provide fascinating information about the evolution of gene systems. The simple question, for example, as to whether the equivalents of the major histocompatibility regions, such as HLA and H2 in mammals, exist in lower organisms, from primitive vertebrates to invertebrates, and, if so, to what extent they retain the same set of genes as has been found in mammals, could illuminate the evolution of this complex genetic region. Can we trace fully the origin of, for example, the immunoglobulin super-gene family right back to its presumed precursor as a primitive intercellular attachment molecule?

The ultimate evolutionary challenge is to be able to analyze the DNA from some really old fossils. There are, however, those that argue that even DNA preserved for 50 to 100 million years in amber may be too oxidized to yield identifiable segments. This lays open to question the whole notion behind *Jurassic Park*! Nevertheless, it seems possible that short sequences may have survived at least enough to be PCR-amplified by the polymerase chain reaction and sequenced so that the mutations introduced by chemical damage can be averaged out to give a consistent sequence. Even short fragments could be overlapped to give longer segments, provided good automated technology has evolved. So, while we may not be able to recreate a dinosaur within the next 40 years, perhaps we may just be able to identify one or a few dinosaur genes and see what they do in say, a lizard. Much more interesting, really, however, would be the possibility of obtaining DNA from the organisms of the Burgess Shale, reflecting the Cambrian explosion.[14] Was this associated with a jump in organismal complexity that came with the development of segmentation? And was this development really due to the evolution of homeobox genes?

CONCLUSIONS

The greatest future practical impact of genome analysis is likely to be on human disease. Perhaps even within the next forty years, we may have learnt how to prevent most cancer and cure the rest, and to do the same for heart disease, mental disease including Alzheimer's, autoimmune diseases such as rheumatoid arthritis and juvenile-onset diabetes, as well as osteoarthritis and other degenerative diseases. In that case, we will have removed much of the disability of old age, but without necessarily extending the life span. The distribution of the quality of life should then become closer to a step function. Namely, people should remain healthy and fit, both physically and mentally, until very shortly before they die. The best, presumably, is to die healthy at a reasonably old age.

But will we ever be able to control and adjust the human life span itself? Will we identify the genes that enable us to live 40 times as long as a mouse? And is there any leeway in going further? If we could increase the human life span at will, how long would we want to live? For one hundred years, for two hundred years, or even more? There may be no limit, but the real challenge is to establish, in some sense, what is best for society and what its members wish.

The social sciences are still the most difficult of all. They require an understanding of human social structures on top of human behavior, which in turn sits on top of biological understanding. Admittedly, just as we do not need to know the details of quantum mechanics to understand most biology at the chemical level, perhaps similarly to understand the social behavior of people we do not need to go much beyond an understanding of their behavior. But this itself entails some understanding of the biology and genetics of behavior. Can we hope that during the next forty years there might be an invigoration of the social sciences from the biological, and, in particular, through what genetics can tell us, in much the same way that the revolution in molecular biology came, to a considerable extent, from the physicists who set an analytical framework to the subject that had not been achieved before. This was accompanied by the structure analysis that gave us the DNA double helix.

Shortly after I went to Oxford, about 20 years ago, as professor of genetics, a sign appeared outside the department as it was being renovated saying "Alterations Department of Genetics." There was, of course, nothing on offer. At that time there was not even a great deal that could be done to identify the alterations you might wish for, let alone to implement them. That situation has, of course, changed dramatically and will continue to do so over the next forty years. We shall, as I have explained, identify and perhaps even understand, many of the genetic differences between individuals, especially as they affect disease, and also the commoner outwardly visible characteristics. Will alterations ever be offered? Will germline therapy ever be accepted? And if so will it only be for the treatment of severe inherited diseases? I have no doubt that germline gene therapy will become feasible, even though it may largely be developed in domesticated animal species. The decision as to whether to allow it, therefore, will become entirely a social and a political one rather than a scientific decision. But the decision will not be made intelligently by a populaion that is not prepared to understand it, namely, does not have an adequate background in science in general, and the science of genetics in particular.

The one sure prediction that, unfortunately, I believe I can make is that our social adaptation will be far too slow to deal with the extraordinary advances that will come from the revolution of genome analysis. To achieve the potential benefits from the fruits of these advances, it is essential to have a population that understands, at least to some extent, some of the underlying science and the nature of the scientific enterprise itself. This is a problem, no doubt faced by Michael Faraday at the Royal Institution more than a hundred and sixty years ago, and it remains with us to this day. In the 1920s and 1930s J. D. Bernal and J. B. S. Haldane strongly promoted the public understanding of science and the scientific enterprise. Some twenty or thirty years later C. P. Snow gave his famous lecture in Cambridge, England on the two cultures, which I, in fact, attended. Now in the United Kingdom, the United States, and elsewhere, there is again a resurgence of interest in promoting the public understanding of science. All scientists should at least take some responsibility for explaining their science intelligibly to the general public. In this way, we might hope that the extraordinary advances in dealing with human disease, improving agriculture and, conceptually in understanding our own evolution, and evolution in general, will be taken advantage of to the full for the greater benefit of mankind.

REFERENCES

1. SRB, A. M. & R. D. OWEN. 1952. General Genetics, 1st Ed. W.H. Freeman. San Francisco.
2. BODMER, W. F. 1986. Cold Spring Harbor Symp. Quant. Biol. 51(P1): P1–13.
3. SAUNDERS, A. M., K. SCHMADER, J. C. BREITNER, M. D. BENSON, W. T. BROWN, L. GOLDFARB, D. GOLDGABER, M. G. MANWARING, M. H. SZYMANSKI, N. MCCOWN et al. 1993. Lancet 342: 710–711.
4. RABIN, M., X. L. WEN, M. HEPBURN, H. A. LUBS, E. FELDMAN & R. DUARA. 1993. Lancet 342: 178.
5. HAMER, D. H., S. HU, V. L. MAGNUSON, N. HU & A. M. L. PATTATUCCI. 1993. Science 261: 321–327.
6. BODMER, W. F. et al. 1993. Ann. N.Y. Acad. Sci. 690: 42–49.
7. BODMER, W. F. 1981. Am. J. Hum. Genet. 33: 664–682.
8. WOLPERT, L. 1991. The Triumph of the Embryo. Oxford University Press. Oxford, England.
9. BUCK, L. & R. AZEL. 1991. Cell 65: 175–187.
10. WINDERICKZ, J., D. T. LINDSEY, E. SANOCKI, D. Y. TELLER, A. G. MOTULSKY, S. S. DEEB. 1992. Nature 356: 431–433.

11. PROFITA, J. & T. G. BIDDER. 1988. Am. J. Med. Gen. **29:** 763–771.
12. CAVALLI-SFORZA, L. L., A. PIAZZA, P. MENOZZI & J. MOUNTAIN. 1988. Proc. Natl. Acad. Sci. USA **85:** 6002–6006.
13. BODMER, W. F. 1993. Proc. Br. Acad. **82:** 37–57.
14. GOULD, S. J. 1989. Wonderful Life. Hutchinson Radius. London.

The World We Have Lost

HORACE FREELAND JUDSON

George Washington University
Department of History
Washington, D.C. 20052

The year 1993 marks not one but two anniversaries in molecular biology, a ruby anniversary and a silver one, anniversaries of the double helix in two different yet in their way complementary senses. The ruby, or fortieth, is that of the discovery of the double-helical structure of DNA early in 1953. The silver, or twenty-fifth, anniversary is that of the appearance, early in 1968, of that now classic, still curious anti-memoir, *The Double Helix*.

That moment in 1953 was a discovery of great beauty, astonishing parsimony, and transcendent explanatory power. We've been hearing about it and its consequences for months this year, for days this week, and still its fascination does not wear off. Yet perhaps we should take a moment to look again at that book, as we read it in 1968 and as we read it today. *The Double Helix* has a great deal to tell us—if we're willing to attend to it—about the world of molecular biology then and about how it has changed to what we see today.

As a publishing phenomenon, *The Double Helix* was instantaneous and dazzling. One index of its power is the outpouring of brilliant reviews it elicited: I can think of few other books this century of which the reviews make great reading in themselves. One remembers especially Erwin Chargaff's waspish, witty denunciation; or George Steiner's burst of imaginative fireworks in *Encounter*, under the pseudonym "F.R.S."; or André Lwoff's extended review in *Scientific American*, magisterial, profoundly responsive, and above all just.[1]

The book was more than the literary *cause célèbre* of 1968. Most of you here remember exactly where you were when you learned of Jack Kennedy's assassination. Some will remember what you were doing when you heard of the first atomic bomb. For molecular biologists, *The Double Helix* was an event of that kind. "I have never seen Francis Crick in a modest mood." That oddly gawky opening sentence—I venture to say that most of you can remember how you first encountered it. I, for one, was in the lobby of a hotel here in Chicago, early in January, with Matthew Meselson and some other friends, when Matt hurried over to the newsstand because he had spotted the first delivery of the issue of the *Atlantic Monthly* that carried the first installment of the book's serialization.

In 1968, we read *The Double Helix* at first more for its whiff of scandal than for enlightenment about the science—though for the discerning Jim delivered a couple of insights into the science, too. Our interest had been whetted by rumors that when Watson circulated a first draft—under the title *Honest Jim*, a nod to Kingsley Amis as well as a bitterly ironical self-perception—it was met with outrage. Some of us had heard that Crick had found it an infuriating invasion of privacy, vulgar, inaccurate, and a gross violation of friendship. Others knew that a score of Watson's scientific community, led by Crick, Linus Pauling—and Ava Helen Pauling—and perhaps Maurice Wilkins, had campaigned stridently to stop the book's publication, and had indeed succeeded in blocking it at Harvard University Press. A few, closest to the titanic struggle, realized that the book's great defender had been Sir Lawrence Bragg, who wrote numerous letters in its support and even travelled to the United States to urge editors and publishers to understand that this was a great book.

Watson revised and revised, toning down and fudging over most of the gamier revelations and characterizations. Crick was not placated, and in the spring of 1967 wrote Watson an incandescent six-page letter attacking the latest draft—sending copies to at least ten others involved, starting with Bragg and Nathan Pusey, the president of Harvard. Crick's letter offered some pretty gamey characterizations, too—but this is not the occasion for quotes.[2]

Anyway, Jim's editor quit Harvard University Press and took the book to Atheneum. *The Atlantic Monthly* bought the serial rights. The book appeared. Crick left Cambridge for Greece for a couple of months to avoid journalists. Like me. For I was working in London, then, as arts and sciences correspondent for *Time*, and when *Time*'s editors in New York woke up to the book and the science I was sent hurrying up to Cambridge to interview everyone who was still at home—which meant Max Perutz, Sydney Brenner, Fred Sanger, and Hugh Huxley. One can fairly say as a footnote to history that *The Double Helix* changed my life, too, setting me on the path that has led me to this podium this afternoon.

From the published *Double Helix*, the discerning reader can divine what must have been some of the things omitted from or fudged in *Honest Jim*. And a careful reading in historical context reveals that behind the seeming ingenuous frankness lies an astonishing complexity of motives and conflicts that the book presents with extreme indirection. I'll save a full deconstruction of *The Double Helix* for another place—but there is one story about that for which you are the perfect audience.

While I was working on the history of molecular biology, we lived for nearly seven years in Cambridge. I frequented the Medical Research Council lab and interviewed many people there, visitors too, and some of them repeatedly. Several times, in conversations with Francis, he told me about this or that, "Well, Odile actually saw more of Rosalind than I did at that time," or "You ought to ask Odile about her impressions of Jim." I hadn't met Odile

Crick, but then one evening at a mid-summer's party in Mark Bretscher's thatched cottage and apple orchard to the north of Cambridge I did meet her and told her what Francis had said. A week or so later, I went around to the Cricks' house. Francis was away, but Odile had a woman friend visiting. We sat in the sitting room and talked. We moved to the kitchen and talked. We went to another sitting room and talked. For some reason, I wasn't using a tape recorder, but rather a notebook. The notebook sat open on my knee, but I wrote nothing down. Odile was telling me the same things Francis had told me—and most often in the very same words. This was a vivid demonstration of something I had long realized, that one of the problems with interviewing is that people often tend to remember not the events themselves but rather the way in which they have recounted those events most recently. I was getting the canonical accounts, the traditional versions, of Rosalind and Jim, and not succeeding in breaking through to fresh memory.

Then I asked, desperately, "Well, what was Jim like in those days?" Odile told the obligatory story about Jim's arriving with a crew cut. My page remained blank. I waited. Then she said, "He ate a lot of ice cream." I wrote down *ice cream*, and waited. She said, "He came around a lot." I had a sudden vision of this skinny post-adolescent, brilliant but socially awkward, yet ingratiating, who hung around to the point, no doubt, of being something of a pest. Then Odile said, "Of course, he was jealous of Francis." I wrote down *jealous of Francis*, and waited. "Of his success." I wrote down *success*. And waited. "Of his social success." I wrote *social success*. And waited. "Of his success with girls."

✌

Re-reading *The Double Helix* today, I find it nostalgic and immeasurably sad. How remote from us, how different from what we know now, that time was. Indeed, though Watson wrote it just fifteen years after the events, already the harsh immediacy of his prose could not conceal the distancing that that prose accomplishes. Watson was himself deconstructing the past, and himself. (Maurice Wilkins has said to me a couple of times that Watson presents himself in the book in the archetype of the Holy Fool—or, better, the Trickster.) Yet, if indeed there was ever a golden age, those early days of molecular biology are surely a candidate. Perhaps there have been others. Morgan's group at Columbia in the decade from 1908, say? Göttingen in the late 1920s? Gowland Hopkins' laboratory at much that same time? For molecular biology, we can specify the period and places. From the 1940s through the early '50s, the small, slowly expanding group around Max Delbrück, who worked with the genetics of bacteriophage and who gathered in the summers at Cold Spring Harbor. From the early 1950s and for the rest of the decade, the Medical Research Council unit in the Cavendish Laboratory at Cambridge, and André Lwoff's garret at the Institut Pasteur in Paris.

What is a golden age like in a science? Watson and Crick pointed out to me the other day that the obvious first requirement is that the science be just aborning, nascent, small. Certainly that small size at the start of a new field is crucial. An important aspect of this is that the people in the new field have arrived there from elsewhere. The small size maximizes the collision frequency, the intensity of intellectual interactions; the variety of starting points maximizes the interplay, the scope and angle of intersection of ideas. In short—and we see this in the arts, too—new intellectual movements spring up between the paving stones of more rigid, established approaches and disciplines.

Again, at the beginning of a new field both concepts and experiments can have a freshness and simplicity that, among much else, may mean that the actual costs of research are low, in absolute terms but especially in relation to the payout.

Yet a golden age in science is more than small size and modest needs. A true golden age is also an age of innocence. It thrives, for a while, in the competitive harsh ocean of the twentieth century, as an island of idealism, and of play, and of, at the same time, an austere devotion to intellectual enthusiasm and openness. The phage group has been called "one of the little communities of intellectual purpose and excitement that constitute the only genuine utopias of the twentieth century."[3] Competition surely there will be, but competition that's more unifying of the tight small community than divisive—competition for the approval, respect, and lively immediate intellectual response of your colleagues.

Listen again to Watson, this time from the essay he wrote about the same time as *The Double Helix* and published in the volume prepared to honor Max Delbrück on his sixtieth birthday, *Phage and the Origins of Molecular Biology*. He's describing his induction into the phage group:

> As the summer passed on I liked Cold Spring Harbor more and more, both for its intrinsic beauty and for the honest ways in which good and bad science got sorted out. . . . Most evenings we would stand in front of Blackford Hall or Hooper House, hoping for some excitement, sometimes joking whether we would see Demerec [the director then] going into an unused room to turn off an unnecessary light. . . . On other evenings, we played baseball next to Barbara McClintock's cornfield, into which the ball all too often went. . . . When August began the Lurias went home to Bloomington because Zella would soon have a child. This left Dulbecco and me even more free to swim at the sand spit or to canoe out into the harbor, often in search of clams or mussels.[4]

Surely, Max Delbrück created the ethos of the phage group in the 1940s—and this was something he brought from the Bohr circle in Copenhagen of nearly fifteen years earlier. Surely, André Lwoff created something very similar at the Pasteur: one has only to look at the photographs of everybody eating lunch together in the attic in the early 1950s to experience an almost physical awareness of the rush of excitement that filled so much of those days there.[5] And we know from *The Double Helix* that the unit at the Cavendish, increasingly biological in the midst of a physics laboratory, was another such island of intellectual play and enthusiasm and dedication to getting the answers (if not always of high idealism).

Oh, but that enthusiasm is infectious, even after forty years and more. Listen to Max Delbrück, in his Harvey Lecture, in 1946. He described the fascination of the field of bacteriophage, "a fine playground for serious children who ask ambitious questions." He went on:

> You might wonder how such naïve outsiders get to know about the existence of bacterial viruses. Quite by accident, I assure you. Let me illustrate by reference to an imaginary theoretical physicist, who knew little about biology in general, and nothing about bacterial viruses in particular, and who accidentally was brought into contact with this field. Let us assume that this imaginary physicist was a student of Niels Bohr, a teacher deeply familiar with the fundamental problems of biology. . . .
>
> Suppose now that our imaginary physicist, the student of Niels Bohr, is shown an experiment in which a virus particle enters a bacterial cell and 20 minutes later the bacterial cell is lysed and 100 virus particles are liberated. He will say, "How come one particle has become 100 particles of the same kind in 20 minutes? This is very interesting. Let us find out how it happens! . . . This is so simple a phenomenon that the answers cannot be hard to find. In a few months we will know. . . ."
>
> Perhaps you would like to see this childish young man after eight years, and ask him, just offhand, whether he has solved the riddle of life yet? This will embarrass him, as he has not got anywhere in solving the problem he set out to solve. But being quick to rationalize his failure, this is what he may answer, if he is pressed for an answer: "Well, I made a slight mistake. I could not do it in a few months. Perhaps it will take a few decades, and perhaps it will take the help of a few dozen other people. But listen to what I have found, perhaps you will be interested to join me."[6]

Delbrück was one of the most seductive intellects of our time: the charm of the man and his ideas and style are palpable here as they were in his Nobel lecture a quarter-century later. But style is surely essential to a golden age. And the style I speak of is an integral component of leadership—which is, of course, another essential. Morgan, Rutherford, Bohr, Delbrück, Lwoff, Crick—these men had styles of leadership that attracted younger scientists of great individuality and ambition, scientists like Monod or Jacob, Watson, or Brenner, who would never have fit comfortably or creatively into more traditional laboratories. André Lwoff, in particular, was a leader and scientific stylist who has been unfortunately neglected in Anglo-American accounts of the origins of molecular biology. His review of *The Double Helix* reveals a man of immense rectitude, compassion, and, indeed, wisdom. His work was wildly original in its day, and, with all that, he remains the one scientist I know who could be deliberately funny in a serious scientific paper.[7]

Contrast these men with those who have risen in molecular biology to be figures of great power today. Once more, this is not the occasion to go into particulars—but look about you. Can you name anyone likely to create the setting for a new golden age?

One more example—small but telling—of the high style in molecular biology. Sydney Brenner once remarked, of the time in the early 1960s when the primary question was the nature of the genetic code, that one attempted to create experiments, to invent experiments, where the absolute minimum of work and of data would produce the maximum insight.

The golden age, the age of innocence, has been followed by our age of brass. Brenner himself, for the past fifteen years, has been driving a project almost without style: his brute-force attack on the total developmental biology of a *Caenorhabditis elegans* nematode. Watson has been for two decades himself the director of Cold Spring Harbor laboratory: consider how he has transmuted it into a powerhouse of big biology, a place where the young Jim Watson could hardly have flourished.

᎒

Three great structural transformations have affected molecular biology over the past twenty years and will operate with increasing and irresistible force. For convenience in introducing them, although they overlap, we can characterize the first of these transformations as internal and the other two as external.

The internal transformation is the steady increase in ethical problems, which is to say, problems of scientific misconduct. Yet again, this is not the occasion nor have we got the time to go into this area in detail; but I must offer one type of misconduct as a demonstration of the forces that are at work. This is plagiarism, or the purloining of intellectual property. On the scale of misconduct, such purloining may appear less heinous than outright fabrication or falsification of data: after all, fabricated or falsified data are untrue, are a betrayal of science itself, while stolen data and ideas have at least a chance to be correct, or why steal them and betray your neighbor? But what plagiarism might seem marginally to lack in gravity it more than makes up in frequency. Theft of intellectual property is by far the most common of all forms of scientific misconduct. Make no mistake: it is rife.

The root of the problem of plagiarism in science is a contradiction inherent in the system of peer review and refereeing. For the fact is, of course, that the persons most qualified to judge the worth of a scientist's grant proposal or the merit of a submitted research paper are precisely those who are his closest competitors. This is simple and obvious and cannot be avoided. Its implications are rarely confronted.

A question: When you receive a paper to review, typically you are required to return the text of the paper with your comments. I won't ask for a show of hands on this—but who among you could say that you never make a photocopy of the paper before returning it? And that the paper is never mentioned in your lab before it appears?

Another question: Have you never had the experience of submitting a paper only to get it back weeks later with comments from one of the referees that oblige you to revise it in petty ways, perhaps to perform several additional experiments or controls – only to see the essential content of your paper appear in some other journal, beating you to it? Lucy Shapiro, at Stanford, said to me cheerfully not long ago that scientists must expect this to happen to them from time to time, but that, fortunately, one is judged on the totality of one's work; an incident of this kind must be shrugged off.

In several notorious cases publicized in recent years, whole tables and paragraphs, whole chapters even, have been lifted. But the subtler form of piracy is the theft of an idea, an insight, a conclusion, while your data and your language remain untouched. A friend who works on the pathology of the human kidney has provided an example. Some years ago, he was scrutinizing a body of data he had accumulated on failure rates in kidney transplants. He realized that the American practice was to keep the kidneys perfused while in transit from donor to recipient. In Scandinavia and the Netherlands, kidneys were not perfused, just shipped quickly on ice. Transplant failure in those countries, he saw, was significantly less frequent than in the United States. A simple series of experiments showed that perfusing damages kidney tissue. He wrote up these observations and submitted the paper to the premier specialized journal. His manuscript came back with niggling objections from one reviewer, who also listed a number of papers that my friend had failed to cite. Ten weeks later, a paper appeared in the *New England Journal of Medicine* making exactly the same case against perfusion, although using an entirely different set of data – comparable data, of course, and no doubt perfectly legitimate in themselves. The principal author of that paper was the man who was principal author on the papers my friend had been scolded for failing to cite.

This problem is inherent in the ways we judge, publish, and fund recent and new research. The pressure to steal could be held at bay so long as molecular biology was small, young, fresh, and led by scientists of dedication and ideals. But in the past quarter-century molecular biology has grown immense: laboratories, projects, journals have all multiplied like cancers. In many laboratories, mentoring has become attenuated. Look again at peer review and refereeing. Once upon a time, these were accepted as necessary and even instructive parts of the doing of science. The meeting of a peer-review panel twenty-five years ago could be intellectually demanding and exciting – a three-day seminar in the latest thinking in one's specialty. Now fatigue has set in, and the tasks seem endless and onerous. The quality and even validity of papers and grant applications have become more difficult to assess. Meanwhile, the results of such assessment have come to seem less relevant. For grant applications in particular, the rank order of the top 20% is increasingly difficult to determine, while the increasingly stringent monetary limits mean that only some among that top 20% are likely to get funded anyway. The combination means that a stochastic element is in part replacing the judgment of one's peers.

Worst of all, however, an alternative mode of science has grown up. It competes for the allegiance of graduate students and post-docs. Those remarkable small self-governing scientific democracies of the golden age, symbolized by the phage group, the garret at the Institut Pasteur, the RNA Tie Club, and the Medical Research Council unit at the Cavendish—such collaborations are steadily being superseded by a different sort of laboratory life and ethos: laboratories that are large and rigidly hierarchical, and an ethos driven by careerism, in which doing science is in large part a way to secure more grants, to get promotions, to gain power. This is an ethos of the 1980s.

੨੦

We are now experiencing two further transformations of the scientific enterprise, which I characterized for convenience as external. One of these relates to the phenomenal growth of molecular biology, and of science generally. But the problem that now envelopes us, conditioning everything else we do, is not one of growth in the simple sense. Rather, it is the transition of the sciences from exponential growth to the steady state. In 1963, Derek de Solla Price, who was an historian of science, published a graph showing that scientific activity had been expanding at an exponential rate for 300 years or more—indeed, that the output of scientific papers was doubling every 15 years. At that rate, Price said, by the turn of the millennium every man, woman, and child in the United States would be spending every working minute doing research and writing it up.[8] Derek Price was a peppery, amusing man, and everyone laughed and said, What a droll idea! Like most assertions of Malthusian limits, this one was taken seriously by few. Yet those limits have been creeping up on us. The tectonic plates deep beneath the enterprise have changed their direction and rate. The transition to a steady state is producing enormous systemic strains, some obvious, some subtle.

For most scientists, the obvious sign of the transition is the shortage of funds and the much increased competition for what funds are available. Along with this, they see an ever-increasing pressure from politicians and government agencies for directed or targeted research: the slogan in 1993 was "national needs." The campaign of the previous director of the National Institutes of Health, Bernadine Healey, to draft a strategic plan for the priorities of the NIH is perhaps the most conspicuous instance. Her plan was beloved of none, written off by most as a political exercise—selling out the citadel of biological science to advance her career. In short, parsimony and political interference: these seem to most of us to be the characteristics of an abnormal situation, antithetical to the practice of science in the mode that has proved so successful since the Second World War. One longs for a return to normality.

Of course, we are not going to get any return to such a nostalgic normality. The Clinton administration came into office with a declared commitment to research and high technology. Yet almost at once we heard from Vice-President Al Gore and from other of Clinton's creatures a renewed emphasis on the practical, industrial, technological exploitation of research—while any increases of funding will be minor and at the margin. Nor will the Republican party be more generous.

These are signs of that transition to the steady state. The transition has other concomitants. One of these is the internationalization of research. Another comprises the intensifying and increasingly complex linkages between university and industry.[9-13]

This last is evident in many fields of science, but is becoming a dominant factor in molecular biology. In 1970, 1972, 1974, the men (and one or two women) who were developing the methods we now call recombinant DNA, or genetic technology, were pulled by the desire to find ways to get at a pure and intractable scientific problem: development and differentiation in eukaryotes, or how the fertilized egg, a single cell, becomes the adult organism of perhaps billions of cells with scores of different functions. I am sure that this was their prime motive, because I was interviewing them at the time. Since then, many of those men (none of the women, so far as I know) have made fortunes in biotechnology. Naturally, money has power to influence the direction of research; but the point for our purpose here is that the linkages between the academic laboratory and industry are changing the career structures of molecular biology.

In molecular biology, at least from the 1970s onwards, the line between so-called "pure" and "applied" research has grown ever more tenuous. It is now vanishingly thin. The trick one laboratory devises to take a tiny step forward in the puzzle of cell development has often turned out to allow manipulations of some other sort of cell for the production of something commercially profitable; indeed, we can no longer predict which kind of advance will come first and in what sort of laboratory, academic or biotechnological. Often today you'll find that the laboratories in the biotechnology company are better equipped and funded than those in the university department across town; these days, much of the work done in them is just as publishable in the journals. Multiple examples of this rapid interchange can be cited in almost any sub-field of molecular biology: consider, for example, the history of monoclonal antibodies.[14]

Similarly, new technical methods facilitating biological research are typically indifferent to the uses to which they will be put. From the 1920s to the present day—including the ultracentrifuge, all the varieties of chromatography and electrophoresis, and protein and DNA sequencing, and on—the list is long of Nobel prizes awarded for inventing technology that today crowds the corridors and work-benches of university labs and biotech companies indistinguishably. The telling recent instance is the polymerase chain-reaction, for multiplying DNA sequences *in vitro*, without the need to put the sequences into cells—cloning without cloning, as it were. The polymerase chain-reaction was invented by Kary Mullis and others at Cetus Corporation, in Berkeley, California. It has been called the most revolutionary new technique in molecular biology in the 1980s. In sum, the technology of genetic experiment and analysis has done more than facilitate research and theory.[15] It has *driven* research and theory. Often the new technology *is* the science.

One crucial consequence for the transition to the steady state is a change in the patterns of work and advancement in molecular biology. The young biologist today sees not one but two career ladders, one academic the other high-tech industrial—and the best of them soon realize that as they ascend they will be able to step from one to the other, either way.[16]

The aspect of the transition to the steady state that is most subtle yet most powerful is the growing emphasis on evaluation and accountability. Consider peer review and refereeing yet again. These are, after all, institutionalized means of evaluating research. The first characteristic of these evaluations is that they are done by scientists. Secondly, especially in the review of grant applications, the evaluation is of inputs. But the new emphasis in advanced management these days, in government as in industry, is the evaluation of outcomes. In the new political order, the managers of social programs, of industrial policy, and of technological and scientific activity of all sorts are already emphasizing this. Outcomes have replaced strategic plans as the new management shibboleth. For those in science the evaluation of outcomes will mean evaluation by non-scientists, evaluation at end-points rather than at the beginnings or in mid-process, evaluation according to new criteria over which scientists will have far less control.

In the midst of a great structural transformation, its longer-term consequences are not predictable in detail—at the microlevel, so to speak. But keep your eyes on such things as the changes in the career structures of the biological sciences, the changes in the aims and conditions and organization of biological research, and the changes in the ways research is evaluated. And the golden age, or even its successor, the grand imperium of science—if these ever existed, they are forever gone. The barbarians are in the city: civilize them if you can, but work with them as you must.

ಈ

The other great structural transformation has only just begun, but is accelerating at an astonishing rate. In it lies a hope of remedy, at least partial, for some of the problems I have listed. This transformation is the shift to electronic publishing, and in particular, right now, the electronic distribution of what amount to preprints.

Awhile ago, in October of 1991, at a conference of editors of scientific journals at Woods Hole, the central topic was electronic publication. The prospect seemed remote. When one of the more daring participants suggested that papers in electronic journals—biomedical and clinical journals, at that—might be issued without peer review, the general response was horror at the thought. Today, we have several biomedical journals on-line. More interesting, we are seeing a proliferation of preprint clubs using the Internet. And hardly a month ago, the *Los Angeles Times* announced that soon after the turn of the year they will publish their product for general readers on-line.[17,18] Indeed, such publication does no more than make widely available what editors and others at major newspapers now consider routine: I know a columnist who lives in Baltimore and who, every night, opens his line to the *Baltimore Sun* and scans not only that paper's edition for the next morning, but goes on to scrutinize the next day's *New York Times*, *Washington Post*, and *Los Angeles Times*.

The most thoughtful and engrossing talk at the Woods Hole conference was given by Joshua Lederberg. You may have read the version that he published in *The Scientist* at the beginning of this year. Lederberg welcomes electronic publication. To be sure, many journals will continue to function best in much their present form, that is, on paper, to be read on airplanes or trains or sitting in the garden. But any journal should be available for electronic browsing or searching. Electronic publication offers the only possible answer to the need to sort out from the ever-more-vast literature all the papers, but only the papers, that are directly relevant to one's work, and then at the same time to record, retain, and keep accessible one's responses—notes, commentary, linkages, inspired ideas—to those papers. Electronic publication, together with more sophisticated search algorithms, may even ease the problem of locating highly specific but elusive matters within the vast literature, what Lederberg called "the exquisite detail needed to take the next step."[19]

Two years ago we could already see that electronic publication can overcome two other substantial problems that ink-and-paper journals now present. The first, of course, is the lag time from submission to publication, at least insofar as this is a product of backlog and of less-than-instantaneous communication among editors, staff, referees, and authors. Papers in on-line journals will be offered electronically as soon as they are accepted, thus getting them to journal subscribers many months before they appear in print. The second problem is the editorial pressure on authors to condense papers, to simplify discussion and conclusions, and, worst, to excerpt the data—pressure that has become routine in certain "hot news" journals, including *Nature*, *Science*, and most notoriously *Cell*. Electronic publication can eliminate forever that problematic parenthetic notation: (data not shown). Indeed, it will go much further. When the *Los Angeles Times* announced its plan to go online, one promise was that readers would be able to request further information about any story—for example, to call up material that may have been dropped for space reasons from the printed version, or to see related articles that may have appeared in regional editions other than theirs. We must imagine that comparable options will quickly develop for readers of scientific papers published electronically. You will be able to scan not just the compressed references to other papers that we now see in the notes, but, if you wish, the abstracts of those papers; soon you will be able to demand the referees' reports and ultimately, perhaps, the raw data.

I go into this detail about the potential of electronic publication because we must recognize how revolutionary it really is. We are not talking of the substitution of one medium for another, of the replacement of the printed page by the screen, with everything else—including editing, refereeing, and readers' correspondence—to go on as before. Lederberg, prophetically, called for a dialectic, or dialogue, between readers, journals, and authors, and saw this as the greatest long-term gain from electronic publishing.

This dialectical mode is already being built. Sydney Brenner remarked to me four years ago that when we deal with genomic sequences, "To publish *is* to perish!" Which is to say, put new sequences on paper and they are lost: they must go directly into the electronic data bases. In February 1993, we all saw the news report in *Science* by Gary Taubes: "Publication by electronic mail takes physics by storm," said the headline, and the article went on at length to recount the burgeoning in the physics community of the practice of communicating a paper to an electronic bulletin board at the same time it is submitted to a journal; the bulletin board makes abstracts available to users, who can then call up full texts at will.[20,21]

At the same time, we must be concerned with the problem of quality control in electronic publication. In some fashion, the speed and potential openness of the network must be adapted to allow certain journals, at least, to preserve—indeed, to enhance—the values that are added by skilled editors and thoughtful expert referees.

What electronic publication will mean for the way we do science is of momentous interest. Of the e-print archives, one Yale physicist was recently quoted as saying that now, "The only thing I use journals for is looking back for papers that came out before the bulletin boards existed."[22] The development of dialogue has hardly begun, but we see already that it is not blinded or anonymous, but open. The time is coming soon when the act of publishing a paper electronically for your colleagues will be but a preliminary step, inviting open criticism, suggestions, rebuttals; the paper so published will be able to go through revisions on the screen, comments and old versions retained for reference; when a colleague calls up the paper six months later it will bring with it indexes of all these responses and changes that have accrued, available to anyone who wants to look. Corrections and retractions will similarly be forever linked to the paper itself.

The communities—already they are being called "collaboratories"—that are coming to use the Internet in this and related ways are, in aggregate, potentially numerous, international, and highly active. Yet the number of scientists in any one such subset may number a dozen, two score, two hundred. The members of any such group, competing and simultaneously collaborating, are each and collectively their own peers, doing their own reviewing. Furthermore, such a group need not be hierarchical; it can be as egalitarian as any community we have known in the sciences.

How large was the group that established the Académie des Sciences in France in the seventeenth century? Or the group that founded the Royal Society at the same period and developed open peer reviewing in the eighteenth? Might electronic publishing possibly be evolving to allow, in this radically new way, the flowering of new golden ages?

ʚ̣ɞ

Francis Crick remarked in Paris six months ago, at a meeting at UNESCO similar in purpose and size to this one, that a chief reason for observing the fortieth anniversary of the discovery of the structure of DNA is that so few of the founders of molecular biology will be alive for the fiftieth. Francis is not here today because he has a touch of influenza. Linus Pauling's prostate is giving him trouble. Fred Sanger never accepts invitations to meetings of this sort. Sydney Brenner often accepts such invitations, and then frequently fails to show up. Erwin Chargaff has said that to write and deliver a paper is now beyond his powers—a grievous loss. André Lwoff is now very old, and I'm told that his memory is not what it was. But there are three others whose absence we must deeply regret. When and wherever we gather later this evening for a drink or two, let our first toast be *to absent friends*—and in particular to three of the great figures of the golden age. To Max Delbrück. To Jacques Monod. And to Rosalind Franklin.

NOTES AND REFERENCES

1. WATSON, J. D. 1980. The Double Helix, Norton Critical Edition. W.W. Norton, New York. This edition reprints Lwoff's and other reviews.
2. JUDSON, H. F. 1979. The Eighth Day of Creation. Simon & Schuster. New York.: 112, 182–183, 202. Unpublished correspondence in the Bragg papers, file on The Double Helix, archives of the Royal Institution, London.
3. FLEMING, D. 1969. Emigré physicists and the biological revolution. *In* The Intellectual Migration. Donald Fleming and Bernard Bailyn, Eds.: 179. Cambridge: The Belknap Press of Harvard University Press. Cambridge, MA.
4. WATSON, J. D. 1966. Growing up in the phage group. *In* Phage and the Origins of Molecular Biology. John Cairns, Gunther S. Stent, and James D. Watson, Eds.: 241. Cold Spring Harbor Laboratory. Cold Spring Harbor, Long Island, NY.
5. JUDSON, *op. cit.*, :350–352, 385–386.
6. DELBRÜCK, M. 1946. Experiments with bacterial viruses (bacteriophages). Lecture delivered on 17 January 1946. Harvey Lectures **41**: 161–192.
7. LWOFF, A. 1953. Lysogeny. Bacteriol. Rev. **7**: 269–337; the epigraphs to individual sections and appendices.

8. PRICE, DEREK J. DE SOLLA. 1963. Little Science, Big Science: 10, 12, 19. Columbia University Press. New York.
9. RIP, A. 1990. An exercise in foresight: The research system in transition–to what? *In* The Research System in Transition. Susan E. Cozzens, Peter Healey, Arie Rip, and John Ziman, Eds.: 387–401.Kluwer Academic Publishers. Dordrecht, the Netherlands.
10. KRIGE, J. The international organization of scientific work. *In* Cozzens *et al.*, *op. cit.*, :179–197.
11. U.S. CONGRESS, OFFICE OF TECHNOLOGY ASSESSMENT. 1991. Biotechnology in a Global Economy. Office of Technology Assessment. Washington, D.C.
12. KRIMSKY, S. 1991. Biotechnics and Society: The Rise of Industrial Genetics. Praeger Publishers. New York.
13. NELSEN, L. L. 1991. The lifeblood of biotechnology: University-industry technology transfer. *In* The Business of Biotechnology. R. Dana Ono, Ed.: 31–75. Butterworth-Heinemann. Boston, MA.
14. JUDSON, H. F. & TONG XUESONG. Unpublished data, March 1994.
15. JUDSON, H. F. 1992. A history of the science and technology behind gene mapping and gene sequencing. *In* The Code of Codes: Scientific and Social Issues in the Human Genome Project. Daniel J. Kevles and Leroy Hood, Eds.: 37–80. Harvard University Press. Cambridge, MA.
16. ZIMAN, J. Research as a career. *In* Cozzens *et al.*, *op. cit.*, :345–359.
17. *Los Angeles Times*, 5 August 1993, :D2, D4.
18. STIX, G. 1994. Extra! Extra! Newspaper publishers reinvade cyberspace. Scientific American **270** (February): 110–111.
19. LEDERBERG, J. 1993. Communication as the root of scientific progress. The Scientist 8 February: 10–11, 14.
20. TAUBES, G. 1993. Publication by electronic mail takes physics by storm. Science **259:** 1246–1248.
21. WULF, W. A. 1993. The collaboratory opportunity. Science **261:** 854–855.
22. TAUBES, G. 1993. E-mail withdrawal prompts spasm. Science **262:** 173–174.

The Biomedical Revolution at 40 Years[a]

DONALD A. CHAMBERS,[a,b]
KENNETH B. M. REID,[c] AND
RHONNA L. COHEN[b]

[a] *Department of Biochemistry (M/C 536), and*
[b] *Center for Molecular Biology of Oral Diseases*
University of Illinois at Chicago
1819 Polk Street
Chicago, Illinois 60612

[c] *MRC Unit on Immunochemistry*
Department of Biochemistry
University of Oxford
Oxford, United Kingdom

INTRODUCTION

The year 1993 marked the fortieth anniversary of the publication of the landmark papers of Watson and Crick, and of Wilkins, Franklin and their collaborators in *Nature*, describing the structure of DNA as a double helix. This series of papers, published in the April 25, 1953 issue of *Nature*, was the focusing event for the biological revolution of the twentieth century. To celebrate, the New York Academy of Sciences, the University of Illinois at Chicago, and Green College of the University of Oxford, sponsored a conference entitled "DNA: The Double Helix–Forty Years, Perspective and Prospective," held in Chicago on October 13–17, 1993. The governor of Illinois, James Edgar, proclaimed the week of the meeting "Biomedical Sciences Appreciation Week." The conference brought together many of the pioneers and current leaders in biomedical science to reflect upon the path biology and medicine has taken since the publication of the original papers describing the double-helical structure. The audience of more than 1,000 persons consisted of scientists and teachers, postdoctoral fellows, graduate and undergraduate university students, and high school students–some of whom came from as far as Berlin and Australia to attend the conference. Of the four public meet-

---

[a] This paper has been adapted from one published in Volume 8 (15): 1219–1226 of the *FASEB Journal* and is reprinted here by courtesy of the Federation of American Societies for Experimental Biology.

ings held in the fortieth anniversary year, to our knowledge this was the largest and best-attended, and its proceedings constitute this volume. Throughout the three days of meetings, the enthusiasm of a scientific festival prevailed, as scholars of all ages raced around the meeting halls, asking their scientific heros for autographs. The conference was organized in eight sections, which will be described in this paper. In addition, there was a conference banquet, which featured James D. Watson and Joshua Lederberg (President of the New York Academy of Sciences) as the principal speakers. At this event Professor Watson was presented with certificates of appreciation by the New York Academy of Sciences, the University of Illinois at Chicago, and Green College, University of Oxford. What follows is a brief recapitulation of the entire convocation.

PART I. THE DOUBLE HELIX: PERSPECTIVE

Professor Gunther Stent of the University of California at Berkeley began the conference with a lecture entitled "The Aperiodic Crystal of Heredity," by making the point that 500 years after the beginning of the Renaissance, a similar renaissance in biology began with the publication of the Watson-Crick "double helix" paper on April 25, 1953. Stent reviewed the early history of genetics, indicating that Aristotle was the first to recognize the concept of heredity's being an informational plan transferred from father to progeny. The Renaissance, although supportive of the physical sciences, produced few great insights into genetics, and it was only with the rediscovery of the 1855 paper of Mendel's concept of the gene, that the science of genetics began to flourish. The early work of Thomas Hunt Morgan and his associates led to the understanding that genes were linearly arrayed on chromosomes, that they could be mutated, and that mutation could supply a molecular understanding of Darwin's concept of natural selection. However, these developments did not focus on the molecular nature of the genetic material. Notwithstanding the publication of the nature of the "transforming principle" by Avery, MacLeod, and McCarty in 1944, scientists were unable to appreciate the fundamental role of DNA as the aperiodic crystal of Avery's paper. In fact, in a ceremonial volume on genetics published in 1950, only two mentions were made of Avery's work: one giving a note of caution by Alfred Mirsky, and another of praise by Joshua Lederberg. What then led to the almost immediate acceptance of the paper by Watson and Crick? Stent argues that a number of new developments opened the pathway to the double helix. These included the development of a concept of a genetic code, first put forth by Erwin Schrödinger in his book *What is Life?*; the establishment of bacteriophage as model systems for genetic investigations by Max Delbrück, and the use of phage by Hershey and Chase to reproduce the work of Avery *et al.* that DNA was the genetic material; the overturn of the tetranucleotide hypothesis by

Chargaff's discovery in 1950 of the equivalence of purines and pyrimidines in DNA; and finally the initial development of DNA X-ray crystallography by Astbury in 1947, followed by the concepts of structural biology and model-building through which protein structure was solved by Linus Pauling in 1951. "What if not Watson and Crick?" Stent hypothesizes that others would have made the discoveries (perhaps Wilkins and Franklin), but more gradually. What made Watson and Crick so unique, Stent suggests, in referring to a view put forth by Peter Medawar, was the all-encompassing dimension of their paper. To quote Medawar, "The great thing about Watson and Crick's discovery was its completeness, its finality. . . . If Watson and Crick had been seen groping toward an answer. . . . [if] the solution had come out piecemeal instead of in a blaze of understanding, then it still would have been a great episode in biological history. But it would not have been the dazzling achievement that it, in fact, was."

PART II. THE PATHWAY TO THE DOUBLE HELIX

Maclyn McCarty, Professor Emeritus at Rockefeller University and one of the co-authors of the "Avery paper," opened this session on the contemporary biology of DNA by reviewing the discovery of DNA as the transforming principle, described in a paper submitted for publication on November 1, 1943, fifty years before this meeting. Although the first observations of DNA's functioning biologically as the transforming principle were made in January of 1941 (and by 1942 the findings were accepted in the laboratory as reproducible), no mention of the work was made to the Rockefeller Institute's community until the laboratory's annual report of 1943, and the manuscript was submitted at the end of the same year. The extreme caution suggested by this timetable, McCarty argues, lay in the need to be careful because of the paucity of information concerning DNA. As late as 1929, P. A. Levine first reported that both deoxyribose and ribose could be found in nucleic acids. At the time of the Avery work, although Albert Matthews, E. B. Wilson, and Emil Fischer had suggested that DNA might be genetically important, most scientists still accepted the tetranucleotide hypothesis of Levine, which implied a structural role for DNA, and agreed that the more diverse nature of proteins would fit them to be carriers of genetic information. Notwithstanding the delay in publication, it was clear that those in Avery's laboratory were cognizant of the significance of their observations. In a letter to his brother, Roy, in 1943, Oswald Avery ascribed to DNA the properties of predictable, hereditary change and spoke to the concept of "gene" and "gene product" in relation of transformation. In his retrospective, Dr. McCarty spoke to the issue of clinical and basic research by reminding us that the impetus for the research on the transforming principle was the need to understand the clinical problem of pneumococcal pneumonia, yet the completion of this work laid the ground for understanding genetics and the immediate acceptance of the Watson-Crick paper, a generic lesson still of major consequence.

Dr. Rollin Hotchkiss, Professor Emeritus at Rockefeller University, whose contributions built upon the work of Avery, MacLeod, and McCarty, defined the decade between 1943 and 1953 in his talk "DNA in the Decade before the Double Helix" as an age in which the concepts of classical genetics (gene linkage groups, mutation, and genetic recombination) were brought together with the material reality of biochemistry (the molecule and its structure and chemical reactions). Three major questions relating to DNA were framed by Hotchkiss for the ensuing decade: (1) Is DNA a specific entity? (2) How do DNAs differ in composition and secondary structure? (3) What is the genetic significance of DNA? Hotchkiss' laboratory itself did much to answer these questions. He developed the first qualitative base analysis of DNA by paper chromatography, enabling him to show quantitatively that the transforming principle was entirely free of protein in that all the glycine found in the DNA preparation could be accounted for by the spontaneous decomposition of adenine. In the course of this work, Hotchkiss also showed that DNA contained the four bases, A, G, C, and T, but uracil could not be detected. In terms of DNA secondary structure, Hotchkiss was one of the first investigators to show the denaturability of DNA and to use the hyperchromic effect to define differences in DNA secondary structure induced by alkali. Finally, as 1953 approached, Hotchkiss and his colleague Harriet Taylor were using genetic bacterial drug resistance to define genetic transformation as the biologic manifestations of a single DNA molecule in a single cell. With this new information in hand, the scientific community was ready to grasp and immediately accept the work of Alfred Hershey and Martha Chase that DNA was the genetic material contained in the bacterial virus, despite the fact that the Hershey–Chase viral DNA preparations were 100-fold less pure than the contested bacterial preparations of Avery. Dr. Hotchkiss concluded by describing the decades before the double helix as an exciting, rewarding time characterized by challenging opportunities, a chance for cooperative fellowship, a growing interest from a wider public, and respect without too much notoriety—attributes, he argues, that make for good science.

Unfortunately, because of illness, Linus Pauling was unable to attend the conference, but in his place, asked Alex Rich of MIT, a former postdoctoral fellow, to describe life in the Pauling laboratory. We are sad to note that Professor Pauling died at the age of 93 on August 19, 1994. Dr. Rich reviewed Pauling's early work on the nature of the chemical bond and ionic crystal structure, and went on to describe Pauling's "growing" interest in the problem of biological specificity resulting from interactions with Karl Landsteiner about antigen/antibody interactions. Pauling's early biological interests were aided and abetted by the fertile environment of Caltech, where he studied with the chair of biology, Thomas Hunt Morgan, the pioneer of *Drosophila* genetics. In the 1930s, Pauling turned his attention to the nature of proteins, studying the magnetic properties of proteins, and with Alfred Mirsky (who went on to study the chemistry and biology of DNA at The Rockefeller Institute) published a paper on the general theory of protein structure, which anticipated the role of hydrogen bonds in maintaining protein structure. In 1940, Pauling turned to biological specificity and with Max Delbrück published a paper on the nature of the intermolecular forces operating in biological processes, arguing that biological specificity results from the existence of molecules with complementary structures. Pauling devised a methodology consisting of developing a strong knowledge of the stereochemical dimensions of molecules coupled with clues obtained from X-ray diffraction to build molecular models (with his colleague Robert Corey) which led to the solution of the α-helical structure of proteins. This method not only resulted in the solution of protein structure, but also it was used by Watson and Crick to solve DNA structure as well. Rich recalled that a critical focus of attention in the Pauling lab was *structure*. The apprehension by Pauling (after discussion with the eminent hematologist, William Castle) of sickle cell anemia's being a function of the difference in structure between the normal and sickle cell hemoglobins allowed the disease to be the first to be understood in terms of molecular interaction, thereby signalling the beginning of molecular medicine. In the pathway to the double helix and to the modern age of molecular medicine, Linus Pauling played a pivotal role in the development of structure–function concepts of the two major biological macromolecules, proteins and nucleic acids.

Dr. Paul Heller, Professor Emeritus at the University of Illinois at Chicago and an eminent hematologist, concluded the session with a paper entitled "Historic Reflections on the Clinical Roots of Molecular Biology." After reviewing the clinical history of sickle cell anemia, Dr. Heller discussed the famous interactions between William Castle and Linus Pauling that resulted in the electrophoretic observations of Pauling and Itano that the molecular basis of sickle cell anemia is an abnormal hemoglobin, resulting in erythrocyte sickling. In terms of the historic record, Pauling's recollection differs from that of Castle in that Pauling remembered overhearing a hallway conversation between Castle and others about the disease at a famous meeting organized by Vannevar Bush to determine the future organization of science in the United States after World War II. Castle, on the other hand, maintained that a face-to-face discussion between Pauling and himself occurred on a train transporting both scientists to the meeting. Heller pointed to this conversation as one of the most productive interactions between basic scientists and clinicians and asserted that such dialogue must become a way of life between scientist and clinician if molecular medicine is to achieve its greatest potential.

PART III. THE STRUCTURE AND SYNTHESIS OF DNA

Dr. Alexander Rich of the Massachusetts Institute of Technology opened this session with a talk about nontraditional DNA structures. Using an historical approach, he described the variations seen in the double helix structure over the period 1953 to the present. He noted that as early as 1953, at Caltech, fiber diffraction patterns of RNA were obtained revealing a helical structure, and with hindsight it was clear that they had been examining double-helical segments of ribosomal RNA. By 1956, now at NIH, Rich and others obtained good evidence that RNA was able to make a double helix. Then a triple-stranded structure was shown to be possible in experiments involving addition of Mg^{2+}, although this was very much a finding ahead of its time since it is only quite recently that great interest has arisen over the formation of triplexes *in vivo* such as is seen in certain DNA regulation events. By the early '60s he had moved on to the question of how DNA makes RNA and the formation of hybrid duplexes. Work on tRNA in 1974 that provided the basis for the three-dimensional structure of that molecule suggested an interesting and complex modification of what the double helix looks like in its "RNA mode." Thus, by 1979, it was considered that most of the varieties of DNA's structure had been identified, but the finding of Z-DNA was an interesting surprise, especially since this type of structure can be seen when certain genes are being transcribed.

Dr. Aaron Klug of Cambridge University continued this session, speaking on protein designs for the recognition of DNA. He reviewed the different types of protein structures that are known to be involved with the regulation of the transcription of DNA with respect to recognition and switching on specific portions of DNA. He showed how different types of protein structure such as "zinc fingers" or "leucine zippers" act as "reading heads," where a small portion of protein structure interacts with a specific DNA sequence. Indeed, one general survey of this area indicated that a helical structural region formed from only 12 amino acids was a common feature of many of these proteins which bound to DNA in a similar way. The leucine zipper structure is composed of repeating leucines plus a region of basic residues that bind the DNA, and the structure relies on the "dimer principle," which makes use of the symmetry of the DNA backbone. The zinc finger motif is very common, but in this case the symmetry of DNA is not used in binding the 30-amino-acid-long segments of protein, which are brought together in a compact mini-domain by zinc atoms. He then brought the proceedings up-to-date by discussing recent data, from Yale and the Rockefeller, which showed that other motifs, such as the A-stranded β-sheets in the TATA-box binding protein, form a pseudo-dimer which transforms the DNA structure, thus showing the beginning of the unwinding of the DNA strands in transcription. Dr. Klug ended his paper with a description of current work on how some of the five different classes of zinc fingers interact with DNA and how combinations of these proteins provide great potential for the regulation of DNA transcription.

Professor Walter Gilbert of Harvard University discussed introns, exons, and the evolution of genes. Up to the 1970s all genes were envisaged as being continuous with RNA polymerase's transcribing in an uninterrupted fashion. It therefore came as a surprise that, while this held true for most bacteria, for all vertebrates and higher plants the situation was much more complex, with their genes being split into relatively small exons (of average size encoding 50 amino acids), which were split by introns, which showed a wide variation in size (30 bases to 100,000 bases). While exons clearly show conservation, there are no general rules governing intron size or function. Professor Gilbert had speculated in 1978 that an important property of introns was likely to increase the rate of genetic recombination between exons and thus accelerate evolution. This concept is displayed by the apparent exon shuffling seen in such protein superfamilies as the immunoglobulin superfamily. It is accepted that introns enhance recombination, but it is not clear how they arise, especially since most bacterial genes do not contain introns. Did the bacteria lose their introns? Or did the higher organisms gain introns by transposition? By analyzing phylogenetic and structural data on ancient genes, such as triose phosphate isomerase, which shows 40 percent sequence identity across all phyla, Dr. Gilbert concluded that the data were consistent with the loss of introns through evolution. This conclusion was supported by the correct prediction that a particular exon would be found in mosquito triose phosphate isomerase. Indeed, because exons, in general, represent compact structures in the proteins, the introns, in his view, are likely to be ancient.

Professor Matthew Meselson of Harvard University asked the question "Why do genomes mix?" He first considered the possible advantages and disadvantages for the mixing of genomes, or sexual reproduction. Clearly, the rapid production of new combinations of genetic material, by, for example, bringing together two advantageous, but rare, mutations could be beneficial, as could the elimination of deleterious mutations, or of retrotransposons, or mobile elements, via sexual reproduction. He has initiated the study of DNA extracted from two species of rotifer: one species (the bdelloid type) is parthenogenic, and therefore has diploid eggs, and the other (the monogonont) reproduces by fertilization of the eggs. Examining for selected genes within the genomic DNA from both species allows conclusions to be drawn about the possible advantages and disadvantages of the two modes of reproduction with respect, for example, to number of mutations seen in the DNA. With respect to the amplified products of a heat-shock protein gene, the monogononts showed only limited heterozygosity (0.1 percent silent divergence), whereas the heat-shock protein gene products from the dbelloid DNA showed many silent mutations (27 percent divergence or two orders of magnitude greater). More data are needed, but an overview suggests that this study will support the view that parthenogenic species have a disadvantage since they represent only about 0.1 percent of all living species and appear to die out quickly.

Professor Howard Temin of the University of Wisconsin at Madison ended the session with a talk on the genetics of retroviruses. He pointed out that the interest in retroviruses is high because of the presence of the HTLV-I and HTLV-II viruses in leukemia and the HIV virus in AIDS; he also pointed to the possibility of using retrovirus-mediated somatic gene therapy for genetic diseases, cancer, and AIDS. He described how the organization of the retrovirus DNA genome gives the virus autonomy from cellular control sequences and how it is relatively easy to manipulate any desired retrovirus vector. He then outlined how errors might arise, via the use of reverse transcriptase, in retroviral replication, and he proposed that the majority of these errors may be related to the reverse transcriptase's bringing about strand transfer. He envisaged that this error-prone mechanism would be an advantage to retroviruses because it generates genetic variation and that HIV might be an example of such mutation-driven evolution. Sadly, Professor Temin died of metastatic adenocarcinoma of the lung not very long after (February 1994) giving this talk. It is clear that recombinant DNA technology owes a great deal to Dr. Temin's pioneering work in the mid 1960s which showed that RNA tumor viruses reproduce by a process involving an integrated DNA intermediate, as well as by his involvement in the discovery of the role of reverse transcriptase in infectious RNA tumor viruses.

PART IV. THE CONFERENCE BANQUET

At the end of the first day of the conference, a banquet was held honoring Watson, Crick, and Wilkins. Although Crick and Wilkins were unable to be present, their contributions are contained in these proceedings.

Sir Crispin Tickell, Warden of Green College, University of Oxford, presented Professor Watson with a certificate of achievement from Green College as Dr. Stanley Ikenberry, President of the University of Illinois, presented a certificate from that institution. Sir Crispin went on to warn us not to become captives of dogmas that do not allow for the progress of civilization. Dr. Joshua Lederberg, President of the New York Academy of Sciences and President Emeritus of The Rockefeller University, presented Dr. Watson with a certificate from the New York Academy of Sciences. In introducing Dr. Watson, Lederberg commented on the question of prematurity in science, previously discussed by Professor Stent, and suggested that the Avery paper had been met with critical skepticism, which is so important in maintaining the integrity of science. Debate and the search for critical corroborative evidence should be welcomed. Lederberg went on to suggest that "to introduce Jim Watson is the oxymoron of all time" and to describe the double helix and its accompanying concept of biological complementarity as the major construct that has "dominated DNA research for the last 40 years, informing every branch of biology and medicine. . . . The duplex is at the root of DNA as an informative molecule and of every experiment involving sequence specificity, enzymatic reactivity, biological specificity, and so forth."

James Watson then responded to a cheering, tumultuous welcome by more than 600 guests at the banquet with the almost stunned look of disbelief that asks "why me?", allowing that many of the audience were equally deserving and still making important scientific contributions, such as Drs. Gilbert, Baltimore, and Temin, among others. Watson went on to describe Chicago as the place that had had the most influence on his life. He grew up on the south side of Chicago, went to South Shore High School, developed a love of books at the 73rd Street library, and a love of birds birdwatching with his father. Watson was introduced to university science and the quest for knowledge at the University of Chicago, which he credits for instilling within him three values: (1) to read original sources instead of textbooks; (2) to understand the importance of ideas and theory rather than the accumulation of fact; and (3) to concentrate on learning how to think rather than on improving memorization skills. In retrospect, Watson believes that the acquisition of these mental habits made him acceptable first to Luria and Delbrück and later to Francis Crick. Watson credits Erwin Schrödinger's book *What is Life?* and the University of Chicago geneticist, Sewall Wright, as the two most important influences on his decision to have "the gene as my life's principal objective." Watson concluded with some powerful statements on the role of genetics in the present and future. Responding to critics of genetic solutions to medical problems, Watson argued that "we are increasingly going to be accused of unwisely playing God when we use genetics to improve the quality of either current or future human life. Partly these accusations reflect the objections of individuals who don't think we have the right to do 'God's' work. But I also sense that sometimes the uneasiness comes from the fear that we might someday use genetic procedures in Hitler-like ways, using our scientific powers to further discriminate against unpopular political and racial groups. But diabolical as Hitler was, and I don't want to minimize the evil he perpetuated using false genetic arguments, we should not be held in hostage to his awful past. For the genetic dice will continue to inflict cruel fates to all too many individuals and their families who do not deserve this damnation. Decency demands that someone must rescue them from genetic hells. If we don't play God, who will?"

PART V. MOLECULAR, CELLULAR, AND INTEGRATIVE BIOLOGY

This session featured Susumu Tonegawa, François Jacob, and Marshall Nirenberg, three Nobel laureates whose Nobel prize work focused on the genetic code and gene regulation. Of interest is that the work discussed by Tonegawa and Nirenberg revealed how their thinking has evolved to study the role of genes and gene products in the nervous system and neurobiology.

Dr. Tonegawa, of the Massachusetts Institute of Technology, opened the session by describing how gene knockout technology has allowed him to study mutant mice lacking kinase genes controlled by Ca^{2+}, either calmodulin-kinase II or protein kinase C γ. Using such mouse mutants, Tonegawa is asking whether these animals show impairment in long-term memory and learning. Early experiments in which different Ca^{2+}-dependent protein kinases were obliterated suggest that these molecules could play key roles in both long-term potentiation and long-term depression and in neurobiological events characteristic of learning and memory. It is the hope of Tonegawa that continued experiments using genetic techniques of gene knockout will elucidate the neurobiological mechanisms involved in learning and memory.

Professor François Jacob of the Institut Pasteur reviewed the history of bacterial genetics, in which he played such a major role, and discussed the dominance of transcriptional regulation to regulatory circuits in both eukaryotes and prokaryotes. In understanding such circuitry, the famous remark made by his collaborator, Jacques Monod, that "what is true of *E. coli* is true of the elephant" gains in contemporary significance. Jacob ascribed the current rapid growth in understanding eukaryotic regulation to the advances in recombinant DNA technology coupled with the uses of yeast and *Drosophila* as model systems. He argued that the additional complexity necessary for eukaryotic gene regulation arises from the multimeric diversity of eukaryotic regulatory proteins, allowing for building aggregations of such proteins which can acquire necessary regulatory specificity through interactions that characterize the protein aggregate. Professor Jacob calls such protein aggregates "aggregulates" and he suggests that the complexities of eukaryotic regulation will be described in terms of the interactions of such molecules.

In the concluding talk of the session, Dr. Marshall Nirenberg of the National Institutes of Health began by describing some of the developments in the nature of the genetic code that have occurred more recently, including the discovery of the twenty-first amino acid, seleno-cysteine, coded for by a specific tRNA with UGA as its anti-codon, dialects in the genetic code (e.g., mitochondria have somewhat different codons), introns, exons, RNA splicing and editing, and the presence of natural suppressor RNAs in eukaryotic cells. He feels that the next challenge in genetic code research may be the explanation of relationships between codons and the specific amino acids they code for. Analysis of such relationships may give rise to the possibilities that more primitive codes have generated the present genetic code. Dr. Nirenberg went on to discuss his present work on the genetic mechanisms that dictate the development of the nervous system in *Drosophila*, in particular the role of the homeobox gene, NK-2, in the development of the *Drosophila* nervous system. The NK-2 gene is the earliest known neural gene regulator in *Drosophila*, and its activation responds to changing fluxes of dorsal-ventral regulatory proteins, both inhibitors and activators. By studying the detailed regulation of the NK-2 gene, and the nature of its protein product, which he hypothesizes regulates the differentiation of neuroectoderm cells into more specialized neuroblasts, Nirenberg hopes to address the nature of the rules that govern the development and function of the nervous system.

PART VI. DNA AND MOLECULAR MEDICINE

Professor David Weatherall of the University of Oxford was the first speaker of this session. In discussing the molecular basis for phenotypic diversity of genetic disease, Sir David used the hemoglobin disorders of sickle cell anemia and thalassemia to demonstrate that in the complex interactions between genotype and environment some phenotypic variations may be explained by the range of underlying mutations in a single gene combined with heterogeneity at a few other loci. Other single-gene diseases have recently been associated with multiple trinucleotide repeats, which in the case of Huntington's disease may be related to the age of onset of disease. However, the basis for heterogeneity of most monogenic diseases, even those such as hypercholesterolemia and Duchenne muscular dystrophy where much is understood about the pathophysiology, is still a puzzle. Understanding polygenic diseases, Sir David cautioned, will remain a major challenge.

Dr. Richard Mulligan of MIT, an expert in retrovirus-mediated gene transfer, described himself as being born scientifically in the era of tumor viruses. He went on to detail that the key event in gene transfer–mediated gene therapy is the ability to specifically transfer genes to the correct part of the genome and to maintain the persistence of the particular transferred gene. At the time of the meeting, 31 of the 40 clinical protocols currently being investigated related to cancer, and Mulligan suggested that future developments in gene-mediated cancer therapy would involve modifying the immunogenicity of cancer cells and activating and/or transferring specific cytokines to the tumor location. Although initial experiments utilizing cytokines such as GMCSF in conjunction with other cellular regulatory molecules are proving successful, much additional work needs to be done to define tumoricidal-effective batteries of biological-response modifiers. The greatest challenge to successful gene therapy, Mulligan feels, is developing the ability to transplant *in vitro* genetically modified cells to recipient patients.

In this session, returning to the theme of molecular mechanisms of behavior discussed by Susumu Tonegawa, Dr. Eric Kandel of Columbia University addressed long-term learning (i.e., days, weeks, and even the lifetime of the organism). He noted that such mechanisms have in common the modulation of pre-existing synapses and covalent combinations utilizing protein kinases. However, long-term memory requires the synthesis within a specific time frame of new, slowly turning-over proteins. Among the questions his laboratory is attempting to answer are which genes are turned on in long-term memory, what underlies the switch from short-term memory, and what underlies the stability of the new proteins. Kandel's co-workers have devised an *in vitro* system (paralleling their *in vivo* experiments with the invertebrate *Aplysia*) in which sensory and motor neurons can be co-cultured. These experiments simulate learning and memory by modulating exposure of the cells to pulses of serotonin. The extension (learning), strength (memory), and retraction (forgetting) of connections and between single neurons and the changes within the cells can be monitored at the molecular level. Repeated stimulation results in the translocation of the catalytic unit of cAMP to the nucleus, where it binds a cyclic AMP response element binding protein (CREBP) and leads to phosphorylation of substrates common to short- and long-term memory. Decreasing the amount of cAMP by injecting extra copies of CREBP directly into the nucleus, increasing endogenous cAMP response elements (CRE), or making new CREs in transgenic mice blocks long-term memory, but does not affect short-term memory. The genes activated by the CREBP are intermediate early genes, ubiquitin hydrolases whose function may be involved in cleavage of the regulatory subunit and transcription factors. In *Aplysia* the CREB protein is a leucine zipper transcription factor having a basic DNA binding domain and an activation domain. This CREBP may in part be the switch between short- and long-term memory. Kandel commented that progress in defining synaptic transmission has evolved slowly. The late 1950s saw the delineation of the nicotinic neuromuscular junction, in the '70s the effects of hormones and second messengers were described, and in the '90s there is the beginning of understanding that neurotransmitters act like growth factors, translocating their signals to the nucleus and making neurobiologists consumers of molecular biology.

Professor Kay Davies of the University of Oxford followed with a detailed discussion of Duchenne muscular dystrophy (DMD). In the muscle sarcolemma, the product of the DMD gene, dystrophin, normally binds to a glycoprotein complex linking the internal cytoskeleton of the muscle cell with the extracellular matrix. The absence of dystrophin results in the disease phenotype. Recent work suggests that increased expression of utrophin, a related autosomally encoded protein localized to a different chromosome, may be able to replace dystrophin in the glycoprotein complex. This potential for gene therapy is currently being explored.

Professor Henry Bourne of the University of California, San Francisco enlarged the domains of molecular medicine in his discussion of the role of signal transduction in biological regulation and in the molecular mechanisms of disease. Focusing on G signaling proteins, he described the centrality of GTPases to biological control of a variety of processes such as protein synthesis, cytoskeletal changes, vision, inflammatory responses, and cancer and other diseases. The determination of the molecular fine structure of the G proteins has resulted in the understanding of a variety of apparently diverse diseases including cholera, acromegaly, and the 30 percent of types of cancer in which the ras G protein has been mutated. All these diseases have in common an altered αGs protein component of the G signaling system such that normal bioregulation of G protein–mediated signal transduction is no longer possible. By determination of the three-dimensional structure of the αG protein unit, Professor Bourne and his colleagues have been able to locate the molecular lesion to a specific arginine moiety. Through the use of recombinant DNA technology they have made constructs of G protein domains which will mimic and/or repair the aberrant signal transduction mechanisms. Such experiments will lead to enhanced knowledge, undoubtedly setting the stage for rational therapeutic intervention to restore proper signal transduction control and thereby eliminate pathology.

Professor Susan Ross of the University of Illinois at Chicago described the use of transgenic mice to study obesity genes. Her laboratory has identified and used an adipocyte enhancer to drive the expression of heterologous genes and alter the function of adipose cells resulting in fat cell tumors. Brown fat cell lines derived from these transgenic mice represent the first model system for metabolic studies of brown fat and genetic regulation of the mitochondrial uncoupling protein (UCP) that is uniquely expressed by brown fat.

Professor Harold Slavkin of the University of Southern California extended the theme of molecular medicine to problems of dental and craniofacial development. For example, a Hox code for first and second branchial arch development has been formulated and a large number of homeobox genes expressed during early craniofacial development have been isolated, sequenced, and mapped. The genetic bases for several craniofacial syndromes have been established and candidate genes for some craniofacial anomalies, such as cleft lip, have been identified. Approximately 70 human genes related to dental tissue disorders, clefting defects, and craniosynostosis have been mapped. Gene therapies are currently being sought for infectious oral diseases and xerostomias as well as developmental anomalies of dental structural proteins such as tooth enamel and dentin.

PART VII. DNA, ONCOGENES, AND CANCER

Sir Richard Doll, in chairing this session, reminded the audience that there is collective pressure on scientists to discover how cancer can be prevented or treated more effectively. Molecular biology is providing a way to classify people according to their susceptibility to particular types of cancer and to understand the molecular basis of neoplasia; both approaches provide the potential for intervention that could reduce risk for the individual. Sir Richard cited the example of cancer of the cervix and its association with the papilloma virus, and the retinoblastoma gene product, the RB protein.

Professor Robert Weinberg of the Massachusetts Institute of Technology began the scientific session with the theme introduced by Sir Richard, that cancer is created by multiple steps, "a succession of four to five distinct stochastic events happening during a person's lifetime." Each step is demarcated by a genetic change conferring growth advantage. The work Weinberg described was focused on tumor suppressor genes that limit normal cell proliferation and, by extension, prevent malignancy. He reviewed the state of understanding of how the cell cycle is controlled by kinase-mediated phosphorylation and dephosphorylation of cyclins and how viral oncoproteins (such as the adenovirus oncoprotein E1A, SV40 large T or human papilloma virus type 16) complex with the retinoblastoma protein, RB, compromising its nuclear binding ability, altering its state of phosphorylation, and neutralizing growth control.

Professor David Baltimore of The Rockefeller University related an anecdote in which, as a college student in the late 1950s, he had occasion to drive Jim Watson to the airport. Watson told him that a virus that could cause cancer had just been discovered and that the virus contained only a small amount of DNA. "That such a virus is able to cause cancer means that a very small amount of genetic information is all that's required to cause cancer . . ." he recalls Watson saying. Baltimore focused on the role of nonreceptor kinases and he presented evidence that suggested that such kinases organize signal transduction proteins in a manner similar to that of the receptor tyrosine kinases. SH2 and SH3 domains are found in many different proteins, as well as being distinct proteins in themselves having no catalytic function other than being adapters which act as linkers between other protein molecules such as a receptor tyrosine kinase and a target protein. He described SH3 as a domain that can bind to sites on other proteins that are linear determinants and that are proline-rich. Individual SH3 regions bind with high specificity to particular linear sequences. The oncoprotein Abl contains an SH3 domain and SH3 binding sites that greatly increase its potential for interaction with signaling systems. The oncoprotein Crk (which is a fragment of an adaptor protein) seems to show marked specificity for binding to those sites in Abl and this may provide an explanation for Crk-mediated cell transformation.

Dr. Harold Varmus, Director of the National Institutes of Health, reviewed the ways that retroviruses capture cellular genes and showed how recent research has moved beyond finding the genes and uncovering their control mechanisms to an understanding of how they function in the context of the whole organism, from development to growth and differentiation, mutation, and cancer. The ability to manipulate the mouse genome through targeted mutations and knockout germ line mutations to engineering transgenic mice is leading to the development of whole-system approaches for investigation of multistep carcinogenesis. He reminded the audience that science is indebted to Peyton Rous, who in 1910 reported the association of a virus with mammary tumors in chickens, the Rous sarcoma virus. The src-gene family now numbers nine tyrosine kinase genes expressed widely with different levels of tissue specificity. The extensive distribution of src would suggest that members of this gene family play a role in normal development. Clues as to their functions have been found indirectly by correlation of tissue expression and associations with certain proteins, such as the association of the tyrosine kinase lck with the T-lymphocyte cell surface proteins CD4 and CD8. Deletion of src in knock-out experiments had little gross effect on the mice other than osteopetrosis and delayed tooth eruption. This surprising finding led Varmus and collaborators to follow the message of developmental biologist Lewis Wolpert and find environmental challenges that will reveal knock-out phenotypes. Src genes were conserved for some good reason, and scientists, Varmus said, needed to identify the right challenge in order to discover their function. Using this approach to investigate functional changes in cells of hematopoietic lineage, they discovered that double homozygous (hck/fgr) knock-out mice had lethally diminished peritoneal macrophage phagocytotic activity which was specific only for certain pathogens. Other src family double homozygous transgenic mice have been shown to have hematopoietic deficits in maturation, growth retardation, and immune complex diseases. This work now needs to be extended to demonstrate the rescue of selective genes in specific tissues. In another approach involving mammary tumorigenesis, Varmus described work on the WNT-1 gene, which is part of a large gene family conserved from insects to *C. elegans* and mammals. WNT-1 plays a role in neural development during embryogenesis and is involved in multistep carcinogenesis. Collaboration between fibroblast growth factor genes and WNT genes has been shown to increase and accelerate the development of mammary tumor in mice, particularly in the absence of a functional p53 gene.

Dr. Ira Pastan of the National Cancer Institute spoke of the desperate need for new therapies for cancer and the approach taken by his laboratory of using cell surface molecules expressed by tumor, but not normal cells, as targets for immunotoxic moieties. They are using a powerful bacterial toxin, *Pseudomonas* exotoxin A, which has been cloned and is readily made in *E. coli*, and whose three-dimensional structure is known. This enables the production of mutant forms of the toxin with different pharmacologic properties that are chemically linked to an antibody directed at the tumor-specific cell surface antigen selected. The immunotoxin enters the tumor cell by endocytosis and is proteolytically cleaved, releasing a fragment targeted to the endoplasmic reticulum, where it translocates to the cytosol, ADP-ribosylates elongation factor 2, irreversibly inhibits protein synthesis, and kills the cells. Because the immunotoxins are recognized by the body as foreign, they elicit the production of neutralizing antibodies within 10 days, thus limiting the use of such agents to adjuvant therapy for patients who have already undergone surgery or radiation. A phase I clinical trial using this approach is currently under way.

PART VIII. RECOMBINANT DNA AND BIOTECHNOLOGY

The opening talk in this session was given by Dr. David Jackson of DuPont-Merck Pharmaceuticals, who discussed the many industrial and commercial openings made possible by the discovery of the structure of DNA, the development of DNA sequencing and the methods for synthesis of DNA which collectively form the basis of modern recombinant DNA technology. He pointed out that even by 1971, by a series of apparently unconnected routes of *basic* research, a vast new area of biotechnology was rapidly developing covering a very wide spectrum of disciplines from animal husbandry and agriculture to environmental sciences, epidemiology, taxonomy, and polymer sciences. Even the entertainment industry, in films such as *Jurassic Park*, has joined the boom, illustrating one route by which the general public can be easily reached and made aware of this important topic. Dr. Jackson emphasized the growth of government spending on R&D and, like other speakers at the conference, equated the impact on society of the ability to "read," "write," and "edit" DNA with that of other, revolutionary scientific findings, such as Copernicus' theories or the development of quantum mechanics.

John Baxter of the University of California, San Francisco then illustrated how advances in recombinant DNA technology had rapidly increased the understanding of thyroid hormone action.

Professor Leroy Hood of the University of Washington in Seattle ended this session by providing insight into the progress being made by those involved in the large-scale DNA sequencing projects that form part of the Human Genome Project initiative. He emphasized the importance of the Project in the understanding of both the evolution of gene families and the roles and structures of the proteins within these families. He considered that extensive studies on multigene families would eventually lead to a much greater understanding of fundamental mechanisms in complex networks such as neutral networks. Hood estimated that within 15 years scientists would be faced with trying to correlate protein structures and functions to the products of as many as 100,000 human genes. This major challenge is exemplified by the immunoglobulin superfamily, which already has more than 200 members that, despite having the same overall protein fold, can show up to 90 percent difference in linear sequence between members and a wide variety of different functions. The future major challenges, Dr. Hood considered, will probably be the analysis of networks, such as neuron networks, which will require interdisciplinary research of a much higher order than is usually seen today. He suggested that future progress in this area will require graduate students to have projects (and supervisors) that bridge at least two quite diverse disciplines in order to bring together topics such as biology, applied mathematics, computer science, and engineering. Hood also emphasized the importance of stimulating and educating the scientists of the future and illustrated how this could, and has been, done by teaching high school students how to carry out genomic sequencing, which allows them to identify their very "own" unique piece of information that will be useful to all those carrying out genomic analysis. These new approaches to thinking about and teaching biology, he considered, should allow us to confidently face future large-scale biology projects.

PART IX. THE DOUBLE HELIX: PROSPECTIVE

The final session of the meeting featured Sir Walter Bodmer, Director-General of the Imperial Cancer Research Fund Laboratories, London, and Horace Freeland Judson, author of *The Eighth Day of Creation: The History of Molecular Biology*.

Sir Walter concluded the scientific portion of the meeting with a consideration of DNA, asking prospectively where genome analysis will lead us. Bodmer suggested that the completion of the sequencing of the human genome will have great ramifications leading to the identification of disease genes and detection of disease susceptibility and a new era of molecular epidemiology. In addition, it is likely not only that there will be widespread somatic gene therapy, but also that the domain of pharmacology will spread to encompass gene regulation and gene products corresponding to receptors, enzymes, signalling proteins, and transcription factors. The biggest challenge will be to analyze gene function. Bodmer predicted that in the next fifty years the greatest impact of the biological revolution will be on human disease. Perhaps cancer, heart disease, autoimmune diseases, and degenerative diseases will be preventable or curable, thus enhancing the quality of life without increasing life span. Bodmer, however, poses an interesting dilemma: suppose we do learn to increase the life span, then to what age and for whom? Who will make the decisions for society? These real questions led Sir Walter to hope that enhanced knowledge of the genetics of behavior could lead to the next dramatic revolutions in understanding. The one sure prediction that Sir Walter could make is that "our social adaptation will be far too slow to deal with the extraordinary advances that will come from the revolution of genome analysis." To overcome this problem, enhanced public understanding of science is essential. Sir Walter concluded that "All scientists should at least take some responsibility for explaining their science intelligibly to the general public. In this way, we might hope that the extraordinary advances in dealing with human disease, improving agriculture, and conceptually in understanding our own evolution, and evolution in general, will be taken advantage of to the full for the greater benefit of mankind."

Horace Freeland Judson, in a paper entitled "The World We Have Lost," pointed out that 1993 was not only the ruby (fortieth) anniversary of the Watson–Crick *Nature* paper, but also the silver (twenty-fifth) anniversary of Watson's autobiography, *The Double Helix*. Contrasting then and now, Judson described the beginnings of the golden age of molecular biology which was characterized by the "excitement, and the free communication among a comparatively small number of early participants." The golden age, Judson argues, has given way to an age of brass resulting from and marked by three great structural transformations. The first transformation is an increase in scientific misconduct, particularly in the theft of intellectual property, or plagiarism. The second is the evolution of molecular biology and science, in general, from a state of exponential growth to a steady-state, thereby creating a funding shortage and overwhelming competition among scientists; this in turn fuels both the rise in misconduct and a movement from academe to industry (as a source of funding), marked by the ethos of the business enterprise. The third great transformation is the shift to electronic distribution of information and electronic publishing. The rapidity of electronic information transfer promises change to the scientific community, not only in terms of classic journal publishing, but, perhaps even more importantly, in the generation of new collaborations, new groupings of scientists, and a loss of hierarchies. The potential of electronic information nets could radically change the organization and sociology of the scientific effort, setting the stage for new golden ages of science.

As a final note: During the two-year period as this conference was being organized, the concern was not infrequently raised that scientists and students of today have little interest in the development of their disciplines, and that, as molecular biology has flowered, they have tended to embrace technology. Accordingly, this hypothesis led to anxiety that the conference would be of little interest and that few would attend. But, gratifyingly, the overwhelming response in numbers of attendees as well as the enthusiasm and interactions of the participants across generations at a meeting that coupled historic overviews with "state of the art" reviews attests to a continuing profound association between intellectuality and technology as both students and researchers contribute to the progress of the biomedical revolution. Continued opportunities to foster this dynamic will assure the flowering of new "golden ages" as molecular medicine proceeds to confront the pathologies of human existence.

Index of Contributors

Alberini, C.M., 261–286
Alexandropoulos, K., 339–344
Apriletti, J.W., 366–389

Baltimore, D., 166–170, 339–344
Baxter, J.D., 366–389
Blake, D.J., 287–296
Bodmer, W., 414–426
Brinkmann, U., 345–354

Chambers, D.A., xiii–xiv, 1–11, 24, 171–173, 413, 441–458
Cheng, G., 339–344
Choy, L., 297–313
Cicchetti, P., 339–344
Crick, F.H.C., 198–199

Davidson, R.L., 355
Davies, K.E., 287–296
Doll, R., 329–330

Fitzgerald, D.J., 345–354
Fletterick, R.J., 366–389

Ghirardi, M., 261–286
Graves, R.A., 297–313

Heller, P., 86–96
Hood, L., 390–412
Hotchkiss, R.D., 55–73, 205–207
Huang, Y-Y., 261–286

Jackson, D.A., 356–365
Jacob, F., 218–223
Judson, H.F., 427–440

Kandel, E.R., 261–286
Klotz, I.M., 46–47
Klug, A., 143–160
Koop, B.F., 390–412

Lederberg, J., 176–179, 182–193

McCarty, M., 48–54

Nguyen, P.V., 261–286
Nichols, R.W., 174–175
Nirenberg, M., 224–242

Pai, L.H., 345–354
Pastan, I.H., 345–354

Ren, R., 339–344
Ribeiro, R.C.J., 366–389
Rich, A., 74–82, 97–142
Ross, S.R., 297–313
Rowen, L., 390–412

Schaufele, F., 366–389
Slavkin, H.C., 314–328
Solaro, R.J., 211–212
Soleveva, V., 297–313
Spiegelman, B.M., 297–313
Stent, G.S., 25–31
Storti, R.V., 83–85

Temin, H.M., 161–165
Tickell, C., 180–181
Tinsley, J.M., 287–296
Tonegawa, S., 213–217

Wagner, R.L., 366–389
Walton of Detchant, Lord, 243–244
Watson, J.D., 194–197
Weatherall, D.J., 245–260

Weinberg, R.A., 331–338
West, B.L., 366–389
Wilkins, M.H.F., 200–204

Subject Index*

ADA deficiency. *See* Immunodeficiency
AIDS, 167, 320, 448. *See also* HIV; Retroviruses
Allergies, 415
Altman, R., 48
Altman, S., 183
Alzheimer's disease, 415, 424
Amelogenesis imperfecta, 314
Antisense RNA, 113–114
ApC/EBP. *See also* CCAAT enhancer-binding protein
 cAMP-dependent expression of, 277–279
 5-HT induction of, 272–277
 and long-term facilitation, 267, 272–277
Aperiodic crystal of heredity, 25–31, 442
Arber, W., 360
Aristotle, 25, 28, 442
Astbury, W. T., 29, 198, 200, 443
Asthma, 415
Atheroma, 257
Aus meinem Leben, 49
Austrian, R., 56, 68
Autoimmune diseases, 424
Avery, O. T., 5, 50, 61, 201. *See also* Avery-MacLeod-McCarty discovery
Avery-MacLeod-McCarty discovery
 and aperiodic crystal, 442
 and Avogadro's number, 4, 185
 development of, 55–57
 and DNA as language, 358
 and genetic role of DNA, 26–27, 29, 48–54, 442
 perspective on, 48–54
 "premature discovery" of, 26–27
 recognition of, 199
 transforming principle as gene, 184

Bacterial genetics, 66, 450. *See also* Molecular biology; Transformation
Bacterial viruses. *See* Phages
Baltimore, D., *21*
Basic leucine zipper, 146–149, 446
Bateson, W., 46
Baxter, J., *19*
Bayev, A., *80*
Beadle, G., 218
Becker muscular dystrophy (BMD), 287–288. *See also* Duchenne muscular dystrophy
Berg, P., 358

Bernal, J. D., 200, 425
Bernheimer, H., 68
Biological information. *See also* Human Genome Project
 categories of, 390–391
 deciphering chromosomal languages, 391–392
 DNA as informational duplex, 190
 T-cell receptor sequence analysis, 395–402
Biomedical revolution
 DNA and molecular medicine. *See also* Biomedical science
 and craniofacial-dental development, 453
 gene therapy, 451–452
 molecular mechanisms of behavior, 452
 signal transduction in biological regulation, 452–453
 single-gene diseases, 451
 transgenic mice use in, 453
 DNA/oncogenes/cancer
 gene manipulation and, 455
 multiple-step nature of cancer, 454
 new cancer therapies, 455–456
 nonreceptor kinases and, 454
 double helix prospective, 457–458
 and integrative biology, 450–451
 pathway to double helix, 443–445
 recombinant DNA and technology, 456–457. *See also* Human Genome Project
Biomedical science. *See also* Oral medicine; Recombinant DNA technology
 double helix and, 182–183, 190–191
 development of molecular genetics, 186–187
 DNA as informational duplex, 187–190
 historical background, 183–186
 mechanistic interpretation and, 192
Biotechnology. *See* Recombinant DNA technology
Bodmer, W., *21*
Boivin, A., 56
Boyer, H., 358, 360
Bragg, L., 428
Brenner, S., 7, *9, 36,* 358

CAMP
 and implicit learning forms, 279–280
 and long-term facilitation, 265–267, 270–271, 277–279
 in short-term memory, 263

* Page numbers in italics indicate photographs.

cAMP response element (CRE)
  in long-term facilitation, 267, 278
  protein binding to, 267
Cancer. *See also specific types*
  adenovirus and, 334–335
  genetic makeup and, 257
  genome analysis and, 424
  immunotherapy for, 329–330, 416
  molecular basis of, 331–337
  multistep process of, 331–332, 454
  new therapeutic agents for, 345–346
  oncogenes and, 332–334, 453–456
  oral forms, 314
  recombinant toxin therapy for, 329–330, 416
  specific antiviral immunization and, 329–330
  tumor suppressor genes and, 332–333
Candida albicans infection, 314
  gene therapeutics for, 321, 323
Caspar, D., *36*
Castle, W., 81, 91–93, 445
Cavendish Laboratory, in molecular biology's golden age, 429, 431
CCAAT enhancer-binding protein (C/EBP)
  and *c-fos* regulation, 267
  and long-term facilitation, 272–276
C/EBP. *See* CCAAT enhancer-binding protein
Cech, T., 183
Cell in Development and Heredity, The, 49
Central dogma
  of information flow, 188–189
  process described, 83–85
  reverse transcription in, 84–85
  RNA-DNA relations in, 70
  Temin's challenge to, 167
  transformation studies and, 186
  Watson's concept of, 7–8
Cervical cancer, 329, 454
Chambers, D., *208*
Chang, A., 360
Chargaff, E., *4*
  appreciation of, 199
  DNA analyses, 58, 60
  and DNA nucleotide variety, 29
  and nucleotide base ratios, 60–61, 185
  overturn of tetranucleotide hypothesis, 442–443
  "science makes the men," 30
Chase, M., *7, 41. See also* Hershey-Chase experiments
Chemosensory deficiencies, 314
Chloroplasts, DNA in, 118–119
Chorionic somatomammotropin genes (hCS), 377–378
Christmas disease, 256
Chromatin, 46, 49
Circuits. *See* Gene regulatory proteins

Cohen, S., 358, 360
Cold Spring Harbor Phage Group. *See* Phage Group
Cold Spring Harbor Symposium for Quantitative Biology, 1961, 220
Colon cancer, 331–332
Corey, R., 77, 121, 445
Coryell, C., 76
Cowan, P., *36*
Craniofacial-oral-dental anomalies. *See also* Oral medicine
  congenital malformations, 314–319
  craniofacial morphogenesis, 320
  gene mapping of, 318–319
Crawford, L., *43*
CRE. *See* cAMP response element
CREB-like binding proteins
  in implicit learning forms, 279–280
  in long-term facilitation, 267, 272
Crick, F. H. C., *2, 33, 34, 36, 43, 45*
  in 1953, 1
  cooperative discovery of DNA, 198–199
  and *The Double Helix*, 428
  and α-helix, 146
  and language of DNA, 358
  and molecular biology, 4, 7
  and molecular biology's golden age, 430
  and RNA double helix, 99
Crothers, D. M., 358
Cystic fibrosis, 256, 322

Darlington, C. D., 200–201
Darnell, J., *43*
Darwin, C., 26, 194
Davies, D., 100
Davies, K., *22*
Davis, R., 359
Delbrück, M., *6, 9, 38, 39, 40*
  on biological specificity, 445
  Bohr's influence on, 200
  gene replication model, 69
  gene stability hypothesis, 28
  and intermolecular forces, 76
  and molecular biology's golden age, 218, 429–431
  and molecular genetics, 4, 28
  and phage infection studies, 186
  phage models for genetic studies, 442
  Watson postdoc with, 99, 195
Dental caries, 314, 321. *See also* Oral medicine
Dentinogenesis imperfecta, 314
Diabetes
  evolution and, 257–258
  obesity and, 304, 309–310
  polygenic factors in, 258
  transgenic mouse model for, 258

SUBJECT INDEX 463

DiGeorge syndrome, 314
DMD. *See* Duchenne muscular dystrophy
DNA. *See also* Central dogma; Gene(s); Nucleic acids
  as aperiodic crystal of heredity, 25–31, 442
  base pairing equivalence rule, 29
  and bioengineering tools, 9, 189–190
  and biotechnology. *See* Recombinant DNA technology
  and cancer, 329–330
  chemistry of, 58–64
  chromatography of hydrolysate bases, 58–61
  denaturability of, 61–64, 187–188
  discovery of, 1–2, 27, 48, 183
  DNA polymerase and, 102, 108
  DNase and, 52–54, 58, 417
  and drug discovery, 190
  duplex structure of, 182–183. *See also* DNA double helix
  and enzymology, 189
  and evolution, 188, 408, 410, 422–423
  four-stranded, 133–137
  and genetic code, 83–85
  and genetic linkage/recombination, 66–67
  as genetic material, 2, 27
    and Avogadro's number, 4, 185
  genetic role of. *See also* Avery-MacLeod-McCarty discovery; Hershey-Chase experiments
    acceptance of, 4, 7
    doubts about, 2, 4, 27–28
    perspective on, 48–54
    Phage Group and, 4, 28
  history of. *See also* Avery-MacLeod-McCarty discovery; Hershey-Chase experiments
    acceptance of central dogma, 186
    Fischer in, 49–50, 443
    historic papers, 13–17
    Miescher in, 27, 48, 183
    to 1954, 65
    phage experiments, 186
  hybridization of. *See* Hybridization
  as informational duplex. *See also* Human Genome Project
    central dogma and, 188–189
    in drug discovery, 190
    engineering tools and, 189–190
    and genetic relatedness, 188
    mutagenesis and repair and, 188
    reverse transcriptase and, 189
    separability of strands and, 187–188
  intraspecific polymorphism of, 188
  as a language, 358, 363–364
  and mechanistic interpretation, 192
  misconception about, 57
  and molecular biology, 7, 9, 83–85

  molecular/genetic study of, 68–69. *See also* Transformation
  and molecular medicine, 451–453
  mutagenesis and repair of, 188
  nature of genetic code, 83
  nucleotide content variability of, 29
  nucleotide equivalence rule, 2, 29
  in organelles, 118–119
  replication of, 83–84, 97, 107
  of retroviruses, 162–163
  RNA-dependent synthesis of, 167–168
  RNA involvement with, 58–60
  and RNA synthesis, 107–109
  sequencing projects, 456
  splicing, 189
  structure of. *See* DNA double helix
  synthesis of, 102, 108–109, 357, 446–448. *See also* Hybridization
  TATA-box and, 150, 272, 446
  transcription of, 143–145, 188–189
  and transformation questions, 57
  as transforming principle, 2, 48–54, 58–61
  triplex (H-DNA), 106
  in tRNA synthesis, 117–118
  viral single-stranded, 109
DNA-binding domains
  hormone receptor, 150–153
  zinc-binding, 155–156
  zinc finger proteins in, 150–155
DNA double helix
  acceptance of, 7
  at atomic resolution, 121–124
  bending mechanism, 133
  Chargaff's question about, 30
  concepts of, 182
  Conference sessions
    banquet, 448–449
    DNA and molecular medicine, 451–453
    DNA/oncogenes/cancer, 453–456
    DNA structure and synthesis, 446–448
    molecular/cellular/integrative biology, 449–451
    pathway to double helix, 443–445
    perspective, 442–443
    prospective, 457–458
    recombinant DNA and biotechnology, 456–457
  conformations of, 191–192
  discovery of, 1, 13–17
  The Double Helix, 30–31, 356, 427–431
  experimental data for, 1–2, 13–17
  gene evolution and, 447
  and gene regulation and morphogenesis, 191
  and genome mixing, 447
  and golden age of molecular biology, 429–432
  higher orders of organization in, 190–191

implications of, 30
and life sciences transformation, 30
Lwoff and, 429–431
and medical diagnosis, 183
and molecular genetics, 357
pathway to, 443–446
perspective, 442–443
Phage Group and, 429–431
prospective, 457–458
questions raised by, 83
in retrospect, 427–439
reverse transcriptase and, 448
scientific ramifications of, 177–178
supercoiling of, 191
what follows, 98–100
X-ray crystallographic analyses of, 29
Z-DNA, 130–133, 191, 446
zinc fingers and, 446
DNA revolution, 69–70
Doty, P., 60, 117
Double Helix, The, 356
  Chargaff review of, 30
  Medawar review of, 31
  reviews and responses to, 427–431
Down's syndrome, 420
Duchenne muscular dystrophy (DMD). *See also* Becker muscular dystrophy
  allelic BMD gene, 287–288
  DMD locus genes, 289–290
  dystrophin-associated glycoproteins and, 289
  dystrophin gene and, 256, 288
  dystrophin protein and, 288–289
  first report of, 243
  gene therapy strategies for, 291
  linkage analysis of, 287–288
  single-gene mouse model for, 319
  utrophin and, 290–291, 452
Dulbecco, R., *43*, 166–167
Dunitz, J., *97*, 99
Dyslexia, 415

Electronic complementarity, 118–120
Engelhardt, V., 80
Enhancer response element (ERE), 267, 272–274, 277
Ephrussi-Taylor, H., 56, 67–68, 70, 444
Epidermal hyperplasia, 320
ERE. *See* Enhancer response element
Evolution
  of biological information transfer and origin of life, 112–114
    RNA transcriptase in, 167–168
  DNA-level polymorphism and, 188
  gene mutation and, 26
  and genetic diseases, 257–258
  genetics and, 26
  genome analysis and, 422–423

genomics and, 408
sequence analyses and, 410

Facilitation. *See also* Learning; Memory
5-HT in, 263–265
long-term
  ApC/EBP and, 267, 272–277
  cAMP-dependent gene expression and, 265–267
  protein synthesis and, 265
  phosphorylation cascade in, 263
  protein/RNA synthesis and, 263–264
Felsenfeld, G., 100
First branchial arch syndromes, 314–315
Fischer, E. O., 49–50, 443
Flemming, W., 49
Fox, M., 68
Fragile X syndrome, 256–257, 415
Franklin, R. E., *3*
  in 1953, 1
  and B-pattern helix, 203
  and DNA diffraction studies, 203
  and DNA structure, 185
  and molecular biology, 7
  recognition of, 198
  and two-chain helix interpretation, 30, 204
  X-ray diffraction studies, 1–2
  and X-ray equipment, 202
Franklin-Gosling paper, 16–17, 441
Fraser, B., 203

Garrod, A., 187
Gene. *See also* DNA; Human Genome Project; Oncogenes
  Aristotle's notion of, 25, 28
  for dyslexia, 415
  for fragile X syndrome, 415
  hCS, 377–378
  hGH, 377–378
  homeotic, 146
  for Huntington's disease, 415
  as information carrier, 28
  linear genetic code, 28
  long-term stability hypothesis, 28
  for male sexual orientation, 415
  nature of, 26–28
  new description language for, 187
  nucleoprotein nature of, 50
  for partial deafness, 415
  replication of, 97
  stability of, 28
  TR effects on, 377–378
Gene expression. *See also* Gene regulatory proteins
  biotechnology and, 107
  cAMP-dependent, 265–267
  in long-term memory, 263

SUBJECT INDEX

molecular biology and, 191
multiple protein factors in, 143–144
Gene mapping. *See also* Human Genome Project
of craniofacial anomalies, 318–319
Gene regulatory proteins
basic leucine zipper, 146–149
and circuits, 220–223
DNA sequence recognition by, 144–145
helix-turn-helix motif, 144–147
homeobox proteins, 221, 224–226
multiple protein factors in, 143–144
operon model, 83–84, 143, 220–223
protein-DNA interactions and, 144–145
β-ribbon motifs, 149–150
TATA-box binding proteins, 150, 272, 446
zinc finger proteins, 150–156
Gene targeting, 213–217
Gene therapy. *See also* Recombinant DNA technology
for DMD, 291
genome analysis and, 415–417
in hemoglobin disorders, 94
in oral medicine, 322
for xerostomia, 321, 323
Genetic code
and neurobiology, 224–237. *See also* Homeobox proteins; Homeotic genes
research on, 450–451
RNA and, 83–84
transcription/translation of, 83–84
transmission of, 83–84
Genetic diseases. *See also* Craniofacial-oral-dental anomalies; Hemoglobin disorders; Molecular medicine
evolutionary selection for, 257–258
genotype/phenotype relationships and, 258
phenotypic diversity of
in human globin disorders, 246–255
molecular basis for, 245–258
in monogenic diseases, 255–256
transgenic mouse models for, 297–310
Genetic material. *See* DNA
Genetic recombination, 66–67
Genetics. *See also* Molecular genetics
DNA genetics by 1954, 66
and evolution, 26
Mendel and, 26
of retroviruses, 161–164
rise of, 26–27
Genome analysis. *See also* Human Genome Project
and agriculture, 417
and autoimmune diseases, 424
conclusions about, 424–425
and development and the brain, 419–420
and disease genes, 414–415
and evolution, 422–423
and gene function, 418–419

and gene sequence function, 417–418
and gene therapy, 415–417
and normal variation, 420–422
and population studies, 422–423
and technology, 417–418
Genomics. *See also* Human Genome Project
and biology, 408
and chromosomal languages, 410
and evolution, 410
and languages of theoretical biology, 410
motifs and, 408–409
and regulatory code, 409–410
Giacometti, G., 97
Gilbert, W., *18, 20, 21, 80, 208*
DNA sequencing, 358
and intron concept, 447
and mRNA synthesis, 84
Watson's view of, 195
Gosling, R., 201–202. *See also* Franklin-Gosling paper
Goulian, M., 359
Griffith, F., 2, 49, 54, 55, 184

Haldane, J. B. S., 425
Harker, D., 78
hCS. *See* Chorionic somatomammotropin genes
Heart disease, 257, 415
α-Helix
and DNA-binding proteins, 146–149
Pauling and, 29, 47, 77
zipper sequence and, 146
Helix-turn-helix motif, 144–146
Helling, R., 360
Hemoglobin disorders. *See also* Sickle cell anemia; Thalassemia
evolutionary selection and, 257
Hb types, 91–93, 246–247, 253
Pauling and, 81, 86, 89, 91–92, 445
sickle cell hemoglobin
discovery of, 88
electrophoretic demonstration of, 92–93
Hb types in, 91–93
hypoxia and, 90–91
Hemophilia, 256
Heredity
aperiodic crystal of, 25–31, 442
theories of, 25–26
Herpes simplex, 314, 321
Hershey, A. D., *7, 41*
and linkage in bacteriophage, 67
and phage group, 28–29
reservations about DNA's role, 177
Hershey-Chase experiments
DNA as genetic material, 358, 444
and "DNA-only" model, 186
phage DNA as genetic material, 4, 7, 67–68
phage in genetic studies, 442

hGH. *See* Human growth hormone
Hibernomas, 301–303
Hippocrates, on heredity, 25
HIV
 in AIDS, 448
 retroviral identity of, 167
 in transgenic mouse line, 320
 vaccine for, 416
HIV infection, 314, 320
Homeobox proteins, 221, 224–226, 226–228
Homeodomains
 helix-turn-helix motifs in, 144–146
 in homeotic genes, 146
 NK-2 domain, 226–227, 227
Homeotic genes
 homeodomain in, 146
 NK-2 homeobox gene, 224–237
 and transcription factors, 221
Hood, L., *22*
Hotchkiss, R. D., *22*, *40*, *208*
 and bacterial genetics, 47
 and DNA denaturability, 443
 and pneumococcal transformation, 56–57
 reflections on DNA revolution, 69–70
 "The Night Before Crickmas," 205–207
5-HT. *See* Serotonin
Human Genome Project. *See also* Biological information; Genome analysis; Molecular genetics
 and craniofacial anomalies, 318–319
 DNA as informational duplex and, 190
 DNA revolution and, 198
 and evolution of gene families, 456
 major objective of, 391
 and motifs, 408–409
 Watson and, 390
Hunter, A., 340
Huntington's disease, 256–257, 415
Hybridization
 of DNA, 187–188
 DNA-DNA, 111
 and double helix RNA, 100–102
 phage RNA-DNA, 117
 RNA-DNA, 58, 109–112, 117
Hypercholesterolemia, monogenic, 255

Ikenberry, S., *208*
Immunodeficiency, severe combined, 322
Immunotoxins
 animal experiments with, 349–350
 completely recombinant molecules of, 351–352
 cytotoxic activity of, 348–349
 future directions, 352–353
 preclinical toxicology studies, 350–351
 tissue culture experiments with, 347–349
 tumor cell lines tested, 349

Jackson, D., *23*
Jacob, F., *9*, *20*, *35*, *37*, *210*
 and *E. coli* Lac system, 219
 and gene regulation, 83–84, 143
 and molecular biology's golden age, 218
Jacob-Monod operon model, 83–84, 220–223
Jordan, P., 76

Kaiser, D., 359
Kandel, E., *22*
Kendrew, J., 29, 200, 218
Khorana, H. G., 186, 358
King, M., *78*
Klug, A., 83–84
Kornberg, A.
 DNA polymerase studies, 108
 DNA synthesis/replication, 358–359
 enzymic replication of DNA, 183, 186
 paradigm shift statement, 364
 and reverse transcriptase, 109
Kossel, A., 48–49
Krauss, M., 56
Kuhn, T., 363–364

Lacks, S., 68
Learning. *See also* Memory
 explicit forms of, 261–262
 gene targeting studies of, 213–217
 implicit forms of
  cAMP-induced gene expression in, 279–280
  CREB in, 279–280
  and reflex systems, 261–264
  sensitization of reflex response, 262–264
 mechanisms in, 261–262, 280–281
Lederberg, J., *23*, *39*, *208*
 and Avery paper, 27, 442
 DNA and Avogadro's number, 4
 and NYAS, 171
Lehman, R., 359
Levene, P. A., 49, 50, 184, 443
Lipodystrophy, 309–310
Lobban, P., 359
Long-term potentiation (LTP)
 induction of, 213
 and learning and memory, 214–217
 mossy fiber, 280–282
 mossy fiber model, 282
 Schaffer collateral, 280–281
 synaptic facilitation mechanisms in, 280–281
LTP. *See* Long-term potentiation
Luria, S., 28–29, *38*, *39*, *43*, 195–196, 218
Lwoff, A., *37*, *40*, *43*, 429–431

MacArthur, I., 47
MacLeod, C. M., 56, 68. *See also* Avery-MacLeod-McCarty
Mandibulofacial dysostosis (MFD), 318–319

SUBJECT INDEX

Marmur, J., 60, 68
Marshak, A., 58
Mathews, A., 49, 443
Mazia, D., 70
McCarty, M., 19, 21, 52–54, 58. *See also* Avery-MacLeod-McCarty discovery
McClintock, B., 42, 69, 187
Medawar, P., 31, 443
Melanoma, 421
Memory. *See also* Learning
  consolidation mechanism, 262
  long-term, 262
    and cAMP-mediated gene expression, 263
    mRNA and, 262–264
    and protein synthesis, 263
  short-term, 262
    and phosphorylation cascade, 263
Mendel, G., 25–27, 186, 442
Mertz, J., 359
Meselson, M., 9
  and double helix replication, 70, 84
  and genome mixing, 447
  and molecular biology's golden age, 218
  and mRNA synthesis, 84
  and recombinant DNA technology, 360
Meselson-Stahl experiments, 70, 84
Messenger RNA (mRNA)
  in central dogma, 83–84
  concept of, 99, 113
  in information flow, 188–189
  isolation of, 111
  and long-term memory, 262–264
  in protein synthesis, 115–116, 143
Methemoglobinemia, 91. *See also* Hemoglobin disorders
MFD. *See* Mandibulofacial dysostosis
Microphthalmia, 319
Miescher, F., 27, 48, 183
Mirsky, A., 6
  on Avery's discovery, 27
  caution on Avery paper, 442
  and protein structure, 445
  protein structure theory, 76
  skepticism about genetic material, 2, 4
  and transforming principle as DNA, 53
Mitochondria, DNA in, 118–119
Molecular biology. *See also* Central dogma; Molecular genetics; Recombinant DNA technology
  beginning of, 25
  and cancer susceptibility, 453–454
  circuits in higher organisms, 220–223
  clinical roots of, 86–94
  of cognition, 261
    explicit learning mechanisms, 280–281
    genetic mechanisms in, 281–283
    implicit learning, 262–264

implicit learning mechanisms, 280–281
long-term facilitation, 265–280
major forms of learning, 261–262
consequences of, 9
control unit in bacteria, 220
and gene therapy, 93–94
golden age of, 218, 429–432
growth of, 9–10
α-helix and, 47
and neoplasia, 454
nuclear hormone receptors, 368–370
operon model, 220
pervasiveness of, 30
phage group and, 7, 28
of phenotypic diversity
  in human globin disorders, 246–248
  in other monogenic diseases, 255–256
  in thalassemias, 248–255
regulatory circuit models, 220–223
scientist pairs in, 218–220
sickle cell hemoglobin and, 86–94
significance of, 10
structural transformations in
  electronic publishing, 434–439
  ethical problems, 432–434
  exponential growth to steady state transition, 434–436
of thyroid hormone action. *See also* Thyroid hormone receptors
Watson-Crick and, 7
Molecular Biology of the Gene, 9–10
Molecular disease, 81. *See also* Genetic disease
Molecular genetics. *See also* Molecular biology; Polymerase chain reaction; Recombinant DNA technology; Transgenic animals
Beadle-Tatum experiments and, 186
and biological sciences unification, 261
commercial enterprises using, 361–362
craniofacial anomalies gene mapping, 318–319
financial impact of, 363
flowering of, 186–187
human biochemical defects and, 187
recombination analysis in, 187
Monod, J., 37, 40. *See also* Jacob-Monod operon model
  and *E. coli* Lac system, 219
  and gene regulation, 83–84, 143
  and molecular biology's golden age, 218
Morgan, T. H., 26, 75, 442, 445
Morrow, J., 360
mRNA. *See* Messenger RNA
Muller, H. J., 26–27, 188, 196
Mulligan, R., 21, 451
Mullis, K., 436
Muscular dystrophy, 322
Myopathies, 314
Myotonic dystrophy, 256

Nature of the Chemical Bond, The, 75
Neurobiology. *See under* Genetic code
Nichols, R., *208, 210*
Nirenberg, M., *20, 22*, 358, 449–450
NK-2 homeobox gene
  and CNS development, 450–451
  distribution during development, 228–230, 233
  expression of, 224–225
    with developmental age, 228–237
    by neuroblast/glioblast, 231–233, 236–237
    pattern of, 233–236
  homeodomain of, 227
  nucleotide sequence of, 226–227
Nobel Laureates
  in chemistry
    Altman, S., 183
    Berg, P., 358
    Cech, T., 183
    Fischer, E. O., 49
    Gilbert, W., *18*
    Kendrew, J. C., 48
    Klug, A., 83
    Mullis, K., 436
    Pauling, L., *78*
    Perutz, M., *77*
    Sanger, F., 358
    Todd, A., 199
  in physiology or medicine
    Baltimore, D., *21*
    Beadle, G., 218
    Cohen, S., 358
    Crick, F. H. C., *2*
    Delbrück, M., *6*
    Dulbecco, R., *43*
    Hershey, A. D., *7*
    Jacob, F., *9*
    Khorana, H. G., 186
    Kornberg, A., 167
    Lederberg, J., *23*
    Luria, S., *38*
    Lwoff, A., *37*
    McClintock, B., *42*
    Medawar, P., *31*
    Monod, J., *37*
    Nirenberg, M., *20*
    Ochoa, S., 167
    Tatum, E., 218
    Temin, H., 167
    Tonegawa, S., *22*
    Watson, J. D., *2*
    Wilkins, M. H. F., *3*
Nuclear hormone receptors. *See also* Retinoic acid receptors; T-cell receptor families; Thyroid hormone receptors
  function, 369–370
  mechanism of action, 221
  superfamily, 368–369
  TR heterodimerization with, 374–375
Nucleic acids. *See also* DNA; RNA bases of
  electronic complementarity of, 118–120
  Hoogsteen-type pairing of, 104, 121, 124
  Watson-Crick pairing, 121–124, 128, 130
  X-ray analyses of, 121–122
  DNA-RNA relations, 107–109
  as genetic material, 50
  early ideas about, 49–50
  skepticism about, 46
  hybridization of, 100–102
    DNA strand-RNA strand, 109–112
    Hoogsteen base pairing and, 104
    and triple-stranded structure, 100–102
  identification of sugars in, 49
  Miescher's description of, 48
  nucleotides and, 49
  tetranucleotide hypothesis and, 50, 184
  triple-stranded, 102–107
  work on composition of, 48–49
Nucleic Acids, The, 61
Nuclein. *See* Nucleic acids

Obesity
  fat cell enhancer and, 299–300
  genetic forms of, 298–300, 304
  and hormonal secretion levels, 304
  induction of, 304–309
  as risk factor, 304, 309–310
  transgenic animal models of, 297–300
Ochoa, S., 167
Oncogenes
  Abl protein and, 339–343
  adapter proteins and, 342–343
  and cancer, 453–456
  molecular basis of, 331
  retrovirus vectors of, 164
  tyrosine kinase and, 339–344
Operon. *See* Jacob-Monod operon model
Oral medicine. *See also* Biomedical science; Recombinant DNA technology
  animal models for human diseases, 319–321
  biomaterials for, 321–322
  conditions included, 314
  craniofacial-oral-dental anomalies, 314, 316–319
  gene therapy opportunities in, 315, 322
  molecular approaches to, 314–316, 322
  perspective, 315–316
  prospective, 322–323
Orgel, Leslie, 33
Origin of life
  Oparin's view, 112
  reverse transcriptase and, 167–168
  RNA and, 112–114
Osteoarthritis, 314
Osteoporosis, 314, 319, 321

SUBJECT INDEX

Pardee, A., 219
Pastan, I., *19*
Paul, B., *43*
Pauling, A. H., 428
Pauling, L., 78, *80*. *See also* Sickle cell anemia
 and antigen-antibody reactions, 76
 and antioxidant vitamins, 81–82
 appreciation of, 199
 and biological molecules, 76
 and chemical bond, 75
 and chemical processes in biology, 75–76
 and DNA structure, 185
 and *The Double Helix*, 428
 and enzyme action, 77, 79
 and gene replication, 79
 gene replication model, 69
 and guanine-cytosine pairing, 121
 and helical DNA structure, 203
 and helical proteins, 202
 and α-helix discovery, 29, 47, 77, 202
 and intermolecular forces, 76
 and molecular evolution, 81
 and molecular structure, 74–75
 The Nature of the Chemical Bond, 75
 Nobel Prizes, 75, 81
 and β-pleated sheet structure, 77
 and protein structure modeling, 29, 443–445
 and protein structure theory, 76–77
 on radiation damage, 81
 Senate Internal Security Subcommittee and, 81
 tribute to, 74–82
 and triple-stranded DNA structure, 98
 Watson-Crick debt to, 79
PE. *See Pseudomonas* exotoxin A
Periodontal diseases, 314, 321
Perutz, M., 77, 200, 218
Phage and the Origins of Molecular Biology, 430
Phage Group, 9
 and DNA's genetic role, 4
 founding of, 28
 and molecular biology, 7
 Temin and, 166
Phages
 and DNA activity, 67–68
 and gene replication, 28
 as model systems, 4
 transformation and, 67–68
Pierre Robin syndrome, 314
Pollock, M., *43*
Polymerase chain reaction (PCR). *See also* Molecular genetics
 and entertainment industry, 362–363, 423
 invention of, 435–436
 and molecular biology, 189–190
Polysomes, 114–117
Pontecorvo, G., *43*
Poortman, Y., *44*

"Premature discovery," 26–27, 176
Protein kinases
 and cancer, 340
 in long-term facilitation, 265, 267, 271, 277–279
 in LTP, 280–281
 and oncogenes, 339–344
 in short-term memory, 263
Proteins. *See also* Gene regulatory proteins; Homeobox proteins
 Abl protein, 339–343
 adapter proteins, 342–343
 α-helix structure of, 29, 47
 specific DNA sequence recognition by, 144–145
 synthesis of, 114–117
 viewed as genetic material, 2, 27, 50, 184
Pseudomonas exotoxin A (PE)
 in cancer therapy, 346–347, 455
 cytotoxic activity of, 347
 endocytosis and, 347–348
 as immunotoxin, 347
Psychoses, 257

Randall, J., 200–201
Ravin, A., 68
Recombinant DNA technology. *See also* Biomedical science; Molecular genetics
 creation of, 359
 development of, 360
 EcoRI restriction enzyme and, 359–360
 and gene regulation, 221
 genetically engineered vaccines, 321
 review of, 456–457
 scientific disciplines using, 360–361
 Watson-Crick model and, 356–357
Regulatory proteins. *See* Gene regulatory proteins
Retinitis pigmentosa, 319
Retinoblastoma, 332–337, 454
Retinoic acid receptors (RXRs), 318, 371–375. *See also* Nuclear hormone receptors; T-cell receptor families; Thyroid hormone receptors
Retroviruses. *See also* HIV; Rous sarcoma virus
 in AIDS, 161, 448
 DNA of, 162–163
 genetics of, 161–164
 in leukemia, 448
 in mediated somatic gene therapy, 448
 as oncogene vectors, 164
 and reverse transcriptase, 161, 189
 RNA of, 161–164
Reverse transcriptase, 108–109
 in evolution, 167–168
 functions of, 161
 and retroviruses, 161, 189
 Temin and, 166–168

Reverse transcription
  and central dogma, 84–85
  error rates in, 161, 163
Rheumatic disease, 257, 314
β-Ribbon motifs, 149–150
Ribosomal RNA (rRNA), 99
Ribosomes, in protein synthesis. See Polysomes
Rich, A., 22, 33, 36, 78, 80, 97, 208
  and nucleic acid structure, 47
  and protein structure, 47
  and protein synthesis, 83–84
  and RNA structure, 446
  tRNA studies of, 83
Rich, J., 97
Rieger's syndrome, 314, 318
RNA. See also Antisense RNA; Messenger RNA; Ribosomal RNA; Transfer RNA
  and DNA structure-function relationship, 7
  double helix structure of, 100–102, 446
  in genetic code transmission, 83–84
  as genetic determinant, 70
  misconception about, 57
  molecular structure of, 99
  and origin of life, 112–114
  and protein synthesis, 107
  of retroviruses, 161–164
  synthesis of
    DNA and, 107–109, 111–112
    hybridization and, 100–102
    and transformation, 58
    triple-stranded, 102–107
  X-ray diffraction study of, 100–101, 103–107
RNA polymerase, 111, 113
  activation of, 143
Roberts, R., 360
Rolfe, R., 9
Rous sarcoma virus (RSV), 166–167, 455
rRNA. See Ribosomal RNA
RSV. See Rous sarcoma virus
Ruben, H., 43
Rubin, H., 166
RXRs. See Retinoic acid receptors
Ryan, F., 171, 173, 176

Sachs, L., 43
Sanger, F., 358
Schizophrenia and depression, 420
Schlenk, F., 46
Schrödinger, E.
  aperiodic crystal of heredity, 28–29
  gene as information carrier, 28
  genetic code concept, 442
  What Is Life?, 28, 196, 442, 449
Schultz, J., 50
Science, public audience for, 70–72
Scientific revolutions, 363–365
Sensitization, 262–264. See also Learning

Serotonin (5-HT)
  and ApC/EBP induction, 272–277
  and long-term facilitation, 265–271, 277–279
  in pre-synaptic facilitation, 263–264
Sgaramella, V., 359
Sickle cell anemia, 445. See also Hemoglobin disorders
  blood picture of, 87–88
  chronology of developments in, 89
  diagnostic test for, 89
  gene therapy trials for, 322
  hereditary nature of, 89–90
  hypoxia and, 90–91
  and selection against malaria, 257
Siminovitch, L., 40
Slavkin, H., 21
Smith, H., 360
Snow, C. P., 425
Sonneborn, T., 58, 196
Spiegelman, S., 43
Spinobulbar muscular atrophy, 256
Stahl, F., 70, 218. See also Meselson-Stahl experiments
Staker, M., 43
Stent, G., 9, 21, 210
  on double helix perspective, 442–443
  in phage group, 24
  and "premature discovery," 4, 56
Stokes, A., 201–203. See also Wilkins-Stokes-Wilson paper
Structure of Scientific Revolutions, The, 363
Sueoka, 60
Sutherland, G., 78
Szilard, L., 4, 41

T_3. See Thyroid hormone
T_4. See Thyroid hormone
TATA-box binding protein, 150, 272, 446
Tatum, E., 218
Taylor, D., 201
Taylor, H. See Ephrussi-Taylor
T-cell receptor families. See also Nuclear hormone receptors; Retinoic acid receptors; Thyroid hormone receptors
  genomics and, 390–392
  human β T-cell receptor family, 403–408
  sequence analysis of, 392–394, 403–405
    non-TCR elements, 399
    regulatory elements, 399–400
    serendipity, 411
    TCR elements, 395–399
T-cell receptor loci
  CδCα region
    conservation of, 395
    α elements sequence analysis, 400–402
    identifying receptor elements, 395–397
    Jα segments, 397–399

SUBJECT INDEX

non-T-cell receptor elements, 399
regulatory elements, 399–400
human β locus
  chromosome translocation in, 403, 405
  evolutionary comparisons of, 405–408
  homology units, 403–404
  sequence analysis of, 403–405
  trypsinogen genes, 403
Temin, H.
  in memoriam, 166–170
  and mutation-driven evolution, 448
  and reverse transcriptase, 167–168
  and reverse transcription, 84–85
  and RSV, 166–167
Tetranucleotide hypothesis, 50, 61, 184
Thalassemia, 94. *See also* Hemoglobin disorders
  evolutionary selection for, 257
  heterogeneity of, 246–247, 255
    α-thalassemia, 252–255
    β-thalassemia, 248–252
  phenotypic diversity of
    molecular basis of, 248–255
    α-thalassemia, 252–255
    β-thalassemia, 248–252
  selection against malaria, 257
  single-gene mouse model for, 319
Thomas, R., 68
Thyroid hormone receptors (TRs). *See also*
    Nuclear hormone receptors; Retinoic
    acid receptors; T-cell receptor families
  aspects of, 366–368
  characteristics of, 370–371
  and DNA conformation, 380
  heterodimerization of, 374–375
  and hGH/hCS, 377–378
  and nuclear hormone receptors, 368–370, 374–375
  properties of purified TRs, 371
  recombinant overexpression of, 371–372
  target gene recognition by, 372–374
  three-dimensional TR-LBD structure, 382–383
  thyroid hormone-induced property changes in, 380–382
  transcription regulation by, 378–380
Thyroid hormones (T3; T4)
  actions of, 456
    on TR properties, 380–382
    in vertebrate tissues, 367–368
  forms of, 366–367
Thyroid response elements (TREs), 368, 372–376, 378–379, 382
Tickell, C., *208, 210*
Todd, A., 199
Tonegawa, S., *22,* 449–450
Transfer RNA (tRNA), 83–84
  backbone of, 125–126

crystallization of, 124–125
high-resolution structure of, 126–129
in protein synthesis, 114, 116
structure of, 446
synthesis of, 117–118
Transformation. *See also* Avery-MacLeod-McCarty
    discovery
  of *Bacillus subtilis*, 68
  continuing thread of, 68–69
  DNA as genetic information, 2
  DNA chemistry and, 58–61
  DNase experiments, 52–54
  as genetic process, 63–67
  growth after 1954, 68
  of *Hemophilus*, 68
  and phage, 67–68
  and pneumococcal encapsulation, 56
  questions about, 57–58
  review of, 68–69
  treatment of DNA extracts, 52
Transforming principle, 27, 46, 48–54
Transgenic animal models
  adipose tissue types, 300–302
  BAT culture line isolation, 303
  with hibernomas, 300–303
  for human oral disease, 319–321
  for human viral diseases, 320
  for obesity, 297–300, 453
  obesity-resistant, 304–309
  for retinoblastoma, 332–337, 454
  uses for, 298
  for Waardenburg syndrome type 1, 319
  for X-linked cleft palate, 319
Treacher Collins syndrome, 314
Trigeminal neuralgia, 314
tRNA. *See* Transfer RNA
TRs. *See* Thyroid hormone receptors
Tuberculosis, 416
Tumor suppressor genes, 332–337
Turner's syndrome, 243

Vaccines, molecular biology and, 321, 416

Waardenburg syndrome type 1, 319
Watson, J. D., 2, *21, 23, 33, 34, 45, 97, 208, 209, 210, 357*
  in 1953, 1
  and ApU structure, 124
  at Caltech, 99
  The Double Helix, 30
  Hershey-Chase experiment influence on, 68
  and Human Genome Project, 390
  laboratory notebook entry, 7–8
  and language of DNA, 358
  meeting with Crick, 29
  meeting with Wilkins, 29
  and molecular biology, 7–9

and molecular biology's golden age, 430
Nobel Prize, 356–357
and Pauling, 196
personal reminiscence, 191–197
and Phage Group, 28
and RNA double helix, 99
tribute to, 171
Watson-Crick double helix. *See also* Central dogma
  1953 paper, 7, 13–14, 30–31, 121, 441
  base pairing in, 122–124, 128, 130
  and helical DNA structure, 203
  history of
    DNA as genetic material, 185–186
    nucleic acids discovery, 183
    protein chemistry developments, 184
    tetranucleotide theory, 184
    TMV crystallization, 184
    transformation studies, 184–185
  Hotchkiss and, 69–70
  modeling methods and, 79
  and molecular biology beginning, 25
  and mRNA concept, 99
  and recombinant DNA technology, 356–357
  and structure of DNA language, 358–359
Weatherall, D., 22
What is Life?, 28, 196, 442, 449
Wilkins, M. H. F., 3, 78
  DNA at King's College, 200–204
  and DNA structure, 185
  and double helix, 30
  and *The Double Helix*, 428
  letter to Crick, 204
  and molecular biology, 7
  recognition of, 198
  and structure of DNA language, 358
  X-ray crystallographic analyses, 29
  X-ray diffraction studies, 1–2
Wilkins-Stokes-Wilson paper, 14–16, 441
Wilson, E. B., 49, 183–184, 443. *See also* Wilkins-Stokes-Wilson paper
Wollman, E., 35, 219

Xeroderma pigmentosum, 322
Xerostomia, 314, 321, 323
X-linked cleft palate, 319
X-ray diffraction studies
  and DNA structure, 29
  of nucleic acids, 14–17, 98
  of RNA, 100–101

Z-DNA. *See* DNA double helix
Zinc finger proteins, 446–447
  first class, 150–153
  hormone receptor DNA-binding domains, 153–155
  other domains, 155–156
Zipper. *See* Basic leucine zipper